"The handbook refreshingly celebrates a carrier bag-full of multiple and insightful engagements with gender in watery contexts. Together with its commitment to highlight differently situated hydrofeminisms, this collection also draws on more than 40 years of shared intellectual and activist streams that questioned harmful water and knowledge authorities, along with their neoliberal extractivist logics, lying at the heart of persistent water-related intersectional injustice, inequitable distributions, pollutions, and precarity. This handbook is a must-read for those contemplating feminist, water, and more-than-human futures."

Bernadette P. Resurrección, *Professor, Global Development Studies, Queen's University, Canada*

"This book fizzes and pops with fresh perspectives, thought-provoking provocations and original insights. The carefully curated collection celebrates the richness and diversity of gender and water research, embracing differences between the varied contributions. Some of the chapters revisit old debates, updating and nuancing our understanding of gender-water dynamics, others venture into emerging areas for gender and water research.

The introductory chapter sets the tone for the collection – written with a light touch and expressing a collaborative, reflective approach to water research, rooted in the authors' personal experiences as scholars and activists. It usefully locates the collection in the last 40 years of gender and water scholarship, and shows that feminist water studies continue to be needed to inform ongoing struggles against injustice and to imagine more just and sustainable futures."

Frances Cleaver, *Professor of Political Ecology, Lancaster Environment Center, Lancaster University, UK*

"The *Routledge Handbook of Gender and Water Governance* weaves together an impressive collection of voices, perspectives, and framings from around the world on feminist water studies. It offers critically important insights to advance interdisciplinary water scholarship as a whole, and in particular to those interested in pluralizing and achieving water justice for all."

Farhana Sultana, *Professor of Geography and the Environment, Syracuse University, New York, Washington D.C., USA*

"Amid the gloom and doom of news on water, this book provides refreshing glimpses of hope. The feminist lens on water expands recognition of the range of water-related issues, from embodied experiences of accessing water to transboundary water cooperation, but also expands our understanding of how to address the challenges through examples of individual and collective agency and activism and creative approaches such as water museums to shape the water heritage and ethics. The understanding of care – for water, for the environment, and for each other – as a central concept and motivation provides a welcome alternative to mechanistic regulatory or economic approaches to motivating people's actions in relation to water. The diversity of action research case studies does not define rigid boundaries, but invites and inspires others to join in."

Ruth Meinzen-Dick, *Senior Research Fellow, International Food Policy Research Institute, Washington D.C., USA*

"A significant contribution to the extensive body of literature on the gendered nature of water from a wonderful collective of established and younger researchers. A comprehensive volume of this nature and coverage has been lacking in the shelves for some time, and this book will fill that gap. Readers will enjoy the book for the timeliness of its topic, its scope and the mastery of the style of the authors selected carefully by the editors. This book is a momentous addition to the growing understanding of women, gender and the resources and the environment. It weaves together contributions from a range of disciplinary scholars and valuable insights from practitioners on a wide variety of themes to draw out the intersecting threads that make gender more visible in thinking about water. I can guarantee the readers they will never see water management in the same way again."

Kuntala Lahiri-Dutt, *Professor, Crawford School of public Policy, ANU College of Asia and the Pacific, The Australian National University, Canberra, Australia*

ROUTLEDGE HANDBOOK OF GENDER AND WATER GOVERNANCE

This handbook provides a comprehensive overview of the field of gender and water governance, exploring how the use, management and knowledge of water resources, services and the water environment are deeply gendered.

In water there is a recognized gender gap between water responsibilities and water rights and bridging this gap is likely to help achieve not just goals of equity but also those of sustainability. Building on a rich legacy of feminist water scholarship, the *Routledge Handbook of Gender and Water Governance* is a collection of reflections and studies that can be used as a prismatic lens into a thriving and ever proliferating array of feminist water studies. It provides a clear testimony of how hydrofeminism has evolved from rather instrumental gender and water studies to scholarship that uses feminist tools to pry open, critically reflect on and formulate alternatives to water development-as-usual. The book also shows how the community of feminists interested in studying water has diversified and expanded, from often white female scholars studying projects and gender relations in the so-called Global South, to a varied mix of scholars and activists theorizing from diverse geographical and political locations – prominently including the body. It is organized into five interconnected parts:

- Part I: Positionality and embodied waters
- Part II: Revisiting water debates: diplomacy, security, justice and heritage
- Part III: Sanitation stories
- Part IV: Precarious livelihoods
- Part V: New feminist futures

Each of these parts brings out the gendered nature of water, shedding light on the often neglected care and unpaid labour of women and its relationship with extractivism and socioeconomic inequalities. The overall aim of the handbook is to apply social science insights to water governance challenges, creating synergies and linkages between different disciplines and scientific domains.

The *Routledge Handbook of Gender and Water Governance* is essential reading for students, scholars and professionals interested in water governance, water security, health and sanitation, gender studies and sustainable development more broadly.

Tatiana Acevedo-Guerrero is Assistant Professor at the Copernicus Institute for Sustainable Development, Utrecht University, the Netherlands. She leads the ERC project titled 'Homescapes make the world we live in,' that takes water as an entry point of an investigation into the homes of the urban South.

Lisa Bossenbroek is a researcher at the iES, Institute of Environmental Sciences, University of Kaiserslautern-Landau, Germany.

Irene Leonardelli is a postdoctoral researcher at the University of Calabria, Italy. She holds a PhD from the IHE-Delft Institute for Water Education, the Netherlands, where she worked at IHE Delft for more than four years as a junior researcher.

Margreet Zwarteveen is Professor of Water Governance at the IHE-Delft Institute for Water Education and the University of Amsterdam, the Netherlands. She is the co-editor of *Drip Irrigation for Agriculture* (Routledge, 2017).

Seema Kulkarni is a senior fellow at the Society for Promoting Participative Ecosystem Management (SOPPECOM).

ROUTLEDGE ENVIRONMENT AND SUSTAINABILITY HANDBOOKS

ROUTLEDGE HANDBOOK OF ENERGY COMMUNITIES AND SMART CITIES
Edited by Maciej M. Sokołowski and Anna Visvizi

ROUTLEDGE HANDBOOK OF GLOBAL LAND AND RESOURCE GRABBING
Edited by Andreas Neef, Chanrith Ngin, Tsegaye Moreda Shegro, and Sharlene Mollett

THE ROUTLEDGE HANDBOOK OF ECOMEDIA STUDIES
Edited by Antonio López, Adrian Ivakhiv, Stephen Rust, Miriam Tola, Alenda Y. Chang and Kiu-wai Chu

ROUTLEDGE HANDBOOK OF URBAN BIODIVERSITY
Edited by Charles H. Nilon and Myla F.J. Aronson

ROUTLEDGE HANDBOOK OF CLIMATE CHANGE IMPACTS AND ADAPTATION STRATEGIES ON INDIGENOUS PEOPLES AND LOCAL COMMUNITIES
Edited by Victoria Reyes-García

ROUTLEDGE HANDBOOK OF INSECT CONSERVATION
Edited by James S. Pryke, Michael J. Samways, Tim R. New, Pedro Cardoso and René Gaigher

ROUTLEDGE HANDBOOK OF PRIVATE LAW AND SUSTAINABILITY
Edited by Marta Santos Silva, Andrea Nicolussi, Christiane Wendehorst Pablo Salvador Coderch Marc Clément and Fryderyk Zoll

ROUTLEDGE HANDBOOK OF GENDER AND WATER GOVERNANCE
Edited by Tatiana Acevedo-Guerrero, Lisa Bossenbroek, Irene Leonardelli, Margreet Zwarteveen and Seema Kulkarni

ROUTLEDGE HANDBOOK OF GENDER AND WATER GOVERNANCE

Edited by Tatiana Acevedo-Guerrero, Lisa Bossenbroek, Irene Leonardelli, Margreet Zwarteveen and Seema Kulkarni

Routledge
Taylor & Francis Group
LONDON AND NEW YORK

earthscan
from Routledge

Designed cover image: Angie Vanessa Cárdenas Roa www.angievanessita.com

First published 2025
by Routledge
4 Park Square, Milton Park, Abingdon, Oxon OX14 4RN

and by Routledge
605 Third Avenue, New York, NY 10158

Routledge is an imprint of the Taylor & Francis Group, an informa business

British Library Cataloguing-in-Publication Data
A catalogue record for this book is available from the British Library

Library of Congress Cataloging-in-Publication Data
Names: Acevedo-Guerrero, Tatiana, editor.
Title: Routledge handbook of gender and water governance / edited by Tatiana Acevedo-Guerrero, Lisa Bossenbroek, Irene Leonardelli, Margreet Zwarteveen, and Seema Kulkarni.
Description: Abingdon, Oxon ; New York, NY : Routledge, 2024. | Series: Routledge environment and sustainability handbooks | Includes bibliographical references and index.
Identifiers: LCCN 2024013052 | ISBN 9780367607586 (hardback) | ISBN 9780367607630 (paperback) | ISBN 9781003100379 (ebook)
Subjects: LCSH: Water-supply—Social aspects—Developing countries. | Water use—Social aspects—Developing countries. | Women—Developing countries—Social conditions. | Water security—Developing countries. | Water rights—Developing countries.
Classification: LCC HD1702 .R68 2024 | DDC 363.6/10901724—dc23/eng/20240531
LC record available at https://lccn.loc.gov/2024013052

ISBN: 978-0-367-60758-6 (hbk)
ISBN: 978-0-367-60763-0 (pbk)
ISBN: 978-1-003-10037-9 (ebk)

DOI: 10.4324/9781003100379

Typeset in Sabon
by Apex CoVantage, LLC

CONTENTS

FOREWORD

I am delighted to have been able to submerge myself in this engaging and meticulously researched *Routledge Handbook of Gender and Water Governance*. In this short foreword, I set out reasons for why the readers should engage in the Handbook's exciting, original and challenging feminist conversations on water.

The first is its methodological daring. The handbook bubbled up from classroom conversations to become a feminist critical thinking hub to be distilled and shared in this handbook. The chapters capture the excitement of the course in Delft as students and teachers co-created ways to explore the materiality of water in its social, political and cultural dimensions, looking at how gendered power relations intersected with other dynamics of class, caste, age, race and ability in water worlds. The conversations which are shared in the handbook's many chapters provide profound gendered insights into watery questions across disciplines, contexts and bodies.

Secondly, the handbook fully acknowledges the histories and genealogies that produced this unique combination of feminist inquiries into water governance which is embodied and embedded in different bodies and places. The book builds on eco-feminism, material feminism, decolonial feminism, queer and feminist political ecology and black ecologies to make visible the gender norms, power asymmetries, multiple social relations that shape today's experiences of watery domains. It does not shy away from critical reflections on the environmental changes the world is experiencing due to the violence of hetero-patriarchal colonial capitalist development. Building on rich and profound feminist theoretical work, the chapters show how gender and water relate in ways that determine different meanings and practices of water, from defining and giving life to threatening health due to lack of access, contamination and toxicity. The handbook interweaves theory with the stories. The experiences and knowledge of the authors based on their reflections of their own direct watery connections of bodies and places are at times surprising and unsettling.

Thirdly, the diversity of the book is astounding. You dive deep into many stories from around the world, beginning with the recognition that we ourselves are watery, porous and not separate from water, it flows through our corporeal selves. Embracing the fluidity of watery lives, the handbook makes visible the violence and damage of colonial capitalist thinking and practices that define water as a resource to be exploited, managed and governed. Instead, it invites readers to engage in gendered, queer and trans concepts and methods in a radical rethinking of gendered hydro-social futures which explores the flowing movements and fluid relations of water with(in) humans, animals, plants, cultures and imaginaries.

Fourthly, from these unsettling and often dismaying stories a theorised feminist water politics emerges. The complex question of how to understand ourselves as bodies of water shifts attention away from humans as the focus of research to look at how humanity is connected to and depends on the watery relations of different ecologies, which are threatened by patriarchal colonial developmentalist extractivist practices which shape and destroy waterscapes in the past, present and future. The handbook challenges us to change how we see gender and water relations and to create new subjectivities that take responsibility for our embedded connections to other watery bodies, within global flows of capitalist hetero-patriarchal power. Finding ways to live well together with other bodies of water requires a new way of understanding and practice.

Fifthly, as part of the feminist politics of radical hope, the handbook surfaces from these deep and profoundly challenging ways of understanding water worlds to propose what practices are needed in the context of hetero-patriarchal capitalist driven colonial climate catastrophe. Listening and learning from the margins will allow us to create new modes of water governance where watery connections can flourish based on a profound shift in feminist ethics, political commitments and activism which the handbook beautifully outlines.

In conclusion, the handbook provides the answers in this shimmering array of imaginative, empirically based studies to its opening query of 'what are feminist water questions?' I congratulate the editors and authors for their cascade of gendered water stories and invite readers to take an exhilarating dive into the water worlds of possibilities they present.

Wendy Harcourt
The Hague, September 2023

ACKNOWLEDGEMENTS

The coming together of this handbook was an endeavour that included many more than those who are formally acknowledged as editors, authors or contributors. Beyond the chapters they contributed, we would like to first of all thank the book's gifted and motivated authors for trusting us throughout the process of its development. Partly because of the COVID-19 pandemic (that left none of us unaffected), putting the book together took longer than we hoped, testing the patience of all involved. We are also grateful for the Water Governance Department of the IHE Institute for Water Education for financial support, thank Leila Rezvani for editing the chapters with a lot of care, and Angie Vanessita for illustrating our cover.

The book project has been enormously inspired by the many international students (from the Global South and the Global North) who shared their experiences and reflections with us during our courses on gender and water governance. While putting together this handbook, their voices and questions always were in the back of our minds. We were keen to create a bundle of texts to guide their intellectual journeys, bringing together different stories, theoretical analysis and methodologies that allow for multiple water worlds to co-exist. After all, similar handbooks have inspired us when we started embarking on scholarly and research adventures. They allowed us to dream of water futures where feminist questions are acknowledged, and where there is no distinction between research and activism. We were equally inspired by the experiences of women and men across the globe who are striving, often on a daily basis, for more equitable and sustainable water worlds. Many of them generously shared their water stories with us. We dedicate the book to them, as well as to the many students, researchers and activists who accompany them in their courageous struggles.

TABLES AND FIGURES

Tables

Figures

CONTRIBUTORS

Mohammed Abubakari is an interdisciplinary Ph.D. candidate at the School of Environment and Sustainability (SENS), University of Saskatchewan, Canada. His current research focuses on the intersections of traditional ecological knowledge systems for agricultural practices focusing on migrant farmers from the north working in southern rural communities in the forest transition zone of Ghana. He has an interest in community engagement, food security and livelihood nexus, local approaches to resource management (land, water, green spaces and forest) and regional and community planning for environmental sustainability. He aims to become a scholar-practitioner in sustainability science through science with society research and practices.

Tatiana Acevedo-Guerrero is an engaged researcher and educator from Santander, Colombia. She currently works as Assistant Professor at the Copernicus Institute, Utrecht University, where she forms part of the environmental governance group. She has experience in interdisciplinary social-environmental studies, learned from schools and universities but also through the environmental and political struggles which she's had the honour to accompany. Her work is situated broadly at the intersection of urban studies and political ecology but she also cultivates interests in postcolonial/decolonial, Caribbean, Colombian and Latin American studies.

Sara Ahmed is Adjunct Professor at the Department of Humanities and Social Sciences at the Indian Institute of Science Education and Research, Pune. She has over 30 years of applied research experience on water, livelihoods and social equity. Sara has been actively engaged in teaching and mentoring young development professionals in India, managing large and complex regional research portfolios on water, food security and climate change in Asia, and advising a range of development organizations and water networks globally. Sara holds a PhD in environmental sociology from the University of Cambridge and is currently on the board of Water Aid, India, and Wetlands International South Asia. She has published extensively on water governance and her last co-edited book is entitled, *Diverting the Flow: Gender Equity and Water in South Asia (2012)*. Apart from founding Living Waters Museum, Sara is one of the vice presidents on the board of the Global Network of Water Museums.

Evelyn Arriagada is an anthropologist and MA in social sciences (Universidad de Chile), a MA in political and social sciences (Universitat Pompeu Fabra) and a Ph.D. candidate at the Institute for Resources, Environment and Sustainability (University of British Columbia). Her dissertation focuses on shifting subjectivities and hydrosocial relationships in contemporary water justice movements in Chile. Her Ph.D. is supported by a Vanier CGS Award, Killam Doctoral Scholarship, Public Scholar Initiative and the National Agency for Research and Development (ANID – Chile) – among other funding. Her most recent co-authored publication explores the experiences of 'lived environmental citizenship' in female rural leaders in central Chile.

Durba Biswas is Adjunct Fellow with ATREE. She received her PhD in environmental economics in 2011 for her thesis on estimating the economic value of canal irrigation farmers in North Karnataka, India. Currently, she is Adjunct fellow with ATREE. Her research is focused on WASH challenges in India, particularly on the outcomes of WASH policies on poor and marginal communities using a multi-disciplinary approach. Biswas has led multiple projects broadly on the areas of water, wastewater and diets with a focus on gender and marginalized communities. She has collaborated as Principal Investigator (PI) and Co-PI on large grant projects with Indian and international organizations and universities. Publications from her research have been published in international journals. Biswas led a multi-disciplinary project on *Groundwater and sanitation nexus in small towns of Bangalore: Case study of Nelamangala*. She was a co-PI in a Scottish government funded project on *global innovation for sustainable rural communities* which aimed to co-design and implement a sustainable and low-cost sanitation and wastewater treatment system in rural Karnataka. Presently, she is co-leading an all-women's research team on a project that aims to build a web-based tool for government schools to identify appropriate, low-cost, greywater treatment systems. Her work spans rural, peri-urban and urban areas across Southern India.

Chantal Victoria Bright is a researcher and children's author with 12 years of combined professional experience in West Africa, Europe and the United States. She is a recipient of the University of Manchester Distinguished Achievement Medal for Postgraduate Researcher, the Royal Geographical Society Frederick Soddy Postgraduate Award and Leverhulme Trust. She is also a Library of Congress Kluge Scholar, Missing Peace Initiative Scholar and a West African Research Center Affiliate Scholar (WARC).

Chantal is pursuing a PhD in geography at the University of Manchester, and her research sits at the nexus of gender-water-peace in fragile and conflict-affected settings from an African ecofeminist perspective. She is a visiting PhD student and lecturer at the Kofi Annan Institute for Conflict Transformation (KAICT), University of Liberia. She holds an undergraduate degree in English and political science (dual course) from Seton Hall University and a graduate degree in environmental management from the University of Pennsylvania. Chantal is a Liberian American who settled and resides in the United Kingdom.

Jessica Budds is Professor of Development Geography at the University of Bonn. Her research explores how environmental policy is influenced by economic and political processes and power relations in the Global South, particularly Latin America, and its implications for and effects on the lives, livelihoods and landscapes of low-income groups. Her main foci have comprised urban water infrastructure and access, water privatization and markets, the increased demand for water for industrial sector growth, and the links between drought responses and capital accumulation. She is co-editor of *Journal of Latin*

American Geography, and a member of the editorial advisory board of *Environment and Urbanization*.

Anyi Castelblanco holds a degree in pedagogy and education with an emphasis in social sciences and is an anthropologist from Colombia. She also has a master's degree in political ecology and alternatives to development. Additionally, she has conducted research on conflicts related to water access and distribution, as well as mining extractivism. Furthermore, she has collaborated with peasant and Indigenous communities on various topics, including territorial control, community environmental monitoring, advocacy and defence of peasant livelihoods, agroecology, women's rights and environmental democratization. She is a co-founder and partner of the Association CAMBIUM: Climate, Environment, Research-Action Uniting Worlds, a local organization interested in building bridges between activism and academia in Colombia and Latin American.

Stroma Cole is a reader at the University of Westminster. She combines her academic career with action research and consultancy. Her research explores the interconnect between tourism, gender and water rights. Stroma has a long history of working on small islands, particularly in Indonesia. She holds a British Academy Knowledge Frontier grant, and manages an interdisciplinary, international team to explore the connections between household water insecurity and gender based violence in Peru and Indonesia. Stroma is most well known for her tourism and water nexus research in a Bali longitudinal action research project. Stroma is an activist researcher critiquing the consequences of tourism development, in particular the overuse of water by hotels and the consequences for local women. Stroma has over 40 publications, is Associate Editor for *Annals of Tourism Research* and on the editorial board at *Journal of Sustainable Tourism* and *Tourism Geographies*.

Qurratul Ain Contractor is an architect and researcher from Mumbai, with a MSc. in water management and governance from IHE-Delft, Netherlands. She is part of the Right to Water movement in Mumbai. Her master's research used a lens of care to map invisibilised contributions of fisherwomen to artisanal fishing economies and ecologies in Tamil Nadu, India. By documenting the ecological knowledge and perspectives of marginalised Indigenous peoples, she hopes to understand diverse forms of nature-society relations that can offer alternatives to the dominant models of development.

Silvia Corredor-Rodríguez is an investigative journalist, anthropologist and specialist in conflict resolution. She is currently working as a senior journalist in *El Espectador* (Colombian oldest newspaper); previously she was a researcher in the Commission for the Clarification of Truth, Coexistence and Non-Repetition, the institution created in the framework of the 2016 Peace Agreement to draft the final report of the Truth Commission, the most detailed and exhaustive report on the armed conflict in Colombia. She also worked as a reporter on the website Rutas del Conflicto (Paths of armed conflict). She has particular expertise in Colombian armed conflict, memory reconstruction, the implementation of the Peace Agreement, human rights, analysis of guerrilla organisations and socio-environmental conflicts in the Colombian Caribbean.

Gabriela Cuadrado-Quesada is Senior Lecturer/Researcher of Water Rights and Justice at IHE Delft, the Netherlands. Her research and teaching activities focus on the legal and institutional dimensions of water at both the local and national levels, the role of community organizations with explicit attention to gender in using legal instruments and activism to promote water/environmental justice, and the connections between water practices and law. Prior to joining IHE Delft, Gabriela was a PhD fellow at the University of New South Wates (UNSW), Sydney, Australia. She has also extensive practical experience working for various non-governmental organizations in Costa Rica and India and as an environmental litigator in Costa Rica.

Amina Dessouki is a PhD candidate at Central European University, Vienna. She holds a master's degree in anthropology from American University in Cairo. She works on issues related to development, gender, water, infrastructure and the urban environment.

Kelly Dombroski is a researcher in community economies, feminist economic geography and political ecology at Massey University in New Zealand. She is the co-editor of the *Handbook of Diverse Economies* with JK Gibson-Graham, and associate editor at Asia Pacific Viewpoint and New Zealand Geographer.

Madeline Donald is a daughter, partner, friend and researcher interested in the interaction of possibilities that humans perceive in relation with plants in their shared, day-to-day and lived-in environments. Currently, she is located at the University of British Columbia Okanagan, learning with riparian habitats in the Okanagan watershed, Syilx territory.

Benjamin Dosu is Community Consultation Officer with the Crown-Indigenous Relations and Northern Affairs Canada, Government of Canada. Benjamin is an environmental planner with an interest in environmental policies and management, focusing on the resource rights of marginalised populations, particularly in rural communities. Benjamin has been extensively involved in community engagement and consultations in rural and remote Indigenous communities in Canada and Ghana, where he has contributed immensely to advancing the interest of marginalised communities in environmental and resource stewardship.

Hind Ftouhi is an agronomist and rural sociologist. She is currently a professor at the Institut National d'Aménagement et d'Urbanisme in Rabat. Her research interests focus on the roles of rural youth in the processes of agrarian and territorial change in Morocco. She is a member of different research projects where she also studies access to and governance of natural resources from a gender perspective.

Jacqueline Gaybor Tobar has a diverse background with expertise in sexual and reproductive health and rights, international development, education and qualitative research. Jacqueline holds a Ph.D. from Erasmus University Rotterdam, where she critically examined menstrual health, international development and menstrual activism. In addition, Jacqueline has extensive experience in teaching and course coordination in the Netherlands, Brazil and Colombia, focusing on gender equality, international relations, human rights law and qualitative interviewing. Jacqueline has authored various peer-reviewed publications on menstrual health and activism, gender and development, WASH and science and technology. In her current role as Senior Technical Adviser at Rutgers Kenniscentrum Seksualiteit in the Netherlands, Jacqueline manages relationships with partners in West and North Africa and advises on strategies for gender equality and sexual and reproductive health and rights. She is also a lecturer at the Erasmus University College, in Rotterdam.

Nikita Gopal is Principal Scientist at the Indian Council of Agricultural Research-Central Institute of Fisheries Technology (ICAR-CIFT), Kochi, Kerala, India. For over 25 years she has been working in the fisheries sector focusing on gender; seafood trade and markets;

technology evaluation; labour and migration; and small-scale fisheries. She is currently the chair of the Gender in Aquaculture and Fisheries Section (GAFS) of the Asian Fisheries Society (AFS).

Maura Hanrahan earned her PhD at the London School of Economics, taught at Carleton University, Memorial University, the University of Lethbridge and worked on Indigenous land claims and rights throughout Canada. She is the author or editor of 14 books in several genres, numerous journal articles and book chapters. Her work has won awards in Canada, the United States and the United Kingdom.

Holly M. Hapke is a geographer and interdisciplinary social scientist based in the United States. Her research projects have examined the impact of technological transformation and globalisation on artisanal fishing communities, fish markets and fisherfolk livelihoods in India; the role of fisheries in food security for the urban poor in India and Ghana; the social economy of dried fish, and the gendered impacts of the blue economy in small-scale fisheries in the Indian Ocean region. Dr. Hapke is currently researcher and Director of Research Development in the School of Social Sciences at University of California, Irvine.

Leila M. Harris is Professor with the Institute for Resources, Environment and Sustainability and with the Institute for Gender, Race, Sexuality and Social Justice at the University of British Columbia. Her work analyses a range of governance and political considerations from feminist, equity and sustainability perspectives – with work that has focused on water politics, governance and justice in a variety of contexts (from Turkey, to Ghana, South Africa and Canada). Recent projects consider lived experiences and equity concerns related to the uneven implementation of the human right to water, ongoing water governance shifts in varied contexts, as well as focus on social, affective and political dimensions of the uneven geography of water access, quality and infrastructures. Recent publications include focus on issues trustworthiness around drinking water, rethinking gender and water research, and enriched engagement with storytelling and narrative inquiry for political ecology research (*Environment and Planning E*).

Cara Jacob is a Ph.D. candidate in the anthropology department at Michigan State University. Her work focuses on urban environmental contamination with emphasis on infrastructural degradation, water insecurity and gender dynamics. Her recent research examines the gendered and racialised impacts of water insecurity in two large urban areas of the United States. Cara is dedicated to the use of feminist methodologies and collaborative, participatory methods that center marginalised community voices.

Amie Jammeh is a researcher from The Gambia and with several years of teaching experience. She is a PhD candidate at the University of East Anglia, Norwich, United Kingdom. Amie's research interest spanned across gender analysis of migration, literacy and education, WASH, agriculture, peri-urban livelihoods and Sub-Sahara Africa's urbanization. She graduated from IHE Delft Institute for Water Education with an master's in sanitation. She received the Best MSc in Sanitation Thesis 2020 award from the Bill & Melinda Gates Foundation for her thesis *The many meanings of menstruation: practices and imaginaries among school girls in Lusaka, Zambia*.

Priyanka Jayakodi is a sociocultural anthropology Ph.D. student at Michigan State University (MSU). She specialises in medical and environmental anthropology. Her research interests include entanglements of health, environment and gender, water justice and water governance, and state violence and social suffering. Her doctoral research examines social and political aspects of water insecurity in the context of chronic kidney disease of uncertain etiology (CKDu) in Sri Lanka.

Tom Johnston is Professor Emeritus, Department of Geography and Environment, University of Lethbridge. His research interests centre on rural geography, resource management, and public policy analysis.

Deepa Joshi is Gender, Youth and Inclusion Lead at the International Water Management Institutes (IWMI). A feminist political ecologist by training, her research has analysed shifts in environmental policies and how these restructure contextually complex intersections of gender, poverty, class, ethnicity and identity.

Seema Kulkarni has over 30 years of experience of working on issues of gender and natural resources. She is one of the founding members of Society for Promoting Participative Eco-System Management, Pune, India (SOPPECOM). She is a senior fellow here and co-ordinating the gender and rural livelihoods activities within the organization. She has co-ordinated various studies and programmes around decentralization, gender and land, water and sanitation. She has published several articles/book chapters around issues of gender, water, sanitation and rural livelihoods. She has been associated with the feminist movement in India. She is currently a member of the National Facilitation Team of Mahila Kisan Adhikar Manch (MAKAAM), forum for women farmers' rights and a core member of the Feminist Policy Collective (FPC). She has made a consistent effort towards bringing in a feminist analysis in research and actions in the areas of water, land and agriculture. She is currently involved in a programme with women farmers on food, nutrition security and water.

Kyoko Kusakabe is a professor of gender and development studies at the Department of Development and Sustainability, School of Environment, Resources and Development, Asian Institute of Technology in Thailand. She has over 20 years of experience in research and teaching gender and development in Asia especially focusing on the Mekong subregion. Her research focuses on gender issues in labour/work, especially on labour migration, garment factory workers, informal employment and agriculture and fisheries. She is currently an executive committee member of Gender in Aquaculture and Fisheries Section of Asian Fisheries Society, and a co-editor-in-chief of *Gender, Technology and Development* journal (Taylor & Francis).

Irene Leonardelli just completed her PhD at IHE Delft Institute for Water Education, in Delft, the Netherlands. She worked at IHE Delft for more than four years as Junior Researcher involved in different research projects, focusing particularly on gender and water. She was a Marie Sklodowska-Curie Fellow of the feminist political ecology network WEGO-ITN, funded through the European Union's Horizon 2020 research and innovation programme under the Marie Sklodowska-Curie grant agreement No. 764908. In her research, she looks at processes of rural agrarian transformation and water governance, from a feminist perspective.

Janwillem Liebrand is Assistant Professor at the International Development Studies group, Department of Human Geography and Spatial Planning, Faculty of Geosciences, Utrecht University, the Netherlands. His research and education focusses on the role of science and technology expertise in promoting development interventions in the field of land, water, food and climate change. He is author of the book *Whiteness in Engineering: Tracing*

Technology, Masculinity and Race in Nepal's Development (2022), and coordinator of the IOS platform Fair Transitions (Institutions for Open Societies) at Utrecht University.

Gayathri Lokuge is Senior Researcher at the Centre for Poverty Analysis, Sri Lanka, and at present leads the research cluster titled 'livelihoods and employment.' Her research and publications focus on fisheries, post-war development and rebuilding, peace and conflict studies, gender and intersectional identities and entrepreneurship studies.

Sandra Manuel is a social anthropologist, assistant professor at University Eduardo Mondlane (UEM) and researcher at Kaleidoscopio Research Institute on culture and public policy. Her teaching, research and policy analysis focuses on the social study of body, gender, sexuality and health themes. Her research questions normative gendered notions looking at the intersectionality of gendered relations and understanding socio-cultural readings of the body and sexuality, specifically in the African context. She is, currently, a member of the editorial board of *Feminist Africa* journal and previously was at the editorial of *Anthropology Southern Africa* journal. Her work at university also comprises empowerment of young women students on sexual reproductive health and rights through action research within a southern African regional partnership. Additionally, she worked as the adviser for the UEM Vice-Chancellor for research and a planning dimension.

Amanda Matabele is a master's finalist in social anthropology at University Eduardo Mondlane. During her studies she became interested in the topics of health, gender and sexuality and joined a research and advocacy group focusing on sexual health and reproductive rights, called Young Women Leaders. Her initial research experiences emerged from such platform. She developed her bachelor's research under the Dengue Water and Households project in 2017, from which she had her first experience with ethnography and fieldwork and developed her skills in data collection and analysis, academic writing and participation and presentation at academic conferences and seminars. Outside the academia she has worked as a researcher for social consultancy firms, mainly in the health sector.

Danícia Munguambe has a master's degree in anthropology with a focus on health and illness from the University Eduardo Mondlane (UEM), Mozambique and a bachelor's in anthropology from the same university. She also has a bachelor's degree in theology from the Seminário Unido de Ricatla in Maputo. She is a social researcher and one of her first steps in social research was in the study on *Dengue, Water and Households: Informing Suppliers and Government Officials in Small Towns* (part of the DUPC2 programme – UNESCO) developed by the Institute of Water Education (UNESCO-IHE) in coordination with the UEM. She is currently carrying out social research at the Mozambique Health Committee (CSM). Her research interests include health, gender and public policies – housing policies in times of COVID-19 in the Mozambican context.

Astrida Neimanis is a writer, researcher and teacher who works at the intersections of feminist and environmental scholarship, often in collaboration with artists, scientists, writers, educators and other community members. With a focus on water, weather and embodiment, their publications include *Bodies of Water: Posthuman Feminist Phenomenology* (2017), the co-edited collections *Feminist, Queer and Anticolonial Propositions for Hacking the Anthropocene* (2021) and *Thinking with Water* (2013), alongside many journal articles, book chapters, blogs, podcasts and catalogue essays. They are currently Associate Professor and Canada Research Chair in feminist environmental humanities at UBC

Okanagan, on unceded Syilx territories, in Kelowna BC, where they are also Director of the FEELed Lab (www.thefeeledlab.ca).

Kathleen O'Reilly is Professor in the Department of Geography and Texas A&M University Presidential Impact Fellow. She has more than 25 years of experience doing research on gender, water and sanitation (WASH) interventions in rural and urban India. She is trained as a feminist geographer, ethnographer and South Asia scholar. Over the course of her career, she has sought an in-depth understanding of internal community and household dynamics as they pertain to access to resources, like water and toilets, for women and socially marginalised groups. Her research highlights the need to understand the complexities of social relations and sanitation policy as they pertain to spatial patterns of inequality in WASH. Her work has been funded by the National Science Foundation and the Bill & Melinda Gates Foundation, among others.

Margarida Paulo has a PhD in social anthropology from the University of Gothenburg, Sweden (2018); master's in social anthropology from the University of Cape Town, South Africa (2004); and bachelor's in social sciences from the Federal University of Rio de Janeiro (UFRJ), Brazil (1996). She has teaching experience at undergraduate and postgraduate levels, academic management, research and extension in Mozambique. She was head of anthropology session at University Eduardo Mondlane (2017–2018). Since 2019, she is Head of the Central Research Department at the Gender Issue Coordination Centre at the same university. Her research interests include health and the informal economy, with gender as a cross-cutting aspect.

Catalina Quiroga is a Colombian anthropologist with a research interest in feminist political ecology and critical geography. So far, her research work has been characterised by collaborative work with the local populations involved. She finished her BA in anthropology at the Universidad Nacional de Colombia. Then, she obtained a master's degree in geography from the Universidad de Los Andes, Colombia. Currently, she is a PhD student at the Human Geography Department at Lund University in Sweden. The aim of her research is to analyse how the interaction between climate change policies and local strategies to overcome socioenvironmental injustices produces landscapes of climate change. She is a co-founder and partner of the Association CAMBIUM: Climate, Environment, Research-Action Uniting Worlds, a local organization interested in building bridges between activism and academia in Colombia and Latin America.

Lucero Radonic is an applied anthropologist working on water governance and environmental change in Latin America and the United States. She focuses on water rights, water (in)security and water conservation policies and practices. Her recent collaborative research with partners in the public and private sectors documents cultural norms and values influencing water use and management in urban areas and seeks to envision strategies for more just climate adaptation. Lucero is Associate Professor in the anthropology department at Northern Arizona University. She is also affiliated with the School of Earth and Sustainability.

Kavita Ramakrishnan is Associate Professor of Geography and International Development at the University of East Anglia. She is an urban geographer who specialises in experiences of belonging, informality and everyday life on the margins. Her research has focused on the lived experiences of eviction and resettlement in India, and refugee provisioning and care in European cities. This work has been published in multiple journals, such as *Antipode*, *CITY*, *Environment and Planning D: Society and Space*, *Environment and Planning E: Nature and Space* and *Humanity*. She has been recently awarded a British Academy Wolfson Fellowship to explore embodied experiences of extreme heat in Delhi.

Binahayati Rusyidi is Associate Professor of Social Welfare and senior researcher at the Centre for Gender and Children Studies, Universitas Padjadjaran (UNPAD), Indonesia. She obtained her terminal degree from School of Social Welfare, University at Albany, New York, United States. Since the last 20 years her research focus has been on child welfare, trafficking of women and children, domestic violence and sexual violence on campus. Her work makes significant contributions to improve policy, intervention programmes and human resources capacities to prevent and tackle child protection and gender-based violence issues in Indonesia. In recent years, she has extended her research interests toward water insecurity and gender-based violence, water governance and gender, green social work and international social work education and practice. Dr. Rusyidi is the recipient of both individual and group's research grants/fellowships from the United Nations organizations as well as Fulbright, the British Academy, Raoul Wallenberg Institute, Taiwan Fellowship, DFAT Australia and Netherlands Fellowship programmes. Dr. Rusyidi also has authored and co-authored the publication of international journal articles and book chapters from her research.

Gabriela Salmón-Mulanovich is a researcher, member of the Institute for Nature, Earth and Energy at the Pontificia Universidad Catolica del Peru. She has a BSc in biology from the Universidad Nacional Agraria La Molina (2000) and a Ph.D. in international health, global disease epidemiology and control track from the Johns Hopkins Bloomberg School of Public Health (2014). Her research applies a one health approach to examine how environmental, social and structural determinants are linked to health and well-being. She is a former Fogarty fellow from the University of Washington and has experience developing epidemiology training modules with Latin American ministries of health and the Pan American Health Organization. In 2018 she was a finalist for the L'OREAL/UNESCO award for women in science in Peru. She has more than 30 publications in peer-reviewed scientific journals and has published other dissemination documents.

Yesaya Sandang is a lecturer at the Faculty of Interdisciplinary Studies-Universitas Kristen Satya Wacana, Salatiga, Indonesia. Yesaya has been conducting research on the intersection of law, tourism and human rights, specifically utilising socio-legal approaches for the past nine years. As an activist, Yesaya maintains affiliations with community development and human rights NGOs in Indonesia. Yesaya earned his PhD from the University of the West of England, Bristol, United Kingdom.

Alexandra Schindler is a postdoctoral researcher specializing in gender and migration with the International Water Management Institute. She has her PhD in cultural anthropology from the CUNY Graduate Center and her master's in gender and women's studies from the American University in Cairo.

Abhay Shukla is a public health physician based in India, working on public and community health issues since the last 35 years. He is a senior consultant with SATHI Pune and is a national co-convenor of Jan Swasthya Abhiyan (People's Health Movement – India). He has co-edited/authored the books *Review of Health care in India*, *Health system in*

India – crisis and alternatives, as well as *Dissenting Diagnosis*, exposing malpractices in the private medical sector in India. He has been involved in drafting of the national 'Patients' Rights Charter' developed by National Human Rights Commission. He has a number of scientific publications, and also writes regularly for *Economic and Political Weekly* and online journals. Abhay has contributed to developing community-based health responses to the COVID epidemic, as well as drafting of official health rights advisories. He is active in campaigns for patient's rights, social regulation of the private healthcare sector, right to health and developing a system for universal health care.

Dacotah-Victoria Splichalova is a community-based water researcher, an MSc in water resources, policy and management (Oregon State University), and a Ph.D. candidate at the Institute for Resources, Environment and Sustainability (University of British Columbia). Her dissertation focuses on reimagining water beyond material dimensions in the Yucatan Peninsula (Mexico) within the Cuxtal Ecological Bioreserve to more fully attend to what we 'know' as water and to incorporate elements such as culture, identity, belonging, sacredness, emotions, psychology, the senses and agency. Dacotah employs participatory action and decolonising research approaches with creative arts-based engagements of community-led storytelling and filmmaking. Working with partner communities, her research aims to co-reveal water's complementary elements to foster community interconnectedness (well-being) and cultural resurgence (guardianship) with water. Doing so offers broader learnings for management and policy initiatives to address water insecurity risks more fully.

Paula Skye Tallman is an anthropologist investigating the causes of health inequalities and formulating strategies to promote environmental and human health. She aims to identify the pathways creating global health disparities and to use this scientific knowledge for action to improve human and ecological well-being. As an undergraduate, she investigated stress biology at Johns Hopkins University and graduated in 2008 with a BA in behavioural biology with honors. The same year, she was awarded a grant from the Johns Hopkins programme in Latin American studies. In 2009 she pursued a Ph.D. in anthropology at Northwestern University and graduated in 2015. Since 2021 she is part of the Department of Anthropology at Loyola University Chicago (LUC) as Assistant Professor of biological anthropology and the Director of the Laboratory for Human Health and History (HHH lab). She has authored over 20 peer-reviewed publications and her work has been supported by more than 15 grants and fellowships.

Nancy R. Tapias Torrado is Barry Pashak Postdoctoral Fellow of the Social Justice Centre at Concordia University. She is also Visiting Fellow at McGill University's Centre for Human Rights and Legal Pluralism, former Postdoctoral Fellow in the Faculté de Science Politique et de Droit (FSPD) at the Université du Québec à Montréal (UQAM), doctorate in sociology (U. Oxford) and a human rights lawyer (LLM U. Essex; LLB/JD U. Javeriana). She has dedicated her professional life to working with and for human rights defenders and vulnerable communities across the Americas, many of them Indigenous peoples.

Rozemarijn ter Horst works as a lecturer and PhD researcher at the Water Resources Management group at Wageningen University since October 2020. She wants to understand how models gain influence, and how we can constructively engage with the power of models in the context of transboundary water conflict. Her work focuses specifically on everyday practices of those who do and engage with transboundary water governance, including diplomats, policymakers, scientists, consultants and donors.

Before her work with Wageningen University, Rozemarijn worked at IHE Delft on water diplomacy and remains affiliated as Guest Researcher Transboundary Water Governance.

Lisa Bossenbroek's research focuses on gender and water/environmental change interactions. She is deeply connected with the people she works with and attempts to do justice to the diverse voices experiencing environmental change. Currently, she is co-leading the project SALIDRAA 2 – salt in the system at iES, Institute of Environmental Sciences, RPTU – Landau, Germany, which focuses on the social and environmental impacts of salinization in the Draa River basin in Morocco. Within this project, she analyses environmental changes from a gender justice perspective and is interested in how local actors develop (alternative) development paths.

Arianna Tozzi is an environmental social scientist whose research sits at the intersection between feminist political ecology, science and technologies studies and environmental humanities. She holds a PhD in human geography from the Universty of Manchester. In her research she uses feminist and postcolonial approaches to unpack the intersecting impact of climate change, development policies and processes of agrarian transformation in the rain-fed areas of Maharashtra. Her field research was supported by the Postgraduate Research Award 03.21, Royal Geographical Society (with IBG).

Yaffa Truelove is Assistant Professor in the Department of Geography and the Program in International Affairs at the University of Colorado, Boulder. Her research focuses on urban and feminist political ecologies of water, urban infrastructure and the governance of Indian cities.

Maitreyi Koduganti Venkata is a researcher with the School of Environment and Sustainability at the Indian Institute for Human Settlement (IIHS), Bengaluru. She heads Water Youth Network (WYN), a Netherlands based youth organization that brings young people work together to become agents of change and ultimately improve the world's water situation. An alumnus from UNESCO IHE Institute for Water Education, her work is situated at the intersection of climate adaptation, water and gender studies. She also works across the themes of urban water governance, climate change adaptation and vulnerability assessment, sustainability studies, gender and Research-into-Use, where she engages in working at the ground level along with local communities across the rural-urban continuum. In addition to this, she is a researcher. Her passion towards action-oriented research and putting research into use led her to developing documentaries and participatory games alongside IIHS, Red Cross Red Crescent Climate Centre (RCRCC), and UNESCO-IHE Delft. She is also engaged in writing several academic articles, blogs and op-eds around water management, climate change, sustainability and resource management.

Özge Yaka received her doctorate in sociology from Lancaster University. She works at the intersection of human geography, gender studies and environmental humanities and is interested in critical, feminist, phenomenological and posthumanist theories of body, agency and subjectivity, especially in the context of grassroots environmental struggles. Her work bridges contemporary (critical, feminist and post-) phenomenology, environmental justice and Indigenous and relational ontologies in developing new perspectives on gender, body and justice in the context of local environmental movements (e.g., socio-ecological justice). She is currently based at the Institute of Geographical Sciences, Freie Universität Berlin.

Margreet Zwarteveen is an irrigation engineer and social scientist, who joined IHE Delft in 2014 to become its professor of Water Governance, within the Water Governance Department. Zwarteveen studies water allocation policies and practices, focusing on questions of equity and justice. Her research centres around how institutions, technologies and markets shape water allocation and regulate water flows, and of different ways to make sense or legitimise these. Zwarteveen uses an interdisciplinary approach, seeing water allocations as the outcome of interactions between nature, technologies and society.

INTRODUCTION

A carrier bag for gender and feminist water research

Tatiana Acevedo-Guerrero, Lisa Bossenbroek, Irene Leonardelli, Margreet Zwarteveen and Seema Kulkarni

This book is the fruit of friendships and long-time collaborations that move across times and continents. We (the editors of the handbook) are women (a category that none of us innocently or easily inhabit) coming from Santander, in north-central Colombia; the lakes and rivers of Utrecht and the so-called 'peat colonies of the North – both in the Netherlands; the mountains in Trento, northern Italy; and Maharashtra in India. We got to know each other through various feminist (action-)research projects on water and in the classroom while teaching a course on gender and water resources management.

The action-research projects became occasions for joint learning about the many connections between gender and water, while also providing much inspiration for inventing and discussing forms of hydrofeminism. We jointly engaged in feminist analyses of agrarian change (in Morocco and India) premised on the intensification of water use (Leonardelli et al., 2022, 2023; Bossenbroek & Zwarteveen, 2015); looked at the irregularity of water services provision through a feminist lens (in Colombia and Mozambique) (Bayona-Valderrama et al., 2020; Acevedo-Guerrero, 2022); and wondered about how whiteness and masculinity color the water profession (Liebrand, 2017; Zwarteveen, 2008). These research collaborations taught us, among many other things, the importance of always situating feminist reflections and actions in the contexts from which they arise. Teaching feminist water insights gave us different opportunities to learn together, with the classroom turning into a joyful and creative feminist critical thinking hub. Together with students, we engaged in joint explorations of the many social dimensions of water, questioning in particular how gender and other power relations based on class, ethnicity, caste, religion or ability shape water worlds. Building on a rich scholarly legacy that connects water with resistance, justice and the environment, one of the aims of the course was to illustrate how water is a key site for the production and (re-)negotiation of social and gender differences and inequalities, using this as a basis for interrogating mainstream definitions of, among others, water security and water governance. In the course, we also presented various strands of contemporary feminist environmentalism, using these to make feminist sense of ecological devastation and crisis or the uneven and inequitable impacts of climate change.

A red thread in our research and education collaborations has been the combining of feminism with water, with many of us engaging in a range of experiments to either make

DOI: 10.4324/9781003100379-1

feminist insights applicable to watery questions or reconcile being a feminist with being a water scholar or activist (also see Resurrección & Elmhirst, 2020). We share a concern about the (lack of) equity and sustainability of prevailing water (and water-related rights, incomes, risks) distributions, and recognize that the distributional questions that form the heart of water governance are always contested and disputed (Zwarteveen, 2015; Zwarteveen et al., 2017). We also share a commitment to activist scholarship: challenging conceptions and practices of knowledge and authority that do harm, and undertaking our work through engaged research and pedagogies that keep analyses of power at their centre but carry a spirit of hope for better ways of living together in a multi-species world. Through our interactions, we learned to appreciate the many differences between us, starting with the very different ways of inhabiting our (also very different) bodies and enacting (or performing) our womanhood. Hence, some of us queer gender through the exuberant celebration of femininity, while others prefer doing this through subtle forms of refusing to act and behave in 'feminine' ways. Differences partly stem from how and where we were educated, disciplined and socialized; are linked to having grown up and worked in very different political landscapes; and stem from engaging with different watery contexts. Such differences co-determine how and where we position ourselves in academic and activist debates and struggles, indeed shaping our feminisms. In navigating and discussing such differences, we gradually discovered how to cherish them as a source of inspiration and reflection.

We have tried to carry this spirit of respectfully staying with differences over to this book, infusing it with similar pleasures of curiosity, generosity, open-mindedness and discovery as those that marked our collaborations. While the book is called a handbook, we made a conscious choice to not use it to establish another 'water and gender' canon consisting of 'must-read' texts and authors or providing a curated introduction to key concepts and theories. We did not want to use the book to mark (and police) the boundaries of an emerging body of 'gender and water' scholarship. Instead, we used it as an opportunity to show the richness and diversity of an expanding field of study and activism. Rather than to simplify and eliminate by foregrounding or prioritizing some accounts or theorizations of gender-water linkages or hydrofeminisms at the expense of others, we want the book to complicate, hesitate and slow down, so as to multiply feminist water voices. When we started with the book project in 2020, we therefore decided to make an explicit effort to include as many different voices in/around feminism (or gender) and water as possible. In particular, we were keen to include a younger generation of researchers and practitioners, those for whom it is more difficult than it ever was to ignore the destruction and harm caused by globalizing processes of exploitative profit-making. We also wanted to bring in new ideas and fresh perspectives by inviting feminist water activists or gender specialists coming from less usual places, or those writing about less familiar topics. Our wide search for possible contributors resulted in positive and enthusiastic responses, many of which eventually resulted in chapters for the book. Contributing authors range from those just starting to work as researchers and scholars to those already having built their career in academia; from those who reflect on their professional practices for the sake of this book, to those who on a daily basis try to make gender-meaningful changes in their water-work through diverse forms of activism.

Each of the chapters tells stories that situate the narrator(s) in specific waters and specific feminisms. In bringing together these stories without necessarily forcing them into one coherent framework, our book brings to mind Ursula Le Guin's carrier bag theory of fiction. This essay that Le Guin wrote in the 1980s proposes replacing hero-centric stories of action and success, starring phallocentric weapons (spears) and linear temporalities of conquest and destruction, with 'carrier bag' stories that foreground containing, receiving and sharing. Perhaps this book indeed resembles a carrier bag, a sling, a shell or a gourd in the sense that it holds a jumble of things together, in sometimes tangled and messy ways. Doing away with a spear's clear trajectory, carrier bags allow testing out different political eventualities and make room for complexity and contradiction, for difference and simultaneity (Le Guin, 1988; also see Leddy, 2019).

While not looking for coherence, we were interested in tracing connections and learning from each other. When the pandemic Covid-19 raged around the world and limited the free movement of people we gratefully learned to make use of advanced communication programmes to continue with the exchange of ideas among ourselves and with the contributors. In May 2022, we organized an online workshop with all the authors of this book. Everyone tuned in from their home office or from another spot, sharing questions, concerns and passions about today's water worlds. How to conceptualize the social relations that govern water? What are feminist water questions? And how does a feminist water perspective allow making visible some of the so far untold water stories related to women, men and their livelihoods? Behind the colorful panoply of small screens on Zoom, the workshop revealed a wide range of different methodologies to do gender and feminist water research, characterized by diverse positionalities, theories, geographies and experiences – once again confirming that there are many ways to combine feminist questions with water questions.

The book builds on a rich legacy of feminist water/irrigation scholarship. Attention to gender in water started in the 1980s and 1990s, often as a response to the gender-blindness of water projects in so-called developing countries. Already in 1973, Hanger and Moris recognized this gender-blindness in their evaluation of the Mwea irrigation system in Kenya. They wrote: 'it is our contention that the unsatisfactory recognition of women's rights and needs within the scheme remains one of the greatest weaknesses of the "Mwea system."' It is our doubt about this central aspect, so important for the long-term welfare of Mwea families, that has led us to question . . . whether the Mwea pattern ought to be replicated elsewhere.' (Hanger and Moris, 1973, p. 244). Some ten years later, Schrijvers wrote about the Mahaweli H irrigation system in Sri Lanka:

> [T]he chronic undernutrition in the Mahaweli H area is a direct result of planning that cuts women off from their productive resources. It is of primary importance that women, who have to provide the daily food for their children and other members of the family, have the means to obtain sufficient food. . . . Research showed that only 35% of the net income of the male farmer (after debts were paid off) benefitted the rest of the household.
>
> *(Schrijvers, 1984, p. 270)*

Drinking water (and later sanitation) projects likewise attracted critique from feminists and gender specialists for their lack of sensitivity to prevailing gender relations, with pictures of women carrying water quickly turning into iconic reminders of the genderedness of water. Early analyses include the book of Christine van Wijk, which provides a beautiful introduction to why attention to gender in water projects is needed while also suggesting useful ways to meaningfully engage with questions of gender in drinking water projects and planning (van Wijk, 1998).

These early analyses provided the inspiration for a wealth of studies that documented and analyzed the genderedness of processes of 'water development.' While many of those were rather instrumental in wanting to improve the design and planning of water projects, there were also studies that provided more profound feminist critiques of development or modernization, for instance by showing how capitalist forms of profit-making rested on and reproduced structures of patriarchy. One of the early books of Vandana Shiva (called *Staying Alive*), for instance, includes a chapter entitled: 'Women and the vanishing waters' (Shiva, 1988, pp. 179–217). In this chapter, she describes drought and water scarcity and the exploitation of women as two effects of the same reductionist knowledge that 'violates cycles of life in rivers.' Shiva posits the scarcity of drinking water and the problems this entails for women as one of the effects of irrigation development (Shiva, 1988, p. 179). She ends her list of descriptions of severely water scarce regions and villages in India as follows:

> The cause of the water crisis and the failure of solutions both arise from reductionist science and maldevelopment working against the logic of the water cycle, and hence violating the integrity of water flows which allows rivers, streams and wells to regenerate themselves. The arrogance of these anti-nature and anti-women development programmes lies in their belief that they create water and have the power to "augment" it. . . . That is why in water management, it is imperative to think and act ecologically, to "think like a river" and to flow with the nature of the water.
>
> *(Shiva, 1988, p. 182)*

Also, Judith Carney's seminal analysis of the Jahally Pacharr project in the Gambia did more than bemoan the lack of gender sensitivity of those who had planned and designed the project. Her analysis placed the project in a broader historical perspective, seeing it as part of a distinct development strategy that was aimed at import substitution and national food security (Carney, 1986, 1988). Carney showed how realizing the project's objectives of agricultural intensification (among others by making two crops per year possible) crucially depended on the labour of women. Yet, because the project vested all land- and water-rights in men, while also targeting all support services and credit to men only, it 'unwittingly legitimized male control over the surplus gained from double-cropping' (Carney, 1998, p. 330). This provoked gender conflict, while also negatively affecting the realization of the project's objectives (see Carney, 1986, 1988, 1992, 1993, 1994, 1996, 1998; Carney & Watts, 1991).

With their analyses, Shiva and Carney paved the way for feminist water studies that not just created recognition for the importance of women as water actors – as users, decision-makers and managers – in projects and policies but also mobilized feminist analytical tools to question these projects and policies. The last decades have seen a proliferation of studies that look at water questions through a gender or feminist lens, including several excellent edited volumes. Hence, there are the early edited volumes by Merrey and Baviskar (1998); Bennett et al. (2005); Ahmed (2005); Coles and Wallace (2005) and Kuntala Lahiri-Dutt (2006) as well as the more recent volumes by Resurreccion and Elmhirst (2008), Zwarteveen et al. (2012), Buechler and Hanson (2015) and the even more recent one by Sehring et al. (2022). There have also been several special journal issues on gender and water, including an early one in (edited by Frances Cleaver, 1998) *Agriculture and Human Values*, one in *Gender, Place and Culture* (2009, p. 16:4 edited by O'Reilly, et al. 2009); one in the *Journal of Gender and Development* (2010, p. 18, edited by Wallace and Porter) and one in *Economic and Political Weekly* (2011, p. 18 edited by Maitreyi Krishnaraj). The most recent one probably is a special issue of the journal *Frontiers*, dedicated to 'Changing the Discourse: From the Rhetoric on Women and Water to a Feminist Water Agenda' (edited by Joshi et al., 2023). The University of Pennsylvania even publishes a journal solely dedicated to gender and water issues (https://watercenter.sas.upenn.edu/journal). There is also an ever-expanding plethora of guidelines and more practical books on how to 'do gender' in water and sanitation projects, many of which originate from donor organizations.

This wealth of studies perhaps indicates that, almost 40 years after the first gender and water studies saw the light, feminism in water continues to be very much alive. It also continues to be needed. This is so, first of all because neo-liberal ideologies have become much more entrenched and hegemonic than they used to be, making feminist critique both more difficult and more urgent. Neoliberal reforms have provoked forms of privatization and commercialization of water with often detrimental effects for (gender-)equity in access. In agriculture, the 'freeing' of land and water from prior investments to allow their exchange on markets has gone accompanied with the flexibilization (or indeed feminization) of agricultural wage labour. The effect has been the replacement of place-based relational webs of responsibility, solidarity and care with externalized forms of accountability and control. Neoliberal reforms also tend to entail the publicly promoted and supported push to increase profits per drop of water, provoking a process of re-allocation of water (and rights to water or water infrastructures) towards what are seen as more productive and more efficient uses and users. This often happens in parallel with a downsizing of public water budgets, something that significantly reduces public capacity to regulate and manage water. Feminist scholars have forcefully demonstrated that the ability of some to access water or accumulate land and water rights is premised on deeply gendered and racialized institutional structures and ideologies. Indeed, the ability to perform and succeed as efficient and productive ('modern') water users, managers, irrigators and farmers hinges on and reproduces the active and continuous interiorization of 'feminized others – including women and nature (Mies, 1986, 1998; Shiva, 1988, see also Ahlers & Zwarteveen, 2009; Harris, 2006, 2009; Ramamurthy, 1991; Sultana, 2009; Ahmed & Zwarteveen, 2014; Zwarteveen, 1997).

At the same time, climate change; the increased awareness of the precarity and finality of water resources; and the loss of biodiversity are further increasing the urgency to re-think '(water) development-as-usual.' Yet, struggles to do this run up against those ever more fiercely protecting distinct heteronormative, white and masculine forms of privilege, something that among others becomes apparent in the rise of gender-related violence (including violence against LHBTQI+ people). Forms of hydrofeminist activism and scholarship are important to help expose how the under-valuation of rivers, lakes and aquifers and that of labour (much of it from women) are connected; show how intersecting relations of gender, class and race (among others) allow making it possible and justifying that some can access water, benefit from water-related incomes or protect themselves against floods and droughts at the expense of others; and systematically question how voice and authority in water continue to be defined in masculine terms.

Feminist insights are crucial in helping articulate how 'living well' is about relating and connecting with others – humans and more-than-humans – much more than it is about economic profit-making or the pursuit of self-interest. This for instance entails learning to recognize and appreciate the importance of the work of care/caring, work that is often done by and associated with women. It also about learning to cultivate the attachments and solidarity needed to protect water, use it frugally and share it fairly. Feminism and feminist scholarship in and around water, in sum, are about more than making sure that water policies and projects also benefit women or contribute to gender equality. Feminism also provides powerful tools to re-think, re-imagine and re-do water in more just and sustainable ways. Here, accounts of women actively organizing to either resist harmful developments or (re-)shape their own watery livelihoods provide powerful inspiration. In the command areas of one of the irrigation projects (Khudawadi village) located in the drought-prone region of Maharashtra, for example, women collectivized and demanded a share in the water from canal irrigation to irrigate the land they had collectively leased (Kulkarni, 2005). In northern India, Dalit women collectivized as *jal sahelis* or water friends and embarked upon the restoration of old and defunct water bodies to ensure drinking water security in their villages. In the Kurdish majority areas in Northern Syria, the Women's Economic Committee of the Rojava movement irrigated more than 100 ha in a former state-run farm. This was in response to the water embargo from Turkey, which hit the water and food security of Northern Syria (Rushton, 2023). In the coastal city of Buenaventura, Colombia, women from the Committee for the Defense of Water and Life defend their right to water. The members of the committee have organized protests and blockades in the streets to draw the government's attention to the poor state of infrastructure and the lack of potable water in most of the city's neighborhoods. Their acts of defending water are not limited to the realm of collective action and overt activism against state and market decisions. Rather, defending water in cities such as Buenaventura is also an everyday gendered practice entailing a repertoire of domestic activities and techniques in order to secure water, store it in and around the home, and keep this water clean (Acevedo-Guerrero et al., 2024). Such stories of women transforming ways of engaging and living with water, and organizing social relations around water, need to be told and disseminated to fuel new imaginaries of what water development is or can be.

This handbook, then, is a collection of reflections and studies that can be used as a prismatic lens into a thriving and ever proliferating array of feminist water studies. It provides a clear testimony of how hydrofeminism has evolved from rather instrumental gender and water studies to scholarship that uses feminist tools to pry open, critically reflect on and formulate alternatives to water development-as-usual. The book also shows how the community of feminists interested in studying water has diversified and expanded, from often white female scholars studying projects (and gender relations) in the so-called Global South to a varied mix of scholars and activists theorizing from diverse geographical and political locations – prominently including the body. The early work of such authors as Vandana Shiva, Maria Mies and Carolyn Merchant continues to be an important source of inspiration, with ecofeminisms forming important threads in ongoing conversations within Feminist Political Ecology (FPE) but also with feminist technoscience studies and environmental humanities. Slowly but steadily, engagement with decolonial and anti-racist politics is starting to enrich this feminist water work, questioning forms of white colonialist privilege and associated forms of epistemic violence. We hope that the book will serve to further open up the field of feminist water studies, providing new fuel to ongoing struggles against enduring forms of marginalization, exclusion and oppression as well as yielding new inspirations for imagining and enacting more just and sustainable water futures.

Outlook of the handbook

The handbook is organised into five interconnected sections: 1) Positionality and embodied waters; 2) Revisiting water debates: diplomacy, security, justice and heritage; 3) Sanitation stories; 4) Precarious livelihoods; 5) New feminist futures. Each of the sections brings out the gendered nature of water, shedding light on the often-neglected care and unpaid labour of women and its relationship with extractivism and socioeconomic inequalities.

Most chapters in the handbook underscore the need for forging emotional connections with the more-than-human world, recognizing this as critical for bringing about a transformative change. The chapters also examine the intricate interrelationship between water governance and women's bodies, drawing on studies done in different regions of the world. Some of the chapters in the handbook bring us back to where we started forty years ago – women's access to water and the burdens they carry due to its lack. From countries in the Global North, like the United States and Canada to countries in the Global South such as Mozambique and Colombia, the question of poor access to water and the disproportionate burden on women highlights the failure of the privatisation model and the withdrawal of the state from welfare provisioning.

The theoretical diversity of the chapters enriches the exploration of new futures while remaining firmly grounded in the everyday experiences of women and the economically marginalised in the context of water. Moreover, the authors explore a number of innovative research methods, such as photo voice, life history, storytelling, creative drawings and virtual museums, to tell their stories around the gendered nature of water.

Section 1: Positionality and embodied waters

In recent years, feminist scholars have shown how waterscapes are always infused with power dynamics: studying gender-water relations means recognizing that one's position determines different ways of accessing, using, experiencing, remembering, narrating and sensing different bodies of water and that these different ways shape water management practices as well as political struggles (Neimanis, 2016; Sultana, 2011; Truelove, 2011). Focusing on how different subjectivities are embodied through more-than-human relations is thus crucial to understanding how inequalities unfold, how they are reproduced and how they can be challenged (Cruz-Hernández, 2016). In this section, the contributors dig into and expand different aspects of this theme from various perspectives, using a wide variety of methodological approaches and proposing cases from around the world.

In her chapter, Özge Yaka develops a body-centered approach to studying gender-water relations. She does so by focusing on women's leading role and radical activism in organizing against small-scale, run-of-the-river hydroelectric power plants (HEPPs) in the rural parts of the East Black Sea Region of Turkey. Memories, history and sense of place are produced and conserved in the material relations between women and the river, and these importantly shape their struggle. Through her analysis, Özge Yaka argues that paying attention to bodily and intimate relations with the more-than-human world is essential for water governance to enact more equitable water worlds.

Janwillem Liebrand, in his chapter, uses the method of life history to investigate the linkages between men, masculinity and professional authority in development, particularly in irrigation engineering. To do so, he focuses on the life story of Baidya, a senior Nepali agricultural engineer. By considering Baidya's life story as a cultural performance and a product of intercultural interaction between individuals, Janwillem Liebrand illustrates how masculinity – intersecting with whiteness and heteronormativity – in water engineering is enacted through everyday interactions, behaviour and practices. This allows him to reflect on his own professional identity as a white Dutch engineer and feminist scholar investigating masculinity in engineering. The author concludes by arguing how life history interviews as a qualitative research method can work as a political tool to deconstruct masculinities, whiteness and heteronormativity in water engineering, yet only if used in a critical self-reflective way that problematizes the positionality of the researcher.

Maitreyi Koduganti Venkata and Gabriela Cuadrado-Quesada bring back attention to the materiality of the body. They focus on bodily experiences and emotions to illustrate the gendered impacts of groundwater salinity and other environmental and chemical contaminants on different community members in the coastal village of Lodhva, Gujarat, India. They rely on different methodologies, like participatory creative drawings and in-depth interviews, to show how the bodies and intimate relations of children and women from lower castes were most affected by groundwater salinity. Through their analysis, the authors suggest that emotions – and their connection to the materiality of the body – are an important entry point to contemplate the everyday struggles of accessing and using groundwater.

From Gujarat on the western coast of India, we move to Chennai on the east. The chapter of Qurratul Ain Contractor provides an ethnographic analysis of the everyday caring practices of women from Indigenous coastal wetland communities in Ennore-Pulicat (Chennai). By identifying different caring practices (family care, care in harvesting food, collective care), the author illustrates the strong interdependence between maintaining wetland health and the health of the wetland communities, blurring the lines between family, community and wetland care. In doing so, she illustrates how the socionatural world is entangled and co-constituted. Through the author's ethnographic accounts, the chapter contributes to challenging the dominant narrative of development as industrial growth, which portrays the wetlands as empty spaces, thereby neglecting the myriad informal economies relying on the wetland ecosystem. In addition, the chapter is a call to recognize and value women's labour and knowledge in wetland conservation.

Silvia Corredor-Rodríguez focuses on Latin America. Her chapter builds on an ethnographic analysis of the everyday water labour practices of women from inland delta townships in La Mojana (Colombia). Although they are surrounded by rivers, communities in this region do not have access to potable water for domestic consumption. Domestic water scarcity is caused by pollution caused by agricultural and livestock waste, the adverse effects of mining on the banks of the Cauca River, and heavy periodical floods. She argues that water access and treatment do not only entail technical processes but also *bodily* ones. Women have built particular relationships with rivers, as well as with rains and groundwater. They have developed different practices and knowledges around these different sources, these experiences are embodied and sensorial. Informed by its taste, smell, feel and colour, they classify water as *fresh*, *thick*, *thin*, *blond* and *clear*, giving it different uses.

With a focus on cities of the Global South, Yaffa Truelove explains why and how recent feminist urban political ecology literature focuses on the body to understand everyday urban water access and inequality. She summarizes three key contributions of this work: 1) how the scale of the body matters in multi-scalar approaches to water, 2) how gender, class, race and ethno-religious relations shape patterns of water inequality and insecurity and 3) how 'everyday' practices and politics help unfold often unnoticed and under-theorized dimensions of water insecurity and inequality. These embodied aspects are central, she argues, to understand how different city dwellers experience urban water insecurity and thus to address such inequalities.

In the final chapter of the section, Kathleen O'Reilly, Kavita Ramakrishnan and Jessica Budds explore when, why, how and by whom infrastructure is maintained, repaired or left to decay. In doing this, the authors delve into the affective labour and political implications of infrastructure's shifting materialities and meanings. The authors document how the decay, maintenance and repair that characterize the life of infrastructures are premised upon the often unremunerated and invisibilized labour of ordinary people, particularly of women, who secure connectivity and flow. This labour is connected to differentiated and situated embodied and affective experiences. Through their analysis, the authors connect the temporal, material, affective and social dimensions of infrastructure – showing the possibilities of such an analysis for social justice.

Section 2: Revisiting water debates: diplomacy, security, justice and heritage

This section engages with discussions on water security, water justice, water diplomacy and water heritage. Water security has been defined as the availability of sufficient quantities of potable water through community preferred methods and the just distribution of water collection labour and water management decisions (Cook and Bakker, 2012). Similarly, water justice connects water availability with the 'socio-technical and legal-cultural determinants of how available water flows are accessed and allocated' (Zwarteveen & Boelens, 2014, p. 144). Water diplomacy, in turn, helps countries settle disputes and cooperate peacefully about transboundary water resources (Sehring et al., 2022). Finally, in an uncertain present characterized by climate change, water heritages can help think about emancipatory futures. A quest to unearth heritage promotes storytelling around water and its intersection with natural and built environments, water wisdom, social justice and livelihoods.

Benjamin Dosu, Mohammed Abubakari, Maura Hanrahan and Tom Johnston explore gender inequalities in domestic household activities focusing on the water security experiences of women and girls in rural Ghana. Little research has been conducted on this topic in rural African settings, especially in rural Ghana, although rural dwellers are five times more likely to be water insecure than those in urban areas. Using a cross-sectional survey of three rural communities, the chapter shows how gender inequalities are entrenched in domestic water collection and use, with women and girls performing almost all the duties. Highlighting the importance of the intersections between gender and age, the authors show how, in these rural communities, boys stop engaging in domestic water collection when they enroll in high school or when they start an apprenticeship.

In a similar direction, Cara Jacob, Lucero Radonic and Priyanka Jayakodi tackle the issue of failing infrastructure, pointing to how it is increasingly becoming a source of water insecurity throughout the United States. They focus particularly on the city of Milwaukee, in the area of the Great Lakes in Wisconsin, where – even though water quantity is not an issue – thousands of people experience water insecurity because the tap water is contaminated with lead. This toxic environment – or so the analysis shows – generates inequalities that need to be understood through an intersectional lens. Firstly, the lead epidemic is clearly racialized as the most affected area is the northern part of the city, where most African American families reside, and which is where the ageing infrastructure is located. Secondly, homeowners (usually white people belonging to the upper-classes) are better able to mitigate and cope with lead in the home than people living in rented houses, for the latter have less agency to modify their homes. Last but not least, the lead epidemic has clear gendered unfoldings as women end up bearing most of the responsibilities related to reducing potential lead hazards in the home due to the water-intensive nature of care work.

Writing from a historical perspective, Chantal Victoria Bright investigates the connections between poverty, water security, war and its aftermath and women's activism in Liberia. Liberia is still recovering from long civil wars that lasted from 1989 to 2003. The prevalence of extreme poverty remains, and although water is abundant throughout the country, water availability is compromised by pollution, poor infrastructures, degraded wetlands and climate change, among other factors. Foregrounding Liberian women's activism and political participation from the 1920s to the feminist peace intervention in 2003, the author suggests how women could play an important role in water governance to reduce inequalities, particularly in relation to accessing drinking water. This is particularly relevant, she argues, as water insecurity disproportionately affects women in conflict-affected countries. Significantly, women could play an essential role in developing strategies for more sustainable water management and for addressing drivers threatening peace and security.

In turn, Stroma Cole, Paula Skye Tallman, Gabriela Salmón-Mulanovich, Binahayati Rusyidi and Yesaya Sandang follow the UN guiding principles to evaluate Indonesia's water governance. Specifically, they show how gender mainstreaming has been incorporated at a national framework level. Yet, they also critically identify some of the reasons why these national initiatives are not translating into increased gender equality in water governance at the local level. These reasons include patriarchy, legal gaps, overlapping mandates, lack of law enforcement, competition with other industries and unequal urbanization in the midst of climate change, which all continue to lead to increased water insecurity for women. The authors conclude by arguing for the need for new women's leadership development and empowerment programmes, human rights impact assessments with a gendered lens and increasing women's control over the local distribution of water.

Lisa Bossenbroek and Margreet Zwarteveen build on extensive ethnographic fieldwork to analyze changing agricultural policies in Morocco's Saïss region and interrogate them from a feminist water justice perspective in order to tease out how prevailing gendered ways of being, working and relating shape and are shaped by new agricultural and groundwater dynamics, in ways that are not always straightforward. They discuss how new farm 'entrepreneurs' and peasant farmers who have the willingness and ability to modernize, both clearly cast as men, mostly benefit from government subsidies. They explain how only a privileged few young men want and can pursue new modern farming identities, mimicking entrepreneurial stereotypes. New modern farms, in turn, have modified the existing gendered organization and meanings of farm work, as they heavily rely on (cheap) wage labour for the labor-intensive activities of planting, weeding and harvesting, much of it done by women. The chapter argues that all rural actors actively navigate existing spaces or create new ones to fuel, co-shape and fulfil emerging desires. Women for example took up and reworked feminine identities according to social position. Those belonging to landowning families and of the older generation assumed a more 'bourgeois' feminine identity, performing supposedly more domestic roles; younger women instead re-invented the domestic domain, coming up with entrepreneurial careers, such as craft-skills or small businesses. Women belonging to landless families, who rely on agricultural wage work to make ends meet, enjoy little personal pride in what they do and many even try hiding their wage-work involvement from the public eye. By documenting these diverse grounded positionings and spaces, the authors hope to pave the way for pragmatic and modest strategies for interrogating and challenging injustices.

Moving to the global scale, Rozemarijn ter Horst reviews a growing body of literature on gender, diplomacy, international negotiations and water. She discusses how diplomacy and engineering are interlinked in international negotiations over water. These are two distinct yet gendered fields, both in terms of the number of men and women that participate, as well as in terms of the gendered norms and practices that influence who participates and how. She argues that through discourses and infrastructures of securitization, states tend to limit the space for the substantial participation of women. When women do participate, it is often tokenistic or merely symbolic. On the other hand, increased cooperation over the course of decades creates space for increased, more real forms of participation by women, both through different policies and through changing norms.

This section ends on a hopeful note, with a reflection on the potential of water museums to exhibit and interpret both tangible and intangible liquid heritages, from ancient artefacts and technologies to strategies to combat water scarcity, pollution and climate change. Sara Ahmed explores the opportunities and challenges faced by the Living Waters Museum. This virtual museum, which is also her life's work, engages with citizens and communities through digital media, storytelling and the creative and performing arts. Launched in India in 2017, this museum portrays the value of building inclusive, collaborative partnerships which address social equity and gender justice through research, design and curation. The chapter shows how, through collaborative endeavours, Ahmed and her team have managed to flesh out the ways in which our water worlds are socially constructed through material realities, rituals and agency.

Section 3: Sanitation stories

This section takes a critical look at literature and developments in the water, sanitation and hygiene sectors (WASH). According to a 2023 report by the World Health Organization WHO, improving access to water, sanitation and hygiene can save 1.4 million lives per year (Wolf et al., 2023). This and other United Nations bodies have stated that WASH is not only a prerequisite to global health but contributes to school attendance and dignity (O'Reilly, 2016). The section first tackles debates around menstrual hygiene management (MHM), which entails access to WASH facilities, affordable and appropriate menstrual hygiene materials, information on good practices and a supportive environment to manage menstruation without stigma.

Jacqueline Gaybor Tobar explores how the WASH development sector's historic focus on product-based interventions frames menstrual health as a compliance issue rather than a systemic one. She documents how this approach reinforces the figure of the West as the saviour of girls and women from the Global South and diverts attention away from combating menstrual stigma. Highlighting the importance of not immediately searching for solutions, she argues for a critical examination of how restrictive sexual and gender norms promote unequal gender relations, impacting people's perceptions of and experiences with menstruation and the menstrual body. The chapter calls for the adoption of thoughtful, nuanced and feminist approaches to menstrual health to develop policies and programmes that are better aligned with the needs and realities of menstruating people and those that can support gender equality and social justice.

Building on similar discussions, Amie Jammeh and Tatiana Acevedo-Guerrero analyze the implementation of MHM programmes in schools. They draw on photovoice research with a group of 22 students living and studying in George and Chawama, two of the fastest-growing neighbourhoods of Lusaka (Zambia). They document how MHM programmes have only translated into the delivery of reusable pads whenever there are international funds available. The chapter argues that, facing multiple uncertainties regarding economic stability, infrastructural malfunctions in neighborhoods and schools, and socio-cultural changes due to widespread unemployment among men, girls were not particularly worried about clothes, pads and stains. This chapter increases understanding of challenges and expectations related to menarche and menstruation by looking through the eyes of students themselves and exploring their contexts and views.

Later on, the section explores the sanitation needs of transgender communities. Durba Biswas reviews available literature and delves into the case of India, where reforms providing gender-neutral and sex-segregated public bathrooms for transgender individuals have met with transphobia, undermining the welfare benefits of friendly sanitation interventions as violence and transgender bathroom use experience are deeply entangled. She argues that the country's traumatic colonial past and cultural and gendered norms acting upon intersections of caste, ethno-religion, class, sexuality, age and ability produce trans experiences which are context specific. She concludes that technical solutions can meet their sanitation needs partly, but a nuanced examination of the barriers to safe sanitation for transgender communities is essential for achieving various sanitation goals.

With a focus on the structural underpinnings of gendered vulnerability, Seema Kulkarni and Abhay Shukla discuss how extractivist development intersects with the exploitation of migrant female labour in sugarcane harvesting, causing irreversible changes in menstrual and reproductive health. The chapter is located in the 'prosperous' western Indian state of Maharashtra that attracts forced labour migrants from water-stressed regions who seek new employment opportunities. It highlights the exploitation of migrant workers and the commodification of women's bodies in both the production and social consumption sectors. Firstly, it calls into question the distorted model of market-driven water and agriculture development that perpetuates such crises in social reproduction. Secondly, it discusses the feminist agency exhibited by the invisible women workers, now advocating for meaningful employment opportunities in their home districts to reduce the need for migration. They are demanding a reassessment of water, agriculture and energy policies to prioritize food and nutritional security, as well as rational, quality health care and ecologically sound local employment. These efforts are challenging regressive social structures and policy frameworks towards improving the lives of women sugarcane harvesting workers and reshaping the terms of their work.

Finally, Kelly Dombroski engages in a critique of water-based, individualized sanitation structures, which contribute to 'flush it away' mentalities and inhibit the clear-minded consideration of alternatives. She shows how global elites, accustomed to not having to deal with their own or others waste, make decisions about water and sanitation infrastructure according to their own embodied norms of waste. She then explores alternatives to water-based sanitation in the minority and majority world, centering the 'soft infrastructures' of embodied practice that have often been the responsibility of women. She focuses on the practice of infant toileting in China and Australia and the implications of such a practice in habituating infants and caregivers to diverse kinds of embodied relationships with human waste. She argues that those doing hygiene and sanitation care work differently (particularly women) should have a prominent role in discussions about what safe and adequate future sanitation systems look like, with an openness to radical change. This way, those who design these systems should rehabilitate themselves to different relationships with waste, which might mean long-term visits to places with mountain outhouses or shared community toilet facilities.

Section 4: Precarious livelihoods

This section delves into the impact of extractivism, capitalism and colonialism on everyday livelihoods. Specifically, it explores the connection between capitalism and patriarchy by analyzing cases of extractivist regimes of (water) resource management. That is, regimes such as industrial agriculture, mining and rapid urban development which extract value without fostering socio-ecological reproduction (Ojeda et al., 2022). The different chapters illustrate how extractivist regimes of water management (re-)produce new socio-economic categories and new forms of marginalization based on gender, poverty, labour, caste, religion, age and origins. The chapters explore the intersection of different forms of exploitation in terms of what they mean for water and gender and examine entry points of change.

The section starts with a critical revision of initiatives proposing the increasing engagement of poor women as frontline agricultural workers. Deepa Joshi, Amina Dessouki and Alexandra Schindler argue that such initiatives fail to pay attention to the contextual challenges of the combined effects of economic, water and ecological crises, alongside persisting patriarchy and increasing poverty. They build on focus group discussions with 170 male and female water-reuse irrigators, followed by in-depth discussions with ten irrigators, to analyze the contentious water-reuse irrigation work for women in the Kafr El Sheikh collectorate, an area north of Cairo in the Nile Delta, Egypt. The chapter explains how women's increasing role in water-reuse irrigation offers limited economic gains, is risk-prone and insecure and is work that is no longer aspirational to anyone, including poor men. Employing a feminist political ecology lens, the chapter documents women's domestic and productive work burdens, their emotional anxiety in working with and consuming crops grown with poor quality water, and the limited scope they have in negotiating the increasing demands on their time and labour and in exploring other livelihood options. The authors conclude by fleshing out the gendered dimensions of water-reuse irrigation, which have roots in the colonial legacy of irrigation, a persisting patriarchal system and the political economy of agriculture in an arid geography.

Continuing with the exploration of agriculture in arid geographies, Lisa Bossenbroek and Hind Ftouhi analyze how the oases in the Draa Valley in southeast Morocco are affected by declining water availabilities. Looking beyond the natural decline of water availability they illustrate how water re-allocation processes and the growth of particular water-consuming sectors in the research area have drastic impacts on local livelihoods, gender relations and identities. The authors illustrate how women used to play an essential role in oasis agriculture. Despite its hardship, their work allowed them to keep busy, fulfil themselves and create and maintain social relationships. It also enabled them to find spaces of relative freedom: in the fields, they used to meet other women, exchange, laugh, gossip and have fun. The authors illustrate how farming has become more difficult due to less water availability and its diminishing quality. They illustrate what this means for women: some are increasingly confined to the private feminine domain, whereas others take up new opportunities to make a living and give sense to their daily lives. In doing so, they renegotiate feminine rural identities and shape the future becoming of the oasis.

From agriculture, the section moves to fishing. Holly M. Hapke, Nikita Gopal, Kyoko Kusakabe and Gayathri Lokuge explore the relationship between gender, water resources and fishing, revealing ways in which political, economic and ecological developments impact water resources and livelihoods. They analyze the entanglement between gendered livelihoods, gendered access to water resources and the norms that govern these resources through an exploration of three case studies. The first case study is from the Vembanad Lake region in Kerala, India, where urbanization and related economic development have polluted the lake, creating competition for access to water. These developments have affected women's access to water resources, which support the small-scale fishing activities in which they have historically engaged. The second case documents how, in Trincomalee, Sri Lanka, military occupation, fear of violence, minority social status and lack of formal rights to lagoons and waterways for fishing and gleaning have deprived women of access to water resources to pursue their traditional livelihood of gleaning for mussels and clams. The third one is the case of Banteay Meanchey province, Cambodia, where agricultural development, namely intensified dry season rice cultivation relying on reservoir-fed irrigation, combined with deforestation, inefficient infrastructure and poor water management practices, have reduced reservoir water levels, adversely impacting fishing and gleaning activities by reducing fish stocks. Women who fish and glean at the edge of the reservoir feel these effects most severely. Each of these cases demonstrates that when water resources are degraded or impacted by economic change, the impacts of such change are felt most acutely by particular groups of women – in these case studies, by poor, elderly women and women from particular castes or ethno-religious groups.

From fishing in South and Southeast Asia, the section moves to mining in Canada, where for over a decade, Wet'suwet'en women have been leading the defense of the Yintah (their ancestral territory) against the construction of megaprojects, including the largest private investment project in the country, the Coastal GasLink (CGL) pipeline. The pipeline crosses over the Wet'suwet'en Yintah and Wedzin Kwa, a sacred and fundamental river for this Indigenous people. Nancy Tapias documents how Wet'suwet'en women-led-mobilization and hereditary chiefs have never granted consent to CGL to work in their territory, calling on federal and provincial authorities and the corporations involved to stop the project in the Yintah. Yet, at the end of 2022, CGL started drilling. Drawing on her learnings and previous work in Latin America, Tapias presents the 'braided action' theoretical framework to study how, in a minimally favourable human rights context, Indigenous women-led mobilizations claiming respect and protection for their ancestral territories and most fundamental rights can harness a mobilization power in 'a braid of action,' achieving 'success' over much more materially powerful forces. She concludes that to live and to act in the Yintah is of fundamental importance for the Indigenous Wet'suwet'en people. Yet, despite the strength of their women-led mobilization the entrenched legacies of a colonial past that persist have made their efforts have little impact on corporate behavior.

The section ends with the chapter by Sandra Manuel, Margarida Paulo, Tatiana Acevedo-Guerrero, Danícia Munguambe and Amanda Matabele telling a gender and water story from urban Mozambique. They argue that most water-related campaigns implemented in the area (including disease prevention programmes to prevent new dengue outbreaks) are designed within colonial and western frames that tend to associate femininity with domestic, unpaid work and masculinity with non-domestic paid work. This is problematic as such frames oversimplify and homogenize complex local gender dynamics, risking to not reach the many people who are responsible for different aspects of water management. The chapter illustrates how in Pemba, there are Makhuwa migrant men who work as domestic employees and are in charge of cleaning, cooking and managing water for domestic uses – tasks that are historically associated with femininity. At the same time, although in Pemba it is men who are historically responsible for earning a salary and providing for their households, many women also contribute to the household's income by selling the extra water stored in the household's cement tank. With this evidence, the authors conclude that post-colonial governmentalities should acknowledge and embrace local gender dynamics and their diversity when developing different water programmes.

Section 5: New feminist futures

The final section of the book proposes some ideas to imagine and co-enact new feminist water futures. The chapters all start from a relational understanding of water inspired by feminist, post-humanist theories and ontologies. This means that they define water and gender in a situated manner, always in relation to specific more-than-human bodies, ecologies and histories. The aim is both political and theoretical: choosing to look at gender and water by foregrounding specific bodies, subjectivities, labours and knowledges – such as those of women, queer and trans- peoples, and Indigenous people – means proposing new and different understandings of water governance, while also providing the basis for re-imagining equitable and sustainable feminist water futures.

In their chapter, Arianna Tozzi and Irene Leonardelli foreground the relational materiality of water – co-emergent between water's biophysical characteristics and its sociocultural situatedness – to unfold gender dynamics in two semi-arid areas of Maharashtra, India: one where agriculture and livelihoods are dependent on monsoonal rainfalls and the other where farmers use wastewater to irrigate their crops. More specifically, they draw attention to three more-than-human elements – the solutes and sediments of water, the organisms of soils and goats – to explore how these contribute to shaping different watery agrarian environments and gendered labours. Their feminist post-human analysis aims to rethink metrics of efficiency and productivity in water governance and infrastructural planning to account for the work different people and their ecologies jointly do in different rural agrarian contexts.

Writing as activists and politically engaged academics, Catalina Quiroga and Anyi Castelblanco call to examine the connection between waters, women's bodies and territories to understand the impacts of climate change. They scrutinize the more dominant debates around water security and water justice. They explain that, frequently, debates around water security do not delve into the production of scarcity, erasing responsibilities in how inequalities come to exist in the first place. For their part, debates around water justice, which have engaged researchers and social movements in Latin America, understand water as part of the territories and the guarantees of a dignified life. They embrace the concept of water-body-territory, which emerged from social movement struggles in Latin America. This concept expresses how water is an integral part of the relationships between territories and the bodies of women. Studying these intricate relations is essential to analyzing causes, consequences and forms of adaptation and mitigation to climate change.

Evelyn Arriagada, Leila M. Harris and Dacotah-Victoria Splichalova review and critically reflect on different bodies of gender and water literature to propose an approach to gendered aspects of water insecurities that highlights non-material dimensions. Firstly, inspired by Indigenous ontologies and epistemologies, as well as by post-human feminism and science and technology studies, they highlight the importance of approaching water as a living being and as a set of relationships. Focusing on the non-material aspects of these relationships means foregrounding senses of community, belonging, fairness or values linked with water. Secondly, inspired by queer and trans theories, they emphasize the need to explicitly consider trans, non-binary and gender-diverse peoples and experiences in relation to water beyond mainstream gender approaches. These two approaches together, they argue, build up a less anthropocentric, patriarchal, colonial way to address gender and water issues: they provide inspiration for more inclusive ethical-political accountability, Indigenous resurgence and care of diverse water bodies/bodies of water.

This framework resonates well with what Madeline Donald and Astrida Neimanis articulate in their chapter. They use anticolonial-feminist-queer-trans theories to propose the concept of 'fringe natures' as an alternative to the dominant conceptualizations of land and water rooted in settler colonial imaginaries. They focus on the Semá:th X_ó:tsa, a body of water located in lower mainland British Columbia that used to be 'a fecund body of water of great seasonal variation in depth, size and temperature' and was transformed into a controlled agricultural prairie for hundreds of years, by settler colonial powers. Yet it re-emerged in the form of a lake after some devastating flooding in 2021. As a porous, shifting and interconnected body of water, Semá:th X_ó:tsa is a fringe nature because it challenges strict land and water divides, similarly to how trans bodies challenge dominant gender binaries. Looking at bodies of water as 'fringe natures' seems particularly important in times of great climate and environmental change and instability, as new ways of governing water need to be developed, some that encompass marginality, transition and movement.

References

Acevedo-Guerrero, T. (2022). Water with larvae: Hydrological fertility, inequality, and mosquito urbanism. *Environment and Planning E: Nature and Space*, 1–18.

Acevedo-Guerrero, T., Camargo, A., Roa-García, C., & Roa García, C. (2024). Defending water and life: Domestic ecologies and social reproduction in Buenaventura, *Gender, Place and Culture*. Upcoming.

Ahlers, R., & Zwarteveen, M. (2009). The water question in feminism: Water control and gender inequities in a neo-liberal era. *Gender, Place and Culture*, *16*(4), 409–426.

Ahmed, S. (Ed.). (2005). *Flowing upstream: Empowering women through water management initiatives in India*. Foundation Books.

Ahmed, S., & Zwarteveen, M. (2014). *Diverting the flow: Gender and water in South Asia*. Gender Equity and Water in South Asia.

Bayona-Valderrama, A., Acevedo-Guerrero, T., & Artur, C. (2020). Cities with mosquitoes: A political ecology of Aedes aegypti's habitats. *Water Alternatives*, *14*(1), 186–203.

Bennett, V., Dávila-Poblete, S., & Rico, N. (Eds.). (2005). *Opposing currents: The politics of water and gender in Latin America*. University of Pittsburgh Press.

Bossenbroek, L., & Zwarteveen, M. (2015). "One doesn't sell one's parents": Gendered experiences of shifting tenure regimes in the agricultural plain of the Sais in Morocco. In C. Archambault & A. Zoomers (Eds.), *Global trends in land tenure reform. Gender impacts*. Routledge.

Buechler, S., & Hanson, A.-M. (2015). *A political ecology of women, water and global environmental change*. Routledge.

Carney, J. (1986). *The social history of gambian rice production: An analysis of food security strategies* [Berkeley ProQuest Dissertations Publishing, University of California], 8717920.

Carney, J. (1988). Struggles over crop rights and labour within contract farming households in a Gambian irrigated rice project. *The Journal of Peasant Studies*, *15*(3), 334–349. https://doi.org/10.1080/03066158808438366

Carney, J. (1992). Peasant women and economic transformation in The Gambia. *Development and Change*, *23*, 67–90. https://doi.org/10.1111/j.1467-7660.1992.tb00446.x

Carney, J. (1993). Converting the wetlands, engendering the environment: The intersection of gender with agrarian change in the Gambia. *Economic Geography*, *69*(4), 329–348. https://doi.org/10.2307/143593

Carney, J. (1994). Contracting a food staple in the Gambia. In P. Little & M. Watts (Eds.), *Living under contract. Contract farming and agrarian transformation in Sub-Saharan Africa*. The University of Wisconsin Press.

Carney, J. (1996). Landscapes of technology transfer: Rice cultivation and African continuities. *Technology and Culture*, *37*(1), 5–35. https://doi.org/10.2307/3107200

Carney, J. (1998). Women's land rights in Gambian irrigated rice schemes: Constraints and opportunities. *Agriculture and Human Values*, *15*, 325–336. https://doi.org/10.1023/A:1007580801416

Carney, J., & Watts, M. (1991). Disciplining women? Rice, mechanization, and the evolution of Mandinka gender relations in Senegambia. *Signs: Journal of Women in Culture and Society*, *16*(4), 651–681.

Cleaver, F. (1998). Choice, complexity, and change: Gendered livelihoods and the management of water (Introduction to the Special Issue). *Agriculture and Human Values*, *15*, 293–299. https://doi.org/10.1023/A:1007511816437

Coles, A., & Wallace, T. (Eds.). (2005). *Gender, water and development*. Routledge.

Cook, C., & Bakker, K. (2012). Water security: Debating an emerging paradigm. *Global Environmental Change*, *22*(1), 94–102. https://doi.org/10.1016/j.gloenvcha.2011.10.011

Cruz-Hernández, D. (2016). Una mirada muy otra a los territorios-cuerpos femeninos. *Solar*, *12*(1), 46–53.

Hanger, J., & Moris, J. (1973). Women and the household economy. In R. Chambers & J. Morris (Eds.), *Mwea: An irrigated rice settlement in Kenya*. Weltforum Verlag.

Harris, L. M. (2006). Irrigation, gender, and social geographies of the changing waterscapes of southeastern Anatolia. *Environment and Planning D: Society and Space*, *24*(2), 187–213.

Harris, L. M. (2009). Gender and emergent water governance: Comparative overview of neoliberalized natures and gender dimensions of privatization, devolution and marketization. *Gender, Place & Culture*, *16*(4), 387–408.

Joshi, D., Nicol, A., Peveri, V., & Mitra, A. (2023). *Changing the discourse: From the rhetoric on women and water to a feminist water agenda*. Frontiers in Human Dynamics, Sec. Environment, Politics and Society.

Krishnaraj, M. (2011). Women and water: Issues of gender, caste, class and institutions. *Economic and Political Weekly*, *46*(18), 30.

Kulkarni, S. (2005). Looking back, thinking forward: The Khudawadi experience with access to irrigation for women and the landless. In S. Ahmed (Ed.), *Flowing upstream: Empowering women through water management initiatives in India*. Foundation Books.

Lahiri-Dutt, K. (Ed.). (2006). *Fluid bonds: Views on gender and water*. Stree.

Leddy, S. (2019). We should all be reading more Ursula Le Guin. *The Outline 2023*. https://theoutline.com/post/7886/ursula-le-guin-carrier-bag-theory

Le Guin, U. K. (1988). The carrier bag theory of fiction. In D. Du Pont (Ed.), *Women of vision, women of vision: Essays by women writing science fiction*. St Martin's Press.

Leonardelli, I., Kemerink-Seyoum, J., Kulkarni, S., Bhat, S., & Zwarteveen, M. (2023). A feminist analysis of women farmers navigating groundwater qualities in Maharashtra, India. *Water Alternatives*, *16*(1), 134–152.

Leonardelli, I., Kemerink-Seyoum, J., & Zwarteveen, M. (2022). Obliqueness as a feminist mode of analysing waterscapes: Learning to think with overflows. *Environment and Planning E: Nature and Space*. https://doi.org/10.1177/25148486221117725

Liebrand, J., & Udas Bhushan, P. (2017). Becoming and engineer or a lady engineer: Exploring professional performance and masculinity in Nepal's department of irrigation. *Engineering Studies*, *9*(2017), 120–139.

Merrey, D., & Baviskar, S. (Eds.). (1998). *Gender analysis and reform of irrigation management: Concepts, cases, and gaps in knowledge: Proceedings of the Workshop on Gender and Water, 15–19 September 1997, Habarana, Sri Lanka* (p. xii, 268p). International Water Management Institute (IWMI).

Mies, M. (1986). *Patriarchy and accumulation on a world scale: Women in the international division of labour*. Zed Books.

Mies, M. (1998). Globalization of the economy and women's work in a sustainable society. *Gender, Technology and Development*, *2*(1), 3–37.

Neimanis, A. (2016). *Bodies of water: Posthuman feminist phenomenology*. Bloomsbury Academic.

Ojeda, D., Nirmal, P., Rocheleau, D., & Emel, J. (2022). Feminist ecologies. *Annual Review of Environment and Resources*, *47*(1), 149–171. https://doi.org/10.1146/annurev-environ-112320-092246

O'Reilly, K. (2016). From toilet insecurity to toilet security: Creating safe sanitation for women and girls. *Wiley Interdisciplinary Reviews: Water*, *3*(1).

O'Reilly, K., Halvorson, S., Sultana, F., & Laurie, N. (2009). Introduction: Global perspectives on gender – water geographies. *Gender, Place & Culture*, *16*(4), 381–385.

Ramamurthy, P. (1991). Rural women and irrigation: Patriarchy, class, and the modernizing state in South India. *Society & Natural Resources*, *4*(1), 5–22. https://doi.org/10.1080/08941929109380739

Resurreccion, B., & Elmhirst, R. (2008). *Gender and natural resource management*. Routledge.

Resurrección, B., & Elmhirst, R. (Eds.). (2020). *Negotiating gender expertise in environment and development: Voices from feminist political ecology* (1st ed.). Routledge. https://doi.org/10.4324/9781351175180

Rushton, S. (2023, March 22). Women's cooperatives overcome water wars and climate drought in Rojava. *Equal Times*. www.equaltimes.org/women-s-cooperatives-overcome?lang=en

Schrijvers, J. (1984). Blueprint for undernutrition an example from Sri Lanka. *Sociologia Ruralis*, *24*, 255–273. https://doi.org/10.1111/j.1467-9523.1984.tb00648.x

Sehring, J., ter Horst, R., & Zwarteveen, M. (2022). *Gender dynamics in transboundary water governance*. Routledge.

Shiva, V. (1988). *Staying alive: Women, ecology and development*. Zed Books.

Sultana, F. (2009). Fluid lives: Subjectivities, gender and water in rural Bangladesh. *Gender, Place and Culture*, *16*(4), 427–444.

Sultana, F. (2011). Suffering for water, suffering from water: Emotional geographies of resource access, control and conflict. *Geoforum*, *42*, 163–172.

Truelove, Y. (2011). (Re-)Conceptualizing water inequality in Delhi, India through a feminist political ecology framework. *Geoforum*, *42*(2), 143–152.

van Wijk, C. (1998). *Gender in water resources management, water supply and sanitation: Roles and realities revisited*. IRC International Water and Sanitation Center. In: Technical Paper Series No. 33-E. www.irc.nl/page/1893

Wallace, T., & Porter, F. (2010). Introduction. *Gender & Development*, *18*(1), 1–10.

Wolf, J., Johnston, R. B., Ambelu, A., Arnold, B. F., Bain, R., Brauer, M., Brown, J., Caruso, B. A., Clasen, T., Colford, J. M., Jr., Mills, J. E., Evans, B., Freeman, M. C., Gordon, B., Kang, G., Lanata, C. F., Medlicott, K. O., Prüss-Ustün, A., Troeger, C., . . . Cumming, O. (2023). Burden of disease attributable to unsafe drinking water, sanitation, and hygiene in domestic settings: A global analysis for selected adverse health outcomes. *The Lancet*, *401*(10393), 2060–2071. https://doi.org/10.1016/S0140-6736(23)00458-0

Zwarteveen, M. Z. (1997). Water: From basic need to commodity: A discussion on gender and water rights in the context of irrigation. *World Development*, *25*(8), 1335–1349. https://doi.org/10.1016/S0305-750X(97)00032-6

Zwarteveen, M. Z. (2008). Men, masculinities and water powers in irrigation. *Water Alternatives*, *1*(1), 111–130.

Zwarteveen, M. Z. (2015). *Regulating water, ordering society: Practices and politics of water governance (Inaugural Lecture)*. University of Amsterdam and UNESCO IHE.

Zwarteveen, M. Z., Ahmed, S., & Gautam, S. (2012). *Diverting the flow: Gender equity and water in South Asia*. Zuuban.

Zwarteveen, M. Z., & Boelens, R. (2014). Defining, researching and struggling for water justice: Some conceptual building blocks for research and action. *Water International*, *39*(2), 143–158.

Zwarteveen, M. Z., Kemerink-Seyoum, J. S., Kooy, M., Evers, J., Guerrero, T. A., Batubara, B., Biza, A., Boakye-Ansah, A., Faber, S., Cabrera Flamini, A., Cuadrado-Quesada, G., Fantini, E., Gupta, J., Hasan, S., ter Horst, R., Jamali, H., Jaspers, F., Obani, P., Schwartz, K., . . . Wesselink, A. (2017). Engaging with the politics of water governance. *WIREs Water*, *4*(6), e1245. https://doi.org/10.1002/wat2.1245

PART 1

Positionality and embodied waters

1
WOMEN'S ANTI-HYDROPOWER ACTIVISM IN TURKEY
Water, environmental struggles, and bodily experiences

Özge Yaka

Introduction

> And suddenly Fatmi stands with her feet in the sea. . . . Wherever she looks, everything is blue, everything is the sea. She can feel it, she can hear it, she can smell it. "You are speechless when you see this beauty," she says. "I love the waves, the sound of the sea, the colors. I love it all." Fatmi wanted to go to the sea. Absolutely, and for a very long time. Now she is there, together with her husband and her son, and it is the first time for all of them.[1]

This is how Peter Münch begins his *Süddeutsche Zeitung* article on a group of Palestinians who traveled to the Mediterranean coast, which is geographically so close – yet so far away (Münch, 2020). The sea is nearby, as it is only 80 kilometers away from the village where Fatmi lived all her life; but it is far away, as it is located within the borders of Israel.

As a Palestinian living in the West Bank, it is not easy for Fatmi to step on Israeli soil. That is why she stands by the sea for the first time in her life, even though she had wanted to do so for as long as she remembers. The long-awaited encounter overwhelms her at first sight, as she is immersed in a multisensory experience of the Mediterranean Sea.

Not all of us risk a dangerous journey like Fatmi to have a first-hand experience of water bodies, but most of us are drawn to oceans, seas, lakes, ponds, rivers, and creeks. We often spend our holidays on the coast or by a lake or river. We visit spas and thermal springs to relax and rejuvenate; we go to great lengths for a home with a sea, lake, or river view. Immersing one's body in water, watching the movements of water bodies, hearing the waves or a river's flow, drinking from a spring – firsthand sensory experiences of water are highly affective and desirable, even though they are all too rarely discussed in a scholarly manner.

Our relationship with water certainly has an instrumental dimension: it is life-giving and life-sustaining, not only for humans but for all living beings. It is therefore understandable that the literature on water is dominated by discussions of water as a vital "resource," a term that focuses on the aspect of utility in the fields of water management, governance, access, development, security, and sustainability. Utilitarian framings of water, however, render invisible the deep, living connection between water and our bodies and our utter

DOI: 10.4324/9781003100379-3

dependence on water, both fresh and salinated, in every facet of our lives and our well-being. The story I will tell you in this essay, the story of East Black Sea women in Turkey who resist the small-scale hydroelectric power plants (HEPPs) that invade their villages and valleys and "steal" their streams, is about the strength of such sensory, affective, and ultimately social connections to water as an elemental relationship, which is mostly ignored by the scholarly literature (for an exception, see Stefanovic, 2020).

Standard arguments about local community struggles against various forms of resource extraction and dispossession (such as mining, oil drilling, fracking, and hydropower) share the tendency to frame water as a utility upon which lives and livelihoods depend. The struggles, it follows, are motivated by this immediate biological and economic dependence, or, in the cases of Indigenous struggles, by belief systems in which water bodies have a central role. I do not deny the importance of livelihoods and belief systems for many local, especially rural and Indigenous, communities fighting to protect water bodies around the world; however, I aim to introduce an alternative framework, drawing on an empirical case of women's activism against HEPPs in Turkey's East Black Sea Region. I argue that fundamental, embodied relations to waterways are central to such activism, and much more meaningful than simple utilitarian reasoning allows.

Gendered activism against HEPPs in Turkey

Run-of-the-river HEPPs are substantially different from hydro dams, as they require little or no water storage. Instead, they capture the kinetic energy of the natural flow of water running down steep slopes. Typically, water from upstream is diverted to electricity-generating turbines by a weir or a pipeline and then is released back into the river's downstream flow. For this reason, they are often presented as an eco-friendly alternative to hydro dams and a less ecologically harmful way of producing renewable energy, despite the emerging evidence of their ecological impacts (Couto & Olden, 2018).

During the 2000s, under the Justice and Development Party (*Adalet ve Kalkınma Partisi*, AKP), Turkey has decided to make extensive use of small-scale, run-of-the-river HEPPs. In accordance with AKP's construction and energy-based growth strategy, sometimes referred to as "bulldozer capitalism" (Evren, 2022), private construction and energy companies have been given "extraordinary latitude to evict villagers, expropriate private land, clear state forests and steamroll normal planning restrictions to meet the target of four thousand hydroelectric schemes by 2023" (Gibbons & Moore, 2011).

HEPPs started to pop up in the remote villages and valleys of the county, especially in the East Black Sea and Mediterranean Regions, and in the East and Southeast Anatolia to a lesser extent, where the small- and middle-sized rivers have naturally steep slopes. By 2007–2008, it became clear that HEPPs were both destroying river ecosystems and natural habitats especially when they are built extensively and without adequate impact assessment, regulation, and monitoring such as in Turkey (Turkish Water Assembly, 2011; TMMOB, 2011; Kurdoğlu, 2016) and dispossessing riverside communities, which were left with dry streambeds when river waters were diverted through pipelines for a number of kilometers. Riverside communities started to organise themselves locally against HEPPs and local movements emerged in many different parts of the country, slowly leading to the formation of regional and national networks of committed activists.

Even though the anti-HEPP movement is not a women's movement *per se*, women shape the specific character and public image of the movement with their bold, committed, and radical activism (Kasapoğlu, 2013; Radikal, 2015; Yaka, 2017).[2] Especially in the rural parts of the East Black Sea Region, where resistance to HEPPs is concentrated,[3] women have been highly visible, protesting in their traditional clothing of shalvars (*şalvar*), long skirts, and headscarves, keeping guard at the construction sites and blocking construction equipment – even physically confronting the military and the police when necessary. Contrary to men who assume a more moderate language, women adopt a radical position, talking about killing and dying to protect the rivers (see Krauss, 1998; Prindeville, 2004; and Bell, 2013 for women's radicalism and commitment in environmental struggles).

Take Selime, a middle-aged woman, as an example whose words I have recorded during our conversation on her terrace in the village of Arılı, East Black Sea Region:

> We, as women, won't allow them to construct a hydropower plant here. We don't even count on men. . . . Bring them on, if they dare, if any brave fellows think they can come here . . . let them try. We will cut them to pieces. We know how to use guns as well. We take the risk. They really shouldn't force us. Don't make people go mad.

Like many others interviewed by different journalists, Selime's courage and commitment is shared by many other women with whom I spoke during my field research. No one is motivated by an immediate economic dependence on river waters, as scholars of feminist environmentalism and political ecology would assume. In fact, the livelihoods in the East Black Sea Region do not depend on river waters, because they are not used for either agriculture or daily household needs: rainfall alone sustains monocultural tea cultivation (and hazelnut crops in the western parts of the region). In other words, rivers are not "natural resources" for East Black Sea women and their struggle could not be captured by terms like "resource conflicts."

Instead, what drives their activism are the intimate corporeal connections that they have with river waters, and the bodily memories of such connections that are anchored in place. When I asked East Black Sea women why they oppose HEPPs, they talked neither about livelihoods nor did they refer to macro-political struggles, technical data or scientific information, as men tend to do (see Yaka, 2019a). Instead, they told me about the river itself and their everyday bodily interactions with it – about how they grew up by the river, how they are used to sleeping with and waking up to the sound of it, how they love immersing their bodies in its waters, and how the sight of its flowing, cascading waters defines the place (the village and valley).

"We live here, in this narrow valley, only with the joy of the river [*derenin neşesi*]. When it is gone, it means we should also go," said Refiye from the village of Yaylacılar. "We live here with the river. We look at the river every day. We cannot live without it," adds Ayşe from Ulukent. Such testimonials are often repeated. "Rivers are our celebrations [*şenlik*]," said Ülker from Aslandere. "I went by the river today before I prayed. I just sat by it and watched it. It was so beautiful, greenish blue. I watched the fish swimming in it. I stayed there for some time and returned to pray." "The river is life and soul [*hayattır ve candır*] for us. I would probably feel utterly empty if I did not hear the sound of the river," reflected Seniye from the same village. (See Yaka, forthcoming, for more examples and a detailed discussion.)

Women similarly talked about the memories of their parents by the river and the sight of their children and grandchildren playing in the same waters where they once played. "I see my mother and my father by this [Arılı] river; every time I look at the river, I remember them," said Semra from the village of Gürsu. "They are my grandchildren," she said. "Look how happy they are in the river," said Nuran, pointing out three young children playing in the [Arılı] river, shouting and playing with joy during our interview (see Figure 1.1).

Figure 1.1 Children playing in the Arılı River.
Source: Photo by the author.

For these women, the river is neither a mere resource that could be utilised and traded nor an abstract political, cultural, or belief-related symbol. The river is a nonhuman, environmental entity (like a neighbour, like a relative they would say) with which they live in a close relationship that is established through constant bodily encounters and sensory-affective connections.

Women's political agency is therefore shaped by the intimate corporeal connections they have with river waters, and by bodily memories of such connections that are anchored in place. In this sense, women's anti-hydropower activism in Turkey's East Black Sea Region illustrates "how agentic properties emerge and endure within corporeal experience" (Coole, 2005, p. 131). The case also demonstrates that the struggles to protect water bodies against energy and infrastructure projects are not always driven by the instrumental use of water to sustain livelihoods; the centrality of water bodies to our world of lived experience, to our *lifeworld*,[4] could also inspire such struggles.

More-than-human lifeworld and socio-ecological justice

To describe the centrality of rivers to the lifeworld (*Lebenswelt*) of riverside communities, especially of women, means recognising the more-than-human character of the places within which our lives unfold. As reflected in the German origins of Husserl's *Lebenswelt* (lifeworld) or Uexküll's *Umwelt* (environment, surroundings,) natural entities such as waterways are constitutive of the world of experience in which self, subjectivity, and agency are formed. Nonhuman entities such as rivers are an essential part of the intersubjective universe of experience, encounter, and interaction. What we call "natural" or "ecological" is, in fact, an element of the "social." (Haraway, 1991; Whatmore, 2002; Latour, 2007).

This notion of *eco-sociality* (Whitmore, 2018) is central to the justice claims of anti-HEPP activists in Turkey, especially in the East Black Sea Region, where rivers are not seen as "natural resources," but as nonhuman entities with which people live. Indeed, living in a riverside village in the region is *living with* a river, as people work, unwind, celebrate, socialise, sing, cry, and fall in love, always in continuous corporeal interaction with river waters. In this sense, their struggle is not for "resources" but for *coexistence* (Escobar, 2011; Larsen & Johnson, 2017). Such coexistence means living with the nonhuman, environmental entities that they value, in places that they constructed free from institutionally sustained destruction, degradation, pollution, toxification, and commodification of ecological systems, habitats, and entities.

The environmental justice (EJ) movement and body of scholarship have famously linked environmental issues to notions of power and inequality. They stress the unequal and unfair distribution of environmental goods and hazards – a distribution that structurally disadvantages Black, persons of color, minority, Indigenous, and poor populations (Mohai et al., 2009). Although I value the EJ movement and scholarship highly, I believe the struggles of the rural East Black Sea communities, especially of women, to protect their rivers, calls for an alternative notion of justice. This alternative notion, which I call *socio-ecological justice*, corresponds to eco-sociality and the right of coexistence mentioned above and signifies the interconnection between the so-called social and ecological realms. In other words, socio-ecological justice maintains that our intrinsic and intimate relations with the nonhuman world are an essential part of our well-being and are central to our demands to pursue a fair and decent life. Thus, what is at stake here is not merely a matter of extending the community of justice to include nonhuman environments, as the concept of ecological justice attempts to do (Low & Gleeson, 1998; Baxter, 2005). It is also a matter of actively incorporating human–nonhuman relational ontologies and corresponding ethical practices into our understanding of justice. (Yaka, 2019b, 2020).

Muhammet Kaçar (2017) reports in the daily newspaper *Hürriyet* about a group of women who blocked the road in the Arılı Valley and who talked to the experts appointed by the court for the HEPP case. One of them told him: "*Dere bizim eşimiz, biz derenin eşiyiz* – The river is our spouse; we are the river's spouse." The word eş in Turkish is used for partner or spouse in everyday life but it also has the connotation of being one's equal, fellow, or companion. *Dere* (small river/stream) can be all of them – a partner, a relative, a brother, a sister, a friend, a neighbor, a relative, a companion, all in all an equal – a non-human person/being/entity with whom one has a lifelong, intimate, sentient, and affective relationship. The case of the anti-HEPP struggles in Turkey demonstrates that the rights and interests of nonhuman nature and "humans-in-nature" are not at odds with one another. Instead, caring for one's own life and caring for the environment can be one and the same, so long as communities perceive themselves as part and parcel of the nonhuman world that surrounds them.

Notes

1 Translated from German by the author.
2 The gender difference in the ways in which men and women frame their opposition to HEPPs and enact their political agencies resonate with the feminist political ecology literature which stresses "real, not imagined, gender differences in experiences of, responsibilities for, and interests in 'nature' and environments" which are "not rooted in biology per se" (Rocheleau et al., 1996, p. 2). They are instead rooted in gendered organisation of everyday life and material practices, especially in the gendered division of labor. The gendered division of labour in the East Black Sea Region, for instance (i.e., the fact that the agricultural work is perceived and performed as an extension of housework and that it is women who work in the fields planting, fertilising, pruning, and harvesting tea plants), provides the material conditions for women to interact with river waters in a routine and habitual manner. As the fields (as well as the houses) are located by rivers within the long, deep valleys that run through the Kaçkar (Pontic) Mountains to the Black Sea in the region, agricultural work puts women in constant connection with rivers. For a detailed discussion, see Yaka (forthcoming).
3 203 HEPPs were built within a decade between 2008 and 2018 in the region, and 123 more were in the project phase.
4 Lifeworld is a phenomenological concept which originated in Husserl's work but is employed by many others, including Jürgen Habermas. It designates the given, self-evident world of lived experience.

References

Baxter, B. (2005). *A theory of ecological justice*. Routledge.
Bell, S. E. (2013). *Our roots run deep as ironweed: Appalachian women and the fight for environmental justice*. University of Illinois Press.
Coole, D. (2005). Rethinking agency: A phenomenological approach to embodiment and agentic capacities. *Political Studies*, *53*(1), 124–142. https://doi.org/10.1111/j.1467-9248.2005.00520.x
Couto, T. B. A., & Olden, J. D. (2018). Global proliferation of small hydropower plants – science and policy. *Frontiers in Ecology and the Environment*, *16*(2), 91–100.
Escobar, A. (2011). Sustainability: Design for the pluriverse. *Development*, *54*(2), 137–140.
Evren, E. (2022). *Bulldozer capitalism: Accumulation, ruination, and dispossession in Northeastern Turkey*. Berghahn Books.
Gibbons, F., & Moore, L. (2011, April 29). Turkey's Great Leap Forward risks cultural and environmental bankruptcy. *The Guardian*. www.theguardian.com/world/2011/may/29/turkey-nuclear-hydro-power-development
Haraway, D. (1991). *Simians, cyborgs, and women: The reinvention of nature*. Routledge.
Kaçar, M. (2017, July 18). Arılı Vadisinde Yola Oturan Kadınlar Bilirkişi Heyetinin Yolunu Kesti [The women block the road to in the Arılı Valley]. *Hürriyet*. www.hurriyet.com.tr/arili-vadisinde-yola-oturan-kadinlar-bilirkisi-40523696
Kasapoğlu, Ç. (2013, October 25). 100 Kadın: Karadeniz'in Dereleri İçin Direnen Kadınları [100 women: The Black Sea women resisting for their rivers]. *BBC Turkish*. www.bbc.com/turkce/haberler/2013/10/131025_findikli_kadinlar_kasapoglu
Krauss, C. (1998). Challenging power: Toxic waste protests and the politicization of white, working-class women. In N. A. Naples (Ed.), *Community activism and feminist politics: Organizing across race, class and gender* (pp. 129–150). Routledge.
Kurdoğlu, O. (2016). Expert-based evaluation of the impacts of hydropower plant construction on natural systems in Turkey. *Energy & Environment*, 27(6–7), 690–703. https://doi.org/10.1177/0958305X16667186
Larsen, S. C., & Johnson, J. T. (2017). *Being together in place: Indigenous coexistence in a more than human world*. University of Minnesota Press.
Latour, B. (2007). *Reassembling the social: An introduction to actor-network-theory*. Oxford University Press.
Low, N. P., & Gleeson, B. (1998). *Justice, society and nature: An exploration of political ecology*. Routledge.

Mohai, P., Pellow, D., & Roberts, J. T. (2009). Environmental justice. *Review of Environment and Resources*, *34*, 405–430. https://doi.org/10.1146/annurev-environ-082508-094348
Münch, P. (2020, September 11). Ein Tag am Meer [One day at sea]. *Süddeutsche Zeitung*. www.sueddeutsche.de/panorama/israel-westjordanland-1.5027731
Prindeville, D. M. (2004). *On the streets and in the state house: American Indian and hispanic women and environmental policymaking in New Mexico*. Routledge.
Radikal. (2015, July 19). *Havva Ana Yalnız Değil: Çevreyi Paçavra Ettunuz... Biz de Sizi Edeceğuz [Mother Havva (Eve) is not alone: You have messed with the environment... we will mess with you]*. www.radikal.com.tr/cevre/havva-ana-yalniz-degil-cevreyu-pacavra-ettunuz-biz-da-sizi-edeceguz-1399189/
Rocheleau, D., Thomas-Slayter, B., & Wangari, E. (1996). *Feminist political ecology: Global issues and local experiences*. Routledge.
Stefanovic, I. L. (Ed.). (2020). *The wonder of water: Lived experience, policy, and practice*. University of Toronto Press.
TMMOB (Union of Chambers of Turkish Engineers and Architects). (2011). *Hidroelektrik Santaller Raporu [Report on the Hydroelectric Plants]*. TMMOB – Türk Mühendis ve Mimar Odaları Birliği. www.tmmob.org.tr/sites/default/files/682384b57999789_ek.pdf
Turkish Water Assembly. (2011). *HEPPs, Dams and the status of nature in Turkey*. https://studylib.net/doc/7537045/hepp-s-dams-and-the-status-of-nature-in-turkey-turkish-w . . .
Whatmore, S. (2002). *Hybrid geographies: Natures cultures spaces*. Sage.
Whitmore, L. (2018). *Mountain, water, rock, god*. University of California Press.
Yaka, Ö. (2017). A feminist-phenomenology of women's activism against hydropower plants in Turkey's Eastern Black Sea region. *Gender, Place and Culture*, *24*(6), 868–889. https://doi.org/10.1080/0966369X.2017.1340873
Yaka, Ö. (2019a). Gender and framing: Gender as a main determinant of frame variation in Turkey's anti-hydropower movement. *Women's Studies International Forum*, *74*, 154–161. https://doi.org/10.1016/j.wsif.2019.03.002
Yaka, Ö. (2019b). Rethinking justice: Struggles for environmental commons and the notion of socio-ecological justice. *Antipode*, *51*(1), 353–372. https://doi.org/10.1111/anti.12422
Yaka, Ö. (2020). Justice as relationality: Socio-ecological justice in the context of anti-hydropower movements in Turkey. *DIE ERDE – Journal of the Geographical Society of Berlin*, *151*(2–3), 167–180. https://doi.org/10.12854/erde-2020-481
Yaka, Ö. (forthcoming). *Fighting for the river: Gender, body and agency in environmental struggles*. University of California Press.

2

MAKING ENGINEERS TELL THEIR STORIES? MASCULINITY, WHITENESS AND HETERONORMATIVITY "AT WORK" IN LIFE HISTORY INTERVIEWS IN IRRIGATION IN NEPAL

Janwillem Liebrand

Introduction

In this chapter, I reflect on the use of life history interviews in researching masculinity among engineers in donor-supported planned development. Between 2009 and 2011, I researched gender and masculinity among irrigation engineers and agricultural experts in Nepal. I relied on a professional-insider position, being male myself and originally trained as an irrigation expert. I used a mixed research strategy, based on multi-sited ethnographic observations (in offices, workshops, the field and at home), literature and documentation research, and life history interviews. I used life history interviews as a method to talk and interact with engineers, gaining detailed information about their careers, reflections and aspirations, while taking note of styles of decorum (behavior, practices, attitudes, clothes, phrases, jokes). In total, I interviewed and interacted with nine senior engineers. The analysis in this particular chapter draws on a set of interviews that I had with the most senior engineer, Huta Ram Baidya (1921–2013). He was known as the first agricultural engineer of Nepal. There, he also was known as an activist and environmentalist who advocated for the cultural and ecological restoration of the ancient river Bagmati Civilization in the Kathmandu Valley (see Rademacher, 2007).

My interest in studying gender dynamics among engineers in irrigation and agriculture is based on the observation that the meanings and implications of masculinity tend to live an uncomplicated life in development. In spite of more than three decades of gender mainstreaming, there is still an overwhelming physical presence of men in fields that require technical expertise (Liebrand, 2022). Through participation in the professional community, I learned that belonging can be easily achieved by displaying norms of manhood. In fact,

DOI: 10.4324/9781003100379-4

the performance of men as engineers and technical experts in development appears to be so normal that it is one of the ways in which professional authority in irrigation and agriculture presents itself as self-evident, neutral and universal (Zwarteveen, 2011). Increasingly, however, the problem of diversity and female underrepresentation in engineering and technical expertise has become more acute and contested. By and large, the application of science and technology has failed to bring the anticipated progress and transformation in development (Gupta, 2007), while there has been a tremendous increase in engineering education at the global scale – in South Asia approximately a 50-fold increase from the 1950s to 2010s, contributing to a growing level of female student enrollment in engineering in this region (Liebrand & Udas, 2017). In this changed context, the "normal" in irrigation and agricultural development is associated with the oft-noted resilience of the professional establishment to change (Lynch, 1993; Laurie, 2005; Ongsakul et al., 2012; Zwarteveen & Rap, 2017).

To expose and challenge the normal in engineering and investigate linkages between men, masculinity and professional authority in development, it is necessary to move away from an analysis of problems *in* irrigation and agriculture and shift the research gaze to the powerful themselves – to those who promote technical expertise and who are responsible for planning new development interventions (Zwarteveen, 2008). Studying masculinities in the engineering profession therefore requires a practice of "studying up". Anthropologist Laura Nader (1972) originally defined studying up as a research strategy to address the "facelessness" of a bureaucratic society and the major institutions that affect everyday lives (p. 288). Feminist philosopher Sandra Harding (2004) in turn discusses studying up as a strategy that involves doing research from the standpoint of marginalized people in society (women, people of color, lower classes), taking their place in society into consideration and questioning established claims of scientific objectivity and truth. Taking inspiration from them, I use studying up to give engineers a face (and gender) and expose the normal as tied to men and masculinity.

In the world of donor-supported planned development, men have historically occupied the applied technical and engineering disciplines. Its roots can be traced back to a project of colonial empire building and a civilizing mission. In South Asia, then called British India, the British used engineering to create new hierarchies in society, between British white men "at the top" and non-white Indian men, in various ranks, "under the top" and "at the bottom" (Liebrand, 2022). After formal colonial occupation ended, these hierarchies were re-validated by two non-integrated streams of development thinking: the humanities and social sciences on the one hand, and the technical and engineering disciplines on the other. This dichotomy of thought was marked by a hierarchically gendered "horizontal segregation" in science, meaning that men occupied the technical disciplines and expert women were concentrated in the less-valued area of social sciences (Gupta, 2007). The start of donor-supported planned development thus picked up where colonialism left off and, in South Asia, the dominant presence of men in engineering and technical expertise still helps to produce particular norms of masculinity in society (Power, 2003; Nair, 2012). And since then, in turn, performances of masculinity in South Asia have helped to reinforce a particular gendered profession in agriculture and irrigation development, making it normal for men to be engineers (SaciWATERs, 2011). These observations may not resolve the theoretical debate in gender studies on why or when to call something or someone masculine, but they enable me to trace some of the associations between gender, masculinity and professional performance in irrigation engineering and technical expertise in agriculture (see Liebrand & Udas, 2017).

In the next section, I elaborate on the conceptual entry points and method of analysis for this research, discussing the use of life history interviews. This is followed by a brief description of the life and career of Baidya. Then, I present an analysis of masculinities in the interviews, paying attention both to the story that is being told and practices of "doing gender". In the concluding section, I reflect on the use of life history interviews for research on masculinities and draw lessons for using this method for further study.

Conceptual entry point and method of analysis

To grasp associations between male engineers, technical authority and normalcy in the profession of irrigation and agriculture, I conceptualize professional performance as "cultural performance". The definition of cultural performance is derived from public drama and staged or theatrical performance, and it can be defined as follows: to show off, to perform behavior as expected, to act according to social norms or to underline an action for those who are watching (McKenzie, 2001; Schechner, 2006; Rap, 2007; Turner, 2008; John, 2008). In this view, expert authority among engineers is not simply an outcome of merit and innate technical skill and knowledge, or of one's personal network used to secure promotion opportunities; it is also an outcome of one's ability to perform "on stage", to create a credible voice for oneself. It is something that engineers must actively assert, cultivate and guard – sometimes in the face of intense opposition (Hilgartner, 2000). In other words, technical expertise comes in specific cultural forms and bodies and is enacted through ritualized practices – or what Butler calls a "stylized repetition of acts" in relation to gender (1999, pp. 178–179). These acts become visible in professional traditions and skills, and are expressed in everyday interactions, behavior and practices.

Faulkner, writing about gender in engineering workplace cultures in the United Kingdom, describes how "engineering and pleasure in technology are (felt and perceived to be) 'gender authentic' options for men and 'gender inauthentic' options for women" (2009b, p. 172). She uses these terms to call attention to a phenomenon that she describes as the in/visibility paradox in engineering, "whereby women engineers are simultaneously highly visible as women yet invisible as engineers, both within their own communities of practice and in the wider world" (p. 172). In the context of South Asia, the in/visibility paradox and women's difficult position in engineering has received a fair bit of attention in feminist-inspired research (Parikh & Sukhatme, 1994, 2004; Gupta, 2007; Nair, 2012). I have argued elsewhere that, because of it, career prospects in Nepal's Department of Irrigation for female engineers remain grim because for them to succeed and belong, they have to reconcile the near incommensurable: a performance of a "lady engineer" with that of a "normal engineer" (Liebrand & Udas, 2017). In comparison, men's position in engineering, from the perspective of in/visibility, has received little scrutiny.

Researching gender authenticity in engineering and the connections between men, masculinity and normalcy poses particular methodological challenges. In contrast to women in engineering who "have a story to tell" about why they made their choices, Faulkner observes men typically provide little or no account of their choice to become one (2009b, p. 173). Male engineers rarely are confronted with this question; they do not need to explain, they are "normal" in the profession and subsequently, they have gained little or no practice in talking about being an engineer and articulating reflections. By implication, short or semi-structured interviews are not always adequate for an in-depth study on men in engineering. Such interview techniques may not offer male engineers sufficient space to talk. Because the interactions are shorter in such interviews compared to lengthy life history interviews, the researcher's focus tends to be steered to what is being said – to words and content – and less to what is being done – to behavior and practices. These shorter interactions run the risk of obscuring how the practice of interviewing itself is mediated by gender norms. Yet doing interviews is "doing gender" – it embodies social interactions that are mediated by norms of gender, race and sexuality, but such real-life dynamics can escape analysis if the focus is exclusively on words and content. The manner in which I got to know about Baidya and selected him for interviews illustrates this dynamic.

In Nepal, I interacted a lot with Nepali students who, at that time, were following a master's degree program on international land and water management at Wageningen University, the Netherlands. This program was supervised by the irrigation and water engineering group in Wageningen, to which I then belonged as a PhD scholar. The Nepali students and I were thus international Wageningen students and engineers, trained in sociotechnical perspectives on irrigation and agricultural development. We felt a sense of belonging that created professional recognition and friendship. It was in this particular context that one of the male students, who was an agricultural engineer by training, told me about Baidya, who he referred to as the first agricultural engineer in Nepal. His enthusiasm immediately aroused my interest. I felt excited about the opportunity to meet this senior Nepali man and engineer, and I looked forward to getting inspiration from this person who, I understood, had more than six decades of hands-on experience in the field. I came to connect with Baidya for interviews through performing gender and race: through interactions of white, European (myself) and non-white, Nepali students, through man-to-man interactions of friendship and recognition among engineers, and through shared ideals of admiration of apprentices for master engineers – for our heroes and living legends in the field.

To capture the value of life history interviews, both in terms of "stories" and "doing gender", I analyze the interviews as cultural performances. This means that the information of the interviews counts as well as how it is said and to whom (Angrosino, 2002; Ritchie, 2003). The interviews were not necessarily about producing accurate historical information, but more about using dramatic techniques to create and enact a credible performance. The question is how Baidya cast himself as a trustworthy professional in my presence, and at occasions when other people were present as well: how did he lay claim to the authority of an engineer; and how did he create a credible voice for himself? I understand cultural performance as also a product of *intercultural* interaction between two people or more. This means that Baidya told things and acted in ways, at least partially, *because of me*. This goes further than the questions that I asked. My identity as a European irrigation expert and white male person, wittingly or otherwise, informed Baidya's ways of acting and answering, and vice versa. His/our/my performance during interviews was deeply *relational* and gendered.

The life history interview with Baidya took place in the period November 2009 to August 2010, and included in total twelve interview meetings, summing up to nearly 15 hours of recorded conversations. Our meetings took place in his house in Kathmandu. We conversed mainly in English. I used his curriculum vitae to trigger our conversations; I generally aimed for open and unguided conversations, following-up on stories that he was happy to share. At the time of the interview, Baidya was 89 years old; he had a clear awareness and wit, but his fragile physique and physical health sometimes limited the length of our interactions. To get his life story "right", I wrote a report on the basis of the interviews, complemented with literature research to contextualize his stories (adding dates, places and project names), and I shared it with Baidya for feedback. This, in turn, facilitated further social interactions and conversations about his life and career.

In the following section, I present a brief overview of Baidya's life and career. It is based on his account, as he told it, but I contextualized and interpreted it with literature references for readers who are not acquainted with the history and development of Nepal. Most of his stories (and my interpretations of his stories) were about the formative years of his life and career as an engineer and agricultural expert, roughly the late 1940s and 1950s and early 1960s. Table 2.1 presents a summarized overview of Baidya's life and career.

Table 2.1 The Life and Career of Huta Ram Baidya

Personal Details	
Date of Birth	10 July 1921 (23 Asar 1978 BS)
Place of Birth	Tripureshwor, Kathmandu, Nepal
Education & Training	
1939	Durbar High School (English), Kathmandu
1947	Agricultural Engineering (Bachelor), Allahabad University, India
1950	Formulation and appraisal of development projects, Asian Training Centre, Lahore, Pakistan, FAO
1953–1954	Student exchange program, Utah, USA, USOM/Nepal
Employment & Activities	
1947–1961	Agricultural engineer, assistant director and chief of agricultural extension, Government of Nepal, Board of Agriculture/Department of Agriculture
1961–1970	Consultant agricultural engineering
1971–1980	Project officer agriculture, US Peace Corps/Nepal
1981–1990	Consultant rural development projects
1990 <	Retired, advocating for the "Bagmati Civilization"
Key Books and Publications	
1960	*Champion farmer of Nepal.* Regional extension seminar, India. FAO
1968	*Farm irrigation and water management: Principles and practices.* Kathmandu: Royal Nepal Academy
1978	*Application of Science/Technology for food and agricultural development in Nepal.* Tribhuvan University, Research Centre for Applied Science and Technology
1980s	*Huta Ram Baidya tri Shakti: land, labor and capital – economic development triangle concept*
1992	*Nepal's efforts in rural development: What role is expected from agricultural engineers?* Agricultural engineering curriculum development workshop, Tribhuvan University
Key Association Memberships and Social Societies	
1950	Nepal Agricultural Association
1990	Nepal Society of Agricultural Engineers
1990	Nepal Agroforestry Foundation
1990	Save the Bagmati Campaign
1991	Nepal Water Conservation Foundation

(Continued)

Table 2.1 (Continued)

Key Awards and Appreciations	
1996	Nepal Society of Agricultural Engineers
1998	Royal Nepal Academy of Science and Technology
2002	Institute of Engineering, Tribhuvan University
2008	Nepal Engineer's Association: Lifetime Achievement Award

Source: Curriculum vitae of Huta Ram Baidya, November 2009.

The life and career of Huta Ram Baidya

Baidya was born in Kathmandu in 1921 as the third son among seven children (four daughters and three sons) in a family that was part of the inner circle of the Rana aristocracy. The Ranas ruled Nepal in an autocratic manner, treating the kingdom as their personal territory and receiving political backing from the British in India (Whelpton, 2005). The size of the ruling class was limited to some 300 families (Shumshere, 1971), and opportunities for education and travel were strictly controlled by the Ranas. Baidya's family belonged to the upper-caste Newar people in the Kathmandu Valley,[1] and his father was a medical doctor for the Ranas.[2] In this capacity, Baidya's father was one of the few, for instance, who traveled to England in 1908, accompanying prime-minister Chandra Shamsher Rana (1863–1929) in his entourage. Baidya's family home was one of the first in Kathmandu in 1924 to receive electricity, at the occasion of the construction of an electricity grid for the palace of the Rana prime minister.

As a son of the inner circle of the aristocracy, Baidya was one of the few who received education at Durbar High School in Kathmandu between 1933 and 1939. This school was exclusively for the elite male children of the Ranas and was the first school in Nepal that followed English ideas of modern education. Unlike his elder brothers, Baidya did not immediately pursue education in India because his father passed away. While his brothers studied in India, he stayed at home with his sisters. Baidya's mother could not read and write, and she was not in a position to provide encouragement for her son and secure connections for his education opportunities. This changed in the early 1940s when the Board of Agriculture, a newly installed body in the Rana administration, made some scholarships available in India for agricultural engineers. Unlike the established disciplines of civil, electrical and mechanical engineering, agricultural engineering was largely unknown at that time and unpopular. Baidya recalled that "no other people [wanted] to go for agricultural engineering; everybody wanted civil, electrical and mechanical engineering", and that he eventually was approached by a government official who asked, "would you like to go for agricultural engineering?"[3] Baidya had preferred mechanical engineering, but due to his lack of options, he accepted the opportunity.

Baidya went to Allahabad University, India in 1942. The Allahabad Agricultural Institute was set up in 1910 by American economists, agriculturalists and missionaries to address the "Christian neglect" of poor Indian people in rural villages. Training at the Institute integrated education, research and extension, a philosophy typical for the US land grant model at that time (Tolar, 2008; Moon, 1998; Frye, 1993).[4] In 1936, the Institute had started courses in home economics, perceiving the important role of women in agriculture, and in 1942 it introduced the first courses on agricultural engineering. In 1947, Baidya was the first from Nepal to obtain a bachelor's degree in agricultural engineering from this institute.

Back in Nepal, Baidya was employed by the Rana administration to lead several projects, but agricultural development was hardly a priority for them. This changed in 1951. In that year, the United States became the first donor in Nepal to sign an agreement for technical cooperation under President Truman's Point IV program. In the same year, the Rana autocracy fell, and a new democratic government came in its place (Whelpton, 2005). The new administration embarked on an ambitious project of modernization in an open country, and with the help of civil servants of newly independent India, a completely new government structure and civil service was created (Shrestha, 2001). The Department of Agriculture (DA) was one of the newly formed institutions. The American Point IV approach advocated technical assistance and the direct transfer of knowledge as a catalyst for development. It typically provided support in the form of technical advisors and equipment. In Nepal, the technical advisors of the United States Operation Mission (USOM) had mostly worked as country agricultural extension agents in the United States; from this experience they brought a focus on rural works and agriculture to Nepal (Skerry et al., 1992).

Training Nepali government officials at US universities was a key component of the USOM human resources program. Baidya was one of the few officials in the government with a background in agriculture, and therefore, eligible for further training in the United States. In 1953–1954, he went to the United States to participate in a student exchange program at the Agricultural College of Utah. There, he was hosted by an American agricultural development officer to learn more about soil conservation and agricultural extension. He also visited Puerto Rico as part of the program. He recalled that he learned "how to work with [farmers]; how to satisfy them by giving some service; how to build trust with farmers".[5] These new experiences had an immense impact on him; the program in India at Allahabad University had included some theory on extension, but in the United States he learned about working with villagers, building trust and serving the needs of farmers.

In Nepal, through the Village Development Project, the USOM supported the organization of the newly formed Department of Agriculture, focusing on designing and setting up a new national rural extension service. In 1958, Harold Dusenberry, chief agriculturalist of the USOM in Nepal, described the rural development institution conceived by the Americans: "This was to be an organization known as the Village Development Service, established to contact village people, to find out their needs, and then to get assistance from various departments which would channel programs through this service" (Skerry et al., 1992, p. 37). Baidya, working for the DA and trained by the Americans, came to play a prominent role in this mission. He was involved in setting up divisions in the DA for agricultural engineering and rural extension, and in the course of the 1950s, he became the first chief of the extension division and later assistant director of the DA. These were prestigious positions in the department, because all USOM projects for farmers, as well as farmers feedback, went through this division.

In 1961, things changed for Baidya in unforeseen ways. In that year, King Mahendra imposed direct rule and instituted a kind of village-based democracy in which power was concentrated in the monarchy (Whelpton, 2005). This move was couched in nationalism: "Key to the legitimization of indigenous [panchayat] control over the state apparatus in Nepal was the doctrine of 'development' – *bikas* – as '*the* national project'" (emphasis original) (Tamang, 2002, p. 314). The king's claim was that Nepal's democratically elected leaders and government were too much under influence of foreign donors, and that they were incapable of leading the country. Baidya had been working closely with the Americans and his position soon became untenable. He recalled about this episode: "I [was] attached to farmers" and "I was acting more democratic than the Panchayat government [could] swallow".[6] He was soon demoted and decided to leave government service.

Baidya continued to work with the Americans and pursued his career as a consultant and engineer in agriculture and rural development. In the 1960s, he wrote a number of technical books on agriculture in Nepal, on subjects like farm irrigation and water management (Baidya, 1968). These were often the first technical books on the subject in Nepal, and they were readily used. In the period 1971–1980, he worked as a project officer for the US Peace Corps in Nepal, supporting American volunteers in agricultural projects; and in the period 1981–1990, he worked as a consultant in rural development projects that focused on basic health needs for people below the poverty line.

In 1990, at 69 years of age, Baidya formally retired as a consultant but remained active. By that time, he had become deeply concerned with the degradation of the urban reaches of the Bagmati River, and he founded the Save the Bagmati Campaign (see Rademacher, 2007, 2009, for an analysis). Having grown up near the sacred Bagmati and being an upper caste member of the ancient Newari-Kathmandu culture, he became a known activist who advocated for the revival of what he called the "Bagmati civilization". In the 1990s and 2000s, his views circulated widely in the media, from Nepali and English language newspapers to radio, television and Internet sources. In this period, he became the eminent spokesperson for his unique conceptualization of environmental change, weaving ideas of belonging, history and nature into a vision of restoration.

Reporting his death on 23 December 2013, at an age of 94 years old, the Kathmandu Post, wrote about him as "a pioneer environmentalist", remembering him for his active involvement in promoting Bagmati clean-up campaigns and civilization works for more than three decades.[7]

Doing interviews, doing gender, doing race, doing sexuality[8]

The construction of the account of Baidya's life and career through interviews was embedded in the everyday politics of gender, race and sexuality. For instance, reflecting on the questions I asked and how Baidya recounted his life, the account above only mentions men – his father, brothers, male teachers and colleagues in India and Nepal, and men from America. The CV that he shared with me mentioned at the top "Baidya, Huta Ram (son of Dr. Ratna Das Baidya)". Women, such as his mother, sisters, wife and daughter, and female colleagues, both Nepali and American, were hardly mentioned in the interviews, and therefore, appeared absent in, and irrelevant for, his life and career. This was partially because I initially refrained from steering the conversations: it was only after explicitly asking that Baidya gave an account of working with an American female home economics advisor and Nepali female village development workers in the 1950s, as well as with Nepali women as beneficiaries of technologies. This is no coincidence: as I have written elsewhere, the recognition of men, without mentioning women, signifies norms of a particular hegemonic patriarchal culture in Nepal (Liebrand & Udas, 2017) and a professional culture of whiteness among engineers in foreign aid and donor-supported development cooperation (Liebrand, 2022). By focusing on men, Baidya mobilized a familiar ethnic-racially coded scale of sex/gender difference that is hegemonic in development, culminating in the manly white man and the feminine white women – the sharper the contrast, the more "advanced" a society is (Markowitz, 2001; Power, 2003). The practice of signifying men and obscuring women in the interviews breathed life into the narrative of his life and career, making it self-evident and normal.

The account above also clearly reveals that the construction of an engineering identity is tied to a project of modernization and foreign collaboration, notably with Western countries, such as the United States. In his stories, Baidya associated engineering and technical skills exclusively with modern education, the application of science, and principles of rationality and universality, as it was practiced in the West. He did not share, for instance, any stories of how farmers in Nepal were builders of their own systems (Dahal, 1997); nor did he mention the delicate practice of constructing rice terraces in valleys and hills, or, for instance, the constructing of pagodas in the beautiful Newar cities of the Kathmandu valley. He simply did not associate these "technologies" with technical skills and ingenuity, or science and engineering. In contrast, he associated engineering with Western science and foreign men. For example, he recalled an FAO mission of foreign experts in the early 1950s to assess the irrigation potential of the Kathmandu valley, and commented that "they were most of the time right", even though he admitted that their work was largely based on estimations and guesswork in the absence of any reliable data (see Figure 2.1).[9] Such an account exposes an admiration and identification with science and foreign, Westernized men and the engineering identity associated with it.

Figure 2.1 Identification with science and Western male technology experts. Nepali government officials and engineers posing at a research station with a member of the FAO mission and/or technician of USOM (person on the far right), presumably in the Kathmandu Valley, Nepal, 30 August 1952. Baidya is among them (third person from the right).

Source: Private collection, S. Theuvenet (1952).

Baidya's account also subtly shows how the life and career of an engineer gains credibility by building on a particular normative conceptualization of the world that is central in mainstream engineering cultures – the stereotyped dualism of technical/social (Faulkner, 2009b). This dualism holds that the technical is valued higher than the social by signifying the technical as rational, universal, asocial and non-political. Throughout the interviews, Baidya referred to himself as an "engineer" and "a technical man", creating an identity for himself as a man of rationality and ethical conduct in applying principles of science. At the same time, interestingly, he also expressed in interviews that he was not just an engineer but an "agricultural engineer". In so doing, he elaborated, he made a distinction between engineers who work with lifeless material, such as concrete, and those who work with the living environment, such as plants, animals and humans.[10] And later in his life, in 2007, acting as an environmentalist and reflecting perhaps a realization of the inescapability of the end of life, he explained to a journalist of *Nepali Times* (2007) that "we cannot control nature [and] must work together with nature to preserve the [Bagmati] river and our [Nepali] culture". Here, he attached greater value to the living environment than to lifeless material, challenging in the end of his long career, the "hardware" engineering view on technology. Yet, he simultaneously legitimized and normalized his views in the interviews by referring to the application of "the technical".

The interactions embodied in the life history interview, between Baidya and me, and occasionally a third person (a Nepali student or Nepali visitor), were encapsulated in politics of gender, race and sexuality – through jokes, or by means of particular practices and small demonstrations. In particular, practices of heteronormativity surfaced in our interactions, intersecting with norms of masculinity and whiteness. These practices littered the space of the interviews with small cultural meeting points and smoothed our interactions in building rapport. Some of these subtle interactions are particularly illustrative for the politics of gender, race and sexuality involved.

I was first introduced to Baidya by a fellow Nepali engineer and male friend, who had informed Baidya about my interest in meeting him, the first agricultural engineer of Nepal. The first interaction I had with Baidya was thus among three male engineers – two non-white engineers from Nepal and one white engineer from the Netherlands. Right after entering a room with some chairs and expressing formal meetings of "namaste", Baidya jokingly asked me whether I had a girlfriend, looking at me with a big smile and a twinkle in his eyes. My Nepali friend and I laughed in response; it was immediately clear to me that this senior Nepali engineer had not lost his wits and knew about the habits in the world of engineering and development, and I assured him with a laugh that I did have a girlfriend. The joke – a small cultural performance of masculinity, whiteness and heteronormativity among the three of us – cleared the air and effectively functioned as a "common cultural meeting point in order to oil the wheels of collaboration", to use the words of Wendy Faulkner (2009a, p. 9). We, as engineers and group of men, embodied a bewildering range of otherness in the space of one room: Baidya was a senior, high-class and upper-caste Newari man; my male Nepali friend was a junior person (then 26 years of age) with a middle class and upper-caste (Chettri) background, and coming from outside the Kathmandu Valley; and I was a white foreigner, from the Netherlands. Some of the differences were signified by our bodies, notably the fragile physique of Baidya, and the healthy build of my Nepali friend and myself, and of course, our racial appearances. Still, the joke and its mobilization of a heteronormative culture cleared the ground for a smooth and joyful first interaction.

Faulkner (2009a), based on research among engineers at offshore oil rigs in the United Kingdom, identifies heteronormativity as a characteristic hegemonic culture among men who identify themselves as engineers. Baidya knew these hegemonic norms by heart and how they were practiced among engineers in the West, having visited the United States for training in the 1950s and worked with white, American technical experts in Nepal. In making the joke, Baidya, wittingly or otherwise, appealed to me with a ritualized American style of masculinity, tailor-made for my young age (then 30 years old) and Caucasian appearance. The essence of the joke was something like "a healthy bloke is just interested in girls" and I perceived it as a friendly welcome, a test of manhood – are you a real man who has sex with a woman. The joke made clear to me, by means of suggestion, that Baidya, earlier in his life, had been a healthy bloke himself, and that we supposedly (had) shared similar, heterosexual desires. The joke subtly supported the idea that there was nothing wrong with this old engineer; he was (or had been) a real man; he was normal, he was authentic, and I just had to say yes to build rapport.

In that first meeting, Baidya was also quick to show to me a plaque of honor, a "lifetime achievement award" granted to him by the Nepal Engineers Association (NEA) in 2008 on the occasion of Engineer's Day.[11] The NEA is the premier national association for engineers in Nepal, established in 1968 and modeled after its foreign examples such as the Institution of Engineers in the United Kingdom. Membership of the NEA is regulated by eligibility criteria – members need to have a bachelor's degree in engineering; new members can only be proposed by existing members etc. – and honorary titles and distinctions such as lifetime achievement awards are only granted to a few, based on principles of personal achievement and merit. By showing and handing the plaque to me for inspection, Baidya signaled to me that he truly was an engineer, a self-made man who had chosen his own career path, based on hard work, innate quality and merit. Similar things happened in interviews with other engineers – showing medals, prizes, photos as proof of personal achievements, being self-made, having innate qualities and technical competence – and I came to realize that such interactions embody characteristic aspects of engineering identities. They are based on a belief in the meritocracy and impartiality of science and technology (Gupta, 2007) and support the idea that one can choose one's own destiny through rational thinking and the practice of science – and that engineering can bring social and intellectual transformation. It bestowed upon Baidya a status of impartiality and neutrality, granting legitimacy to his identity as an engineer and serving humanity.

In other meetings, Baidya often picked up tools or gave little demonstrations of his inventions to illustrate his experiences and explain what engineering is about. In his view, an engineer is a person who knows how to use tools. He stated, for instance, "how can you be an engineer, unless you are interested in . . . tools, . . . and use tools", and he commented that the story of Robinson Crusoe was his favorite book, because it taught the reader, according to him, how to survive as a man on an inhabited island through the use of simple technology, rational thinking and inventive methods.[12] While telling this, for instance, he picked up a hammer to dramatize his story, suggesting that he was a man with hands-on technical competence. On other occasions, he showed me one of his inventions while talking, such as a model flower that could be unfolded and assembled for extension purposes or the design of a smokeless stove (*chulo*). The "ritualistic displays of hands-on technical competence" among men engineers have been described as a part and parcel of an "engineering culture" in the United States (McIlwee & Robinson, 1992, p. 139). The demonstration of this "nuts and bolts" identity and gratification in working with tools is noted to be typical for engineers (Kleif & Faulkner, 2003; Faulkner, 2007, 2009a). Baidya so passionately talked about hands-on technical skills, while dramatizing them with a tool in his hand, that his whole identity as an experienced man and engineer appeared to me truly normal and authentic.

His stories and demonstrations with tools helped to breathe life into the idea that the performance of an engineer is based on innate qualities, and inborn technical skills and knowledge. By saying that "the morning shows the day", Baidya explained that he had a supposedly natural interest as a boy to play with tools, recalling how he enjoyed himself as a child with a microscope, binoculars and a carpentry toolbox that his father had brought from England.[13] He also explained that he had most enjoyed his education at Allahabad University when working on machinery. In telling these stories and tinkering with tools, Baidya created an origin story – he became an engineer because of his innate interest as a boy in tools. In such a performative representation, the quality of rationality and technical skill is located as inherent to the male subject – as inborn to boys.

The examples above, of stories and interactions in interviews, illustrate that it is through everyday interactions, behavior and practices that normalcy and gender authenticity for men in engineering comes to exist. Put differently, the examples show that there is nothing "natural" about men being engineers – masculinity, whiteness and heteronormativity within engineering in development need to be actively cultivated among men of different ethnic and racial backgrounds, often in the face of opposition and competing narratives. To illustrate this point: Baidya recalled that his father initially wanted him to become a doctor rather than an engineer, and as mentioned earlier, Baidya's preferred choice for a scholarship had never been in agricultural engineering. Instead of working with tools, Baidya initially made a career as a civil servant, working with pen and paper, and acting as a white-collar engineer. In his life, as the description of his career history makes clear, he was equally an extension worker, activist and environmentalist as much as an engineer.

Baidya undoubtedly also practiced various other forms of masculinity, notably those of a caring father and a Nepali upper-caste man. He was married and the father of three children, including one daughter, and he was an elite member of the Newari community in Kathmandu. However, because I focused in the interviews on the professional life of Baidya and how masculinity is tied to engineering, these practices of masculinity have not been touched upon in the analysis.

Concluding reflections

The use of life history interviews for researching masculinity among engineers in planned development demonstrates that the production of knowledge on gender, race and sexuality in the irrigation profession – and water governance at large – can never be disconnected from either the practice of engineering, or the researcher and the research method itself. In the world of agriculture and irrigation development, performances of masculinity are a constituent part of social encounters among engineers, including those that take shape in practices of doing interview research. Doing a life history interview involves building rapport and a relationship between the interviewer and the interviewee. Trust is built through everyday interactions, and therefore cannot be disconnected from everyday practices of doing gender, race and sexuality. To put it simply: Baidya said certain things and acted in ways, at least partially, because of me. My professional identity as an engineer and white man from the West informed Baidya's responses, how he articulated stories and behaved in the interview. My selection of Baidya for a life history interview and my presence made the interview an interaction between male engineers, in which Baidya sought to create a credible voice for himself as an engineer from Nepal vis-à-vis an engineer from the West. Coming from Nepal and knowing the social practice of an engineering culture in international development cooperation by heart, based on more than six decades of professional experience, Baidya cast himself as an engineer with particular stories and interactions: making jokes, tinkering with tools, presenting a "nuts and bolts" identity, emphasizing his achievements as a self-made man and discussing the technical intricacies of his inventions. All these subtle performances helped to portray Baidya's identity as a male engineer as truly normal and authentic.

Doing interviews came thus at a cost: it worked to expose and problematize masculinities in engineering but it also re-enacted and reinforced those norms. It literally provided a space to breathe *new* life into performances of masculinity, whiteness and heteronormativity, re-enacting the idea that science and technology are fundamentally linked to the West, stereotyped dualisms of technical/social, and essentialist ideas that technical skills are inborn to boys. Because of my active participation in the interviews, in which we performed man-to-man interactions, it became difficult for me to scrutinize what feminist scholars have identified as a "strange silence" on the dominance of men in the irrigation engineering profession (Zwarteveen, 2008, p. 111). In hindsight, the interviews with Baidya can be considered to have achieved the opposite objective – they maintained a discursive invisibility of men and masculinity in the irrigation and agricultural profession, normalizing the bond between men, masculinity and engineering, and legitimizing men's authority and voice in technical expertise as self-evident and natural.

That being said, the method certainly achieved its more practical objective, which was to make a male engineer talk. This is an objective both for feminist-inspired research into masculinity and the method of "studying up". The particular resilience of masculinities in irrigation and agricultural engineering in development poses methodological challenges for research. To scrutinize an embedded and taken-for-granted association, even conflation, of men with organizational power, authority expertise and prestige in engineering (cf. Collinson & Hearn, 1996; Zwarteveen, 2008), it is a *prerequisite* for analysis that men themselves start giving an account of their experiences. Letting them explain is not just a necessary first step to hold men accountable for their participation in unequal gender relations; but also, more importantly, to provide the material (their stories, their behavior and practices) for researchers to conduct a critical analysis of masculinity.

The life history interview with Baidya produced a wealth of material for research, not just in terms of accurate historical information but also in everyday interactions of doing gender, race and sexuality – the subtle cultural performances and dramatic techniques that he used to cast himself as a trustworthy professional. The interview gave space for Baidya to reflect on his life and career, and it allowed me to reflect on my own role and contribution to the interviews in terms of embodiment and performances of masculinity. By capturing the rich diversity of his experiences and laying bare ambiguities in these experiences, the life history interview with Baidya prompted me to start deconstructing normalcy in the engineering profession. To recall, Baidya originally did not want to become an engineer: he also acted as an administrator, extension worker, activist and environmentalist, and he contested the dominant "hardware" perspective of technology in engineering. These ambiguities are key for feminist research, because they provide the critical source material for promoting alternative visions on development.

Acknowledgment

I sincerely thank Huta Ram Baidya for participating in the interviews, and I thank Joyce Huisman-Theuvenet for sharing photos of her father's life and work in Nepal. I also thank Wendy Harcourt and John Cameron for early comments on a presentation that I gave on the basis of this chapter in 2012 at the Institute of Social Sciences, and I thank the editors for their critical feedback on drafts of this chapter.

Notes

1 The Nepali caste division diverged from the four-fold *varna* (social order) model in India, which had four occupational categories: i) Brahmin priests (Bahun); ii) Kshatriya warriors (Chhetri); iii) Vaisya farmers and traders; and iv) Sudra laborers. Instead, the Nepali version has five categories to accommodate the "tribal" groups. The Newar, who claim to be the original inhabitants of the Kathmandu Valley formed itself a conglomeration of different castes spanning the entire range from upper-caste tiers to so-called untouchables or Dalits (Höfer, 2004; Gellner et al., 2008).

2 The Baidya caste were traditionally Ayurvedic physicians; the word "baidya" literally means a physician in Sanskrit, and they occupied a place in the Nepali caste hierarchy alongside Bahuns and Chhetris.

3 Tape 12, 7 July 2010, 28:00 to 29:30.

4 The land-grant model involved colleges in the United States that were originally endowed with land by the state and set up to teach practical subjects such as agriculture, science and engineering, as opposed to the general orientation towards liberal arts in institutions of higher learning (Knowles, 1985).

5 Tape 5, 15 December 2009, 1:27:20 to 1:28:10.

6 Tape 5, 15 December 2009, 1:30:04 to 1:31:45.

7 See: http://kathmandupost.ekantipur.com/news/2014-01-04/bagmati-clean-up-to-remember-baidya.html, visited 1 April 2019.

8 Under Western eyes, in the world of international development cooperation, issues of caste and ethnicity basically fall under "doing race". Under colonialism, race was the dominant framework to conceive of and classify people within the global hierarchy of civilization (Mohanty, 1984). To date, Hindu caste systems continue to pose a problem for the conceptual apparatus of Western social sciences (Gellner et al., 2008). By implication, post-colonial scholars argue, when Western authors use terms like "ethnicity" and "caste", they tend to confirm the idea of cultural difference and use it as stand-in for "race" Cooper, 1996; White, 2006). Therefore, since I am from the West, I prefer talking about doing race.

9 Tape 8, 29 December 2009, 00:10 to 15:00. Saliently, Baidya recalled that the irrigation expert of the FAO was an Indonesian man. This person was S. Theuvenet. He wrote the first country report on irrigation for Nepal (Theuvenet, 1953), and he originally came from Dutch Indonesia. In the Netherlands, I came into contact with Theuvenet's daughter who shared her father's photos of his work in Nepal (see Figure 2.1).

10 Tape 12, 7 July 2020, 29:00 to 30:49.

11 In the Nepali calendar, this was engineer's day at 3 Shrawan 2065 BS.

12 Tape 12, 7 July 2020, 24:30 to 25:00.

13 Tape 4, 10 December 2009; 09:10 to 10:30; and tape 12, 7 July 2020, 19:45 to 21:00.

References

Angrosino, M. V. (2002). Conducting a life history interview. In M. Angrosino (Ed.), *Doing cultural anthropology: Projects for ethnographic data collection* (pp. 33–44). Waveland Press.

Baidya, H. R. (1968). *Farm irrigation and water management. Principles and practices*. Royal Nepal Academy.

Butler, J. (1999). *Gender trouble. Feminism and the subversion of identity* (2nd ed.). Routledge.

Collinson, D. L., & Hearn, J. (Eds.). (1996). *Men as managers, managers as men: Critical perspectives on men, masculinities and management*. Sage Publications.

Cooper, F. (1996). Modernizing bureaucrats. In F. Cooper & R. Packard (Eds.), *International development and the social sciences: Essays on the history and politics of knowledge* (pp. 64–92). University of California Press.
Dahal, N. (1997). A review of Nepal's first conference on agriculture. *Water Nepal, 5*(2), 149–164.
Faulkner, W. (2007). Nuts and bolts people. Gender-troubled engineering identities. *Social Studies of Science, 37*(3), 331–356. https://doi.org/10.1177/0306312706072175
Faulkner, W. (2009a). Doing gender in engineering workplace cultures. I. Observations from the field. *Engineering Studies, 1*(1), 3–18. https://doi.org/10.1080/19378620902721322
Faulkner, W. (2009b). Doing gender in engineering workplace cultures. II. Gender in/authenticity and the in/visibility paradox. *Engineering Studies, 1*(3), 169–189. https://doi.org/10.1080/19378620903225059
Frye, D. T. (1993). Linking institutional missions to university and college archive programs: The land-grant model. *American Archivist, 56*, 36–52. www.jstor.org/stable/40293712
Gellner, D. N., Pfaff-Czarnecka, J., & Whelpton, J. (Eds.). (2008). *Nationalism and ethnicity in Nepal*. Vajra Publications.
Gupta, N. (2007). Indian women in doctoral education in science and engineering. A study of informal milieu at the reputed Indian Institutes of Technology. *Science, Technology and Human Values, 32*(5), 507–533. https://doi.org/10.1177/0895904805303200
Harding, S. (2004). A socially relevant philosophy of science? Resources from standpoint theory's controversiality. *Hypatia, 19*(1), 25–47. https://doi.org/10.1111/j.1527-2001.2004.tb01267.x
Hilgartner, S. (2000). *Science on stage: Expert advice as public drama*. Stanford University Press.
Höfer, A. (2004). *The case hierarchy and the state in Nepal: A study of the Muluki Ain of 1854*. Himal Books.
John, G. S. (Ed.). (2008). *Victor turner and contemporary cultural performance*. Bergahn Books.
Kleif, T., & Faulkner, W. (2003). I'm no athlete [but] I can make this thing dance! Men's pleasures in technology. *Science, Technology and Human Values, 28*(2), 296–325. https://doi.org/10.1177/0162243902250908
Knowles, J. (1985). Science and farm women's work: The agrarian origins of home economic extension. *Agriculture and Human Values, 2*(1), 52–55. https://doi.org/10.1007/BF01534993
Laurie, N. (2005). Developing development orthodoxy: Negotiating masculinities in the water sector. *Development and Change, 36*(3), 527–549. https://doi.org/10.1111/j.0012-155X.2005.00422.x
Liebrand, J. (2022). *Whiteness in engineering: Tracing technology, masculinity and race in Nepal's development*. Himal Books and Social Science Baha.
Liebrand, J., & Udas, P. J. (2017). Becoming an engineer or a lady engineer: Exploring professional performance and masculinity in Nepal's Department of Irrigation. *Engineering Studies, 9*(2), 120–139. https://doi.org/10.1080/19378629.2017.1345915
Lynch, B. (1993). The bureaucratic tradition and women's invisibility in irrigation. In *Proceedings of the 24th chacmool conference* (pp. 333–342). University of Calgary Archaeological Association.
Markowitz, S. (2001). Pelvic politics: Sexual dimorphism and racial difference. *Signs: Journal of Women in Culture and Society, 26*(2), 389–414. http://doi.org/10.1086/495598
McIlwee, J. S., & Robinson, G. (1992). *Women in engineering: Gender, power, and workplace culture*. State University of New York Press.
McKenzie, J. (2001). *Perform or else: From discipline to performance*. Routledge.
Mohanty, C. (1984). Under Western eyes: Feminist scholarship and colonial discourses. *Boundary, 2*(3), 333–358. https://doi.org/10.1057/fr.1988.42
Moon, S. M. (1998). Takeoff or self-sufficiency? Ideologies of development in Indonesia, 1957–1961. *Technology and Culture, 39*(2), 187–212. https://doi.org/10.2307/3107044
Nader, L. (1972). Up the anthropologist: Perspectives gained from studying up. In D. Hymes (Ed.), *Reinventing anthropology* (pp. 284–311). Pantheon Books.
Nair, S. (2012). *Women in Indian engineering. A preliminary analysis of data from the graduate level engineering education field in Kerala and Rajasthan*. Occasional paper no. 58. Centre for Women's Development Studies.
Nepali Times. (2007, July 13). An unholy holy river. Pollution and encroachment have destroyed the Valley historic river. Issue no. 357.
Ongsakul, R., Resurreccion, B., & Sajor, E. (2012). Normalizing masculinities in water bureaucracy in Thailand. *International Journal of Public Administration, 35*(9), 577–586. https://doi.org/10.1080/01900692.2012.661177

Parikh, P. P., & Sukhatme, S. P. (1994). Women engineers in India: Present scenario and current trends. *Journal of Scientific and Industrial Research*, *53*, 233–247. www.jstor.org/stable/4414501
Parikh, P. P., & Sukhatme, S. P. (2004). Women engineers in India. *Economic and Political Weekly*, *39*(2), 193–201.
Power, M. (2003). *Rethinking development geographies*. Routledge.
Rademacher, A. (2007). Farewell to the Bagmati civilisation: Losing riverscape and nation in Kathmandu. *National Identities*, *9*(2), 127–142. https://doi.org/10.1080/14608940701333746
Rademacher, A. (2009). Restoration and survival: Remembering the Bagmati civilisation. In A. Guneratne (Ed.), *Culture and the environment in the Himalaya* (pp. 167–185). Routledge.
Rap, E. (2007). Cultural performance, resource flows and passion in politics: A situational analysis of an election rally in Western Mexico. *Journal of Latin American Studies*, *39*, *595–625*.
Ritchie, D. A. (2003). Conducting interviews. In *Doing oral history. A practical guide* (pp. 84–109). Oxford University Press.
Schechner, R. (2006). *Performance studies: An introduction*. Routledge.
Shrestha, M. K. (2001). *Readings in Nepalese public administration*. Educational Enterprise.
Shumshere, P. J. B. R. (1971). Towards an integrated policy of national integration. In P. Pradhan (Ed.), *Aspects of development administration* (pp. 40–45). Tribhuvan University, Centre for Economic Development and Administration.
Skerry, C. A., Moran, K., & Calavan, K. M. (1992). *Four decades of development. The history of U.S. assistance to Nepal 1951–1991*. United States Agency for International Development.
South Asia Consortium for Interdisciplinary Water Resources Studies (SaciWATERs). (2011). *Situational analysis of women water professionals in South Asia*. SaciWATERs. www.soppecom.org/pdf/SituationalAnalysisofWomenFULL7.pdf
Tamang, S. (2002). Dis-embedding the sexual/social contract: Citizenship and gender in Nepal. *Citizenship Studies*, *6*(3), 309–324. https://doi.org/10.1080/1362102022000011630
Theuvenet, S. (1953). *Report to the government of Nepal on irrigation*. Expanded Technical Assistance Programme. Food and Agriculture Organization (FAO), United Nations.
Tolar, R. L. (2008). The land-grant model in U.S. higher education. *Journal of Siberian Federal University Humanities and Social Sciences*, *3*, 408–412.
Turner, V. (2008). *The ritual process*. Aldine Transaction.
Whelpton, J. (2005). *A history of Nepal*. Cambridge University Press.
White, S. C. (2006). The "gender lens": A racial blinder? *Progress in Development Studies*, *6*(1), 55–67. https://doi.org/10.1191/1464993406ps127oa
Zwarteveen, M. Z. (2008). Men, masculinities and water powers in irrigation. Contours of a new research agenda. *Water Alternatives*, *1*(1), 111–130.
Zwarteveen, M. Z. (2011). Questioning masculinities in water. *Economic and Political Weekly*, *XLVI*(18), 40–48. www.jstor.org/stable/41152342
Zwarteveen, M. Z., & Rap, E. (2017). Guest editor's introduction: Engineering masculinities in water governance. *Engineering Studies*, *9*(2), 75–77. https://doi.org/10.1080/19378629.2017.1363019

3
GENDERING GROUNDWATER SALINITY
A study of Lodhva, Gujarat, India

Maitreyi Koduganti Venkata, and Gabriela Cuadrado-Quesada

Introduction

The salinization of groundwater is connected to different aspects such as climate, geologic features, and practices of water use (Misra and Mishra, 2007). More than 15% of the total area in India is impacted by inland groundwater salinity, especially in the states of Rajasthan, Haryana, Delhi, Karnataka, Uttar Pradesh, Tamil Nadu, and Gujarat. Arid coastal regions are especially prone to saline groundwater because the salts that accumulate in the soil due to rainfall quickly evaporate, often in combination with sea water, which then intrudes into the aquifers as result of groundwater extraction. One state that has been particularly affected is Gujarat, located in the northwest of India, which has saline groundwater in more than 12% of its territory, notably impacting coastal zones (Saha & Ray, 2019; Misra & Mishra, 2007).

It is important to note that saline groundwaters in Lodhva are likely to contain other environmental or chemical contaminants (Bhagat et al., 2022; Bhagat et al., 2021). The use and consumption of such groundwater has serious health implications for many communities living in these states, including an increased occurrence of kidney stones, urinary bladder infections, and skin-related infections (López & Hoppe, 2010). Furthermore, a broad range of technical studies are available which assesses the impacts of groundwater salinity on agriculture, mining, health, and ecology (Foster et al., 2018; Li et al., 2018; Khan et al., 2011; Jolly et al., 2008). However, very little is known about how the salinity of groundwater may turn into a daily struggle, especially how it interacts with human bodies and affects the emotions of people exposed to its daily use (Sultana, 2011; Bondi, 2005; Davidson & Milligan, 2004; Sultana, 2009).

Drawing on feminist political ecology, this chapter emphasizes how paying attention to bodies and emotions in the context of resource struggles could provide insight into how people are differently impacted by groundwater salinity, not only in terms of (physical) access to water but also how it affects their bodily awareness and how they engage in intimate relations (Brown & Gershon, 2017; Coole, 2013). This chapter analyzes how groundwater salinity affects bodily experiences and emotions in the coastal village of Lodhva, located in the south of Gujarat, by narrating the experiences of different community members who are being affected by groundwater salinity. The analysis highlights the disparate daily struggles

DOI: 10.4324/9781003100379-5

community members have with saline groundwater, how it impacts the way they view their own bodies and emotions, and how they engage with others (e.g., parents, spouses, friends, doctors) to deal with their bodily experiences and discomforts.

How was the study conducted?

Data collection

This study is based on three months of ethnographic fieldwork in Lodhva, from November 2019 to January 2020, during which we observed and interviewed many of the inhabitants of Lodhva to understand their daily struggles with groundwater. We used a qualitative approach, conducting 40 semi-structured interviews and five focus-group discussions, in addition to daily observations and transect walks. Respondents for semi-structured interviews were selected through a stratified random sampling method based on an intersectional understanding of gender, which accounts for the ways that other social identities such as caste, age, and class intersect with and modify gender identities and experiences.

In addition, the views of school children, both boys and girls between the ages of 12 and 17, were documented through visual methods, including participatory creative drawing (Galman, 2009). Since the local school hosted classes from 6th grade to 12th grade, children in this age range were selected. They were asked to depict the situation of water in their village through drawings. A total of 100 such drawings were received, of which 19 were interpreted in collaboration with the participants, thereby ensuring reflexivity and a comprehensive understanding of their depiction (see Figure 3.1).

Figure 3.1 Collection of drawings by children.
Source: Authors.

Table 3.1 List of Interviewees and Interview Codes

#	*Type of Stakeholder*	*Interview Code*
1	Teachers	Interview T
2	Village Headman	Interview VH
3	Talathi (Block Development Officer)	Interview TL
4	Industries	
	Ambuja Cements Labour Union	Interview AC
	Director of Works – Sidhee Cements	Interview SC
5	Health Inspector	Interview HI
6	Bore-Well drillers	Interview BD
7	Water Operator	Interview WO
8	Dominant-Caste Households	
	Unmarried Women (between 18–22 years)	Interview AH_A
	Married Middle Aged Women (between 27–40 years)	Interview AH_B
	Elderly Women (45 years and above)	Interview AH_C
	Middle Aged Married Men (between 27–40 years)	Interview AH_D
9	Non-Dominant Caste Households	
	Unmarried Women (between 18–22 years)	Interview HH_A
	Young Married Women (22–27)	Interview AH_B
	Married Middle Aged Women (between 27–40 years)	Interview AH_C
	Elderly Women (45 years and above)	Interview AH_D
	Middle Aged Married Men (between 27–40 years)	Interview AH_E
	Young Boys (less than 16 years)	Interview AH_F

To complement this data, five in-depth interviews were conducted to flesh out how gendered impacts of saline groundwater intersect with caste, age, and class. These in-depth interviews focused on the lives of non-dominant caste women, men, and children. The study also included fieldwork observations and conversations with medical staff, schoolteachers, industrial actors, and government officials. Empirically grounded data was triangulated with available secondary literature, including academic articles, government reports, policy documents, and newspaper articles. All data-collection methods fed into gaining a deeper understanding of demography, livelihood activities, gender roles related to water, and the daily struggles of water use and access in Lodhva, as described in the following sections.

Data cleaning and analysis

Empirically grounded data was anonymized from the initial stages of data analysis. Evidence from these interviews was coded as "Interview", followed by codes representing different types of respondents, as listed in Table 3.1. When covering experiences related to sexuality from children, extra care was given to research ethics including the prior consent (of guardians) and full confidentiality of the interviewees.

Feminist political ecology of water and emotions

Drawing on Crenshaw's (1989) fundamental work on intersectionality, feminist political ecology understands gender as it intersects with other axes of social differentiation, such as class, caste,[1] and age (Ahmed & Zwarteveen, 2014; Truelove, 2011). For instance, in many of the rural areas in India, accessing water is highly differentiated, meaning households belonging to dominant-caste and/or with higher incomes tend to have greater access to and control over water. Furthermore, the quantity and quality of water access and use within a household is dependent on its geographic location (Crow & Sultana, 2002).

Grosz (1994) and Harcourt (2016) foreground bodies as sites of cultural meaning, social experience, and political resistance. They indicate that a person's body tends to be marked by their economic and social class, which in turn shapes their lived experiences. Alaimo et al. (2008) and Alaimo (2010) argue that women express themselves and their emotions with their bodies and that these bodies have different manifestations of feelings and experiences, including pain and pleasure. They also have diseases that are subject to medical interventions which may or may not cure those bodies. Therefore, there must be a way to talk about these bodies and the materiality they inhabit (Alaimo, 2010, p. 4). This argument also incorporates ideas from feminist political ecology scholarship, which proposes useful ways to grapple with and make sense of such differences and differentiations. Feminist political ecology scholars have shown how gendered identities and relations affect decision-making practices related to natural resources and how inequities are reproduced and materialized in daily resource struggles (Truelove, 2011; Harris, 2006). Building on these works, the works of Sultana (2011) foreground that resource use and access also play out through emotional geographies; that is, emotions that are produced by and based in particular people and places tend to be fluid and relational (Bondi, 2005; Sharp, 2009; Thien, 2005; Tolia-Kelly, 2006). Women traveling longer distances to defecate in fear of being seen by men, or non-dominant caste households shamed by upper class ones exemplify how emotions like fear or shame play an important role in how people respond to resource struggles. This chapter explores the intersection of gender, caste, class, and age, to understand how saline groundwater (when used and ingested) produces disparate bodily changes, emotions, and experiences through skin contact. In doing so, the chapter focuses on the materiality of water to unpack how its quality reflects, and perhaps reinforces, power relations based on gender, caste, age, and class.

Multiple facets of Lodhva

Demography and hydrogeology

Lodhva is in the southern coastal region of Gujarat (see Figure 3.2). Characterized by semi-arid and dry sub-humid climate, this part of Gujarat receives a little less than 1,000 mm rainfall per year, mostly during August and September. This region is covered by deep black coastal alluvium soil that contains relatively shallow aquifers; however, some regions are underlain by deep basaltic aquifers.

Based on the proximity to the sea, the local population[2] divided Lodhva into saline and less saline regions. The central built-up area of the village falls within the saline region, while the surrounding hamlets, locally known as *wadis*, fall within the less saline regions. The local villagers[3] maintain that as one moves away from the sea, the salinity falls; hence, the wells in *wadis* are a major source of water for human consumption. In times of scarcity and failure of formal water supply, the households from the village access drinking water from these wells. In addition to household consumption, groundwater – through tube wells and/or boreholes – is used to irrigate more than 60% of the cultivated area. Groundwater is also used by the cement industry and by the local municipalities that pump water to the nearby towns of Somnath and Veraval.[4] This indicates a heavy reliance on the shallow and sometimes deep aquifers of this region.

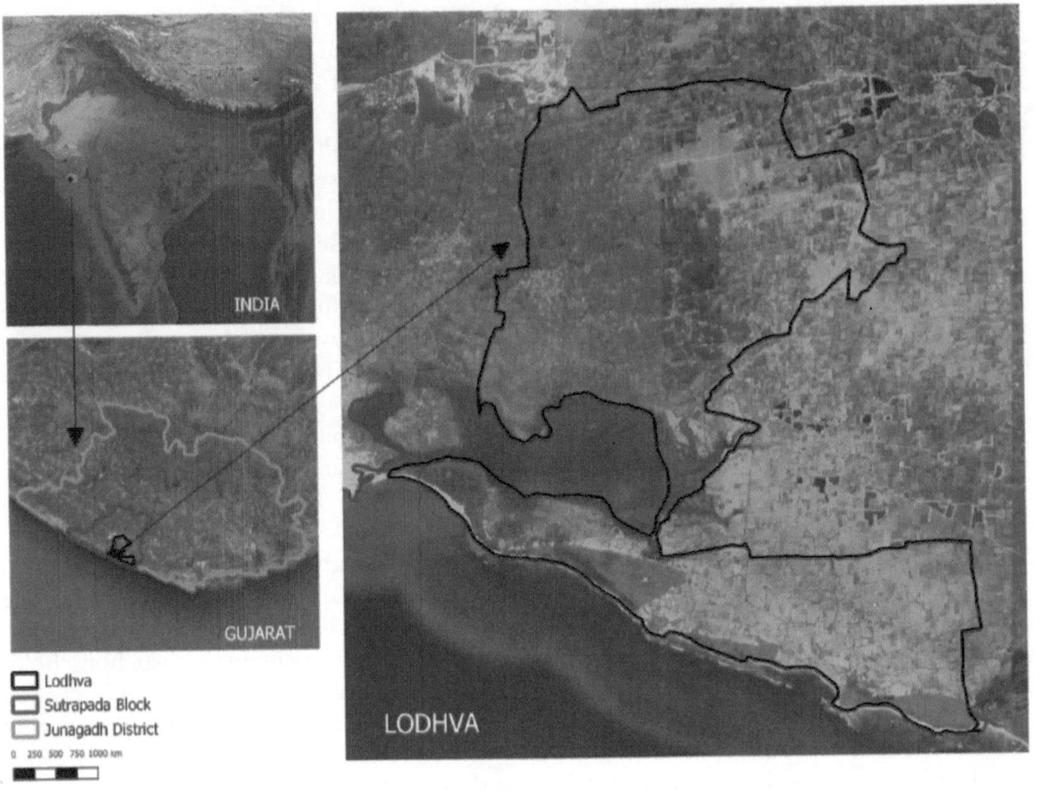

Figure 3.2 Map showing the location of Lodhva.

Source: Authors.

Livelihood activities

With agriculture being a dominant occupation, the major crops in this region are groundnut, millets, wheat, fodder crops and vegetables like onions, eggplant, and potatoes.[5] Only groundnuts are a substantial cash crop in this region and exported to the nearby towns of Somnath, Veraval, and sometimes larger cities like Diu. The rest of the crops produced are mostly consumed within households and the excess is sold in local markets. Farming is largely practiced by dominant-caste households. In turn, non-dominant caste households are engaged as casual labor in the nearby cement industries, fish packaging companies, and sugarcane farms in nearby areas.[6] In addition to this, many community members explore alternative livelihoods like running grocery shops, flour mills or food stalls, and activities like tailoring or welding within the village.

Waterscapes

In Lodhva there are no rivers or (irrigation) canals flowing through the village, nor there are freshwater water bodies like ponds or lakes. However, the village headman and some elderly villagers recollect that several decades ago, Lodhva used to house several freshwater lakes, whose waters were used for drinking and cooking purposes. Over time, with variable

Figure 3.3 Women collecting water near standpipes.
Source: Authors.

rainfall and agricultural run-off draining into the lakes, most of them dried up and became unfit for consumption.[7]

Lodhva received its first piped water connection from the state in 2010. It is sourced from the Umrathi dam built on Hiran River, located 45 kilometers from the village. Access to water significantly depends on caste, among other factors. About 80% of the village, belonging to dominant castes, have private water connections in their households to access this water.[8] The remaining 20% of the population, belonging to the non-dominant castes, use communal standpipes to access water. In general, one standpipe is shared by three to four households and there are about thirty such standpipes throughout the village.

Water from the dam is primarily used for drinking and cooking purposes. Water for washing, cleaning, and bathing is sourced from wells and boreholes.[9] Accessing groundwater is often dependent on the ownership of land and the capacities of individuals to invest in a borewell (Prakash, 2005). This implies that economically well-off people, who often are from dominant castes, enjoy greater access to groundwater by owning individual wells and boreholes. On the other hand, community members belonging to non-dominant castes access water through collective wells or have to get access from bore-holes owned by community members with better positions.[10] Notably, farmers with large landholdings most often belong to dominant castes, while farmers with lesser landholdings and landless farmers belong to non-dominant castes.

Caste and class inequalities

Interwoven with the notion of purity and pollution, discrimination based on the caste system has prevailed in India for over 3000 years. Despite its abolishment, the caste system continues to be a deeply rooted social construction within the fabric of the country, causing discrimination, marginalization, and sometimes deprivation, for instance in access to water resources for the non-dominant castes (Dutta et al., 2015). Segregation is more pronounced in rural areas than in urban areas because people in rural areas continue to hold on to the traditional and cultural values that legitimize caste-based segregation (Joshi, 2011; Tiwari & Phansalkar, 2007). A similar segregation was observed in Lodhva. A dominant portion of the households are Hindus, while less than ten households are Muslims (Census of India, 2011). While contemplating the divisions of caste, several categorizations within the Hindus were identified, of which the four main castes in order of hierarchy are *Ahirs*, *Kolis*, *Prajapati/Kumbhar*, and *Dalits*.[11]

The *Ahirs* are the dominant and most influential caste, who enjoy the benefits of land and asset ownership, as well as better access to services. This puts them highest in the hierarchy of power relations in the village. Next in the hierarchy of power and influence are the *Koli* (also known as Other Backward Castes, or OBC) who seem to appear as the second most influential caste in the village. At par are the castes of *Waniya*, *Mochi*, and *Kumbhar*. The third category in the caste hierarchy are the *Dalits* (also known as Harijans or Scheduled Caste). Categorized as backward castes (by the Government of India), *Dalits* often perform jobs that entail providing services to the other castes in the form of labor. They often engage in "dirty" tasks like cleaning latrines and sewers or clearing away waste.[12] Due to this, they are ostracized and segregated from other social groups and are called the "untouchables" (Tiwary, 2006). This practice of "untouchability" does not allow the Dalits to enter the spaces of the other castes because they are deemed low and inferior by the dominant castes. The very construction and location of houses in Lodhva is based on caste divisions, wherein the *Ahirs* and *Kolis* live in the central parts of the village, which have relatively better access to resources and services, while the *Dalits* lived on the extreme edges of the village with little access to the center of the village, shops, and other activities.[13]

Gender relations

Discussions and observations with respondents revealed that the relations between men and women are governed by a complex intermingling of customs and religious norms (Mehta, 2016; Truelove, 2011). The sociocultural constructions of accepted masculine and feminine roles (Ahlers & Zwarteveen, 2009) have placed men in the role of breadwinner and limited the role of women to fetching water and taking care of the household. Such a construction has also limited the scope of girls pursuing higher education. The village does not have a college. Therefore, while young boys are allowed and/or motivated to pursue higher education outside the village in bigger cities; young girls are often not allowed (by their family members) to travel outside the village to pursue higher education.[14] Such a norm ends up limiting the role of young girls (mostly between the age of nine and seventeen) to household responsibilities. Furthermore, when girls marry, they leave their maternal homes, and stay with their husbands and in-laws, where they bear the responsibility of taking care of the household.

A combination of gendered caste, age, and class inequalities coupled with differential access to water indicate that water struggles are more than material challenges (Wallace &

Coles, 2005; Batra, 2004). These observations form the basis of the analysis in the following sections, that highlight how saline groundwater impacts the way community members view their own (and others') bodies and emotions.

Everyday life, bodies, and emotions

Shubh's burning eyes – The authors met Shubh for the first time near the borehole, when he was fetching water with his mother and elder sister. He is seven years old and lives with his parents, grandparents, and an elder sister in the *harijan basti*.[15] He enjoys attending school where he has many friends to play with. While indicating his interest in attending school, he adds:

> I also like school because I get sweet water to drink. Have you ever tasted the water from our matka?[16] It is not so sweet. Sometimes I fill this water in my bottle and give it to my mother.

Shubh sees his mother give up her share of sweet water to make sure that her family gets more. His recounting of the joy of tasting sweet water and wanting to give it to his mother indicates a gendered experience wherein his mother often consumes saline water. However, at times when he accompanied his mother to fetch water, he seemed sad. When we asked the reasons, he explains:

> To fill this water, I need to walk for 40 minutes. And while walking I cannot use my slippers because they can break easily. The stones hurt my leg a lot. But that is okay. While bringing the pot back, the water keeps dripping on my head because I cannot balance it. And while it drips, it enters my eye and burns. And the burning doesn't stop. It just gets worse day by day. I cannot see well in class. I need to sit in the front desk of my classroom to see the blackboard clearly. But my friends do not sit here with me, and they make fun of me when I sit in the first desk. Only padhaku[17] children sit on the front desk. But I wanted to be 'cool.' So, I end up sitting in the last desk and my eyes burn even more because I cannot see the black board from there. I cry so much but whom do I tell? My mom puts cold water on my face, but that also burns.

His woes echo the struggles of several children of his age, validated by the health inspector, who prescribes eye drops as a temporary remedy to this condition. He advises parents to exempt children from fetching water because "longer exposures to salty water tend to dehydrate the eyes, due to which eyes water and give a burning sensation".[18] This is generally not accepted by several parents because when children are excluded, much of the load falls on women.[19] Furthermore, when questioned further about who should be exempted from filling water, the health inspector responded:

> Of course, the boys! They have more responsibility than the girls. Don't you think so? They need to earn well and serve the family; hence their vision is more important.

The health inspector's narrative depicts how the gendered division of labor is deeply embedded in common sense and everyday life. The concerns of men are acknowledged more than those of women and girls, therefore women and young girls are excluded from accessing health services in a timely manner.

Saroj's embarrassment – Saroj lives in the *harijan basti* with her three sisters, parents, and grandparents. We met her for the first time near the public bore, watching everyone filling water. She is a sixteen-year-old girl whose routine (identical to several other young girls) entails fetching water, collecting firewood, cleaning the house, and washing the clothes. Saroj indicated that her constant intimacy with saline water through these activities caused itching and rashes on her skin (see Figure 3.4). She narrates:

> Nobody notices my pain! My body itches so much. Mother and I are responsible for bringing and storing water, but all the water is khara.[20] We have very little sweet water, which we carefully use only for drinking. For all the other purposes, we use khara water which gives me white patches of dry skin on my hand. From here (pointing towards her hands) itching starts and it's uncontrollable. These white patches are on my hands, legs, my head, almost everywhere. I keep scratching them all the time.

The health inspector stated that it is common that very few women come to him for treatment or medical advice. Most often it is their mothers-in-law or husbands who explain the symptoms based on which he provides medication. When we inquired about the underlying reason, he replied: "Madam, do you discuss an unmarried girl's personal problem openly with other men? It is not in our culture to talk about menstruation in the open. You know it, why are you asking me then?"[21] The need to screen and shelter a female body with white patches and wounds is a culturally derived social construct and a gendered norm that mandates women to maintain privacy about their bodies (Joshi et al., 2011). Such a privacy has largely normalized and silenced the suffering of Saroj. She could not scratch publicly because it is socially inappropriate for her to do so. She remarks:

> As girls we are not supposed to scratch our private parts in public, but sometimes I cannot stop. I feel so embarrassed doing this. During my menstrual cycles, these rashes worsen in my private parts. When I use khara water to clean my private parts, they start to severely itch that lasts for a whole day. That is why I do not attend school when I have periods. The boys stare at me and my own friends laugh, which makes me feel very bad. They call me dirty and uncultured if I itch in public. I say this to my mother, but she ignores it saying that she also faces it and that nothing can be done about it.

In response to her situation, Saroj tends to carefully monitor and discipline her body and emotions by bathing just once in two days using saline groundwater (see Figure 3.5). Her sense of shame and embarrassment – arising from itching – tends to regulate her routines, wherein Saroj overlooks her emotion and continues her daily chores instead of challenging these norms and seeking help from the health inspector.

Through the many narratives encountered,[22] it was apparent that women are involved in dealing with the needs of other family members, as in the case of Shubh and Saroj, wherein they share their concerns with their mothers. However, it is also important to note that mothers, or in general married women, also face difficulties in their intimate relations with their husbands.

Sitaben's painful pleasure – Twenty-four-year-old Sitaben lived in the *harijan basti* along with her husband, parents-in-law, and her three-year-old son. Quite like Saroj's routine, Sitaben walks to the public bore-hole in the early morning, every day, to get three to four pots of water for washing and bathing purposes. She notes:

Since childhood we are constantly in contact with this water more than our men, because we are the ones washing, cleaning, and cooking. The khara pani gives me rashes on my hands, head, and all over my body. It gives me rashes in my private parts, and I scratch. But that is not the problem. Over time they become wounds, which take a long time to heal. When my husband touches me during lovemaking, he feels disgusted. I cannot stop using the water, but I cannot stay away from my husband too. I want to have smooth skin like you all have didi.[23] I am sure my husband will love me then! It hurts me a lot.

Figure 3.4 Drawing from one of the participants indicating skin problems arising due to saline water use.

Source: Authors.

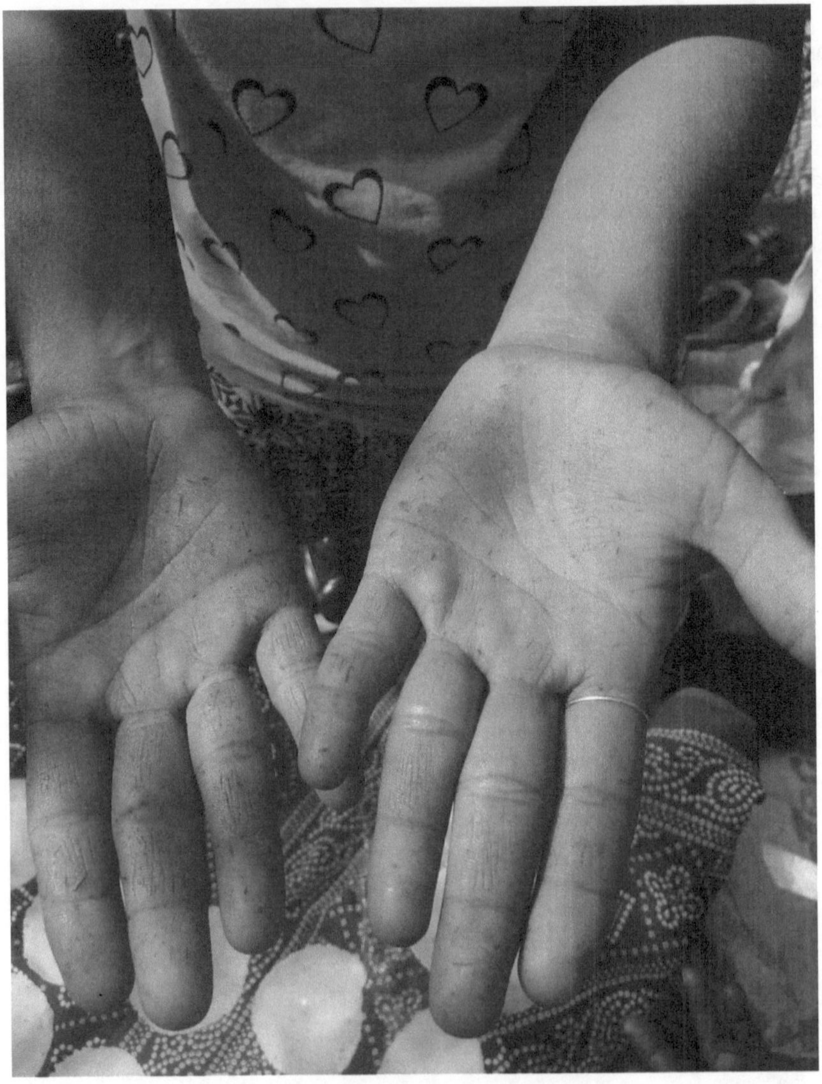

Figure 3.5 Rashes and scratches on the hands of a young woman.

Source: Authors.

Her concerns display how the touch of her husband, and the interest he takes in the different parts of her body, is of utmost enormous importance in the beauty, position, and attitude of her body (Weiss, 2013). It also indicates that such emotions centered on specific bodily features, like beauty, can be damaging and destructive, when women see themselves as unworthy and ugly, as perceived by Sitaben and many other women[24] in Lodhva. When asked about how she treated her infections, Sitaben laughs and says, "Do you expect me to tell the doctor that I have wounds in my private parts? I would suffer myself rather than get embarrassed telling him. I do not even tell my mother; how would I tell the doctor? The only people I share this with are my friends, who face similar things".

Her narrative, like many others,[25] is indicative of women expressing their emotions around sexuality and pleasure privately, either to their friends, like Sitaben did, or to their family members, like Shubh and Saroj, or sometimes not at all. However, one notices how only some emotions are expressed and articulated, like joy and sorrow. But emotions around bodies and wounds are not freely expressed as they become a source of shame or social discomfort and expressing them would perhaps reflect negatively on the person expressing them, as in the case of Sitaben. Therefore, who can express which emotions largely depends on identity and social context as well as gender. Several married and young unmarried women echo similar concerns regarding sexual tensions that seemed to be rooted in their intimate relation with saline water. However, such bodily experiences are not just confined to women, men also experience such impacts and emotions.

Pravesh's disgust – In his mid-30s, Pravesh works as a daily wage laborer at a cement factory. He lives in the *harijan basti* with his wife, two daughters, and his parents. When he was enquired more about the bodily impacts of saline groundwater on women, he notes:

> My wife does not look beautiful anymore with the rashes on her hand. I feel bad that she must endure this, but I don't feel like going to her or touching her. I went to the doctor to get her a cream too. But she never applies it. I tell her to stop using khara water, but she doesn't. And this creates a big fight between us. I come home tired, and I need some peace and love. She does not give it to me anymore. She keeps scratching herself all day, I feel repulsive. It disgusts me.

His narrative reveals the image men tend to form of a woman's body, which is not only the physical form of her body but also by the way their bodies are experienced and emotionally invested, in this case through love-making activities.

When Pravesh was asked how he dealt with his bodily impacts of using saline groundwater, he responds, "I get it cured from the doctor. I use the cream he gives me. But what I also do is while bathing, I mix khara water with some sweet water, so that my rashes are not much".

His recounting demonstrates that though men also experience bodily impacts and emotions, their needs are given more importance than that of the women, highlighting that the usage of water is gendered and allows men to exercise power over women and their bodies (Meah, 2014). This also tends to be mediated by gendered practices of labor. Despite being deeply invested in the management and use of groundwater, the needs of women are given minimal importance; much of this discounting of women's needs arises from patriarchal relations (Mehta, 2016).

On the other hand, it was interesting to note that as women grew older, they seem to have more agency to negotiate and express their needs and emotions and sometimes advocate for the younger women too. Elderly women or mothers-in-law tend to face fewer restrictions in the public space, owing to which they are in a better position to negotiate with the health inspector and spend more time interacting with community members, because they pass on the burden of accessing and using saline water to their daughters-in-law.

Gangabhen's joy – We first met Gangabhen at the health inspector's clinic, where she was advocating for her daughter-in-law's health. She was demanding an ointment to cure rashes on her daughter-in-law's hand, so that her daughter-in-law can play and take care of her three-month-old son. Gangabhen is sixty years old and lives in the *harijan basti* with her husband in a small hut. She has a son and a daughter. Her son is married while her daughter is pursuing ninth grade. Over 60 years of her time, Gangabhen has negotiated for her space in the village and tends to face fewer restrictions because she has surpassed all the struggles and discriminations. Hence, her neighbors hold a very high regard for her and address her as *Daai ma*.[26] In her younger days, fetching groundwater was the best part of her day. As she highlights,

> I was the queen of my desires. Nobody was there to order me or question me. I did my work at my own pace. It is so joyful when people do not boss you around. It feels like you have got wings, don't you think?

Accessing these wells was a getaway and a source of joy for her where Gangabhen enjoyed being free, not bound by social norms. But she also notes:

> The water gave me rashes, itches, and so much more. It made me look so ugly, but I could not stop. I used to cry. Sometimes it made me very anxious too! What if my husband left me because I am no longer beautiful?

Ganagbhen points out that most men give importance to a woman's physical appearance and want their women to be beautiful, while minimizing their contact with saline water. Owing to this, being attractive for their husbands bears utmost importance for most women, like Sitaben. Being a dutiful wife also means trying to please the aesthetic desires of their husband. However, on the contrary, their men were not ready to shoulder the responsibility of doing household chores that involve saline water. This becomes problematic, because it is not possible for women to both use saline groundwater and be (seen as) beautiful (owing to the rashes and itching saline water produces). Many times, not being beautiful enough becomes a reason for young girls to not get married, as expressed by Saroj, or a reason for the separation of a married couple, or sometimes a justification for adultery.

Conclusion

Through these five stories, this chapter sheds light on how salinity and environmental and other chemical contaminants in groundwater affect bodies, emotions, and intimate relations, and these effects are differentiated along gendered lines. We found that the struggles to access, use, and distribute groundwater are mediated through bodily experiences, particularly through skin contact – burns, rashes, wounds, and sores in eyes, hands, legs, and private organs. Empirical data presented in this chapter indicates that the hardships, experiences, and labor relations that were experienced differently by men, women, and children intersected with their gender, caste, age, and class and consequently, with their position in the household. The data demonstrate how they responded to their changing bodies and changing relations. One can see how young children from non-dominant castes, like Shubh, could not attend school because the saline water affected his vision. Young girls, like Saroj, expressed that continuous use of saline groundwater caused itching and deep rashes on their bodies and private parts, but they refrained from itching in public, fearing shame and embarrassment. They tended to overlook their emotions and continue with their daily chores. Married women like Sitaben indicated that deep rashes in private parts altered sexual relations with their husbands. She, probably like many other women, felt ugly, unworthy, and deprived of physical love, care, and pleasure from her husband. Having said that, for some women, especially the elderly like Gangabhen who have negotiated for their voice and agency with time, accessing water from the wells was a source of joy – a getaway from the daily chores and social norms. These five narratives clearly depicted how the materiality of the body, bodily experience, and emotions are intimately tied to spaces where groundwater is saline, and these spaces are made up of social relations that intersect with space and time. Furthermore, the narratives suggest that emotions are relational – in this case to social structures and saline groundwater – reinforcing that emotions are felt through relationships with other people and with groundwater itself.

This chapter explored how emotions are frequently connected to the materiality of the body: in this case, emotions were tied to the bodily experiences and sensations of interaction with saline water. Much of these emotions manifested through place and time, thereby serving as an important entry point to contemplate the everyday struggles associated with accessing and using groundwater. This study sheds light on the bodily impacts of using saline water, which in turn played a key role in shaping our respondents' personal relations with spouses, parents, children, friends, and medical doctors.

Notes

1 Traditionally sanctioned norms, like untouchability, play a critical role in allocating water resources, sometimes more than the natural availability itself (see Dutta et al., 2018).
2 From Interviews VH, AH_C, HH_D, HH_E.
3 From Interviews WO, VH, BD, AH_C, AH_D, HH_E.
4 From interviews AC, SC, WO, BD.
5 From Interviews VH, AH_A, AH_B, AH_C, AH_D, HH_A, HH_B, HH_C, HH_D, HH_E.
6 From Interviews VH, AH_A, AH_B, AH_C, AH_D, HH_A, HH_B, HH_C, HH_D, HH_E.
7 From Interviews VH, WO.
8 From Interviews WO, AA_D, HH_E.
9 From Interviews WO, BD, AA households, and HH household.
10 From Interviews HH_A, HH_B, HH_D, HH_E.
11 From Interviews VH, TL, AA, and HH household.
12 From Interviews HH_A, HH_B, HH_C, HH_D, and VH.
13 From Interview WO.
14 From Interviews AH_A, AH_B, HH_A, HH_B, and HH_C.
15 Hindi term for settlement.
16 Hindi term for an earthen pot to store water.
17 Hindi term for nerd.
18 From Interview HI.
19 From Interviews HH_C, HH_D, HH_E.
20 Hindi term for saline water.
21 From interview HI.
22 From Interviews AH_A, AH_B, AH_C, HH_A, HH_B, and HH_C.
23 Hindi term for sister.
24 From interviews AH_B, AH_C, HH_B, and HH_C.
25 From interviews AH_B, HH_B.
26 Hindi term for motherly figure.

References

Ahlers, R., & Zwarteveen, M. (2009). The water question in feminism: Water control and gender inequities in a neo-liberal era. *Gender Place and Culture*, *16*(4), 409–426. https://doi.org/10.1080/09663690903003926

Ahmed, S., & Zwarteveen, M. (2014). Gender and water in South Asia: Revisiting perspectives, policies and practice. In M. Zwarteveen, S. Ahmed & S. R. Gautam (Eds.), *Diverting the flow: Gender equity and water in South Asia* (pp. 3–30). The University of Chicago Press.

Alaimo, S. (2010). *Bodily natures: Science, environment, and the material self*. Indiana University Press.

Alaimo, S., Hekman, S., & Hekman, S. J. (Eds.). (2008). *Material feminisms*. Indiana University Press.

Batra, L. (2004). Pani ki kahani: Shehri garibon ki zubaani. *Final report*. Hazards Centre.

Bhagat, C., Kumar, M., Mahlknecht, J., Hdeib, R., & Mohapatra, P. K. (2022). Seawater intrusion decreases the metal toxicity but increases the ecological risk and degree of treatment for coastal groundwater: An Indian perspective. *Environmental Pollution*, *310*, 119771. https://doi.org/10.1016/j.envpol.2022.119771

Bhagat, C., Mohapatra, P. K., & Kumar, M. (2021). Unveiling the extent of salinization to delineate the potential submarine groundwater discharge zones along the North-western coast of India. *Marine Pollution Bulletin*, *172*, 112773. https://doi.org/10.1016/j.marpolbul.2021.112773

Bondi, L. (2005). Making connections and thinking through emotions: Between geography and psychotherapy. *Transactions of the Institute of British Geographers*, *30*(4), 433–448. https://doi.org/10.1111/j.1475-5661.2005.00183.x

Brown, N., & Gershon, S. A. (2017). Body politics. *Politics, Groups, and Identities*, *5*(1), 1–3. https://doi.org/10.1080/21565503.2016.1276022

Census of India. (2011). *Office of the registrar general & census commissioner, India*. https://censusindia.gov.in/

Coole, D. (2013). The body and politics. In G. Waylen, K. Celis, J. Kantola & S. L. Weldon (Eds.), *The Oxford handbook of gender and politics* (p. 165). Oxford University Press.

Crenshaw, K. (1989). Demarginalizing the intersection of race and sex: A Black feminist critique of antidiscimination doctrine, feminist theory and antiractist politics. *University of Chicago Legal Forum*, *140*, 139–167.

Crow, B., & Sultana, F. (2002). Gender, class, and access to water: Three cases in a poor and crowded delta. *Society & Natural Resources*, *15*(8), 709–724. https://doi.org/10.1080/08941920290069308

Davidson, J., & Milligan, C. (2004). Embodying emotion sensing space: Introducing emotional geographies. *Social & Cultural Geography*, *5*(4), 523–532. https://doi.org/10.1080/1464936042000317677

Dutta, S., Behera, S., & Bharti, A. (2015). Access to drinking water by scheduled castes in rural India: Some key issues and challenges. *Indian Journal of Human Development*, *9*(1), 115–132.

Dutta, S., Sinha, I., & Parashar, A. (2018). Dalit women and water: Availability, access and discrimination in Rural India. *Journal of Social Inclusion Studies*, *4*(1), 62–79. https://doi.org/10.1177%2F2394481118774487

Foster, S., Pulido-Bosch, A., Vallejos, Á., Molina, L., Llop, A., & MacDonald, A. M. (2018). Impact of irrigated agriculture on groundwater-recharge salinity: A major sustainability concern in semi-arid regions. *Hydrogeology Journal*, *26*(8), 2781–2791.

Galman, S. A. (2009). The truthful messenger: Visual methods and representation in qualitative research in education. *Qualitative Research*, *9*(2), 197–217.

Grosz, E. (1994). *Volatile bodies: Toward a corporeal feminism*. Indiana University Press.

Harcourt, W. (2016). Body politics. In N. A. Naples, R. C. Hoogland, M. Wickramasinghe & W. C. A. Wong (Eds.), *The Wiley Blackwell encyclopedia of gender and sexuality studies* (pp. 1–3). Wiley Blackwell.

Harris, L. (2006). Irrigation, gender, and social geographies of the changing waterscapes of south-eastern Anatolia. *Environmental and Planning D: Society and Space*, *24*, 187–213. https://doi.org/10.1068%2Fd03k

Jolly, I. D., McEwan, K. L., & Holland, K. L. (2008). A review of groundwater – surface water interactions in arid/semi-arid wetlands and the consequences of salinity for wetland ecology. *Ecohydrology: Ecosystems, Land and Water Process Interactions, Ecohydrogeomorphology*, *1*(1), 43–58.

Joshi, D. (2011). Caste, gender and the rhetoric of reform in India's drinking water sector. *Economic and Political Weekly*, *46*(18), 56–63.

Joshi, D., Fawcett, B., & Mannan, F. (2011). Health, hygiene and appropriate sanitation: Experiences and perceptions of the urban poor. *Environment and Urbanization*, *23*(1), 91–111. https://doi.org/10.1177%2F0956247811398602

Khan, A. E., Ireson, A., Kovats, S., Mojumder, S. K., Khusru, A., Rahman, A., & Vineis, P. (2011). Drinking water salinity and maternal health in coastal Bangladesh: Implications of climate change. *Environmental Health Perspectives*, *119*(9), 1328–1332.
Li, P., Wu, J., Tian, R., He, S., He, X., Xue, C., & Zhang, K. (2018). Geochemistry, hydraulic connectivity and quality appraisal of multilayered groundwater in the Hongdunzi Coal Mine, Northwest China. *Mine Water and the Environment*, *37*(2), 222–237.
López, M., & Hoppe, B. (2010). History, epidemiology and regional diversities of urolithiasis. *Pediatric Nephrology*, *25*(1), 49–59. https://doi.org/10.1007/s00467-008-0960-5
Meah, A. (2014). Reconceptualizing power and gendered subjectivities in domestic cooking spaces. *Progress in Human Geography*, *38*(5), 671–690. https://doi.org/10.1177%2F0309132513501404
Mehta, L. (2016). Why invisible power and structural violence persist in the water domain. *Power, Poverty and Inequality*, *47*(5), 31–42. https://doi.org/10.19088/1968-2016.165
Misra, A. K., & Mishra, A. (2007). Study of quaternary aquifers in Ganga Plain, India: Focus on groundwater salinity, fluoride and fluorosis. *Journal of Hazardous Materials*, *144*(1–2), 438–448. https://doi.org/10.1016/j.jhazmat.2006.10.057
Prakash, A. (2005). *The dark zone: Groundwater irrigation, politics and social power in North Gujarat*. Orient Blackswan.
Saha, D., & Ray, R. K. (2019). Groundwater resources of India: Potential, challenges and management. In P. K. Sikdar (Ed.), *Groundwater development and management* (pp. 19–42). Springer.
Sharp, J. (2009). Geography and gender: What belong to feminist geography? Emotion, power and change. *Progress in Human Geography*, *33*(1), 74–80. https://doi.org/10.1177%2F0309132508090440
Sultana, F. (2009). Fluid lives: Subjectivities, gender and water in rural Bangladesh. *Gender, Place and Culture*, *16*(4), 427–444. https://doi.org/10.1080/09663690903003942
Sultana, F. (2011). Suffering for water, suffering from water: Emotional geographies of resource access, control and conflict. *Geoforum*, *42*(2), 163–172. https://doi.org/10.1016/j.geoforum.2010.12.002
Thien, D. (2005). After or beyond feeling? A consideration of affect and emotion in geography. *Area*, *37*(4), 450–454. https://doi.org/10.1111/j.1475-4762.2005.00643a.x
Tiwari, R., & Phansalkar, S. J. (2007). Dalits' access to water: Patterns of deprivation and discrimination. *International Journal of Rural Management*, *3*(1), 43–67.
Tiwary, R. (2006). Explanations in resource inequality: Exploring scheduled caste position in water access structure. *International Journal of Rural Management*, *2*(1), 85–106. https://doi.org/10.1177%2F097300520500200105
Tolia-Kelly, D. P. (2006). Affect – an ethnocentric encounter? Exploring the "universalist" imperative of emotional/affectual geographies. *Area*, *38*(2), 213–217. https://doi.org/10.1068%2Fa43300
Truelove, Y. (2011). (Re-) Conceptualizing water inequality in Delhi, India through a feminist political ecology framework. *Geoforum*, *42*(2), 143–152. https://doi.org/10.1016/j.geoforum.2011.01.004
Wallace, T., & Coles, A. (Eds.). (2005). *Gender, water and development*. Berg.
Weiss, G. (2013). *Body images: Embodiment as intercorporeality*. Routledge.

4

MAPPING WATER CARE PRACTICES

The case of Ennore-Pulicat wetlands in Chennai, India

Qurratul Ain Contractor

Introduction

> As women, we do a lot more than just cook fish – from catching fish, to curing it, auctioning and marketing it and adding value to the fish catch in myriad ways, women are the backbone of the fisheries economy of the region. That said, food is important. In food, there is tradition and the hand-me-down wisdom of our women ancestors.
>
> *(Women of Pazhaverkadu, 2021)*

On 9 September 2021, fisherwomen from *Pazhaverkadu* (Pulicat) wrote a letter to their Chief Minister inviting him to their village to attend a seafood festival on 16 October 2021 – UN World Food Day. Referring to themselves as the "backbone of the fisheries economy in the region," they highlighted the many roles women play in artisanal fishing. Sharing the science of fishers with the Chief Minister, the women explain that "fish are not found everywhere at sea": parts of the ocean floor that have *Paar* (rocks/reefs) and *Seru* (mud) make the best fishing grounds. These are the fertile areas that are threatened by the expansion of ports and industry on the coast. Making a promise to "please the palate" and "relax the mind" of the Chief Minister with Pulicat's specialties and signature recipes, the women reminded him of his electoral promise to protect their wetland from encroachments that threaten their lives, living spaces and livelihoods.

This chapter uses a lens of care to highlight everyday practices of women from these artisanal fishing villages that are otherwise made invisible by patriarchal and caste-ist social structures. These care practices form crucial contributions to the fishing economy and ecology. In documenting fisherwomen's everyday care practices, I also explore how they experience threats to wetland ecology and coastal morphology, such as flooding, encroachments and industrial pollution. These threats, posed by "development" pressures, are driven by industrial economic interests, and supported by the state's view of coastal wetlands as empty and a waste of space that needs to be made productive through the building of ports and industries. Documenting perspectives of fisherwomen challenges this view. Their everyday care practices reveal the many ways in which women derive value from the wetland by handpicking

DOI: 10.4324/9781003100379-6

Figure 4.1 Graphical representation of Ennore-Pulicat wetlands against the backdrop of port and industrial infrastructure in North Chennai, showing caring practices of women from Indigenous fishing communities.

Source: REFRAMED photographers | Drawing made in collaboration with M. Dhanya, March 2022.

fish, prawns and crabs from its shallow mudflats. It also brings attention to the ecological knowledges held by women fishers on account of their gendered and caste-defined roles in the fishing economy. In doing so, it brings forward the crucial role played by women in wetland care, and the need to recognize and value women's labor for effective wetland conservation.

Finally, perspectives of fisherwomen highlight the links between environmental and social justice. In addition to the degradation of wetland habitat, increased rates of coastal erosion and accretion due to large industrial and port infrastructure built on coastal wetlands are increasing incidences of flooding in north Chennai. This is bringing public attention to the need for wetland conservation in the pursuit of flood protection. By collaborating with climate activists in Chennai to organize seafood festivals, fisherwomen effectively link questions of Indigenous livelihoods and food security to those of climate adaptation.

A threatened wetland-scape

Pulicat lake, located 60 km to the north of Chennai, Tamil Nadu, is the second largest brackish water lake in India and a protected sanctuary for migratory birds (see Figure 4.2). It is fed by three rivers and drains into the Bay of Bengal through an estuary that is joined by Kosasthaliyar, the largest of Chennai's rivers. The Buckingham canal, built in the 1800's to transport commodities such as fuel, salt and dried fish, connects Pulicat lake to the Ennore estuary. This complex network of rivers, streams and canals forms the Ennore-Pulicat coastal wetlands. The wide interface of sweet and salt water offered by this wetland system makes it an extremely fertile habitat for fish and prawns (shrimps), hosting a thriving artisanal fishing economy (see Figure 4.3). Additionally, they offer effective flood defense along the coast to the north of Chennai.

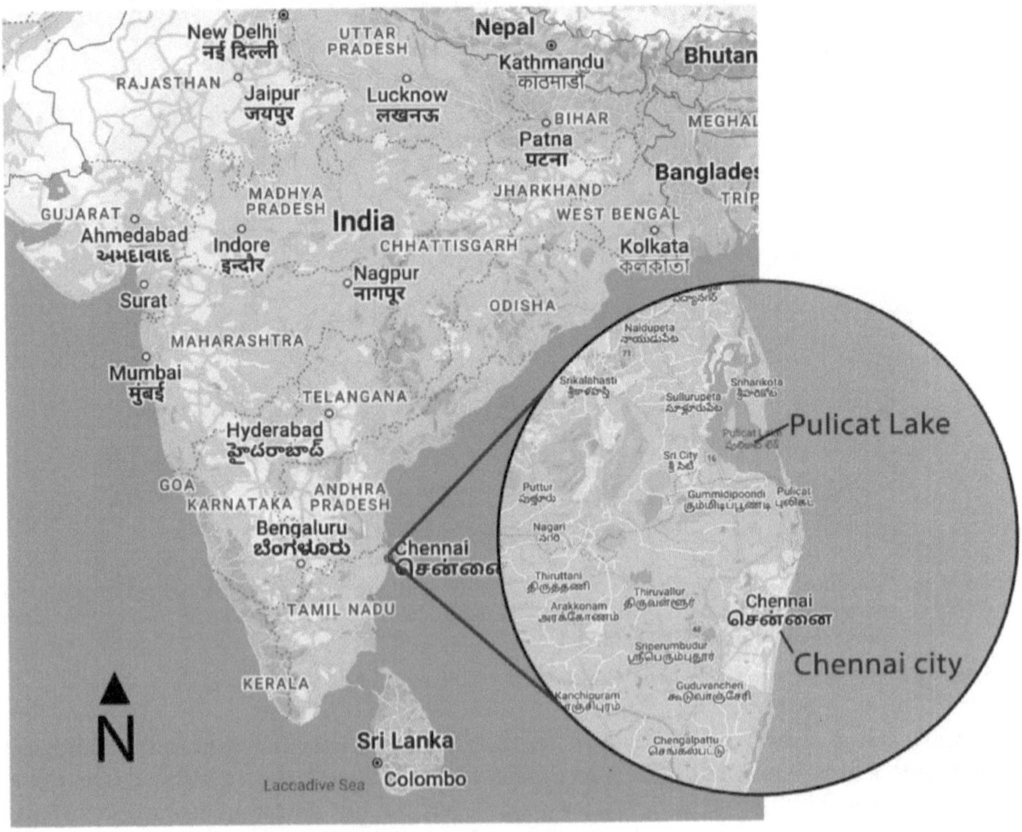

Figure 4.2 Location of Chennai, India.

Source: Author.

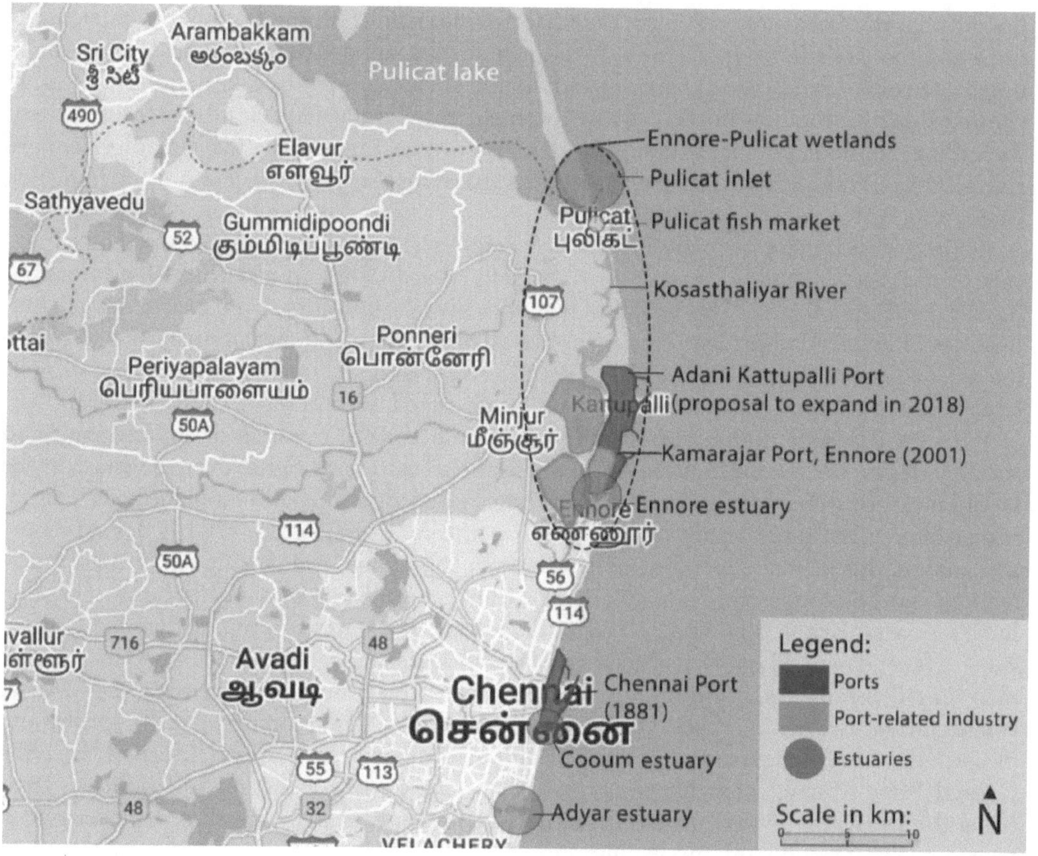

Figure 4.3 Location of ports and estuaries in Chennai.
Source: Author.

However, developments on Chennai's wetlands bring forward contestations in the ways different actors understand the wetland's value. Jayaraman (2017) describes actors involved in the ongoing conversion of wetlands into real estate to include both Tamil Nadu state and central Public Sector Units. Encroachments include industrial installations belonging to the union government's Bharat Petroleum and Hindustan Petroleum and a power plant and coal ash dump by NTECL; a venture involving both state and central governments. Additionally, the Kamarajar Port Ltd. (KPL) in Ennore, owned by the government of India, obtained permission from the State Coastal Zone Management Authority to construct car parking terminals, coal yards and free trade warehouse zones at the expense of several hundred acres of the Ennore Creek. These expansions have been permitted despite evidence that North Chennai is facing shoreline erosion/accretion related problems after the introduction of this port in 2001 (Pandian et al., 2004). Yet, the expansion of port infrastructure along North Chennai's coast continues unabated at the cost of its wetlands.

In November-December of 2015, Chennai saw a devastating flood that claimed over 400 lives and caused enormous economic damage (Narasimhan et al., 2016). However, flood mitigation measures taken by the state government of Tamil Nadu have particularly impacted the city's poor. In 2016, a Parliamentary Standing Committee on Home Affairs noted the role of land cover changes, such as the encroachment of lakes and riverbeds, in causing massive flooding in the city. In response, the Tamil Nadu State government used the excuse of floods to evict urban poor from their dwellings on the margins of Chennai's rivers, without doing anything to check the encroaching illegal construction business which is usurping water bodies (Jayaraman, 2017).

In 2018, a subsidiary of Adani Ports and Special Economic Zone Ltd. proposed an expansion of the cargo handling capacity of Kattupalli port, located on a barrier island north of Kamarajar (Ennore) port. This proposal met with severe opposition from coastal fisher communities of Ennore-Pulicat who expect livelihood destruction and displacement due to resulting coastal erosion (Rupakumar, 2021), as well as environment activists who fear that "fragile coastal habitats within this massive area will be damaged beyond repair" (Aves, 2020).

For over a decade, Indigenous fishing communities, with the help of environment and social activists, have been fighting to establish their customary rights over the coast in the face of increasing encroachments, pollution from waste dumping and the dilution of coastal regulations in the interest of "development" projects. Kumar et al. (2014) describe how artisanal marine fishing depends heavily on beach space, and each fishing village has clearly marked out commons with boundaries not only on land but also in the ocean. Though India's Coastal Regulation Zone 2011 requires maritime states to prepare detailed Coastal Zone Management Plans that document the land-use, incorporate a long-term housing plan for the fishing communities, identify common properties of fishing and other coastal communities, as well as check violations to fisherfolks' traditional rights to spaces and livelihoods, these clauses are not enforced. In the absence of this crucial information in the Coastal Zone Management Plans, Kumar et al. (2014) note the spread of built-up areas along the coast, involving privatization, encroachment and dispossession of fisherfolk of their rights to coastal land and livelihood.

This phenomenon demonstrates that the wetland ecosystem is inherently social, as described by Budds and Zwarteveen (2020), who point toward the bidirectional nature of society-nature interactions in which biophysical and human processes are co-constituted to produce social nature. The wetlands are shaped by power relations favoring industrial and state actors over Indigenous fishing communities, who depend on wetland health to make a living. These power imbalances are evident in the state's Coastal Zone Management Plans, which have failed to document traditional rights and customary uses of the coastal commons in planning coastal development. This failure has resulted in a view of the coastal commons as empty space that must be made productive through industrialization.

Caring for wetlands

The lens of care is useful to uncover wetland values made invisible by dominant views of the natural environment as separate from society, as well as women's contributions to wetland economy and ecology made invisible by dominant patriarchal structures. Focusing on "practical engagements of contributing to restoring, sustaining, or improving something" (Mol & Hardon, 2020, p. 233) the concept of care frames how different social groups, including local communities and citizen collectives, collaborate to further their respective visions of environment protection. I further use Tronto's (1994) definition of care to analyze fisherwomen's everyday practices. Viewing caring as a "species activity," Tronto (1994, p. 103) proposed that care:

> [I]ncludes everything that we do to maintain, continue, and repair our 'world' so that we can live in it as well as possible. That world includes our bodies, ourselves, and our environment, all of which we seek to interweave in a complex, life-sustaining web.

Using the lens of care to document women's everyday tasks sparked an inquiry into understanding what it means to care in the Ennore-Pulicat wetlands.

In response to scholarly critique of the ambiguous definition of the word "care," which has been described as "without semantic boundaries" (Piras & Zanutto, 2010, p. 82), Mol et al. (2010) argue that the un-bounded meaning of the word care encourages curiosity in a researcher to discover ways in which care is practiced. If one is to find out what an apple is, then one needs to have a sense of where to look for it, but not start out with a clear definition of an apple (Mol et al., 2010). Encouraged by this call to explore practices and "move along with the way words are being used" (Mol et al., 2010, p. 85), I focus on the participation of women in campaign activities to understand the complex ways in which their communities depend on wetland health. I documented ways in which women show care for wetland ecology while performing daily tasks. Methods of showing care presented knowledge of wetland ecology acquired through specific daily interactions with the wetland, shaped by diversity in identities of different community members.

I choose a feminist political ecology approach which explains how gender, in relation to other aspects of socio-economic identity such as class and caste, shapes access to and control over natural resources, as well as specific every-day tasks. Bringing attention to gendered and caste-affected everyday practices prevented romanticizing ideas of community or Indigenous knowledges, and instead focused on valuable knowledge about the wetland in a fast-changing, fragile environment. Shining light on power struggles within wetland communities draws links between ecological sustainability and social justice (Joshi et al., 2021). Further, the development of unlikely alliances in the course of the campaign, such as those between different caste groups, different religious communities and across class and gender barriers hinted toward the power of care to fuel innovation in wetland protection strategies.

Methodology

This research was conducted as part of my Master's studies in water management and governance from IHE-Delft, Netherlands.[1] In September 2021, I first established contact with the Coastal Resource Centre, a research, media and advocacy organization that supported the food festivals as part of the on-going wetland protection campaign in Chennai called "Stop Adani Save Pulicat." By December, I arrived in Chennai and began shadowing the network of solidarity organizations working with fisher communities to conserve coastal ecology. Participating in and observing various campaign activities, such as citizen tours of pollution sites in the wetland and a photography exhibition to create a platform for citizen engagement, provided a historical understanding of the site-context from a community perspective. Through conversations with members of a local NGO in Pulicat, I gained an understanding of the history of NGO-driven women empowerment and livelihood protection programs in the region. These include a network of women's self-help groups, members of which campaign for wetland protection by organizing seafood festivals, and the rehabilitation and protection of tropical mangrove ecosystems for enhancing fisher livelihoods.

Due to the stark difference in caste identities[2] of women involved in these two activities, focus of the research was expanded to include those involved in the rehabilitation of mangrove ecosystems. These women belong predominantly to marginalized castes and tribes and make their living by hand-picking prawns and crabs from the shallow inland mudflats of the Ennore-Pulicat wetland system.

Fieldwork in Pulicat began in January 2022, and lasted for six weeks. Pulicat town is surrounded by fishing villages, some located along the coast and others along riverbanks. Each village has a distinct dominant caste identity and dedicated fishing rights called *paadus*, that regulate when and where different villages fish, except for Irular tribal communities who typically do not have specific fishing rights. I interviewed thirty-one women from nine villages,[3] including women who perform different tasks in the fishing economy, as well as those who participated in the seafood festival, and in the rehabilitation and protection of mangroves. They belong broadly to the traditional fishing community and the Scheduled Castes and Scheduled Tribes (in this case, the Irular tribe). Of these, the traditional fishing community holds a dominant social position in Pulicat.

Data collection was done primarily using two methods: semi-structured interviews and participant observations. Additionally, documentation was done of women performing fishing-related tasks through photographs and videos. Semi-structured interviews were conducted to allow interviewees space to express their perspectives and relationship with the wetland. In search of understanding what it means for women to practice wetland-care, an active interviewing method was adopted to treat participants as meaning co-creators (Hathaway et al., 2020). While the presence of a translator made it challenging for interviews to resemble everyday conversations, effort was taken to build rapport with interviewees and facilitate an interactive dialogue with them. As different fishing related activities are done in different parts of the wetland, various sites of participant observation were chosen to understand the interaction of women with the wetland and village environment. This allowed the observation of use of spaces, as well as interaction of women with each other and other community members.[4]

Care adds value

Adopting a lens of care to document lived experiences and perspectives of women from wetland communities brought attention to not just their valuable, life-supporting labor, made invisible by patriarchal social systems and a lack of state recognition but also to the immense life-supporting value offered by the wetland ecosystem.

Analyzing interactions with women in Pulicat, I found that Tronto's depiction of care as a species activity best describes the host of gendered and caste-influenced tasks they perform to facilitate life on the wetland. Though I initially set out to find how women of Indigenous wetland communities show wetland-care through fishing-related practices and participation in wetland protection campaigns, I found that women's everyday tasks were guided primarily by care for their families, followed by their communities. However, in viewing the continuation of their own lives in close connection with the maintenance and repair of the wetland ecosystem, these everyday caring practices extend benefits to the wetland directly or indirectly. This interdependence between wetland health and health of wetland communities blurs lines between family or community-care and wetland-care, such that one bleeds into the next. Below is a list of care practices iterated by women in semi-structured interviews.

Family-care

When asked what they do, some women replied that they "are at home." I met Pushpam, aged 60, at the Pulicat fish market where she was cleaning fish for customers. She said, "I'm at home. Just like that I come here to clean fish." It was nearly noon, and I was sitting with a row of fish cleaners under the slim shade of plastic tarpaulins erected along the edge of the fish market. To unpack the meaning of being "at home," interviewees were asked to describe their typical day, from the time of waking till they sleep. Pushpam then shared that she wakes up every day at three a.m. to begin the tasks at home. Most interviewees, including Pushpam, described their day to begin with cleaning the *vaasal* or threshold of their home by sprinkling water and scrubbing cow dung on it to make the *kolam*. A *kolam* is a geometric pattern drawn by women at the house entrance, traditionally with rice powder. After this daily ritual, Pushpam begins cooking for her daughter who "goes to work," making coffee for her and heating water for her bath, before she comes to clean fish at the market.

All the interviewees described their household tasks to include bringing water home from the source, cooking for the family, cleaning the house and dishes, laundry, collecting firewood for cooking and heating water for baths and childcare tasks like bathing, feeding and dressing. These caring practices are oriented toward maintaining health and hygiene of themselves and their families, without which Indigenous communities would perish along with their Indigenous knowledge of wetland ecology and cultural roots in it.

Post-harvest care

Women from fishing communities are involved in multiple post-harvest tasks to process and add value to the fish catch. These include picking fish out of nets, bringing fish from the boats to the market, selling fish at the market, auctioning fish and cleaning fish for customers. On a good day, Pushpam may clean fish and prawns for up to thirty customers, earning thirty rupees per kilogram. Anju, another fish cleaner, aged 40 years, told us that over forty women come to the Pulicat fish market every day to earn a living by cleaning fish and prawns. This work is often done by single women from traditional fishing communities who need to support their families without the presence of a male family member who could catch fish. Anju single-handedly supports her two children after the death of her husband by spending her mornings as a daily-wage sanitation worker before coming to the market to clean fish.

The informality of work affects earnings at the fish market where all customers may not agree to pay the standard price for fish cleaning, as this price is informally decided. Her fellow fish cleaner, Divya, described the humiliation faced by fish cleaners when interacting with customers,

> Whatever they (customers) have in hand, they give five or ten rupees (EUR 0.06–0.11). . . . We say it's not enough for us, but they say, 'If you can clean, you clean! They talk in a bad way. We bear it and keep cleaning. What else can we do? If we leave this we have no other way. We have no other jobs.
>
> *(Divya, Fish cleaner)*

Despite these irregularities, the Pulicat fish market offers precious opportunities to women to make a livelihood. While interviewing Anju, a woman nearby remarked, "Even a small child can earn ten rupees (a small amount of money) here in Pazhaverkadu (Pulicat). Without this village we cannot survive at all. There is so much value here!" Agreeing with her about the economic opportunity offered by the fertile wetlands, Anju, who works two jobs to support her family, replied that indeed Pulicat is a *nalla uur* (good village).

The fish market is operated almost entirely by women who spend their days and nights making fish from Pulicat's wetland available to customers from the region. My translator and I met Mina on our first night-time exploration of the Pulicat fish market. She is a prawn auctioneer from the Karaiyar street, also called the "Muslim street" of Pulicat town. In the fourth week of our stay in the town, Dhanya and I had become accustomed to the streets around our home going silent by nine p.m. To our collective surprise, the vicinity of the fish market was bustling with women moving baskets of prawn and fish around. Mina, who works at night from eight p.m. till approximately five a.m. at her family's shop near the fish market, explained her task while weighing and examining a prawn catch. As a broker, Mina's task involves setting an acceptable selling price for the fisherperson while also finding a suitable buying price for the merchant. Mina took a pause between answering interview questions to cater to a fisherman with a basket of prawns. She spread out the prawns on the floor for the buyers to examine the size, color and freshness, while the fisher stood silently waiting for his catch to be valued. As some of the prawns crawled around in demonstration of how fresh the catch was, she negotiated a selling price for them, taking care to ensure the satisfaction of the fisher. While weighing the prawns on a heavy manual scale, she explained:

> The merchants will demand the rate at which it sells for them. For the people who are catching (the prawn) and coming, they would also want a particular rate. They are catching with so much difficulty, right?
>
> *(Mina, Fish auctioneer)*

Several of Mina's customers are women who make a living by buying fish at the market and travel several hours a day to interior villages to re-sell it, thereby bringing nutrition from the wetland to a wider community. While women doing these jobs remain undocumented, such that government entities do not recognize their labor rights or keep records of the number of women doing these jobs, these tasks add value to the fish catch, thereby increasing family income and contributing to the fishing economy.

Hand-picking or "scrambling"

This specialized task is done by women from Scheduled Caste communities and the Irular tribe. The performance of this task reflects the low status of prawn hand pickers in caste and gender hierarchies within fishing communities of Ennore-Pulicat wetlands. It involves careful movements through the floodplains, avoiding thorns and harmful fish, to collect prawns and crabs from the wetland using bare hands. This requires enacting multiple caring practices, including maintaining a keen experiential knowledge of wetland morphology through daily interactions with it, caring for accompanying women as this work is done in groups to increase resilience to threats faced in the wetland (snakes, other animals, heat stroke), and caring for the continued propagation of prawns and crabs by choosing to not pick them from breeding grounds.

Prawn hand-pickers describe the action of picking prawns as *tadavu*, which my translator instinctively translated to "scrambling." Scrambling is defined as "to move or climb quickly but with difficulty, often using your hands to help you," or "to compete with other people for something there is very little of" (Cambridge Dictionary). Both these descriptions of the word are fitting to the motions made by women hand-pickers while catching prawns. Early on a foggy morning in the first days of the Tamil month of Thai (starting mid-January), I followed two women as they picked prawns in the shallow ditches near the banks of Kosasthaliyar river. Squatting on their haunches, women graze the floor beneath the shallow water feeling for the prawns. As they sense their sharp tentacles, they pick and deposit prawns in a woven cane back-pack, while taking a step further in the water in a squatting position, making a sweeping gesture with their arm to maintain balance and continue grazing the next patch of soil. Slowly, they made their way forward, occasionally picking out thorny branches fallen on the floor beneath them, squatting for hours at a stretch.

Ruby, a hand-picker from a Scheduled Caste village, explained the importance of knowing wetland morphology as well as seasonal changes in ecology in order to find prawns:

> You need to know the slope, the low points and high points. You need to know where there are thorns, because you can't go there. There might be a lot of Kelthi fish also (that sting) so we don't go scramble there. Where there are prawns there will be a ditch. It all depends on seasons. One day you can catch a lot of prawns, one day you won't get them at all. . . . Wherever there are prawns, there will be small-small ditches. The sand will be like a flower.
>
> *(Ruby – Prawn hand-picker)*

When I asked her to explain what she means by "sand like a flower," Ruby pointed toward soft sand near where we were sitting and generously offered some more tips on prawn catching:

> It will be similar sand like this but in a ditch where the prawns can come and stay. Prawns will come and lie down straight. In places where it is marshy you won't get prawns. Only this kind of sand is good for the prawn (points). In the marsh the prawn won't lie down. If you scramble for prawns in the marsh, they run away. If the prawn is in this kind of sand, even if one prawn runs away you can catch the other.
>
> *(Ruby – Prawn hand-picker)*

Explaining the importance of catching prawns in groups, Ruby explained how women care for each other while practicing their profession:

> If someone isn't able to search for prawns in the sun and falls down, we protect each other and bring each other home. If you go alone there will be boars and snakes. You might get scared of this. . . . If we go as a group, we can be support for each other.
>
> *(Ruby – Prawn hand-picker)*

Experiencing the dangers of being in the water, Ruby also holds knowledge about other wetland species such as snakes. Listing local names of snakes, she casually described the dangers they pose, as well as how to avoid them:

> There is Nalla Pambu (good snake), Tanni Pambu (water snake), Saaru Pambu . . . it just sleeps inside water. If it feels like clay, then we know it is a snake and we just let it be.
>
> *(Ruby – Prawn hand-picker)*

She told us that though not all the snakes are poisonous, in case of a snake bite, one must visit the hospital to ensure that the bite does not get infected. This is important since they cannot afford to have days without earnings. Ruby's matter of fact and sometimes humorous accounts of life as a prawn catcher reveal her familiarity with the wetland ecology, as well as the challenges of living and working in it.

The practice of handpicking prawns is gendered and caste-influenced, such that women hand-pickers are not allowed to use nets for catching prawns on account of being women, and women from the traditional fishing community do not handpick prawns at all. However, women hand-pickers, predominantly from the Scheduled Castes and Tribes, hold valuable knowledge about the wetland ecosystem by virtue of this work and support their families through the value it generates. They also bear witness to the impact of industrial pollution on wetland ecology and can articulate it in detail. Continued sustenance of hand-pickers and their families is testimony to the provisional services offered by the wetland. Therefore, this caring practice incorporates both family-care as well as wetland-care and adds to food security for the people of Chennai.

Reducing waste and growing food

Women from fishing communities, irrespective of caste, are also involved in preserving leftover fish, which further adds economic value to the fish and increases food security. They make *karuvaddu* (dry fish) by salting and sun-drying it. This is a specialized task passed on through generations by women. Drying fish is a traditional method of avoiding waste and is an integral part of the daily diet of coastal communities. It is either sold for a reduced price or consumed by fisher families. I interviewed Rupam, from a coastal, traditional fishing village, on the sand covered street outside her home as she cleaned and salted fish with her daughter-in-law.

> No one has bought the fish, so we got it back (home). Since we got it back, we have to make karuvaddu (dry fish). Instead of leaving it to waste, we prefer putting salt in it. If someone asks we sell it for 200–300–500 rupees (a reduced price).
>
> *(Rupam, traditional fishing community)*

Savi, a professional fish dryer explained the many steps of this special technique which she learnt from her mother:

> We buy fish, then using the Aravamanai (special type of knife used to cut fish) we remove the gills. We wash it in water. Then we put it in this curved container. We put salt and shake it, shake it. Then pour it in the tub. The next day morning, we bring it and wash it in water. We broom the floor and we put the fish and dry it. Same way the prawn, we buy it from the boat, we wash it here and then we dry it.
>
> *(Savi, fish dryer)*

Savi is 45 years old and has been involved in drying fish since her husband abandoned her 25 years ago. She was interviewed while gutting fish behind the Pulicat fish market, along the banks of the Kosasthaliyar river, where the sight of fish and prawns sun-drying on long pieces of cloth is common.

Moreover, the task of packing prawns and fish in cartons with ice is primarily done by women. In the absence of cold storage, this task is crucial to ensure that the fish catch does not spoil on the way to markets further away in Chennai. Poorni, from a traditional fishing village, packs prawns in ice and takes them to sell in Chennai with eight or nine other women every night from eight p.m. till four a.m. In her early 60s, she is one of approximately 500 women who pack prawns in ice and sell them in markets further away. She described herself as an independent businesswoman who must take financial risks, "Some days I might earn 500, some days I might lose 1000 rupees. That's how the business is."

Some of the women I interviewed are also involved in growing food through agriculture, apart from participating in the fishing economy. In a coastal traditional fishing village, I came across women growing a local edible root in their kitchen gardens for feeding their families and for sale. Being a local crop, this root is suited to coastal ecology and has low water consumption. Women from Scheduled Caste villages also work on farms owned by upper caste landlords to plant and harvest rice for half the year. This is a seasonal agricultural practice that responds to freshwater availability in a wetland of fluctuating salinity.

Making visible "the backbone of the fishing economy"

The previously described everyday caring practices make women "the backbone of the fishing economy," as well as illustrate and add value to wetland ecology, the foundation of the fishing economy. Barbier (2019), a proponent of valuation frameworks that value nature in economic terms, argues that the failure to consider the various values and benefits provided by coastal wetlands when deciding their current and future development and use is the core problem behind global decline in coastal wetland ecosystems. He claims that though valuing the goods and services of coastal wetlands is essential to their wise management, large gaps exist in knowledge of these benefits, leading to inadequate estimates of important values needed to make management decisions. However, documenting perspectives of fisherwomen shows that their everyday practices emerge from this very knowledge of wetland benefits. This brings to question *whose* knowledge shows large gaps in understanding the benefits provided by coastal wetlands. Valuing ecosystem services of coastal wetlands is linked with valuing the knowledge and care practices of fisherwomen and other marginalized wetland communities, such that one follows the other.

The lack of recognition of women's care practices is not incidental. Several barriers come in the way of associating value with their labor. Within their own families and fishing communities, patriarchal social systems diminish their voice, such that women are discouraged from speaking out in the presence of male village elders, participating in public life or village administration. As a result, women have few avenues to express their perspectives or gain recognition for their efforts to support the fishing economy. This was illustrated by the lack of recognition women offered to their own labor, with many interviewees describing themselves as just "being at home" or showing low confidence in their ability to answer my interview questions. An NGO worker with fifteen years of experience engaging women's self-help groups in Pulicat explained:

> If you see amongst fishermen . . . women shouldn't cross over men and say anything. Village restrictions. If ten men are talking together, women shouldn't be there. They'll say (to the women), 'You shouldn't say anything!' – it is like that here. . . . Despite all this, there are women who go against it and speak.
>
> *(Devi, NGO staff)*

Women's caste and gender identities also play a role in valuing their own labor. In traditional fishing communities, boat-owning fishermen receive government-issued identity cards that make them eligible for compensation during emergencies and subsidies on fuel. Though women from traditional fishing communities do not receive government recognition themselves, they are able to access state support through the recognition awarded to the men in their families. However, as state recognition is tied to boat ownership, women from Scheduled Caste and Irular communities, who hand-pick prawns without tools, remain completely unaccounted for by the state. These women also showed a greater reluctance to express their views publicly.

Caring as groups

Fisherwomen in Pulicat have a history of organizing in self-help groups called *sangams*, formed with the intention of financial inclusion of women, through which women could save collectively and access credit. However, these groups also function as social support systems for women to share family concerns and encourage each other to challenge gender norms that prevent them from participating in public life. In describing practices of care as inherent parts of human species-being, Jennings (2018) notes that focusing on care shifts away from the individualism of self-reliance to an individuality validated by reliance on others. I find the collectivization of women in *sangams* resonant of Jennings practices of care, that are called forth where a common background condition of fragility, vulnerability, insufficiency and mortality take center stage in the lives of persons in need and those around them. By collectivizing as groups women in a shared condition of vulnerability to a harsh patriarchal society[5] are able to offer care to each other in the form of emotional, social and economic support. Strengthened by these collectives, women in *sangams* have also been successful in extending care to their communities by joining voices to demand improved access to infrastructure such as roads, healthcare, water access and education for their children.

Women's *sangams* have participated in wetland-care and protection through seafood festivals. While community counter-mapping of the coastal commons has served to bring recognition to customary land rights and uses of wetland communities (Kumar et al., 2014), the seafood festival hosted by women from Pulicat served to highlight the bounty of the wetland and their cultural connection to it, in addition to its significance for food security for the region. The festival aimed to inspire state leaders and urban residents to abandon an extractive approach to wetland-use, and instead appreciate the value being offered by the wetland ecosystem in the form of shared interests such as food security and flood protection. By garnering public support and solidarity, the festival worked to empower knowledge claims of the fisher community's maps of the coastal commons, which represent their "water truth" (Boelens, 2014). This water truth sees the wetland as a living system, fertile and bountiful and worthy of protection from pressures of industrial development.

Voluntary efforts by *sangams* of women hand-pickers to rehabilitate and protect mangrove ecosystems also demonstrated their ecological knowledge, as well as the interdependence between the health of wetland species and livelihood of women. Preparing beds and planting mangrove saplings requires strenuous labor and the organization of working groups. The sustained nature of care required to successfully rehabilitate mangroves included long-term attention in the form of guarding against grazing cattle of other wetland communities, protection from male fishers who uproot the saplings to tie their nets and ensuring that saplings received sufficient water. The long duration of this work, which requires regular interventions in response to unpredictable elements, indicates the nature of care necessary to rehabilitate wetland species in the face of deteriorating habitats.

Environmental care must be maintained over long periods of time and is therefore offered by women who have a sustained concern for wetland ecology due to livelihood dependency.

Additionally, their ability to monitor mangrove rehabilitation sites was aided by their daily physical presence in the mudflats of the Kosasthaliyar river for hand-picking. Therefore, the continued presence of women hand-pickers in the Ennore-Pulicat wetlands itself is essential to facilitate daily interaction with the wetland and its species, allowing women to not only look after mangrove saplings but also monitor release of industrial effluents and its effect on prawns, crabs and fish.

Contested wetland discourses

The degradation of Ennore-Pulicat wetlands is a story of deep contestations between opposing views and understandings of the wetland.

Nityanand Jayaraman from the Vettiver Collective, a solidarity organization working to support community-led environment protection campaigns, described how the *poromboke*, which are "reserved for communal use . . . lands that are not amenable to being privatized . . . that cannot be bought or sold or built upon" have been seen by the state as "wasteland, which has to be converted to worthy land by the act of industrialization." By this logic, Jayaraman says that the use of the *poromboke* has been converted into heavy manufacturing, thermal power plants and port infrastructure, at the expense of not just ecological health but also the livelihood and food security of diverse wetland fisher communities. He described the government's vision for development as to "see open space, open earth, open land as worthless. To see economies, cultures that are dependent and arise from open land as also worthless." He urged me to see the "complex connection between land, livelihood, culture and identity" that becomes evident when exploring the many ways in which people work with the land in Ennore-Pulicat wetlands, which he calls a "mosaic of ecosystems and cultures." Describing the variety of ecosystems and livelihoods within coastal wetlands, Jayaraman displayed a complex understanding of the wetland, built through sustained interactions with local communities. This view is distinct from that of industrialists who subscribe to an economic system that does not recognize the myriad informal economies arising from the wetland ecosystem.

Karan Adani, Chief Executive officer of Adani Ports and SEZ Ltd., expressed his views on the value of the Kattupalli port to support economic growth at the Tamil Nadu Global Investors Meet, 2019 in Chennai, where 146 Memorandums of Understanding worth INR 3.4 lakh crore (over EUR 41.1 billion), were signed between investors and the government of Tamil Nadu (Janardhanan, 2019):

> We acquired Kattupalli port, and today we are proud to state that since the acquisition of the port three years ago, it has seen an increase in cargo volume of over 200%. We expect to be able to make this port one of the largest in the country, with a capacity to handle over 300 metric tonnes of cargo and make this a key enabler to support the growth ambition of Tamil Nadu.
>
> *(Adani Group, 2019)*

On the other hand, referring to pollution, coastal erosion and flooding due to increasing port and industrial infrastructure in the Ennore-Pulicat wetlands, Laxmi from a coastal traditional fishing village expressed the fear that her community might be exterminated in the possible expansion of the Adani Kattupalli port. Describing the port as a knife hanging over her head, she said:

> If the port comes, there will remain none of these villages, none of us, no fish habitat . . . we don't know if the fisherfolk will even exist as a powerful community. The Adani port comes, and 50,000 people won't be there (anymore). The (government) survey mentions that this is a barren land unfit for people to live. . . . This is a place where only birds and animals live.
>
> *(Laxmi, traditional fishing community)*

Yet, apart from birds and animals, the Ennore-Pulicat wetlands remain populated with diverse fishing communities, living with and off its birds and animals, earning a livelihood from its fertility and maintaining strong cultural ties with its diverse land formations and ecologies. A denial of their presence on and sustenance from the wetland disregards their relationship with the wetland ecology.

Reframing wetlands through care

The dominant narrative of "development" remains oblivious to the ample gifts offered by the wetlands, attaching more economic value to the port than the wetland itself. By making visible women's everyday contributions to wetland economies, the concept of care brings forward the ecological knowledge that enables them to derive their livelihood from Ennore-Pulicat wetlands. This reframes the view of Chennai's coastal wetlands from mere empty land to a valuable, life-supporting ecosystem.

Caring practices of women from fisher communities serve to counter the dominant narrative of development by highlighting life-supporting services offered by the wetland. Seafood festivals hosted by fisherwomen in collaboration with solidarity groups bring attention to the role of the wetland in maintaining local and Indigenous food security and cultures. They highlight not only the rich biodiversity of the wetland but also the roles of fisherwomen in maintaining its ecology, their contributions to local economies and their cultural knowledge that are crucial to practicing wetland care. Engaging with wetland care by attending these festivals serves to broaden the collective imagination of Chennai's residents, creating a platform to interact with diverse wetland carers, dreaming an alternative to the dominant narrative of growth and development that threatens these wetlands, and celebrating instead a culture of care.

Notes

1 The thesis was mentored by Andres Verzijl and supervised by Prof. Margreet Zwarteveen.

2 Caste is a rigid form of social differentiation and hierarchy that one is born into. It limits or defines a person's occupation, access to resources as well as technology. Caste hierarchies are localized and differ with place and communities in India, such that a certain caste considered low in one context may not have the same meaning in another context. This case study refers to caste identities in the context of fishing communities in Ennore-Pulicat wetlands. Within these fishing communities, caste, in conjunction with gender identities, shape access to fishing technology, such as nets and boats. They also shape access to various wetland ecologies, and therefore water quality and wetland species, such as picking prawns in brackish backwaters or using nets to fish in the sea. Read more: Chakravarti, U. (2009). *Gendering Caste Through a Feminist Lens.* STREE, Kolkata.

3 All names of interviewees have been changed to protect their identity.

4 Combining field notes, reflections and notes from discussions with the translator, initial observations and perceptions were noted to summarize experiential findings from the fieldwork. Data themes or "codes" were designed to specifically answer research questions. Qualitative data analysis software MAXQDA was used to code interview transcripts according to these themes. Apart from codes designed as per initial assumptions based on literature review, more codes were added as observed in the interview data to understand patterns in perceptions of different women. All interviews were conducted with the help of a translator (Tamil to English).
5 Manifestations of this harsh patriarchy were described by women fishers in accounts of domestic violence, struggles to assert the freedom of expression on public platforms, unequal access to public and private services (such as financial inclusion), as well as unequal access to technology, such that women fishers from Scheduled Caste and Scheduled Tribe communities have limited access to tools for catching prawns.

References

Adani Group. (2019, January 23). *Karan Adani, CEO, Adani Ports and SEZ at Tamil Nadu Global Investors Meet 2019* [Video]. Adani Group YouTube Channel. www.youtube.com/watch?v=tuj2jJdtZpc&ab_channel=AdaniGroup

Aves, Y. (2020). A Pulicat story: The lagoon that protects a city. *Sanctuary Asia*, *40*(10).

Barbier, E. B. (2019). Chapter 27 – The value of coastal wetland ecosystem services. In G. M. E. Perillo, E. Wolanski, D. R. Cahoon & C. S. Hopkinson (Eds.), *Coastal wetlands* (2nd ed., pp. 947–964). Elsevier.

Boelens, R. (2014). Cultural politics and the hydrosocial cycle: Water, power and identity in the Andean highlands. *Geoforum*, *57*, 234–247. https://doi.org/10.1016/j.geoforum.2013.02.008

Budds, J., & Zwarteveen, M. (2020). Retheorizing ecosystem services as cultural landscapes: Co-constitution, power relations, and knowledges. *International Journal of Environmental*, *16*(1), 41–59. http://doi.org/10.18848/1832-2077/CGP/v16i01/41-59

Cambridge Dictionary. *Definition of scramble*. Cambridge University Press. https://dictionary.cambridge.org/dictionary/english/scramble

Chakravarti, U. (2009). *Gendering caste through a feminist lens*. STREE.

Hathaway, A. D., Sommers, R., & Mostaghim, A. (2020). Active interview tactics revisited: A multigenerational perspective. *QSR*, *XVI*(2), 106–119. http://doi.org/10.18778/1733-8077.16.2.09

Janardhanan, A. (2019, January 25). Tamil Nadu global investors meet 2019: MoUs worth Rs 3.4 lakh-crore signed, says CM. *The Indian Express – Chennai*. https://indianexpress.com/article/business/economy/tamil-nadu-global-investors-meet-mou-edappadi-k-palaniswami-5554520/

Jayaraman, N. (2017, September 3). Why Chennai – like other Indian cities – will definitely flood again. *Hindustan Times, India*. www.hindustantimes.com/analysis/why-chennai-like-other-indian-cities-will-definitely-flood-again/story-KVxgIt0zyctNFO70EklYGJ.html

Jennings, B. (2018). Solidarity and care as relational practices. *Bioethics*, *32*(9), 553–561. https://doi.org/10.1111/bioe.12510

Joshi, D., Gallant, B., Hakhu, A., De Silva, S., McDougall, C., Dubois, M., & Arulingam, I. (2021). Ramsar convention and the wise use of wetlands: Rethinking inclusion. *Ecological Restoration*, *39*(1–2), 36–44.

Kumar, M., Saravanan, K., & Jayaraman, N. (2014). Mapping the coastal commons – Fisherfolk and the politics of coastal urbanisation in Chennai. *Economic & Political Weekly*, *XLIX*(48), 46–53.

Mol, A., & Hardon, A. (2020). Caring: A fluid concept for adaptable engagements. In J. R. Bowen, N. Dodier, J. W. Duyvendak & A. Hardon (Eds.), *Pragmatic inquiry: Critical concepts for social sciences* (pp. 233–245). Routledge.

Mol, A., Pols, J., & Moser, I. (2010). Author's response. *Tecnoscienza – Italian Journal of Science & Technology Studies*, *2*(1), 83–86.

Narasimhan, B., Bhallamudi, S. M., Mondal, A., Ghosh, S., & Mujumdar, P. (2016). *Chennai floods 2015 – A rapid assessment*. Interdisciplinary Centre for Water Research, Indian Institute of Science.

Pandian, P. K., Ramesh, S., Murthy, M. V. R., Ramachandran, S., & Thayumanavan, S. (2004). Shoreline changes and near shore processes along Ennore Coast, East Coast of South India. *Journal of Coastal Research*, *20*(3), 828–845.

Piras, E. M., & Zanutto, A. (2010). What exactly are "care" and "practice"? Some notes on concepts and their boundaries. *Tecnoscienza – Italian Journal of Science & Technology Studies*, 2(1), 79–82.
Rupakumar, S. (2021, March 11). *Why are Ennore residents protesting against Adani's Kattupalli Port expansionproject?* DowntoEarth–Environment,India. www.downtoearth.org.in/news/environment/why-are-ennore-residentsprotesting-against-adani-s-kattupalli-port-expansion-project-75894
Tronto, J. (1994). *Moral boundaries: A political argument for an ethic of care*. Routledge.
Women of Pazhaverkadu. (2021, September 9). Invitation to a traditional Tamil seafood feast [Blog post]. *Story of Ennore*. https://storyofennore.wordpress.com/2021/09/09/dear-chief-minister-visit-pulicat-fora-traditional-tamil-seafood-feast/

5

WOMEN'S BODILY EXPERIENCES

Accessing and treating water in the Colombian Caribbean[1]

Silvia Corredor-Rodríguez

Introduction

La Mojana inland delta in northern Colombia is an area of about 5,600 km^2 where three major rivers, the Magdalena, Cauca, and San Jorge, converge. It has 405,625 inhabitants and 83.8% of the population is classified as poor. In addition, in the context of Colombia's armed conflict, confrontations in the region have affected local governance and community development (UNDP, 2019). Being "one of the most important freshwater reservoirs in the country" (Camacho, 2018, p. 60) La Mojana is annually flooded despite having seasons of drought and precipitation. During seasons of precipitation, La Mojana's network of freshwater marshes, swamps, and streams flood adjacent lands (see Figure 5.1) (Camargo, 2021, p. 83).

Although they are surrounded by rivers, communities in La Mojana do not have access to potable water for domestic consumption. As a local resident told me (to the author of this paper), living here means "having water without really having it." One of the reasons for this domestic water scarcity, is pollution by agricultural and livestock waste, and the adverse effects of mining on the banks of the Cauca River (Aguilera, 2004). Moreover, multiple floods due to recent dam and dike breaches have deeply affected local rivers,[2] livelihoods are vulnerable to climate dynamics and in 2010, heavy floods resulted in a regional disaster affecting 750,000 people in the context of La Niña[3] (see UNDP, 2019). On top of these issues, access to water and sanitation infrastructure in rural areas is still poor throughout Colombia.

As an investigative journalist covering the implementation of the Peace Agreement in Colombia, I visited La Mojana during the period between 2016–2019.[4] During these years, I found that the communities located along the banks of the San Jorge River had particular relationships with water: not only with the river but also with rain and groundwater. This chapter is based on participant observation in the townships of *El Torno, Venecia 1, and La Mancha 1*, in the municipality of San Marcos. Women in these three townships have developed different practices and knowledges around water provision and treatment. They use hoses, motor pumps, zinc tiles, plastic and cement storage tanks, clay pots, and buckets. Likewise, they use nets, pieces of cloth, chlorine tablets, and filters to purify these waters.

DOI: 10.4324/9781003100379-7

Figure 5.1 Communities on the banks of the San Jorge River.

Source: The author.

I thus explore and analyze the subjective gendered experiences amongst women from these communities, as they struggle to access and treat water in a context of inequality and precariousness.

Working at the intersection of journalism and ethnography, I conducted seventeen ethnographic interviews and directed a drawing workshop with women, inquiring about their daily practices with and around water. The chapter analyzes their daily practices and experiences in a changing climatic, socio-political, and cultural environment – as well as their ways of using and treating different types of water.

The chapter argues that securing water for domestic use does not only entail technical processes but also *bodily* ones. In fact, sensorial experiences inform water's classification, uses, and treatment. That is, knowledges around water are inscribed in the body, which is "a conscious, experiential, acting and interpreting entity" (Esteban, 2013, p. 25). The body operates as a source of knowledge for decision-making regarding water. These bodily processes are lived mainly by women, who in places like La Mojana are in charge of fetching water for drinking purposes and for water-related domestic activities, such as cleaning, washing, cooking, as well caring for children and the ill (Martínez & Minaverry, 2008). In this way, women's bodies are mediators between the home and different sources of water. It is important to mention that in La Mojana there is no idea of "water" in the singular, but rather community members talk about many waters, including rivers, rain, and groundwater. Women's experiences vary depending on the type of water they are dealing with.

In documenting women's embodied experiences of waters and their different ways of treating them, I aim to reflect on women's water work and the consequences it has on their bodies (Sultana, 2014), as well as on their communities well-being and health (Viveros, 1995). By focusing on the knowledges, practices, and bodily experiences of women in relation to different water uses and treatments, this chapter makes a contribution to studies on water quality and water management from a gendered and embodied perspective (Ortiz, 2018; Leonardelli et al., 2023).

For this purpose, in the following section, I will elaborate on how women distinguish between water qualities as the basis for deciding which water to use for which purpose. This process informs water treatment and storage. A second section explores how women experience aches and discomfort while performing everyday water work. The presence of physical aches is naturalized to the point of being incorporated into women's daily lives. The last section proposes some concluding reflections on the chapter.

Women assessing water quality at different sources: sensorial experiences determining water treatment and storage

The women of El Torno, Venecia 1, and La Mancha 1 have similar routines in their homes. The day starts very early, around five a.m., building a fire to prepare coffee. Then they check on the animals, prepare and serve breakfast, and eventually they start cooking lunch. After lunch, women's tasks focus on cleaning the house, washing clothes, feeding animals, and when evening falls, preparing dinner. Around seven p.m., when work in the kitchen is done, women have some time to take a bath and rest.

Family care in La Mojana, meaning activities such as cooking, tidying up, cleaning, and doing the laundry are largely performed by women (mothers and – sometimes – older daughters). Local families are usually composed of around six members, including children, and other relatives that live in the home. Women also take care of domestic pigs, chickens, and ducks. Men, in turn, spend most of their days outside the home, taking animals to graze, fishing, and working in small-scale agriculture. For this reason, women are most frequently in charge of securing water for the house. This requires high physical effort which has long-term physical consequences on women's bodies (see Sultana, 2014).

Accordingly, water-related work performed by women can be defined as care work, where women's body becomes "a space always ready to carry and receive the other" (Lagarde, 2001, p. 382; see also Orlove & Caton, 2010). The responsibility of care falls on women as a result of a "natural" gender division of labour (Viveros, 1995), which assigns particular ideas about women and what is called "the feminine," as opposed to "the masculine." By assuming this role, women incorporate into their daily lives a series of socially accepted experiences and habits (Lagarde, 2001).

The activities and tasks that women carry out throughout the day are all related to water. In order to perform them, in both El Torno and La Mancha 1, women collect and store groundwater. Both townships have an underground well, working with a motor pump, which is turned on in the morning for one or two hours and is connected to the houses through hoses. Women are in charge of placing the hoses served by pumps into storage containers. Wealthier families hire the services of a man who turns on the motor pump, fills the different households containers and delivers them to the respective home. In this case, women do not have to place hoses. In Venecia 1, families have their own private underground well, so they can decide when to turn on the motor pump and for how long.

However, in all three townships, women frequently have to supplement groundwater with water they collect from the river or harvest rain. Some families seek other types of water due to quantity issues. But for the most part, women look for other types of water because of the quality. Women in all three townships believe groundwater has poor quality and they use it mainly to wash clothes and dishes. They say: "if it foams when mixed with soap, it is good enough for washing."

For a significant part of the population of El Torno, the river continues to be the main source of drinking water. Despite having access to groundwater, many feel that its taste and smell are bad and therefore water is harmful (see also, Pangare & Idris, 2012). Luz is a resident of El Torno and consumes water from the San Jorge River. She explains that groundwater is "green, slimy, and gives stomach aches." Luz walks to the San Jorge River, which is close to her home. She must take trips to the river alone because her 12-year-old daughter does not know how to swim, and Luz is afraid that something will happen to her. She has a 200-liter plastic closable tank in which she stores water "to maintain it in good conditions." To fill this tank, she has to make around five trips to the river, filling 20 one-liter buckets a day. When she is finished, she buys two tablets of chlorine and mixes them with water using circular movements. Because of the amount of water she stores, she must wait almost 20 minutes to make sure she added the correct dose of chlorine. She verifies this by looking at the color of the water (it has to be transparent) and by trying its taste (it has to be *fresh*, "not bitter"). When I ask how exactly she understands if the water is drinkable or not, she explains that it is a matter "of trial and error."

For Amelia,[5] another resident of El Torno, fetching water for household consumption is a task that requires a lot of effort (*esfuerzo*). In her opinion, the best water is that of the river because it is "*living*, it runs, it does not breed parasites. It is not *thick*." The appearance of *thick* water is cloudy, due to river sediments. River water is the one she would like to drink and use for everything. However, fetching water at the river requires a lot of work. She leaves at six a.m. so that the heat of the sun does not make it more difficult to transport buckets. She has to walk about 100 meters to the San Jorge River and make around twenty trips to bring back enough buckets of water. Figure 5.2 portrays the different storage

Figure 5.2 Water containers. Drawing by Amelia, 2019.

Source: The author.

containers in Amelia's house: a clay pot, 20 and 60-liter tanks that supply the kitchen, a smaller tank for the bathroom, and a concrete tank.

Once she arrives home, she fills the clay pot, because it will keep the water cold, and this water will be used for drinking. When storing it, she places a piece of cloth over the mouth of the clay pot to filter out any dirt that water may contain from the river. To fill the clay pot, she has to go and fill six buckets of 20 liters. Afterward, she collects 12 buckets to supply the bathroom tank and six more for other kitchen-related activities. This whole process takes more than three hours in the rainy season and a little longer in the dry season: as the river level drops, she often needs to walk further away to find water. Fetching groundwater is much easier than going to the river, she explains. She only has to open the hoses, fill the containers and store them in and around the home. However, when she feels fit enough and has time, she still prefers to go all the way to the river because "water from the river is tastier than groundwater . . . if I would live by the shore, I would always drink from the river."

Miriam, another resident of El Torno explains that since "the water from the river *flows*, it is the best of all." She stores this water for drinking and cooking and uses groundwater only for all other household chores. Miriam's house is only 20 meters from the river, so the effort in collecting water is less burdensome than it is for Amelia, and she has the help of her 16-year-old daughter. They do not have a set schedule because they go to the river as soon as the home's stored water runs out. They, however, prefer to collect water during the early morning when "the river is less polluted."

Miriam and her daughter collect five buckets of 20 liters each daily, a task that takes between ten to 15 minutes. This water is stored in a plastic tank in the kitchen and is enough for approximately one month, as they only use it for drinking. To keep it *"fresh and clear,"* characteristics that women associate with potability, they add chlorine tablets and tightly close the storage tank. Miriam explains how groundwater is more prone to "causing problems." In this way, while they use the concept of *"fresh"* to refer to water's taste, they use *"clear"* to refer to its visual aspect (the absence of sediments). All of these experiences and ideas inform women's decision-making about which water to use for which purpose: cleaning, cooking, and drinking. Although she and her daughter carefully close the concrete tank where they store it, this water is quickly polluted. "It tends to get *thick*", Miriam says. "It is because of parasites," she adds.

Women from the township explain how, during the rainy season, the river becomes "*blond*" (yellow, like sand and river sediments). Since the river overflows its bank, it grows closer to the homes and this way it is easier to access water. On the contrary, during the dry season, the water becomes "*clear*" and recedes, as a large sandy terrain forms around the river and women have to walk a little further.

Besides river and groundwater, communities in Venecia 1 and La Mancha 1 harvest rainwater as a complement for their households. Women of these two townships use a zinc tile to direct rainwater into storage tanks. In the mouth of each tank, they put a piece of cloth that works as a strainer. According to Marta, a resident of Venecia 1, this infrastructure must be set the night before, to take advantage of the downpours of the early morning. She places a wire around the tank to secure the fabric. After rainwater is collected, some women purify it using filters.

Informed by its taste, smell, feel, and color, women classify water as *fresh*, *thick*, *thin*, *blond*, and *clear*. Based on this assessment, they use it for different purposes. In order to secure good quality river water for drinking, women have to invest more work, walking to the river and carrying water containers. After they collect water from the river, women usually treat it with chlorine tablets: they put the water into plastic and cement storage tanks, clay pots or buckets, they add the tablet and wait until water becomes *clear*, that is, until it looks transparent. Only at this point it is suitable for consumption.

Significantly, throughout the implementation of the peace agreements, the communities of La Mojana have been targeted by different state programmes. In community training organized by state agencies, practitioners recommended boiling the water. However, women do not always follow this recommendation. They used to boil water only when there is a visitor who is not used to La Mojana's water or someone in the house is sick. Luz from El Torno, for instance, mentions that this is because boiling water "changes water's taste." However, she does boil water when a member of the household becomes sick "to ensure that water does not contain anything dangerous." After all, boiling water means an additional effort for women as they need to go and search for firewood (see also Sultana, 2010). In recent years, some households also started using water filters provided by the United Nations Development Programme (UNDP) and by the Red Cross as part of emergency programmes implemented in times of heavy flooding.

In the middle of a watery landscape, such as La Mojana, pollution contributes to the difficulties women experience when securing and treating drinkable water. In this particular context, understanding everyday practices and knowledges is key to understand the relations between communities and water (Orlove & Rassmusen, 2017). Women in these townships have a bodily relation with different waters (Le Breton, 2002), not only by carrying different types of water through distances and storing it (which requires physical effort) but also by using their senses to *know* and classify them (see Leonardelli et al., 2023).

Women suffer physical exhaustion (Lagarde, 2001; Das, 2008) after securing and treating water and complying with the daily needs of the household (Sultana, 2010; Wutich and Brewis, 2014). These physical consequences, which emerge in the narratives of women as they talk about aches and pain related to their water work, tend to be often neglected in water programmes, policy efforts and mainstream water research. For this reason, in the next section, I focus on women's bodily experiences (Esteban, 2013) in order to highlight the value of their work and their role as main managers of water in their homes (Sultana et al., 2016; Sultana, 2011; Leonardelli et al., 2023).

Women experiencing aches and discomfort while performing everyday water work

The experiences of women from El Torno, Venecia 1, and La Mancha 1, account for the embodiment of decisions regarding water. The presence of physical aches, which shows the relationship between the individual and the environmental, is naturalized to the point of being incorporated into women's daily lives, since it is a product of a necessary task, or social practice (Esteban, 2013; Le Breton, 2019; Viveros, 1995). In the home, women face two types of aches: one generated by water labour (fetching and storing water), and one caused by water pollution. The latter entails water-borne diseases such as diarrheal disease and skin rashes (see also Lahiri-Dutt, 2015).

Women from La Mojana decide which water to use for different purposes also on the basis of these aches and discomforts. These are not easy decisions for women have to decide which pain and discomfort to avoid the most (aches that result from water labour or aches related to polluted water). In fact, fetching high quality water often requires more effort and physical wear, as it flows further from the home. This is the case of Amelia, who continues to prefer the water from the San Jorge River, even after the installation of a groundwater well next to her house in 2006. Although she does not collect the same amount of water from the river as she used to, walking back and forth from the river continues to cause her body aches. Amelia feels pain in her knees, shoulders, and back because she lifts the bucket from the floor, puts it on her knee and from there she lifts it on her shoulders and then on her head. She does this several times as she needs to walk back and forth from the river several times before filling her drinking water tank, taking about three hours. During the dry season, she makes extra efforts as she has to walk through the dry sandy terrain that forms around the river carrying the weight of 20 water liters.

Amelia explains how her family started experiencing skin rashes every time they used groundwater for bathing. To counteract this situation, she started treating the water with a few drops of "*límpido,*" a strong disinfectant detergent that is popular in Colombia. Moreover, she explains how "groundwater cannot be directly drank or used for cooking because it causes diarrhea." For this reason, she treats it with chlorine tablets. The bodily experiences that Amelia and her family went through after they started using groundwater were key to subsequent decisions on which water to use for different purposes and how to treat them.

Miriam and Luz, who also collect water from the San Jorge River, experience similar body aches, especially during the dry season, when water is five meters further away from the usual shore. The type of aches that Amelia and other women from El Torno experience when fetching river water shows "how the reality of the body refers to unconscious, social, cultural, and individual meanings" (Le Breton, 2019, p. 50). That is, Amelia and other women prioritize the taste and wellbeing that they associate with river water even if fetching water from the river means significant physical effort and pain in the back, knees, shoulders, and legs.

Luz, another woman from the township, lives only two meters away from the San Jorge River. Unlike the other women from El Torno, Luz explains how, in order to access river water during the dry season, she has to use a shovel and create a sort of stairway. She confesses that the dry season is particularly hard for her, especially because of the pain she feels on her feet, due to the weight of the buckets. For her, this effort "is worth it since river water is tastier, sweet, and does not harm her or her family members." She further adds that "the strong current of the river keeps it free from pollution." Furthermore, she bases her decision to drink water from the river on experience. She has spent her entire life in El Torno and she has always drank river water: she knows that it quenches her thirst and never caused any diseases.

In the case of Venecia 1 and La Mancha 1, women's bodily experiences of pain are not caused by fetching river water. This is because these townships depend mainly on groundwater and rainwater. For this reason, everyday physical efforts are related to the movement of the hoses that operate as technological devices (see López, 2016) to transport and distribute groundwater. Fetching and storing groundwater require a lot of effort even if hoses "become an extension of the human body" (Verbeek, 2006, p. 365) as they "carry" the water to the different containers. Moreover, due to the density of groundwater (how "*thick*" it is), cleaning storage containers is challenging and tiring for women. Miriam, for instance, must clean sediments from inside the concrete tank frequently and these cleaning activities take a toll on her back. "My lungs already hurt from washing the tank to make the water *clear*," she explains.

Women's bodies become a source of knowledge (García, 2013) for making decisions related to water consumption and supply. Likewise, in the quest to avoid water-borne diseases, women prolong aches in their own feet, legs, knees, shoulders, and back. Most of them do not seek medical advice because they assume it as being a "natural pain." Therefore, these aches sometimes become diseases (Le Breton, 2019) that are frequently underestimated because women understand them as consequences of their role as water providers.

Conclusions

By studying women's practices and knowledges around water in La Mojana, it becomes clear how water access and treatment are entangled with bodily experiences and gender roles. The variety of classifications and uses associated with different water sources helps understanding how water properties, values and meanings are not fixed or generalizable (Barnes & Alatout, 2012; Orlove & Caton, 2010). Furthermore, the use of motor pumps, zinc tiles, hoses, tanks, filters, chlorine tablets, pieces of cloth, and nets suggest how bodily and sensorial assessments of water quality determine different water treatments and uses (Bønnelykke, 2016).

Understanding water access and treatment from a bodily perspective highlights how assessing water quantity and quality entails not only technical processes but also cultural, sensorial, and embodied ones. Likewise, integrating a gender perspective "that allows us to understand the social and cultural patterns through which . . . natural resources are appropriated, used and managed" (Arellano, 2003, p. 83), offers the possibility of taking the bodily experiences of women seriously. This is important because by recognizing their roles as water managers and providers (Sultana et al., 2016; Sultana, 2018) their contributions and knowledges are vindicated and can be more seriously considered when it comes to understanding the intersections of gender, development, and the environment (Arellano, 2003; Leonardelli et al., 2023), as well as in the water policy arena interested in daily practices of water management in rural areas of the Global South.

Ethnographic methods allowed me, as an investigative journalist, to conduct more in-depth, empathetic interviews that, in turn, allowed women from La Mojana to tell their own stories from their own perspectives. In this way, the needs of the inhabitants can be more readily considered, amidst the implementation of Colombia's peace agreement that touches on topics such as rural reform and broad political participation.

Notes

1 This chapter was translated from the Spanish by Tatiana Acevedo-Guerrero.

2 In August 2021, water level rise of the Cauca River broke a dam in San Jacinto del Cauca causing a major flood that affected more than 150,000 people in 11 municipalities.

3 La Niña is a weather pattern that occurs in the Pacific Ocean. In this pattern, strong winds blow warm water at the ocean's surface from South America to Indonesia. As the warm water moves west, cold water from the deep rises to the surface near the coast of South America (see https://wmo.int/).

4 Among other things, the Agreement included topics of rural reform, political participation, the end of hostilities, solutions to the production of illicit drugs, the rights of victims, and the mechanisms of implementation and verification (see Badrán, 2023).

5 All names have been changed.

References

Aguilera, M. (2004). La Mojana: Riqueza natural y potencial económico. En *Documentos de Trabajo sobre Economía Regional, No. 48, Centro de Estudios Económicos Regionales (CEER)*. Banco de la República.

Arellano, R. (2003). Género, medio ambiente y desarrollo sustentable: un nuevo reto para los estudios de género. *Revista de estudios de género: La Ventana*, *17*(2), 79–106. Universidad de Guadalajara.

Badrán, F. (2023). The signing of the peace agreement in Colombia. Old wine in new skins: Implications for national security and organized crime. *Trends in Organized Crime*. https://doi.org/10.1007/s12117-023-09497-y

Barnes, J., & Alatout, S. (2012). Water worlds: Introduction to the special issue of social studies of science. *Social Studies of Science*, *42*(4), 483–488.

Bønnelykke, M. L. (2016). The affects of water – the materialized morality of wells, pipes, and pumps in Tarawa, Kiribati. *Society & Natural Resources*, *29*(6), 668–680. https://doi.org/10.1080/08941920.2015.1107791

Camacho, J. (2018). *En el camino a la Gloria se pasa por la Mancha*. Instituto Colombiano de Antropología e Historia.

Camargo, A. (2021). Stagnation: Waterflows and the politics of stranded matter in La Mojana, Colombia. In Franz Krause and Mark Harris (Eds.), *Delta life exploring dynamic environments where rivers meet the sea*. Berghahn Books.

Das, V. (2008). *Veena Das: Sujetos del dolor, agentes de dignidad* (1st ed.). Pontificia Universidad Javeriana, Instituto Pensar; Universidad Nacional de Colombia, Sede Medellín, Facultad de Ciencias Humanas y Económicas; Universidad Nacional de Colombia, Sede Bogotá, Facultad de Ciencias Humanas, Centro de Estudios Sociales.

Esteban, M. L. (2013). *Antropología del cuerpo: Género, itinerarios corporales, identidad y cambio*. Bellaterra.

García, N. (2013). La autoetnografía. Una experiencia de corporalidad en la investigación sociológica. In *VII Jornadas de Jóvenes Investigadores. Instituto de Investigaciones Gino Germani, Facultad de Ciencias Sociales*. Universidad de Buenos Aires, Buenos Aires.

Lagarde, M. (2001). *Los cautiverios de las mujeres: Madresposas, monjas, putas, presas y locas*. UNAM.

Lahiri-Dutt, K. (2015). Scaling down: Researching household water practices. *An International E-Journal for Critical Geographies*, *14*(3), 639–651.

Le Breton, D. (2002). *Antropología del cuerpo y modernidad*. Nueva Visión.

Le Breton, D. (2019). *Antropología del dolor* (1st ed.). Metales Pesados. https://doi.org/10.2307/j.ctvvh86m7

Leonardelli, I., Kemerink-Seyoum, J., Kulkarni, S., Bhat, S., & Zwarteveen, M. (2023). A feminist analysis of women farmers navigating groundwater qualities in Maharashtra, India. *Water Alternatives*, *16*(1), 134–152.

López, M. (2016). *Paisajes hídricos en disputa: agua, poder y fragmentación urbana en Medellín*. Confiar.

Martínez, A., & Minaverry, C. M. (2008). Las Cuestiones De Género Y El Derecho Humano Del Agua. *La Aljaba, Segunda Época. Revista de Estudios de La Mujer*, *12*, 257–259.

Orlove, B., & Caton, S. C. (2010). Water sustainability: Anthropological approaches and prospects. *Annual Review of Anthropology*, *39*, 401–415.

Orlove, B., y Rassmusen, M. B. (2017). *Virtual issue-anthropologists exploring water in social and cultural life: Introduction*. American Anthropologist. http://anthrosource.onlinelibrary.wiley.com/hub/journal/10.1111/(ISSN)1548-1433/exploring-water.html

Ortiz Yam, I. (2018). El manejo del agua a través del tiempo en la península de Yucatán, de Mónica Chávez Guzmán (editora). *Revista Colombiana De Antropología*, *55*(1), 316–319. https://doi.org/10.22380/2539472X.581

Pangare, G., & Idris, L. (2012). Water and health security. En H. Bigas (Ed.), *The global water crisis: Addressing an urgent security issue* (pp. 76–83). Papers for the InterAc on Council. UNU-INWEH.

Sultana, F. (2010). Living in hazardous waterscapes: Gendered vulnerabilities and experiences of floods and disasters. *Environmental Hazards*, *9*(1), 43–53. https://doi.org/10.3763/ehaz.2010.SI02

Sultana, F. (2011). Suffering for water, suffering from water: Emotional geographies of resource access, control and conflict. *Geoforum*, *42*(2), 163–172. https://doi.org/10.1016/j.geoforum.2010.12.002

Sultana, F. (2014). Gendering climate change: Geographical insights. *The Professional Geographer*, *66*(3), 372–381. https://doi.org/10.1080/00330124.2013.821730

Sultana, F. (2018). Gender and water in a changing climate: Challenges and opportunities. En C. Fröhlich, G. Gioli, R. Cremades & H. Myrttinen (Eds.), *Water security across the gender divide* (pp. 17–33). https://doi.org/10.1007/978-3-319-64046-4_2

Sultana, F., Mohanty, C. T., & Miraglia, S. (2016). Gender equity, citizenship, and public water in Bangladesh. En D. McDonald (Ed.), *Making public in a privatized world* (pp. 149–164). Zed Books.

UnitedNationsDevelopmentProgramme–UNDP.(2019,February13). *USD117millionwillbeinvestedin environmentalprojectMojana: Climateandlife.* www.undp.org/es/colombia/comunicados-de-prensa/usd117-million-will-be-invested-environmental-project-mojana-climate-and-life

Verbeek, P.-P. (2006). Materializing morality: Design ethics and technological mediation. *Science, Technology, & Human Values*, *31*(3), 361–380.

Viveros, M. (1995). Saberes y dolores secretos. Mujeres, salud e identidad. En M. León de Leal, L. G. Arango & M. Viveros (Eds.), *Género e identidad: Ensayos sobre lo femenino y lo masculino* (1st ed., pp. 149–168). Ediciones Uniandes: Facultad de Ciencias Humanas: TM Editores.

Wutich, A., & Brewis, A. (2014). Food, water, and scarcity: Toward a broader anthropology of resource insecurity. *Current Anthropology*, *55*(4), 444–468.

6

EMBODYING THE URBAN POLITICAL ECOLOGY OF WATER

Three analytical approaches to urban water insecurity

Yaffa Truelove

Introduction

Naila, a Muslim woman living in an informal settlement of Delhi, India rises on a foggy morning in 2016 to begin her daily search for water. While her husband begins his 12-hour shift at a local construction site and their son attends school, Naila waits for a tanker truck carrying state water to arrive outside the lanes of the settlement. She balances this gendered household labor with her own paid work, cleaning three middle-class homes across the street from her house. Hoping to fill two to three jerry cans of water, after two hours of waiting with her empty water containers on the side of the road, Naila becomes convinced that this day will mark the second time in a row that the tanker has failed to deliver vital water needed to get her family through the day. Naila considers walking two kilometers to a neighboring settlement where other Muslim families are willing to share access to their tubewell water, but the time and physical labor of carrying water back to her settlement is daunting. She instead opts to ask one of her employers to fill a few buckets of water from the home where she works. In exchange, Naila's employer asks her to work late that day for extra unpaid hours, a barter which Naila feels she cannot refuse, given the water needs of her household. In the late afternoon when Naila returns home, she first prioritizes the water for her son to drink and later gives it to her husband for bathing, while also saving the leftover bathing water for cleaning his clothes. She decides her own bathing and clothes-washing will need to wait for yet another day and is careful to minimize her own water consumption to ensure that the needs of her husband and son are first met. While Naila's gendered everyday water practices help to enable her family's water security within the complex matrix of Delhi's unequal water supply, improvements in her family members water security often come at the expense of her own. Even at the household scale, one family members relative water security can be based on another's perpetual water insecurity, revealing highly incongruent embodied experiences of urban water in the everyday.

Despite the importance of city dwellers unequal everyday experiences of water insecurity, dominant approaches to water security tend to forego analyses of embodied water

DOI: 10.4324/9781003100379-8

experiences and the connection between water insecurity and patterns of gender/class/race differentiation and inequality. Emerging from frameworks attentive to human and environmental security (Kaplan, 1994; Starr, 1991), and later taken up within Integrated Water Resources Management (IWRM) (see Cook & Bakker, 2012), interdisciplinary approaches to water security have historically privileged national and regional scales of analysis, and to some extent rural areas over urban. While employing diverse conceptualizations of water security, such perspectives align in their analytical attention to scales that "hover above" the scale of the body, whether it is the watershed (the focus of hydrologists, for example), the nation-state (the focus of political scientists), or even the community (the focus of practitioners and policymakers) (Bakker, 2012, p. 925; Lankford et al., 2013). There has thus been a gap in studies which connect the ways that patterns of water in/security across global, national, regional and community scales unequally shape the experience of water and its governance at the bodily scale.

This chapter is based on a previously published article (Truelove, 2019) and provides an overview of emerging scholarship that advances embodied approaches to urban water in/security, inequality and infrastructure. This new literature is broadly informed by political ecology studies of water, which critique depoliticized approaches to water scarcity and inequality and, in relation to cities, give attention to the socially differentiated experiences of urban water and its infrastructures. The chapter details how recent interventions to bring feminist and embodied considerations to water's urban political ecology analyze the site and scale of the body as critical for understanding everyday urban water access and inequality. Drawing from these frameworks, this chapter examines three contributions of an embodied urban political ecology approach for addressing water in/security. These include analytical approaches that give attention to 1) the scale of the body within multiscalar approaches to water, 2) intersectionality and gender/class/race/ethno-religious relations in shaping patterns of water inequality and insecurity and 3) "everyday" practices and politics, in relation to both governance and citizens, that help reveal often unnoticed and under-theorized dimensions of water insecurity and inequality. Embodied approaches to urban water insecurity are poised to expand and deepen work on the everyday politics and lived experiences of insufficient, insecure, and unequal water that profoundly shape city life for urban dwellers.

Politicizing everyday urban water in/security

Political ecologists critique dominant approaches to water in/security as being overly technocratic, depoliticized and environmentally deterministic (Loftus, 2015). While early approaches focused on security and control of water at the level of the nation-state, there has been a more recent movement to mainstream water security into policy spheres as "a question of human security . . . [that] emphasizes a concern for the most vulnerable in the world" (Loftus, 2015, p. 351), particularly the plight of the world's poor. For example, the Millennium Development Goals (MDGs) to improve drinking water access links water security to the quantity and quality of water available for entire populations, while the current Sustainable Development Goal six aims to ensure safe and sufficient water for all by 2030. Although such frameworks are productive in bringing attention to the widespread water disparities for the urban and rural poor globally, the focus nonetheless remains on improving water access and security through predominately technocratic and engineering fixes to mitigate "natural" occurrences that include the impacts of drought and flooding on growing populations. As Loftus (2015) demonstrates, such approaches risk naturalizing the causes of water scarcity and insecurity and undermining the role of power and politics in shaping highly uneven waterscapes for differing social groups. Thus, while bringing the scarcity of potable water for poor populations to the forefront of addressing water security, insecurities in water access nonetheless tend to be explained as a product of environmental events and insufficient technocratic fixes, rather than deeply enrolled in political, cultural and economic processes that actively construct insecure and unequal waterscapes.

Bringing a political approach to water access and in/security in cities more specifically, urban political ecologists (UPE) reveal how social and natural factors combine and co-constitute each other (hence the word socio-natural) to shape water's urbanization, resulting in networked water supplies that often exclude significant portions of the population. In cities as diverse as Guayaquil, Mumbai, Accra and Flint, this body of work demonstrates that water scarcity and everyday insecurity are as much (or more) a product of socio-political processes as ecological occurrences such as drought or groundwater depletion (Swyngedouw, 2004; Loftus, 2015; Truelove, 2016; Kooy, 2014; Furlong & Kooy, 2017; Millington, 2018; Ahlers et al., 2014; Ranganathan, 2016; Jepson, 2012). While such studies do not always utilize the term water security directly, urban political ecologists expand analyses of water in/security, inequality, and infrastructure through attention to the ways social power relations combine with material and ecological factors to shape the unequal flow and regulation of water in cities. These studies reveal that the widespread exclusion of the urban poor and other groups from equitable, secure and sufficient water results not simply from overpopulation and declining water tables, or a "naturally" produced scarcity, but one that is concurrently socially and politically manufactured. Generally, such studies focus on political economy and overarching processes shaping the city, devoting less attention to the scale of the body and the everyday politics of control that are forged in lanes and neighborhoods as residents respond to inadequacies of the networked supply (Truelove, 2016, pp. 144–145).

To fill this gap, feminist political ecology focuses attention to the bodily scale of water-related inequalities, demonstrating how wider political, discursive, and material factors that co-produce urban water are situated and unevenly embodied. This work gives attention to unequal power relations in shaping the ways water is unevenly governed and experienced along intersecting gender/class/race lines, consequently producing intra-household dimensions of water inequity and insecurity (Sultana, 2011; O'Reilly, 2006; O'Reilly & Dreibelbis, 2018; Thompson, 2016; Truelove, 2011). In an effort to bridge such feminist and situated approaches within the canon of urban political ecology, Doshi calls for scholars to "embody" UPE by "connecting socio-natures of . . . resource distribution with the intimate, meaningful and power-laden embodiments of such flows among differently situated groups" (Doshi, 2017, p. 126). Drawing from these frameworks, the remaining sections of the chapter turn to an examination of three contributions of an embodied urban political ecology approach for addressing water in/security.

Analyzing the scale of the body within multi-scalar approaches

Embodied approaches to urban water in/security first and foremost aim to examine how water's socio-nature affects bodies differently and to further account for the myriad human and non-human factors that produce this inequality. This type of analysis draws from feminist scholarship that examines the body as a locus of experience and scale of inquiry where larger power structures are situated (Mountz, 2017). For example, feminist geographers illuminate how processes that traverse the global, national, urban and household scales are situated and unevenly experienced at the scale of the body (Nagar et al., 2002; Silvey, 2004). Drawing from these insights, embodied approaches to water in/security recognize that analyses that are only attentive to regional, national or even urban scales can give an appearance of water access and availability that fails to hold up when the bodily scale is taken into account, consequently excluding important dimensions of insecurity and inequality. For example, while cities such as Delhi, Lilongwe and Amman may appear to provision sufficient and/or secure aggregate water flows at the scale of the city, zooming in to the scale of the body reveals how differing urban, regional and local processes unequally channel water to city dwellers, with particular residents receiving sub-standard and insufficient potable water and encountering detrimental consequences associated with procuring water despite per capita averages (Alda-Vidal et al., 2018; Mustafa & Talozi, 2018; Truelove, 2018). Thus, attention to embodied forms of water in/security reveals that an exclusive focus on national, regional or urban scales of water security can mask important dimensions of insecurity and inequality experienced by city dwellers in everyday life.

Attention to the bodily scale is also critical for demonstrating how social, material and ecological factors interact in complex ways to create differing patterns of water inequality and insecurity for urban residents. For example, Millington's (2018) work on Sao Paulo's 2015 water crisis illuminates residents uneven lived experiences of water emergencies and insecurity, which is not only produced by complex political and ecological processes across the region but also partially mediated by residents differentiated abilities to store water in small-scale reservoirs commonly referred to as "caixas d'água." In a similar vein, scholars such as Swyngedouw (2004), Anand (2018), Ranganathan (2016, 2018) and Peloso et al. (2018) powerfully demonstrate how the experiences of urban water crises are rarely evenly distributed, but rather are nested in unequal social and political power relations that consequently enable some residents within the same city to escape water insecurity while others experience its embodied effects chronically. Furthermore, discourses of crisis can perversely serve to render invisible chronic patterns of water insecurity and scarcity experienced by marginalized residents through narrating crisis as exceptional rather than normative (Anand, 2018; Peloso et al., 2018). For example, Ranganathan's analysis of Flint, Michigan's austerity measures and resulting poisoned water supply shows how discourses of crisis justify and escalate structural violence on marginalized bodies, specifically those of African Americans (see also Anand, 2018). Thus, an embodied approach helps to reveal city dwellers overlapping social, material and ecological experiences of urban water insecurity, whereby one person's exceptional experience of water insufficiency at a particular moment in time co-exists with (and is often connected to) another's ongoing water crisis.

Intersectionality and water inequality and insecurity

Building from political ecology approaches that view water and social power as mutually constitutive, scholarship on water and intersectionality uncovers the complex ways that urban water inequality and insecurity are tied to intersecting gender/class/race/ethno-religious axes of power. Intersectionality is a concept analytic that stems from black, multicultural and postcolonial feminist scholarship that gives attention to the ways gender operates in conjunction with race, class and other power structures in shaping social relations and everyday experiences of inequality (Crenshaw, 1989). Taken up in a growing number of gender and water studies and feminist political ecology more broadly (Mollett & Faria, 2013; Sultana, 2011; O'Reilly, 2006; Truelove, 2011; Thompson, 2016; Harris, 2009; Zwarteveen et al., 2014), analyses of intersectionality in relation to water illuminate how gender/class/race and other power geometries emerge in conjunction with the socio-natural transformation of the urban waterscape. These studies thus demonstrate how intersectional power relations, subjectivities and social differentiation emerge in relation to water and its urban infrastructures.

Intersectionality offers an important lens for examining water vulnerabilities and inequities as they emerge across complex social configurations that affect city dwellers unequal embodied experiences. For example, several studies in South Asia (particularly India and Bangladesh) illuminate how gender intersects with class, caste and other differences in ways that mediate everyday water access, environmental knowledge and unequal patterns of resource control (Sultana, 2011; Joshi & Fawcett, 2005; Truelove, 2011; Birkenholtz, 2013). For example, in Delhi intersectional gender, class, caste and ethno-religious politics shape who controls local water sources, whose bodies tend to be allocated the gendered roles of procuring and managing household water, and political negotiations and water claims-making strategies (Truelove, 2021). In some settlements, intersections of poverty, gender and ethno-religious axes of power compound poor Muslim men's unequal experiences of water insecurity and urban exclusion in Delhi, as they lack both access to local reliable water sources and forms of political representation that are offered to more elite Hindu groups (Truelove, 2016). In Bangladesh, Sultana (2011) demonstrates how women's experience of water stress is congruently shaped by age, class, marital status and poverty, producing additional dimensions of water-related inequity and hardships associated with arsenic contamination. This work collectively brings attention to how water in/security and inequality is intimately tied to overlapping (gender, class, religious and racial) power relations that shape differing forms and levels of water precariousness and hardships for residents who share the same neighborhoods, lanes and even households. This work illuminates the need for approaches to water insecurity to further pinpoint how water's materiality (see also Thompson, 2016), diverse infrastructures and everyday governance interact with social hierarchies to produce unequal socio-environments. Furthermore, an intersectional approach reveals how particular gender/class/race/ethno-religious groups situated knowledge of the waterscape may have emancipatory potential for producing more equitable water access (Loftus, 2006). Without taking intersectionality into account, analyses are at risk of overlooking critical dimensions and spaces of water inequality and insecurity that disproportionately impact particular social groups in the city at the intra-household scale, as well as how these injustices might be transformed.

Everyday practices and micropolitics

In many of the worlds cities, residents must discern and navigate increasingly fluid and complex constellations of everyday governance that shape their unequal experience of, and access to, vital water resources. As Bapat and Agarwal (2003, p. 74) document in relation to water in Mumbai, "anyone can take charge of water and collect money," indicating the diverse delivery configurations and water practices that constitute everyday forms of access in the city. Embodied approaches to urban water in/security give attention to everyday practices of both citizens and political authorities, and their "intentional or unintentional political implications" (Ortner, 1984, p. 3) in shaping patterns of water insecurity and inequality. As everyday practices to access and control make water flow enroll people in differing environmental and social relations, these relations and micropolitics can either enable or constrain forms of water security and equity. Furthermore, attention to everyday practices and micropolitics reveal often unnoticed and under-theorized dimensions of water insecurity, including the variety of relationships, spaces, networks and political arrangements residents rely on to secure water access, and the differing forms of dis/empowerment and inclusion/exclusion that result.

For example, by bringing inequalities that result from everyday practices to the forefront of analyses, this growing literature reveals that interventions that solely focus on quantity and quality of water, or lowering its pricing, may nonetheless do little to improve water security or precariousness for city dwellers (Loftus, 2006; Von Schnitzler, 2008). For example, Loftus's (2006) work on water meters in Durban reveals how the infrastructure of the meter shapes everyday rhythms and water practices for poor populations in inequitable ways that bleed into everyday life. As households wait for "free" metered water that flows at low pressure and recycle water so that scant quantities can be stretched to all manner of tasks, the state's provisioning of "free" water nonetheless creates rippling inequalities that shape poor residents everyday routines, relations with the state and ability to pursue other uses of time. Research on everyday water practices of poor settlers in Delhi shows how the move from "illegal" to "legal" water sources level increasing physical risks for women (including harassment and drawing water from dangerous and fast-flowing canals) that amplify their embodied hardships despite access to higher quantities of "legal" water (Truelove, 2011). These everyday water practices and politics profoundly impact emotion and affect, producing not just material struggles but psycho-social stress and anxiety (Strang, 2014; Sultana, 2011; Wutich & Ragsdale, 2008).

Attention to everyday practices also reveals the complex negotiations that residents, informal authorities, and officials of the state engage in to make water flow to particular urban spaces. This scholarship reveals that even the urban poor within a city experience highly uneven water access and provisioning that is shaped by maneuverings of pipes, pumps, wells, forms of expertise and political pressure (Anand, 2011; Björkman, 2018; Kundu & Chatterjee, 2021; Peloso & Morinville, 2014; Birkinshaw, 2018; Birkenholtz, 2010; Jepson & Brown, 2014). If such micropolitics are bypassed within approaches to water in/security, the poor become grouped together as uniform recipients of unequal and insecure water flows, rather than citizens that experience differing forms and gradations of access, control and socio-political power in relation to the urban waterscape. Finally, a number of studies demonstrate how the uneven practices of water and its governance are connected to differentiated urban subjectivities and unequal urban citizenship that further concretizes forms of embodied exclusions (Anand, 2017; Swyngedouw, 2004; Von Schnitzler, 2008), demonstrating the inseparability of water from what Lefebvre terms the "right to the city" itself (Lefebvre et al., 1996).

Conclusion

Embodied approaches to urban water in/security, inequality and infrastructure bring attention to the body as a critical scale of analysis and site of lived experiences and practices, in order to reveal how water policies and provisioning strategies contribute to wider patterns of urban and social differentiation. This body of work indicates a need to expand analytical approaches to water in/security in order to encapsulate multiple forms of insecurity and precariousness related to (and emerging from) city dwellers unequal embodied experiences of water. Through analyzing the socio-natural and socio-material processes and practices that produce unequal embodied experiences of water, this literature pushes both academic and policy work to go beyond quantity/quality approaches to water to more fully integrate the ways water is tied to a host of additional inequities and vulnerabilities, including life opportunities, income-generating activities, the unequal production of urban space and uneven patterns of social differentiation and rights to the city.

Note: This chapter is based on the previously published article: Truelove, Y. (2019) Rethinking Water Insecurity, Inequality and Infrastructure through an Embodied Urban Political Ecology. *WIREs Water*, *6*(3), 1–7.

References

Ahlers, R., Cleaver, F., Rusca, M., & Schwartz, K. (2014). Informal space in the urban waterscape: Disaggregation and co-production of water services. *Water Alternatives*, *7*(1), 1–14.

Alda-Vidal, C., Kooy, M., & Rusca, M. (2018). Mapping operation and maintenance: An everyday urbanism analysis of inequalities within piped water supply in Lilongwe, Malawi. *Urban Geography*, *39*(1), 104–121. https://doi.org/10.1080/02723638.2017.1292664

Anand, N. (2011). Pressure: The politechnics of water supply in Mumbai. *Cultural Anthropology*, *26*(4), 542–564. https://doi.org/10.1111/j.1548-1360.2011.01111.x

Anand, N. (2017). *Hydraulic city: Water and the infrastructures of citizenship in Mumbai*. Duke University Press.

Anand, N. (2018). "Water crisis", Spotlight on series. *International Journal of Urban and Regional Research*. www.ijurr.org/spotlight-on-overview/parched-cities-parched-citizens/water-crisis/.

Bakker, K. (2012). Water security: Research challenges and opportunities. *Science*, *337*(6097), 914–915. https://doi.org/10.1126/science.1226337

Bapat, M., & Agarwal, I. (2003). Our needs, our priorities; women and men from the slums in Mumbai and Pune talk about their needs for water and sanitation. *Environment and Urbanization*, *15*(2), 71–86. https://doi.org/10.1177/095624780301500221

Birkenholtz, T. (2010). "Full-cost recovery": Producing differentiated water collection practices and responses to centralized water networks in Jaipur, India. *Environment and Planning A*, *42*(9), 2238–2253.

Birkenholtz, T. (2013). "On the network, off the map": Developing intervillage and intragender differentiation in rural water supply. *Environment and Planning D: Society and Space*, *31*(2), 354–371. https://doi.org/10.1068/a4366

Birkinshaw, M. (2018). "Water mafia" politics and unruly informality in Delhi's unauthorised colonies. In L. Leeson (Ed.), *Water power: Creativity and the unlocking of community knowledge* (pp. 189–203). Routledge.

Björkman, L. (2018). The engineer and the plumber: Mediating Mumbai's conflicting infrastructural imaginaries. *International Journal of Urban and Regional Research*, *42*(2), 276–294. https://doi.org/10.1111/1468-2427.12526

Cook, C., & Bakker, K. (2012). Water security: Debating an emerging paradigm. *Global Environmental Change*, *22*(1), 94–102. https://doi.org/10.1016/j.gloenvcha.2011.10.011

Crenshaw, K. (1989). Demarginalizing the intersection of race and sex: A Black feminist critique of antidiscrimination doctrine, feminist theory and antiracist politics. *University of Chicago Legal Forum*, *1*, 139–168.

Doshi, S. (2017). Embodied urban political ecology: Five propositions. *Area*, *49*(1), 125–128. https://doi.org/10.1111/area.12293

Furlong, K., & Kooy, M. (2017). Worlding water supply: Thinking beyond the network in Jakarta. *International Journal of Urban and Regional Research*, *41*(6), 888–903. https://doi.org/10.1111/1468-2427.12582

Harris, L. (2009). Gender and emergent water governance: Comparative overview of neoliberalized natures and gender dimensions of privatization, devolution and marketization. *Gender, Place and Culture*, *16*(4), 387–408. https://doi.org/10.1080/09663690903003918

Jepson, W. (2012). Claiming space, claiming water: Contested legal geographies of water in South Texas. *Annals of the Association of American Geographers*, *102*(3), 614–631. www.jstor.org/stable/23275546

Jepson, W., & Brown, H. (2014). "If no gasoline, no water": Privatizing drinking water quality in South Texas colonias. *Environment and Planning A*, *46*(5), 1032–1048. https://doi.org/10.1068/a46170

Joshi, D., & Fawcett, B. (2005). The role of water in an unequal social order in India. In A. Coles & T. Wallace (Eds.), *Gender, water and development* (pp. 39–56). Berg.

Kaplan, R. (1994). The coming anarchy. *Atlantic Monthly*, *273*, 44–76.

Kooy, M. (2014). Developing informality: The production of Jakarta's Urban waterscape. *Water Alternatives*, *7*(1), 35–53.

Kundu, R., & Chatterjee, S. (2021). Pipe dreams? Practices of everyday governance of heterogeneous configurations of water supply in Baruipur, a small town in India. *Environment and Planning C: Politics and Space*, *39*(2), 318–335. https://doi.org/10.1177/2399654420958027

Lankford, B., Bakker, K., Zeitoun, M., & Conway, D. (Eds.). (2013). *Water security: Principles, perspectives and practices*. Routledge.

Lefebvre, H., Kofman, E., & Lebas, E. (1996). *Writings on cities* (Vol. 63, No. 2). Blackwell.

Loftus, A. (2006). Reification and the dictatorship of the water meter. *Antipode*, *38*(5), 1023–1045. https://doi.org/10.1111/j.1467-8330.2006.00491.x

Loftus, A. (2015). Water (in) security: Securing the right to water. *The Geographical Journal*, *181*(4), 350–356. https://doi.org/10.1111/geoj.12079

Millington, N. (2018). Producing water scarcity in São Paulo, Brazil: The 2014–2015 water crisis and the binding politics of infrastructure. *Political Geography*, *65*, 26–34. https://doi.org/10.1016/j.polgeo.2018.04.007

Mollett, S., & Faria, C. (2013). Messing with gender in feminist political ecology. *Geoforum*, *45*, 116–125. https://doi.org/10.1016/j.geoforum.2012.10.009

Mountz, A. (2017). Political geography III: Bodies. *Progress in Human Geography*, *42*(5), 759–769. https://doi.org/10.1177/0309132517718642

Mustafa, D., & Talozi, S. (2018). Tankers, wells, pipes and pumps: Agents and mediators of water geographies in Amman, Jordan. *Water Alternatives*, *11*(3), 916–932.

Nagar, R., Lawson, V., McDowell, L., & Hanson, S. (2002). Locating globalization: Feminist (re) readings of the subjects and spaces of globalization. *Economic Geography*, *78*(3), 257–284. https://doi.org/10.2307/4140810

O'Reilly, K. (2006). "Traditional" women, "modern" water: Linking gender and commodification in Rajasthan, India. *Geoforum*, *37*(6), 958–972. https://doi.org/10.1016/j.geoforum.2006.05.008

O'Reilly, K., & Dreibelbis, R. (2018). WASH and gender: Understanding gendered consequences and impacts of WASH in/security. In O. Cumming & T. Slaymaker (Eds.), *Equality in water and sanitation services* (pp. 96–109). Routledge.

Ortner, S. (1984). Theory in anthropology since the 1960s in Dirks. In N. G. Eley & S. Ortner (Eds.), *Culture, power, history* (pp. 372–411). Princeton University Press.

Peloso, M., & Morinville, C. (2014). "Chasing for water": Everyday practices of water access peri-urban Ashaiman, Ghana, *Water Alternatives*, *7*(1), 121–139.

Peloso, M., Morinville, C., & Harris, L. (2018). "Water scarcity beyond crisis: Spotlight on Accra", Spotlight on series. *International Journal of Urban and Regional Research*. www.ijurr.org/spotlight-on-overview/parched-cities-parched-citizens/water-scarcity-beyond-crisis-spotlight-on-accra/

Ranganathan, M. (2016). Thinking with Flint: Racial liberalism and the roots of an American water tragedy. *Capitalism Nature Socialism*, *27*(3), 17–33. https://doi.org/10.1080/10455752.2016.1206583

Ranganathan, M. (2018). "Beyond 'third world' comparisons: America's geography of water, race and poverty", Spotlight on series. *International Journal of Urban and Regional Research*. www.ijurr.org/spotlight-on-overview/parched-cities-parched-citizens/beyond-third-world-comparisons/

Silvey, R. (2004). Transnational migration and the gender politics of scale: Indonesian domestic workers in Saudi Arabia. *Singapore Journal of Tropical Geography*, *25*, 141–155. https://doi.org/10.1111/j.0129-7619.2004.00179.x

Starr, J. (1991). Water security: The missing link in our mideast strategy. *Current World Leaders*, *34*, 571–588. https://doi.org/10.1080/02255189.1992.9669480

Strang, V. (2014). Fluid consistencies: Material relationality in human engagements with water. *Archaeological Dialogues*, *21*, 133–150. https://doi.org/10.1017/S1380203814000130
Sultana, F. (2011). Suffering for water, suffering from water: Emotional geographies of resource access, control and conflict. *Geoforum*, *42*(2), 163–172. https://doi.org/10.1016/j.geoforum.2010.12.002
Swyngedouw, E. (2004). *Social power and the urbanization of water: Flows of power*. Oxford University Press.
Thompson, J. (2016). Intersectionality and water: How social relations intersect with ecological difference. *Gender, Place & Culture*, *23*(9), 1286–1301. https://doi.org/10.1080/0966369X.2016.1160038
Truelove, Y. (2011). (Re-) Conceptualizing water inequality in Delhi, India through a feminist political ecology framework. *Geoforum*, *42*(2), 143–152. https://doi.org/10.1016/j.geoforum.2011.01.004
Truelove, Y. (2016). Incongruent waterworlds: Situating the everyday practices and power of water in Delhi. *South Asia Multidisciplinary Academic Journal*, *14*. https://doi.org/10.4000/samaj.4164
Truelove, Y. (2018). Negotiating states of water: Producing illegibility, bureaucratic arbitrariness, and distributive injustices in Delhi. *Environment and Planning D: Society and Space*, *36*(5), 949–967. https://doi.org/10.1177/0263775818759967
Truelove, Y. (2019). Rethinking water insecurity, inequality and infrastructure through an embodied urban political ecology. *WIREs Water*, *6*(3), 1–7. https://doi.org/10.1002/wat2.1342
Truelove, Y. (2021). Gendered infrastructure and liminal space in Delhi's unauthorized colonies. *Environment and Planning D: Society and Space*, *39*(6), 1009–1025. https://doi.org/10.1177/02637758211055483
Von Schnitzler, A. (2008). Citizenship prepaid: Water, calculability, and techno-politics in South Africa. *Journal of Southern African Studies*, *34*(4), 899–917.
Wutich, A., & Ragsdale, K. (2008). Water insecurity and emotional distress: Coping with supply, access, and seasonal variability of water in a Bolivian squatter settlement. *Social Science & Medicine*, *67*(12), 2116–2125. https://doi.org/10.1016/j.socscimed.2008.09.042
Zwarteveen, M., Ahmed, S., & Gautam, S. (Eds.). (2014). *Diverting the flow: Gender equity and water in South Asia*. Zubaan.

7

THE TEMPORAL FRAGILITY OF WATER INFRASTRUCTURE

Conceptualizing the gendered, affective labor of maintenance and repair

Kathleen O'Reilly, Kavita Ramakrishnan, and Jessica Budds

Introduction

In the 1990s, drinking water infrastructure for villages in Rajasthan, India, promised to reverse caste inequality through access to clean water via public standposts. Materially consisting of pipes, concrete stands, and brass taps, the public standposts and their intersection with social actors – contractors, Village Water Committees (VWCs), and caste-based neighborhoods – were expected to deliver time and labor savings in obtaining water, particularly for women. In practice, however, construction decisions were dominated by elites, who dominated VWCs. The VWCs ceased to maintain standposts as they began to decay. Taps were stolen, pipes leaked, and concrete crumbled. Maintenance increasingly fell to individual users, who sealed faucets with torn strips of cloth and collected leaking water with pots. Chips, cracks, and disintegrating concrete were not repaired. Dominant caste households capitalized on the decay to illicitly connect hoses to the standposts that they ran to their houses, depressurizing the whole system and leaving the most marginalized castes in the outermost areas with no or low flows. Women on the social and geographical margins compensated by walking to other standposts in dominant caste neighborhoods. These women and their families refrained from complaining about the illicit connections to the offending households, the VWCs, or the state, since silence was the means by which they preserved the power relations that enabled them to ask dominant caste women for water in times of need (see O'Reilly, 2006; O'Reilly & Dhanju, 2014).

This example illustrates the distinct socio-political struggles, socio-ecological arrangements, and embodied experiences that emerged during phases of infrastructural decay, repair, and maintenance. The changing material form of the standposts shaped social and institutional relations, physical decay, and particular kinds of repair work, which in turn both enabled the uneven distribution of water that differentially affected women's bodies and lives and served to sustain dominant power relations.

Recently, scholarly debates have attended to the processes generated by ongoing material ventures to build and improve infrastructure – as well as the promise that infrastructure holds for improving social worlds (Anand et al., 2018; Harvey, 2018; Larkin, 2013). Both

DOI: 10.4324/9781003100379-9

the social processes and the promise of infrastructure are demonstrated by our example of infrastructure changing over time, and how its shifting materiality is central to the production of social difference – such as intersections of gender, race, caste, class, and sexuality – power relations, and access to resources, like water. As Anand et al. (2018) observe, it is in the "different phases of infrastructure's life span – design, financing, construction, completion, maintenance, repair, breakdown, obsolescence and ruin – [where] one can see the operation of multiple temporalities and trajectories" (p. 18). Our example shows that these temporalities are more than merely moments in time; they are predicated upon, and produce, different material conditions, social norms, and gendered labor.

In this chapter, we explore the dynamic phases of infrastructure and their associated material and affective conditions, to highlight the *temporal fragility* of infrastructure and its contingent social relationships. We focus specifically on the decay, maintenance, and repair that characterize such phases of infrastructural life, on the grounds that these specific temporal phases shed light on the labor enacted by ordinary people, especially women, to secure connectivity and flows, which often goes unremarked or understudied, but which encapsulates embodied and affective experiences (Doshi, 2017; Truelove, 2019). We seek to center the body in our analysis of affect: "*what is at work*: [of] what resonates through bodies as a result of historical imbrications of material relations and [of] what these resonations can *tell us* about those relations" (Dawney, 2011, p. 599; italics in original). By attending to affect, we trace infrastructure's ability to evoke the sensorial and aesthetic, such as longing, anguish, and aspiration (Larkin, 2013; Limbert, 2001) – which are by their very nature temporal and multiple, depending on where infrastructure is in its life course and how people view it (Anand et al., 2018). For instance, the imminent construction of a new road can revive a community's confidence that they matter to the state in terms of development, while suspended or seemingly never-ending construction may evoke despair (Harvey, 2018). The various temporalities of material infrastructure can, each in their own and in overlapping ways, elicit affect, mediated by the positionality of the subject and imbricated with materiality, which in turn, changes with an infrastructure's life phases. Here, *temporal fragility* of infrastructure is intertwined with a range of embodied, situated, and sensory experiences. In turn, affective responses, such as despair over interminable delays to road construction, may inspire human labor and ingenuity to maintain, repair and/or allow infrastructure to decay, thus altering its life phase and affective attachments. In many ways, then, the temporal connects the material, affective and the social aspects of infrastructure, and so attention to temporality gives us the full weight of infrastructures in flux, and how they are mutually configured by social relations over space and time.

Our chapter explores the following two questions: when, why, how, and by whom is infrastructure maintained, repaired, or left to decay? Furthermore, what are the affective, labor, and political implications of infrastructure's shifting materialities and meanings? We seek to push calls to be specific about the infrastructure under examination, acknowledging and foregrounding the agency of materials, but always keeping this in a taut relationship with how affective responses, social institutions, and political forces remake said materials. Here, we examine the dynamics of infrastructural decay, maintenance, and repair in relation to one key dimension (of many) that have animated infrastructure scholarship: gendered, embodied labor.

The chapter proceeds as follows. The first section argues for thinking beyond infrastructure's *configurations* (Lawhon et al., 2018) and introduces our theorization of infrastructure's *temporal fragility*. We address how the material and social dimensions of infrastructure have been conceptualized in the literature, and how these inform our thinking vis-à-vis infrastructure over space and time. Our second section then proceeds to demonstrate what decay, maintenance, and repair as specific temporal-material phases of infrastructural life both obscure and reveal about one specific dimension: embodied labor. We argue for an understanding of infrastructure as dynamic materiality that co-constitutes certain affective conditions and types of embodied labor. We conclude in the last section by presenting our argument that decay, maintenance, and repair illuminate the *temporal fragility* of infrastructure and its possibilities for social justice.

Mapping infrastructure's temporality, materiality, and contingent social relations

Explicit consideration of an infrastructure's life brings together the changes to an infrastructure's material form over time and the (often unequal) embodied labor that is embedded in these transformations. Life phases identified in the literature include destruction, decay, ruination, repair, maintenance, and rebuild (Anand et al., 2018; Humphrey, 2005; Martínez, 2017; Simone, 2004). While these terms are often used to capture infrastructure not "in order" or "working to standard," collapsing these phases, or ignoring their particularities, means missing how materiality in these various phases is connected to infrastructural labor, and how fluidity and transitions between decay and repair mobilize particular affective responses and actants. For instance, writing on dam repair along the River Nile, Barnes (2017) finds that repair and maintenance are not necessarily synonymous (see also Henke, 2007), and she carefully chooses the latter to reflect informants' situated articulations. The social embeddedness of infrastructure renders understandings of repair and maintenance more complex; as Graham and Thrift (2007) contemplate, "what [is] the 'thing' that is being maintained and repaired: is it the thing itself or the negotiated [dis]order that surrounds it or some 'larger' entity?" (p. 4).

We take inspiration from Lawhon et al.'s (2018) consideration of infrastructure as "parts of geographically spread socio-technologic configurations" which bring together "relations, capacities and operations, entailing different risks and power relationships" (p. 722). However, we adopt configuration here to specifically map infrastructure's temporality alongside changes in material form, enabling us to see how each (in)decision, the affective responses and gendered labor associated with it address decay or repair, while allowing us to think about who cares for infrastructure and how this translates to equitable and just access. Our use of configuration is intentional: it is differently defined than Lawhon et al.'s (2018) "heterogeneous infrastructure configurations" so that infrastructure's temporal instability and contingent social relationships can be drawn together *and* deconstructed analytically to disclose the ethical choices made regarding decay, maintenance, and repair and the conditions under which those choices are made.

The embodied, affective labor of repair and maintenance

The value of repair and maintenance cannot be overstated, even if such work is often in the background, or if those who perform such work are overlooked, given its ongoing nature in the face of ever-encroaching decay. In an oft-cited essay on repair, Jackson (2014) proposes "broken world thinking," an ontological, methodological, and empirical approach consisting of two opposing forces – a world on the verge of falling apart and a world that is maintained and restored – that are held together by repair. Instead of consigning repair to a lesser importance than innovation, those who fix and repair see "different worlds" by virtue of fixer knowledge. The ethical also emerges in "broken world thinking" by giving weight to infrastructural labor over time (Jackson, 2014, 2015). Here, Mattern's (2018) work is also incisive, pointing to care(ful) but also exhausting, strenuous, and unremunerated labor that sustains maintenance across a variety of disciplines and fields, and the scales it collapses, from the intimacy of the home to large, public infrastructures. She concludes with the following questions: "Who gets to organize the maintenance of infrastructure and who then executes the work? Who gets cared for at home, and who does the tending and mending?" (Mattern, 2018, n.p.). Those who do the critical work of repairing need care too, bridging the affective with the material and furthering a feminist political commitment to excavating and tending to those who mend. The work of maintenance and repair and those who do it are often out of sight, and yet, is critical to the functioning of infrastructure and its contingent social relations (and in fact, may be creating those social relations so that infrastructure functions). As such, re-centering the care of infrastructure, and the care of those who care for it, may provide important avenues for justice and social change.

While materiality is important to our theorization as discussed, our interest in temporality goes beyond crumbling stone or corroding pipe over time, and instead looks to the ways in which materiality creates, sustains, and is in turn shaped by, affective encounters: from coping with infrastructure demolition and decay, managing expectations when infrastructures do not function as "normal," and to everyday exclusions that infrastructure (in need of repair) creates. Furthermore, multiple temporalities that emerge through an infrastructure's life help us think of living among, between, and within emergent and ongoing phases.

Phases of infrastructural life can be episodic, overlapping, and/or continuous, opening up multiple trajectories of (in)activity, (in)decision, acceleration or delay. Consider repairs made by the urban poor to housing. The decision to shift from *kaccha*, or informal building materials, to more *pukka* (permanent) forms in India, is deeply connected to security of tenure, which is often dependent on provision of other infrastructural connections by the state (Ramakrishnan, 2014). A tarpaulin roof may be replaced by tin, proper doors installed, and walls painted if the immediate or short-term threat of eviction and displacement is considered minimal. Such decisions can gather speed or decelerate due to shifting political climates, environmental degradation, and threats of development and/or forces of gentrification. As Bhan (2019) writes on the Indian context, "[y]ou inhabit and build both incrementally and simultaneously: brick by brick, one layer at a time, moving forward but also sometimes falling behind" (pp. 645–646), thus suggestive of temporalities beyond the linear. Gupta (2018) suggests that conceptualizing infrastructure through a "teleological timeline" resulting in completion misses the various social and political interruptions that affect its installation. Housing among the urban poor and its resultant trajectories demonstrate that an "end" to infrastructure is illusory, and instead, even after the basic structure is constructed, ordinary people may prolong the labor of incremental additions (Lemanski, 2020) or utilize materials of lesser quality during times of (greater) economic uncertainty.

Temporal fragility invokes the visceral and sensory experiences that are generated in between and during phases of an infrastructure's life. If material forms themselves are unstable and unpredictable over time, so too are the surrounding affects that entangle people and infrastructures; simultaneously, the labor that decay, repair, and maintenance demands/evokes shifting, fleeting, and multiple affects. But this labor is often punishing and exploitative. Doshi (2017) argues that struggles over nature exact a bodily toll, and that attention to social reproduction, visceral experiences, and subjectivities can redress the under-theorization of physical embodiment, particularly in urban geography. Here, the ability to "rescript" or "refashion" shifting materialities is intimately linked to the body, and the ways in which affect and infrastructural labor are mutually co-constituted through the physical, psychosocial, and sensorial, have bodily implications. Dawney (2011) offers a definition of affect as entangled material relations resonating through bodies, and in turn, these resonations inform us about the histories and strength of such relations. Building on this, we explore the multiple affective registers at play as affect moves across and connects infrastructural life, labor, and materiality to the body: this includes the physicality and gendered toll; the barriers and risks involved and the anticipation of what is to come when labor is performed, compounding feelings of neglect and marginality, or moving people in more positive ways.

In order to amplify the affect that moves through and between laboring bodies, we thus insist on attention to embodiment, similar to Doshi (2017), a term we use to capture not only visceral experiences and the physical body but also, as Mountz (2018) explains, "where power operating at larger scales is understood" (p. 762). "[T]he shared *relations* that format the materialities of city, [global] capital and body as one" (footnote p. 501, italics in original) come into view through Chalfin's (2019) examination of infrastructural configurations, consisting of affects (such as hope and desire), the physical labor of waste-sorting, and interaction with material decay – in this case, a dumpsite. What becomes visible through the lens of embodiment are configurations of decay, maintenance, and repair that encompass affective responses to physical labor and materiality. Sultana (2020) also recognizes the embodied, affective experience of interfacing with infrastructure. In the center of Dhaka, the embodied labor constantly expended by poor women to access water supply is one suffused with stress and a heightened sense of exclusion from full urban citizenship. Living in slums that are surrounded by more affluent neighborhoods, these women inhabit bodies that labor under conditions of socio-spatial and intersecting inequalities – bodies that become sites where political agency is formed when women devote their unpaid labor to community organizations to secure water (see also Alda-Vidal et al., 2023). Sultana identifies this unpaid, gendered labor as the "externalization of costs" onto women that occurs when men will not pay to secure water access and women must compensate to ensure quality and quantity of water, through the creation of social networks. In Lilongwe's (Malawi) informal settlements, women collaborate to help each other get unreliable water, *and* these collaborations can serve to ease women's psychosocial stress due to strictly gendered norms regarding its collection (Adams, 2023). Here, an excavation of how exactly infrastructural decay, maintenance, and repair maps onto embodied labor can also show how it reshapes the physical and affective life-worlds of individuals facing intersecting inequalities (see also Truelove & Ruszczyk, 2022).

For Schwenkel (2015), affect, labors of maintenance and repair, and gender became inextricably linked during a period of infrastructural breakdown in state-neglected, socialist-era housing in Vietnam. The loss of hydraulic pressure led to a breakdown in water infrastructure, serving not only to reinforce gendered divisions of labor but also to rework gender in important ways. Schwenkel argues that a "collective ethos" of maintenance emerged among both men and women due to this physical "labor of care" that included resource redistribution. Returning to our earlier discussion of infrastructural care (Jackson, 2014, 2015), Mattern (2018) argues that "care" may have less to do with the reproduction of societal value and more to do with the affective responses to maintenance. This calls for understanding where care (as a labor practice) fits with maintenance and repair, and who decides what deserves care, when, and at what juncture (cf. Martin et al., 2015).

The new connections, redistributions, and subjectivities that emerge when maintenance work upends certain gender and social relations may be one mode of understanding infrastructural labor and care. Anand et al. (2018) similarly note that the investment of "labor and care into everyday maintenance and repair make more-than-human assemblies of infrastructure that are generative of differentiated materializations of rights, resources and aspirations . . ." (p. 12). An apt illustration of rights materialized through infrastructural labor comes from Frederick's (2014) study of solid waste collectors in Dakar. Shifting government policies meant that waste infrastructure became displaced more firmly onto bodies and marginal groups such as poor women and youth. Fredericks (ibid) argues that "the materiality of the labor process matters" (p. 539) as waste-workers bear the brunt of demanding physical work, the threat of resultant diseases, and the stigma of working with garbage. Through mobilization, the collectors achieved visibility of their work in the city and shaped a local understanding of its worth, placing pressure on the city government to create safer infrastructure. Fredericks's situated ethnography demonstrates that affects – in this case, stigma and worth – are an important aspect of infrastructural configurations, revealing how subjectivities are reconfigured and political power increased for those on the margins, or a "vibrant politics of refusal to be refuse" (p. 533).

And yet, it is not always the case that infrastructural systems in need of maintenance or repair are tended to or cared for. What happens when care for infrastructures is suspended and, instead, decay is embraced? The materiality of labor is not solely relegated to the realm of maintenance and repair, as decay also requires certain affective and embodied engagements. Again, Schwenkel's (2015) work is illustrative. Residents held the state ultimately responsible for repair, and thus they ignored issues of mold and water seepage, covered up holes and cracks, and tolerated putrid smells from water leaks. As the earlier case of gradual housing maintenance and repairs among the urban poor also showed, choosing not to build, or to let decay, materially symbolizes the expectations ordinary people hold of the state for service provision and the forms of infrastructural integrity they considered acceptable (see also Menon, 2023; O'Reilly & Budds, 2023). More spectacular forms of infrastructural decay include solid waste workers strikes in Dakar, that "deploy the power of dirt to creatively subvert ordering paradigms" (Fredericks, 2014, p. 542), providing a stark, sensory confrontation with government officials through the multiple affective and political effects of rotting garbage.

Highlighting the entanglement of embodied labor practices and their affects with infrastructural decay, maintenance, and repair, also subverts commonly held assumptions about the knowledges "required" to fix infrastructure. Star (1999) argues that infrastructure is a set of learned practices to which outsiders must become accustomed; it is bound by "conventions of practice"; and, it is fixed modularly, so that "[n]obody is really in charge of infrastructure" (p. 382). Thus, although there are communities of practice that loosely delineate infrastructure and its use, repair and maintenance work remains specific to one part of the system (or at a specific scale), given infrastructure's inherent complexity. So-called "experts" may be ultimately "responsible" for repair work, especially in lieu of the state, but uncertainties over materiality, actual scale of usage due to illicit activities, and extent of informal improvisations made to systems by users, may make technical fixes superfluous or the ability to control an infrastructural system difficult. This informality of repair resonates with the work of Alda-Vidal et al. (2023), who trace the gendered labor associated with fragile sanitation infrastructures. Women in the poor areas of Lilongwe enact the labor of repair and maintenance, especially preventative measures to stave off infrastructure failure. In their gendered roles as those who must protect their families, women undertake repairs and care as non-experts, while paradoxically being blamed by city officials for causing infrastructure breakdowns to occur. For these women, infrastructure failure, or the need to prevent it, drives an intensification of gendered labor or increased financial expenditures in order to secure connectivity and flows in the decaying system. "Maintenance here," Alda-Vidal et al. (2023) write, "follows the logic of endurance and survival, of quick fixes and short-term adaption." Social pressures to maintain a healthy family as a good homemaker motivate women to self-exploit in their efforts to avoid sanitation infrastructure failure in the domestic sphere. Women are expected to know and anticipate what is needed to keep infrastructure functioning, while gender norms stipulate that they are not likely to be the ones making decisions that would prevent infrastructure failures. The dynamic materiality of infrastructure is imbricated in the normalization of women's daily labor to support life under precarious conditions.

Those who repair and maintain also help us think through infrastructure beyond design to the forms of work that are normally hidden (Jackson, 2015), and to the instances where repair and maintenance constitute an important part of livelihoods (Cross & Murray, 2018). For Truelove and Ruszczyk (2022), it is women's bodies that compensate for missing or poorly functioning infrastructure, and in so doing play a critical role in the building, development and maintenance of cities. Although women's bodies may be invisible or overlooked at the scale of the city, it is women performing infrastructural labor at the household scale that secure critical resources like water and simultaneously subject them to "slow infrastructural violence" through loss of time and income opportunities (Truelove & Ruszczyk, 2022). Women acting as infrastructure – through physical labor and networks of care that help critical resources reach their households – face emotional and physical harm that reinforces intersecting inequalities. Our conceptualization of infrastructure as the temporal and material entangled with affect, in turn, calls for attention to infrastructural ethics, and the uneven responsibilities and burdens, as the labor of repair and reassembly marks some bodies more than others (cf. Fredericks, 2014).

In sum, we draw on scholarly works that relate to infrastructural phases of decay, maintenance, and repair to argue that embodied labor unsettles notions of who cares for infrastructure over time, even if – and sometimes especially – this work is seemingly mundane. Taken together, infrastructure's shifting configurations demonstrate how at varied points and scales of an infrastructure's life course, some bodies are able to thrive, whereas other bodies are diminished. As the examples we have gathered show, often it is the bodies of women and girls that suffer when infrastructures fail, need maintenance, or must be compensated for. Affective responses, such as fear, frustration, and aspiration, permeate these configurations, demonstrating how inequality and difference are in turn shaped by what infrastructures in flux demand of communities and individuals.

Conclusions

These social, political, and temporal realities and potentialities inform our conceptualization of infrastructure's *temporal fragility*, offering a fruitful pathway to recognize the intimacies, histories, and imaginaries that connect laboring bodies to infrastructure. *Temporal fragility* still invokes "temporal openness" (Gupta, 2018) and means that practices of repair and maintenance often shift in anticipation of emergent or transformed infrastructural states. While fragility is an inherent paradox of infrastructure, as noted by Star (1999) who writes that the "system is necessarily fragile . . . depending on local and situated contingencies" (p. 387), it is simultaneously a window into infrastructural life and gendered labor-worlds. *Temporal fragility* then, is not about describing a particular state or condition, but rather observing, mapping, and responding to the continual repatterning of infrastructural forms, relations, and social worlds. We argued earlier that the power of thinking with infrastructural configurations lies in what it can define, hold to account, and bring to the fore: configurations gesture to specific material forms and their ongoing processes of decay, maintenance, and repair, in turn drawing our attention to the affective conditions and embodied labor that marks these phases, and unsettles people's relations with the state, the city, their neighborhood, and their household. *Temporal fragility* is intimately linked to our usage of configurations because it speaks to the emergent and dynamic nature of infrastructural materiality – as people rework, resist, and reject infrastructural determinacy and official scripts on how infrastructure should and can be used (Lemanski, 2020).

While the focus of our chapter has been on the ways in which people *do* realize, encounter, challenge, and even accept, fragility vis-à-vis decay with routines of repair and maintenance, the emphasis on fragility highlights the imbrication of care(ful), affective, and embodied labors with material form. Thus, the intimate engagement ordinary people have with infrastructure rests upon its inherent dynamic nature – its *temporal fragility* – in any given context. Through an infrastructure's life phases, we are able to see the *longue durée* of infrastructure, speaking to ways in which decay, maintenance, and repair have marked and continue to mark bodies and subjectivities. We take seriously the need to identify the life phases of infrastructure (though not the forms of infrastructure under investigation itself), not least to understand how power and difference become reinforced and reinscribed on material forms, livelihoods, and landscapes (see also Ramakrishnan & O'Reilly, 2023). Here, thinking with configurations becomes productive as a way to center temporality but also to account for the relational dynamics between power and labor.

Our reworking of Lawhon et al.'s (2018) term "configurations" to include infrastructure's *temporal fragility* enhances its conceptual utility by enabling the placement of the temporal and material side by side, thereby revealing a shifting, parallel relationship that can account for the contingent politics of affective responses and physical labor. *Temporal fragility* is integral to our usage of configurations because it encompasses shifts in cooperation, contestation, and resistance – within a context of systemic power relationships – and dynamic materiality. Over infrastructures life phases, processes of decay, maintenance, and repair can be used to trace the multiple meanings associated with these phases and the embodied labor used to make infrastructure function as suits immediate purposes, wider political agendas, and alternative socio-ecological orders. As infrastructure's materiality changes over time, infrastructure can be enrolled in political change, and as embodied experiences and struggles for access shift, openings created by infrastructure's *temporal fragility* can be leveraged. For if infrastructure's materiality is central to the production of social difference, then, infrastructure's *temporal fragility* is also central to new configurations of infrastructural labor and affective responses that anticipate social justice. Decay, maintenance, and repair delineate the repatterning of power geometries, and associated embodied and affective experiences.

Taking this forward, the task that arises from discussing infrastructure through this lens is to more fully engage with temporality and materiality as situated and emergent opportunities to reimagine political claims and futures. The global COVID-19 pandemic has made ethical engagement all the more pressing, especially given the ways in which some infrastructures have been left to decay by various state actors, leading to a prioritization of certain lives over others but also through the emergence of new, progressive political demands for infrastructural care and repair. *Temporal fragility* offers an important framing for how we understand infrastructures experienced in the everyday, on the brink of "failure," or during repair. Returning to the above question of who decides what deserves care, and at what juncture, our framework attends to ever-changing reworkings, where significance is given to the maneuvers, maintenance, and manipulations that stymie or halt decay. Of course, social actors from state agencies to individuals can decide that infrastructures do not deserve care and supported decay can generate positive affects – related to agency and resistance – thus demonstrating an indeterminacy in ways of perceiving and engaging with infrastructure. And yet, *temporal fragility* makes infrastructure more epistemologically grounded even as its materials change, rendering embodied labor, affective encounters, and political processes more visible.

Note: This chapter has been adapted from a previously published article: Ramakrishnan, K., O'Reilly, K., and Budds J. (2021) The temporal fragility of infrastructure: Theorizing decay, maintenance and repair. *Environment and Planning E: Nature and Space* 4(3), 674–695.

References

Adams, E. A. (2023). "Why should a married man fetch water?" Masculinities, gender relations, and the embodied political ecology of urban water insecurity in Malawi. *Social & Cultural Geography*. https://doi.org/10.1080/14649365.2023.2183245

Alda-Vidal, C., Lawhon, M., Iossifova, D., & Browne, A. L. (2023). Living with fragile infrastructure: The gendered labour of preventing, responding to and being impacted by sanitation failures. *Geoforum*, *141*, 103724.

Anand, N., Gupta, A., & Appel, H. (2018). *The promise of infrastructure*. Duke University Press.

Barnes, J. (2017). States of maintenance: Power, politics and Egypt's irrigation infrastructure. *Environment and Planning D: Society and Space*, *35*(1), 146–164.

Bhan, G. (2019). Notes on a Southern urban practice. *Environment and Urbanization*, *31*(2), 639–654.

Chalfin, B. (2019). Waste work and the dialectics of precarity in urban Ghana: Durable bodies and disposable things. *Africa*, *89*(3), 499–520.

Cross, J., & Murray, D. (2018). The afterlives of solar power: Waste and repair off the grid in Kenya. *Energy Research and Social Science*, *44*(2018), 100–109.

Dawney, L. (2011). The motor of being: A response to Steve Pile's "Emotions and affect in recent human geography". *Transactions of the Institute of British Geographers*, *36*(4), 599–602.

Doshi, S. (2017). Embodied urban political ecology: Five propositions. *Area*, *49*(1), 125–128.

Fredericks, R. (2014). Vital infrastructure of trash in Dakar. *Comparative Studies of South Asia, Africa and the Middle East*, *34*(3), 532–548.

Graham, S., & Thrift, N. (2007). Out of order: Understanding repair and maintenance. *Theory, Culture and Society*, *24*(3), 1–25.

Gupta, A. (2018). The future in ruins: Thoughts on the temporality of infrastructure. In N. Anand, A. Gupta & H. Appel (Eds.), *The promise of infrastructure* (pp. 62–79). Duke University Press.

Harvey, P. (2018). Infrastructures in and out of time: The promise of roads in contemporary Peru. In N. Anand, A. Gupta & H. Appel (Eds.), *The promise of infrastructure* (pp. 80–101). Duke University Press.

Henke, C. (2007). Situation normal? Repairing a risky ecology. *Social Studies of Science*, *37*, 135–142.

Humphrey, C. (2005). Ideology in infrastructure: Architecture and Soviet imagination. *Journal of the Royal Anthropological Institute*, *11*(1), 39–58.

Jackson, S. (2014). Rethinking repair. In T. Gillespie, P. Boczkowski & K. Foot (Eds.), *Media technologies: Essays on communication, materiality and society* (pp. 221–239). MIT Press.

Jackson, S. (2015, September 24). Repair. Theorizing the contemporary: The infrastructure toolbox. *Cultural Anthropology*. https://culanth.org/fieldsights/repair
Larkin, B. (2013). The politics and poetics of infrastructure. *Annual Review of Anthropology*, *42*, 327–343.
Lawhon, M., Nilsson, D., Silver, J., Ernstson, H., & Lwasa, S. (2018). Thinking through heterogeneous infrastructure configurations. *Urban Studies*, *55*(4), 720–732.
Lemanski, C. (2020). Infrastructural citizenship: The everyday citizenships of adapting and/or destroying public infrastructure in Cape Town, South Africa. *Transactions of the Institute of British Geographers*, *45*(3), 589–605.
Limbert, M. (2001). The senses of water in an Omani town. *Social Text*, *68*(19), 35–55.
Martin, A., Myers, N., & Viseu, A. (2015). The politics of care in technoscience. *Social Studies of Science*, *45*(5), 625–641.
Martínez, F. (2017). Waste is not the end. For an anthropology of care, maintenance and repair. *Social Anthropology*, *25*(3), 346–350.
Mattern, S. (2018, November). Maintenance and care. *Places Journal*. https://placesjournal.org/article/maintenance-and-care/
Menon, S. (2023). Class, caste, gender, and the materiality of cement houses in India. *Antipode*, *55*(2), 574–598.
Mountz, A. (2018). Political geography III: Bodies. *Progress in Human Geography*, *42*(5), 759–769.
O'Reilly, K. (2006). "Traditional" women, "modern" water: Linking gender and commodification in Rajasthan, India. *Geoforum*, *37*(6), 958–972.
O'Reilly, K., & Budds, J. (2023). Sanitation citizenship: State expectations and community practices of shared toilet use and maintenance in urban India. *Environment and Urbanization*, *35*(1), 238–254.
O'Reilly, K., & Dhanju, R. (2014). Public taps and private connections: Neoliberal water governance and the re-production of caste distinction in north India. *Transactions of the Institute of British Geographers*, *39*(3), 373–386.
Ramakrishnan, K. (2014). Disrupted futures: Unpacking metaphors of marginalization in eviction and resettlement narratives. *Antipode*, *46*(3), 754–772.
Ramakrishnan, K., & O'Reilly, K. (2023). Mapping affective infrastructures otherwise. *Dialogues in Human Geography*, *13*(1), 91–94. https://doi.org/10.1177/20438206231151431
Schwenkel, C. (2015). Spectacular infrastructure and its breakdown in socialist Vietnam. *American Ethnologist*, *42*(3), 520–534.
Simone, A. (2004). People as infrastructure: Intersecting fragments in Johannesburg. *Public Culture*, *16*(3), 407–429.
Star, S. L. (1999). The ethnography of infrastructure. *American Behavioral Scientist*, *43*, 377–391.
Sultana, F. (2020). Embodied intersectionalities of urban citizenship: Water, infrastructure, and gender in the global south. *Annals of the American Association of Geographers*, *110*(5), 1407–1424.
Truelove, Y. (2019). Rethinking water insecurity, inequality and infrastructure through an embodied urban political ecology. *WIREs Water*, *6*(3), e1342.
Truelove, Y., & Ruszczyk, H. A. (2022). Bodies as urban infrastructure: Gender, intimate infrastructures and slow infrastructural violence. *Political Geography*, *92*, 102492.

PART 2

Revisiting water debates

Diplomacy, security, justice, and heritage

8

HOUSEHOLD WATER SECURITY EXPERIENCES OF WOMEN AND GIRLS IN RURAL GHANA

Benjamin Dosu, Mohammed Abubakari, Maura Hanrahan, and Tom Johnston

Introduction

In Sub-Saharan Africa (SSA), gender inequalities tend to be grounded in the culture, norms, and societal expectations of communities (Adjiwanou & LeGrand, 2014). Patriarchal formations, that is, context specific masculine forms of domination "based on the figure of the providing father" (Ojeda et al., 2022, p. 150), tend to be deeply entrenched in rural SSA, where women and girls are tasked with the bulk of or all domestic duties (Jeil et al., 2020) including food preparation, water collection, cleaning, laundry, and waste collection and disposal.

In rural Ghana, women are symbolically linked to three elements of traditional culture at the household level – "earth," "fire," and "water" – respectively representing childbearing and associated responsibilities, food preparation to meet household needs, and water collection for household use (Mazrui, 2014). Some have shown how the status of a rural Ghanaian woman depends largely on her ability to live up to her primary responsibilities for childbearing and home management (Boahene, 2013). Focusing on the element of "water," women and girls are the primary water collectors and domestic users in most rural communities (Lucier & Qadir, 2018; Harris et al., 2017). Water is woven into the daily schedules of women and girls and is inextricably connected to their domestic responsibilities (Van Houweling, 2016). This makes gender norms significant in determining water access, collection, use, and management, as well as the potential outcomes of water projects (Sultana, 2009). Water security in rural Ghana is labor-intensive mainly due to fragile water infrastructure and limited local capacity to provide, maintain and manage safe and reliable potable water services (Sun et al., 2010; United Nations Development Programme [UNDP], 2015). This situation tends to exacerbate women's conditions and struggles, hampering their welfare and well-being. For these reasons, it is essential to understand rural domestic water practices through a gender lens.

This chapter presents survey data gathered in collaboration with the National Community Water and Sanitation Agency (CWSA) and two District Assemblies (decentralized agencies at the local level). Both District Assemblies – Sunyani West District and Sekyere Kumawu District – are rural districts in Ghana with poor water coverage and were randomly selected from the database of the CWSA. The purpose was to explore gender inequalities in domestic household activities focusing on the water security experiences of women and

DOI: 10.4324/9781003100379-11

girls in rural Ghana. Water security is defined as the availability of sufficient quantities of potable water through community-preferred methods and the active involvement of men and women in water collection activities and water management decisions (Goldhar et al., 2013; Global Water Partnership, 2000). Little research has been conducted on this topic in rural African settings, especially in rural Ghana, although rural dwellers are five times more likely to be water-insecure than those in urban areas (Baur & Woodhouse, 2009).

The following section describes the methodological approaches used to capture and analyze the data that feeds the chapter. This is followed by the results section, which uses the household as the unit of analysis and focuses on routine domestic gendered roles in water collection and management. The results section is presented, based on these domestic gender roles and how they affect women and girls compared to men and boys. The chapter concludes with some recommendations on how the entrenched gender inequalities in domestic water security could be dealt with in rural contexts.

Materials and methods

This chapter is based on information gathered in three rural communities. Although coverage does not guarantee water security, the existence of data on rural water coverage offered the best possible means of identifying communities with or without a drinking water facility (Dosu, 2021a; Dosu & Hanrahan, 2021). We selected communities with challenges in drinking water security according to their Medium-Term Development Plans[1] (2015–2019): Esereso and Wabrease were selected in Sunyani West District (Bono region), and Wioso was selected in the Sekyere Kumawu District (Ashanti region) (Figure 8.1).

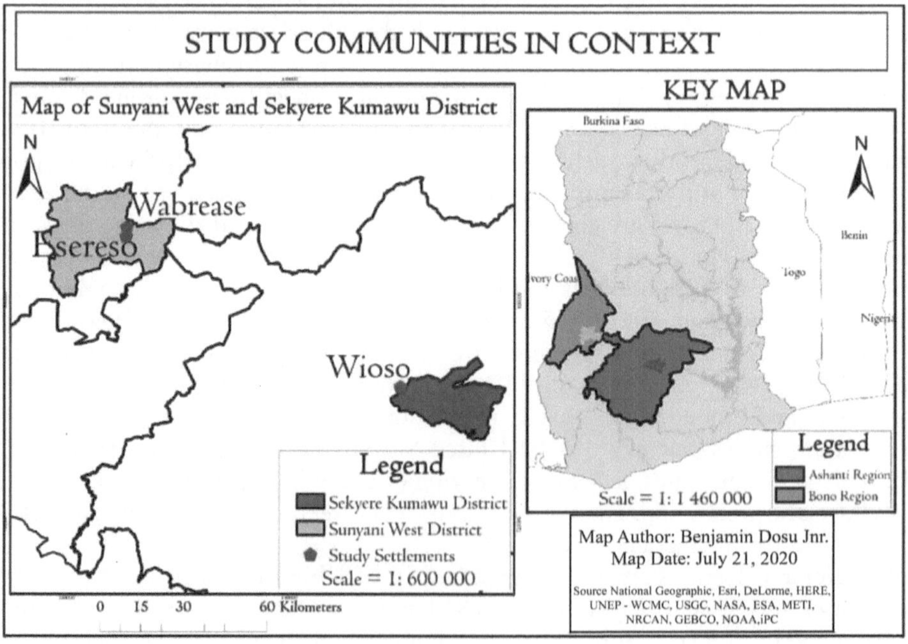

Figure 8.1 Locations of the study communities.

Source: The authors.

These three communities are composed of 276 households, making up a total population of 1,428. A census approach was deemed fit for recruitment purposes. This approach provided an opportunity for all households to be sampled to achieve a desirable level of precision and helped to eliminate sampling error (see Israel, 1992). To be included in the sample, potential participants had to have been permanent residents of one of the study communities for at least 12 months before the date of the survey and had to be at least 18 years old, the age of majority in Ghana (Dosu, 2021b; Dosu & Hanrahan, 2021). The participants were selected to try to ensure gender balance. Household data were collected through a cross-sectional survey in two phases. Phase 1 involved the deployment of a common survey instrument. The survey instrument contained both closed and open-ended questions and focused on demographic and socio-economic household characteristics and experiences of water (in)security. Data from 158 households accounting for a total number of 847 individuals were captured using the Census and Survey Processing System (CSPro), a data collection software.[2]

Phase 2 entailed face-to-face, in-depth interviews with 19 selected households. These were conducted to detail the survey data (after Staddon et al., 2018) and were recorded after obtaining participants' consent. Although a fifth of the survey households (31) were targeted for the in-depth interview, 19 of them consented to participation. The selection of 19 households was based on key issues relevant to the study: gender participation in water collection, female-headed households, and men living alone were identified during phase one. This consideration gave a snapshot to identify and compare in a more descriptive manner how gender permeates and manifests in household water (in)security experiences. On this basis, the interviews focused on experiences in water collection, use, and management.

The data obtained were analyzed using both quantitative and qualitative methods. The IBM SPSS Version 25 was used to support quantitative data analysis. This made it possible for descriptive statistics to be used to examine the socioeconomic characteristics of the participants. To look for differences and associations, the sample data were subjected to inferential statistical tests. For qualitative data, NVivo 12 Pro software was used. The sorting of data enabled coding into categories based on what constitutes household water security. Recorded data were transcribed and presented in the form of discussions, either as a standalone or in support of the quantitative analysis.

Results

The communities that participated in the study – Esereso, Wabrease, and Wioso – are diverse in terms of ethnicity. Although the Akan people[3] are dominant in all three, due to their geographical location, the survey identified other ethnic groups, such as Frafra, Dagaati, and Lobi. Esereso has two hand pump boreholes but only one remains operational. This operational borehole which the Sunyani West District Assembly constructed is unreliable as it breaks down frequently.

Wabrease has one reliable handpump borehole (also constructed by the Sunyani West District Assembly) but owing to the large number of households that rely on it, including households from neighboring hamlet villages, the water source is heavily taxed. The study gathered that the use of Wabrease's handpump borehole exceeds the Ghanaian Government's maximum threshold of 300 people per handpump borehole (Ministry of Environment, Science and Technology, 2011) by a margin of about 140 people.

Wioso has two handpump boreholes provided by the Sekyere Kumawu District Assembly but only one serves as the community's primary source of drinking water and was found to have utilization rates exceeding government guidelines by a margin of about 100 people. The second one is defunct and unused by the community members. The three communities are predominantly peasant and subsistence farmers, relying on seasonal incomes primarily from selling surplus outputs from farm yields. As a result, they cannot afford water, which affects the extent to which they can access safe water from the handpump boreholes that are functional in the communities.

Household water collection in all three communities is done throughout the day and depends mainly on water needs during a particular time of the day. However, most households (61%) collect water at dawn and in the evenings so that water is available for daily chores, meal preparations, and washing at the end of the day. The collection of water at dawn and in the evenings was associated with the risks of snake and scorpion bites, as attested by the respondents. In 79% of the households, only women aged 19 years old and above participate in water collection. Furthermore, about 81% of all girls in the sampled households, between the ages of six and 18, participate in water collection activities compared to half the number of boys in the same age category.

The studied households spend an average of 108 minutes per trip, covering 975 meters per round trip on average for water collection. In addition, households do five trips per day on average for water collection. Over 46% of the households exceed this average.

The study found that while men and boys sometimes use bicycles and motorcycles for water collection, women and girls collect water by walking and carrying loads of water on their heads. Based on observations made in the field, a wide array of vessels was used to haul water, with most of the collection containers having a capacity between 15 to 30 liters. Most female study participants said they would not ride a bicycle or motorcycle to collect water mainly because they do not know how to ride one.

The study also found that about a quarter of the households meet the basic requirement of water access within the 30 minutes benchmark as prescribed by the World Health Organization (see WHO/UNICEF, 2017). The quantity of water required usually depends on the household size, as a large household size tends to consume more water. With an average household size of five, an average of 250 liters is required for each household per day to meet basic water requirements (see World Health Organization, 2011).

To ascertain whether women spend a significant amount of time collecting water in comparison to men, a statistical test – One-way Analysis of Variance – was used to evaluate the time difference between gender participation and daily household hours spent on water collection between adult women and men (aged 19 and above). The independent variable, households water collection by gender, had three levels: households with only men's involvement, households with only women's involvement, and households with both men and women. The mean and standard deviation values for each group are shown in Table 8.1.

The test results, $F\ (2, 154) = 4.164$, $p = .017$, show a statistically significant difference between the hours spent on water collection by women compared to men. Since the variances among the three groups were equal, Tukey's test was used for post hoc comparisons to evaluate the differences among the means. This identified a statistically significant difference between the average time spent by households with only women's participation and that spent by households with both men's and women's participation in water collection. In addition, the analysis of the daily distance traveled for water collection found a statistically

Table 8.1 Descriptive Statistics for One-Way ANOVA

Type of Analysis	*Test Used*	*Variables*	*Frequency*	*Mean*	*SD*
Gender involvement in water collection (Age 18>)	One-Way ANOVA	Men's involvement	8	.4823	.13
		Women's involvement	124	.7084	.23
		Both men's and women's involvement	25	.7525	.263
Total			**157**	**.7039**	**.24**
Gender distance (in metres) Covered in water collection		Men	8	.488	.16
		Women	124	.646	.27
		Both men and women	25	.751	.32
Total			**157**	**.655**	**.28**

Source: Field Data (2019).

significant difference, F (2, 154) = 4.164, p = .047, in the daily distance in kilometers covered by women for water collection as compared to men in the same households. In summary, women spend more time retrieving water and covering a longer distance for water collection than men.

The study recorded an average of 12 hours daily per household for domestic chores, including the time on water collection. The analysis further shows that for every household, girls (six to 18 years) and women (19 years and above) spend an average of five hours each on domestic activities, including daily water collection.

A Pearson r correlation which was applied to examine the relationship between household water collection time (M = 7.1, SD = 6.8) and time for domestic chores (M = 12.4, SD = 8.2), shows a significant positive correlation, r = .85, p < .001 (two-tailed), indicating that water collection time is associated with the overall time spent on domestic activities. Roughly 72% of the variability in time for domestic activities is predicted by knowing the time for household water collection, or vice versa. As women and girls spend more time on water collection, the average time allocated for daily domestic activities tends to increase.

Two factors influence men's participation in water collection; the first is household status, so men who live alone are required to collect water on their own. The second factor is occupation, with some men participating in water collection for farming purposes. There were mixed views regarding men's involvement in water collection. While some women wanted to be supported in domestic chores, including water collection, others argued that it was their responsibility to carry out those duties. Some women regarded those who allowed their spouses to be involved in water collection as irresponsible.

Since boys usually "graduate" from water collection activities during their teenage years, the study also assessed gender involvement in water collection by children aged six to 18 years. This age category represents the school-going ages from basic to high school in Ghana (Ghana Statistical Service, 2013). A chi-square test was conducted to evaluate the differences in gender participation in water collection between boys and girls (aged six to 18). The results show a statistically significant difference between the participation of boys and girls in water collection $\chi 2(1, N = 333) = 37.077$, p < .001. A follow-up test was also conducted to assess the statistical difference between age groups within both boys and girls in water collection. The study categorized respondents into age groups six to ten, 11 to 15, and 15 to 18 years (see Table 8.2).

Table 8.2 Boys and Girls Participation in Water Collection by Age Group (Aged 6–18 Years) Cross-tabulation (N = 333)

Participation in Water Collection		*Age Group*			*Total*
		6–10	*11–14*	*15–18*	
Boy-child participation	Count	48	28	13	89
	Expected count	33.1	27.0	28.5	89.0
Girl-child participation	Count	42	46	38	126
	Expected count	46.9	38.2	40.9	126.0
Boy-child no participation	Count	23	21	44	88
	Expected count	32.8	26.7	28.5	88.0
Girl-child no participation	Count	11	6	13	30
	Expected count	11.2	9.1	9.7	30
Total	Count	124	101	108	333
	Expected count	**124**	**101**	**108**	**333**

Source: Field Data, 2019.

These groupings were based on the age categories of primary, junior, and senior high schools in Ghana, respectively. A significant chi-square statistic was obtained, $\chi2(6, N = 333) = 32.378, p < .001$. Follow-up z-tests of column proportions using the Bonferroni correction found that boys who are not involved in water collection are significantly higher among the age group 15 to 18 years as compared to the rest of the age groups between boys and girls, $\chi2(6, N = 88) = 16.843, p = .01$. The analysis implies that though boys participate in household water collection, girls tend to participate more than boys. In addition, boys, unlike girls, are excluded from water collection after they turn 15 years. Despite their significant involvement in water collection and use, women have limited participation in rural water management and decision-making. Even though community water management committees are mandated to include at least one-third of female membership, the study found that women's roles are limited to mobilizing other women for maintenance purposes (e.g., sweeping and weeding around water facilities) with limited influence in major management decisions.

Discussion

Women and girls are primarily responsible for water access and use in most rural households and walk to and from the source carrying water in containers placed on their heads (Masanyiwa et al., 2015). As Van Houweling (2016) noted, being a good wife in some traditional societies in SSA is tied to the ability to perform domestic tasks, including providing water for the household. Similar to other studies (e.g., Baguma et al., 2013), this research found that household size usually determines the number of daily trips required for daily water collection. With an average of five persons per household, women, and girls undertake about five trips per day, covering over 3.7 kilometers for water collection. Since distance is associated with time, an estimate of one hour and 48 minutes per trip implies that households with five people require 9 hours to collect the daily water requirement.

The study further reveals that water collection time for women and girls is correlated with the time for other domestic work, implying that improvement in water access does not affect gender roles in water access and use but rather provides additional time for women to be involved in other domestic activities. Aligned with other studies (Harris et al., 2017; Karim et al., 2013), this study identified men's involvement in water to be primarily associated with income-generating activities such as irrigated farming and other productive uses or by those who live alone. Although the study found that men were primarily involved in income-generating activities such as farming, the study also revealed that women play active roles in these activities despite being limited by the involvement of domestic activities.

Furthermore, men using bicycles and motorcycles significantly reduce the water collection burden compared to women (Masanyiwa et al., 2015). Generally, men consider their involvement in water collection for domestic uses and household chores as extending "a helping hand" to their spouses rather than as a responsibility (Van Houweling, 2016). Besides, Wallace and Coles (2005) argue that others often disparage men who help in water collection and other household-related tasks, which disincentivizes men's involvement in such activities. Given this, the study found that some women do not encourage men's participation in water collection.

Nauges' (2017) and Van Houweling's (2016) findings suggested that in many rural communities, boys, just like men, exclude themselves from domestic water collection activities in their teenage years. This study went further, finding that such exclusion generally occurs between the age of 15 and 18 years old, when they become enrolled in either high school education or an apprenticeship. Women and girls connection with water limits their geography and place in the public sphere. It also places them in a position of marginalization in the sharing of environmental challenges in cases of drinking water insecurity. As Mukherjee et al. (2017) argue, the involvement of women and girls in low-status, time-consuming work such as water collection limits their opportunities vis-à-vis education and paid work compared to men and boys. According to Nauges (2017), for example, a 50% reduction in water-hauling time in rural Ghana leads to a 13% increase in school attendance for girls. Girls who stay in school are older on average when they marry and can look forward to a brighter economic future and improved health (Hallfors et al., 2015).

The labor-intensive nature of water collection has also been associated with a host of health-related problems, including headaches, fatigue, and body pain, linked to musculoskeletal damage, soft tissue damage, and early onset of degenerative bone disease (Hanrahan & Mercer, 2019; Masanyiwa et al., 2015). In addition, water collection has other risks, which can cause physical and psychological harm to women and girls. While previous studies identified risks such as sexual harassment and abuse (Harris et al., 2017; Wali et al., 2020), this study identified snake and scorpion bites as the major physical risk faced by early morning or late evening water collectors.

International policy recommendations state that spaces should be created for women to play a central role in water provision, management, and safeguarding. For instance, the International Drinking Water and Sanitation Decade (1981–90) and the International Conference on Water and the Environment in Dublin (1992), among others, recognize the need for greater participation of women in water provision and management (Task Force on Gender and Water, 2006). However, this is not just about ensuring a certain percentage of women's participation in water management bodies but actual involvement and participation (Crider & Ray, 2022). Encouraging women's involvement in water management through increased knowledge may reduce water insecurity. Water-related local associations for women may improve knowledge about water resource management during water shortages. Increased investment in infrastructure expansion, such as the provision of more boreholes and rehabilitation of defunct ones, will help improve access to safe water. Sensitizing men to assist in water collection activities can help reduce women's workload. Also, through decentralized government intervention, women could be provided with bikes and driving lessons to ease water collection activities. In addition, household rainwater harvesting can save time on domestic duties (Baguma et al., 2013). Although this may not be feasible in droughts, it could be encouraged as a complementary measure during the rainy season when water collection activities may not be safe and can be hampered by heavy downpours.

Conclusion

The study has shown that the time allocated for water collection activities is associated with the time for other domestic activities. This means that water insecurity, caused by periods of scarcity, may widen inequality between men and women, putting women and girls at a greater disadvantage (Lucier & Qadir, 2018). This makes it imperative to pay attention to meeting basic water access needs in order to make a meaningful effort to advance gender equality. However, the human right to water and sanitation obliges all states, including those in the SSA, to ensure that everyone has access to basic services, including access to improved water within 30 minutes roundtrip, including waiting time. As this study documents, about three-quarters of the households in the study communities do not meet this criterion of access to basic services, which extends to about 844 million people worldwide (WHO/UNICEF, 2017).

Any plan to improve access to safe water in rural Ghana should not only take into consideration the time and distance, as shown in this study, but also the need to ensure the reliability of supply. Unreliable water access means that rural households must find alternative water sources that are more likely to be contaminated, require walking long distances, and require more time for water collection, with adverse impacts on women and girls as primary domestic water collectors. Indeed, as reported here, while some women called for men's participation in water collection, others expressed no concern regarding men's lack of involvement in water collection activities. Regardless of the different positions, this chapter argues that better water access would likely remove barriers associated with water collection freeing time and energy for women and girls. In addition, the chapter suggests that efforts to spearhead women's active involvement in household water management and decisions are a requisite. Women's participation is necessary given that interventions are more successful and sustainable when women are actively involved in every stage of planning, design, and implementation (Task Force on Gender and Water, 2006; Wali et al., 2020). This means that future policy decisions on rural water management should not only include women's participation by the numbers on water and sanitation management committees but also ensure that their voices are heard and incorporated in water management decisions. Ghana's achievement of gender equality by 2030 may not be possible if barriers facing women, such as those identified in this study, concerning domestic water management are not factored into policy analysis and implementation.

Notes

1 These are district level plans that are developed every four years to guide the development of districts. They include community level plans that provide a snapshot of the development needs of communities under a particular district and are developed based on a guideline issued by the National Development Planning Commission (NDPC), Ghana.

2 The choice for CSPro (among other data collection software) was due to our familiarity and expertise in its use.
3 Akan is the largest ethnic group in Ghana, making up 47.3% of the population of Ghana. They comprise the Bono, Asante, Adanse, Twifo, Asen, Fante, Akuapem, Akyem, Akwamu, Kwahu, Sehwi, Awowin, Nzima, and Ahanta (Ghana Statistical Service, 2013, p. 61).

References

Adjiwanou, V., & LeGrand, T. (2014). Gender inequality and the use of maternal healthcare services in rural sub-Saharan Africa. *Health & Place*, *29*, 67–78. https://doi.org/10.1016/j.healthplace.2014.06.001

Baguma, D., Hashim, J. H., Aljunid, S. M., & Loiskandl, W. (2013). Safe-water shortages, gender perspectives, and related challenges in developing countries: The case of Uganda. *Science of the Total Environment*, *442*, 96–102. https://doi.org/10.1016/j.scitotenv.2012.10.004

Baur, P., & Woodhouse, M. (2009). Enhancing private sector in rural water supply: An action-oriented study. *The International Journal of Human Resource Management*, *24*(8), 1671–1684.

Boahene, L. A. (2013). The social construction of sub-Saharan women's status through African proverbs. *Mediterranean Journal of Social Sciences*, *4*(1), 123–131. https://doi.org/10.5901/mjss.2013.v4n1p123

Crider, Y. S., & Ray, I. (2022). Water and development: A gender perspective. In *Oxford research encyclopedia of environmental science*. Oxford University Press.

Dosu, B. (2021a). Doing fieldwork, the Ghanaian way: The dilemma of conducting research in rural Ghana. *The Canadian Geographer/Le Géographe canadien*, *65*(4), 487–498. https://doi.org/10.1111/cag.12688

Dosu, B. (2021b). What defines rural water security? Identifying the contributing factors to household water security in Ghana. *Water Utility Journal*, *29*, 15–28.

Dosu, B., & Hanrahan, M. (2021). Barriers to drinking water security in Rural Ghana: The vulnerability of people with disabilities. *Water Alternatives*, *14*(2), 453–468.

Ghana Statistical Service. (2013). *2010 population and housing census*. National Analytical Report. https://statsghana.gov.gh/gssmain/fileUpload/pressrelease/2010_PHC_National_Analytical_Report.pdf

Global Water Partnership. (2000). Towards water security: A framework for action. *GWP Secretariat, Stockholm*. www.gwp.org/globalassets/global/toolbox/references/towards-water-security-a-framework-for-action.-mobilising-political-will-to-act-gwp-2000.pdf

Goldhar, C., Bell, T., & Wolf, J. (2013). Rethinking existing approaches to water security in remote communities: An analysis of two drinking water systems in Nunatsiavut, Labrador, Canada. *Water Alternatives*, *6*(3). www.water-alternatives.org/index.php/all-abs/228-a6-3-8/file

Hallfors, D. D., Cho, H., Rusakaniko, S., Mapfumo, J., Iritani, B., Zhang, L., & Miller, T. (2015). The impact of school subsidies on HIV-related outcomes among adolescent female orphans. *Journal of Adolescent Health*, *56*, 79–84. https://doi.org/10.1016/j.jadohealth.2014.09.004

Hanrahan, M., & Mercer, N. (2019). Gender and water insecurity in a subarctic Indigenous community. *The Canadian Geographer*, *63*(2), 211–224. https://doi.org/10.1111/cag.12508

Harris, L., Kleiber, D., Goldin, J., Darkwah, A., & Morinville, C. (2017). Intersections of gender and water: Comparative approaches to everyday gendered negotiations of water access in underserved areas of Accra, Ghana and Cape Town, South Africa. *Journal of Gender Studies*, *26*(5), 561–582. https://doi.org/10.1080/09589236.2016.1150819

Israel, G. D. (1992). *Determining sample size*. University of Florida Cooperative Extension Service, Institute of Food and Agricultural Sciences, EDIS.

Jeil, E. B., Abass, K., & Ganle, J. K. (2020). "We are free when water is available": Gendered livelihood implications of sporadic water supply in Northern Ghana. *Local Environment*, *25*(4), 320–335. https://doi.org/10.1080/13549839.2020.1744118

Karim, K. R., Emmelin, M., Resurreccion, B. P., & Wamala, S. (2013). Water, women and marital violence in a Bangladesh village. *Global Water: Issues and Insights*, *113*. https://globalwaterforum.org/2013/03/25/water-women-and-marital-violence-in-a-bangladesh-village/

Lucier, K. J., & Qadir, M. (2018). Gender and community mainstreaming in fog water collection systems. *Water*, *10*(10), 1472. https://doi.org/10.3390/w10101472

Masanyiwa, Z. S., Niehof, A., & Termeer, C. (2015). Users' perspectives on decentralized rural water services in Tanzania. *Gender, Place & Culture*, *22*(7), 920–936. https://doi.org/10.1080/0966369X.2014.917283

Mazrui, A. A. (2014). *The politics of gender and the culture of sexuality: Western, Islamic, and African perspectives*. University Press of America.

Ministry of Environment, Science, and Technology – Town and Country Planning Department. (2011). *Zoning guidelines and planning standards*. www.luspa.gov.gh/media/document/ZONING_GUIDELINES_final_DESIGN.pdf

Mukherjee, S., Lusigi, A., Kamwendo, E., & Bonini, A. (2017). Inequality, gender and human development in Africa. *UNDP Africa Policy Notes* (No. 2063–2018–617). https://doi.org/10.22004/ag.econ.267647

Nauges, C. (2017). Water hauling and girls' school attendance: Some new evidence from Ghana. *Environmental and Resource Economics*, *66*(1), 65–88. https://doi.org/10.1007/s10640-015-9938-5

Ojeda, D., Nirmal, P., Rocheleau, D., & Emel, J. (2022). Feminist ecologies. *Annual Review of Environment and Resources*, *47*, 149–171. https://doi.org/10.1146/annurev-environ-112320-092246

Staddon, C., Rogers, J., Warriner, C., Ward, S., & Powell, W. (2018). Why doesn't every family practice rainwater harvesting? Factors that affect the decision to adopt rainwater harvesting as a household water security strategy in central Uganda. *Water International*, *43*(8), 1114–1135. https://doi.org/10.1080/02508060.2018.1535417

Sultana, F. (2009). Fluid lives: Subjectivities, gender and water in rural Bangladesh. *Gender, Place and Culture*, *16*(4), 427–444. https://doi.org/10.1080/09663690903003942

Sun, Y., Asante, F., & Birner, R. (2010). Opportunities and challenges of community-based rural drinking water supplies. *International Food Policy Research Institute*. https://ebrary.ifpri.org/utils/getfile/collection/p15738coll2/id/4237/filename/4238.pdf

Task Force on Gender and Water. (2006). Gender, water and sanitation: A policy brief. *UN Water*. www.un.org/waterforlifedecade/pdf/un_water_policy_brief_2_gender.pdf

United Nations Development Programme. (2015). Human development report 2015: Work for human development. *United Nations Development Programme*. https://hdr.undp.org/content/human-development-report-2015

Van Houweling, E. (2016). "A good wife brings her husband bath water": Gender roles and water practices in Nampula, Mozambique. *Society & Natural Resources*, *29*(9), 1065–1078. https://doi.org/10.1080/08941920.2015.1095377

Wali, N., Georgeou, N., Simmons, O., Gautam, M. S., & Gurung, S. (2020). Women and WASH in Nepal: A scoping review of existing literature. *Water International*, *45*(3), 222–245. https://doi.org/10.1080/02508060.2020.1754564

Wallace, T., & Coles, A. (2005). Water, gender and development: An introduction. In A. Coles & T. Wallace (Eds.), *Gender, water and development* (pp. 1–20). Routledge.

World Health Organization. (2011). Guidelines for drinking-water quality, 4th edition. *World Health Organization*. www.who.int/water_sanitation_health/publications/2011/dwq_chapters/en/index.htm

World Health Organization and United Nations International Children's Emergency Fund. (2017). Progress on drinking water, sanitation and hygiene: 2017 update and SDG baselines. *World Health Organization*. www.who.int/publications/i/item/9789241512893

9

TOXIC HOMES, TOXIC WATER

Housing, segregation, and gendered responsibilities for household water insecurity in the American Rust Belt

Cara Jacob, Lucero Radonic, and Priyanka Jayakodi

Introduction

The city of Milwaukee, WI, United States, sprang up around the confluence of three rivers that spill out into the broadest natural bay on the western side of Lake Michigan, one of the five Great Lakes of North America. It is difficult to comprehend the possibility of water insecurity, often thought of in terms of hydrological scarcity, when you are regularly confronted with a sparkling lake that holds 1,180 cubic miles of freshwater (Wanamaker, n.d.). Yet less than three miles inland, in Milwaukee's urban core, thousands of people are experiencing water insecurity every day due to concerns that their tap water, which is pumped from Lake Michigan, may be contaminated with lead. In April 2022, in the heart of the predominantly African American North Side area of Milwaukee, a parent advocacy organization called COLE Parents Lead[1] hosted a "Lead Awareness Block Party." As attendees served bratwursts and hotdogs and painted children's faces, COLE volunteers distributed water filtration pitchers and lead safety kits. Coalition member organizations handed out information about the consequences of lead poisoning, potential sources of lead, and available resources. In partnership with the health department, they also offered testing for elevated blood lead levels (EBLLs) to the first 30 children under the age of six to arrive at the event. The location of the party was no coincidence: the North Side has disproportionately high rates of childhood lead poisoning. It also has a disproportionately high percentage of low-income, minority residents, and over 40 percent of households are single female-headed households (Department of City Development, 2000). These demographics are reflected in the attendees, who are mostly African American women and children. The few men present were largely community volunteers posted at the food tables or representatives of the participating organizations. Far from just a reflection of the neighborhood composition; it is indicative of the broader gendered and racialized nature of water insecurity and lead poisoning in the city.

This chapter draws on grassroots narratives to discuss the gendered responsibilities assumed by urban residents dealing with racialized water insecurity in Milwaukee. Household water insecurity (HWI) is the lack of access to clean, safe, affordable, and sufficient water necessary to thrive, and can result from either social or environmental factors

DOI: 10.4324/9781003100379-12

(Wutich & Ragsdale, 2008). In this case, participants' narratives reveal that water insecurity is born from living in a toxic environment; specifically, from living with unknown lead levels within the city's water delivery infrastructure and the intimate infrastructures of their homes. Concerns about exposure to corroded lead pipes and chipped, lead-painted walls often blur together in this environment. Thus, we begin by embedding people's perceptions and experiences of household water insecurity within a framework of housing precarity (Clair et al., 2019), fostered by historical racial inequalities. As our case study shows housing precarity can create conditions "where certain populations become differentially exposed to injury, violence, and death" (Butler, 2009) due to their living arrangements. We then shift attention to the dynamics within the household to explore how the responsibilities people take on due to water insecurity and lead exposure are gendered. Our findings resonate with the literature showing that women are disproportionately burdened with the responsibilities of care work (Crow & Sultana, 2002; Duignan et al., 2022; Hanrahan & Mercer, 2019), and that this labor is both gendered and racialized (Cairns, 2021). The racialized and gendered aspects of water insecurity in this study speak to the importance of taking an intersectional approach (Crenshaw, 1989; Hill Collins & Bilge, 2016) to understand how multiple axes of oppression converge to impact everyday hydro-social relations in late industrialism[2] (Harris et al., 2017; Mollett & Faria, 2013).

An intersectional lens for urban HWI

In 2015 world leaders at a UN Summit adopted the 17 Sustainable Development Goals, which includes the need to "ensure availability and sustainable management of water and sanitation for all" (UN Department of Economic and Social Affairs, 2022). Since then, there has been increased attention to gender parity, and more specifically to the role of women in accessing water for drinking and sanitation. Parallel to this global development agenda, research that examines the relationship between gender and HWI has blossomed. Most research focuses on rural households in low-income and lower-middle income countries (e.g. Crow & Sultana, 2002; Silva et al., 2020), with only a handful thus far addressing gender dimensions of water insecurity in high-income countries (e.g., Eichelberger, 2018; Jepson & Vandewalle, 2016; Radonic & Jacob, 2021). This is not surprising given prevailing narratives portraying access to water in high-income countries as universal, clean, affordable, and trustworthy (Meehan et al., 2020a). In recent years, however, activists and researchers have highlighted the myths of universal household water security in the United States (Meehan et al., 2020a, p. 1) by drawing attention to issues such as water shutoffs in African American neighborhoods in Detroit, Baltimore, and Philadelphia (Kornberg, 2016; Gaber et al., 2021), water contamination in Flint (Radonic & Jacob, 2021), and chronic water disconnection experienced by Indigenous communities (Eichelberger, 2018) and unhoused people (DeMyers et al., 2017).

In this context, this chapter brings into conversation three bodies of work to advance our understanding of the gendered responsibilities over water insecurity in urban areas of high-income countries. The first body of work is built around the housing-water nexus, a term introduced by Meehan et al. (2020b). This emerging scholarship correlates water insecurity to housing precarity, arguing that disparities in water access in US cities result from the conjunction of housing policies, infrastructural investments, and water management systems driven by systemic racism and racial segregation. In the 50 largest US cities – including Milwaukee – households suffering from water insecurity resulting from a lack of piped water access are disproportionately likely to be low-income, headed by a person of color, renting, and spend more than 30% of their income on housing costs (Deitz & Meehan, 2019; London et al., 2021; Meehan et al., 2020b; Pierce & Jimenez, 2015). In this chapter we draw on the concept of the housing-water nexus to highlight how water insecurity in Milwaukee has been racialized through historical processes of structural racism and redlining in the American Rust Belt.

The second body of work we engage with is the literature on gender and HWI, which draws attention to how gender roles and gender norms in specific societies shape access to and responsibilities for water management (Wutich, 2009; Wutich & Ragsdale, 2008). We draw specifically on perspectives from feminist political ecology (FPE) to understand how access to water is based on different social positions. As Rocheleau et al. (1996) outline, FPE acknowledges that resource access is embedded in complex contexts where gender interacts with class, race, and other social positions to shape one's experience with, and in, the environment. Early work by Crow and Sultana (2002) examined how gender and class intersected to shape access to safe drinking water in households in rural Bangladesh, finding that poor women were doubly disadvantaged: first, by their household's lack of material resources and second, by the low priority their communities placed on women's reproductive work, knowledge, and responsibilities. More recently, Harris et al. (2017) advanced an intersectional analysis to water access. Through examples from urban South Africa where labor differences between males and females within household are less pronounced, they illustrate the importance of moving away from viewing gender as a variable that creates a monolithic experience of environmental inequality for all women (or men) within a society. Instead, they highlight the ways in which gender, class, caste, and race intersect to produce multi-faceted relations of power and access for individuals (see also Heynen, 2018; Mollett & Faria, 2013; O'Reilly et al., 2009; Truelove, 2011).

This brings us to the third body of work, which we believe is critical to understanding HWI in the urban United States – intersectionality and black feminist scholarship, which allow us to examine how gender norms and gender roles function in the specific context of the United States. Intersectionality, as we use it here, was conceptualized by black feminist scholars as a framework that calls for us to consider how all axes of a person's social identities *intersect* – it is about the nodes where these axes meet, not the axes alone (Cho et al., 2013; Crenshaw, 1989; Hill Collins & Bilge, 2016). This framework reminds us that an intersectional approach to analyzing gendered responses to household water insecurity must acknowledge that, due to distinctly different lived experiences, gender roles and expectations for African Americans and other minorities in the United States are not necessarily the same as the "traditional" gender roles for middle- and upper-class, white Americans, which are frequently used as the default comparison point in studies of gendered divisions of labor (Carr et al., 2021; Hill Collins, 2006). African American gender roles as they exist today are deeply inscribed by the historical legacy of slavery and structural racism in America and influenced by issues of class. In the years after slavery was abolished, African American women commonly worked outside the home due to economic necessity, as discriminatory hiring practices and pay prevented African American men from earning enough to support their families (Harley et al., 2002; Jones, 1985; Shaw, 1996). Over time, the practice of African American women working outside the home came to be seen as a positive social norm and an integral part of good mothering, encouraged and valued by both men and women (Dow, 2019; Shaw, 1996). This has facilitated a less rigid approach to contemporary gendered divisions of labor in African American families, with relative equity in household decision-making and expectations for working outside the home (Dill & Johnson, 2002; Dow, 2019). However, despite the (comparatively) greater value placed on women's workforce participation, the burden of daily care work is not divided equally, and rests primarily on African American women (Dow, 2019; Shaw, 1996). Using the housing-water nexus, feminist political ecologies of water, and intersectionality we are able to understand how experiences of HWI in deindustrialized urban areas of the United States are inextricably tied to race, class, and gender.

Photovoice methodology

Given the power dynamics at play in this context, it was important that our research benefited the community in a tangible way and that they had a say in how the project was conducted. Thus, this chapter is based on research from a photovoice project conducted during the spring of 2022. Photovoice was chosen because it grew from a feminist tradition[3]: it was conceived of as a way to conduct research "by and with women" instead of "on women," with the explicit goal of centering participant voices and experience (Wang & Burris, 1997). A key element of photovoice is that it provides community members with a photo exhibit for use in education and advocacy. Photovoice aims to elicit participants' understanding of a given issue by asking them to take pictures representing the issue at hand. After taking pictures, they are prompted to discuss what the images mean to them, either as a group or individually through interviews. Participants then write their own narratives or work with researchers to produce them based on their own interviews. When paired with interviews, photovoice has the potential to offer nuanced, highly detailed descriptions for a small number of cases in a real-world context. This collaborative, qualitative study adds ethnographic detail to our limited understanding of gendered water insecurity in the urban United States.

The Milwaukee photovoice project was guided by the question "How has the lead epidemic affected your daily life?" which was developed during a preliminary planning meeting that included the lead author and several organizers and participants from COLE Parents Lead. Participants were recruited from the member list for COLE Parents Lead and the Hephatha Lutheran Church congregation. Nine individuals participated in the photovoice. All have been affected by the lead epidemic, either personally or through their families. Seven people identified as African American women, one as an African American man, and one as a multi-ethnic Latina woman whose family includes multiple races and ethnicities. Each participant was asked to take 15–20 photos answering the guiding question. Following the work of Radonic et al. (2021), in-depth ethnographic interviews were conducted to discuss peoples experiences of the lead epidemic, as a supplement to the discussion of the photos.

Racializing infrastructural decay in the American Rust Belt or, how the legacy of segregation and urban disinvestment became an issue of water insecurity

While this research is focused on present day Milwaukee, the lead epidemic has its roots in the history of structural and individual racism. In this section we provide this historical context and illustrate the importance of an intersectional framework in the study of urban water insecurity by tracing the connection between this history and urban infrastructural decay. Milwaukee is in the Rust Belt, a term used to refer to large and mid-size deindustrialized cities clustered around the Great Lakes[4] (Hackworth, 2019; Sugrue, 1998). Cities in this region grew on the back of a single type of industry (most often steel, rubber, or automobiles), and later experienced steep economic decline when the industry was lost (Hackworth, 2019; Rubado, 2019; Sugrue, 1998).

The Rust Belt saw an economic boom during WWII and in the immediate post-war years. This was a time when life across the southern United States was organized around Jim Crow laws, a racial caste system which relegated African Americans to second-class status by local laws and social norms. When the war increased industrial job opportunities in the north, millions of African Americans from the south settled across the Rust Belt (Jones, 2010; Sugrue, 1998). While the Rust Belt was free of Jim Crow laws, it wasn't free of institutional or individual racism: new African American residents were often restricted to the oldest parts of cities through redlining, underemployment, and the threat of violent attacks from white residents (Jones, 2010; Sugrue, 1998). As a result, housing conditions in predominantly African American neighborhoods were already deteriorating, even as African American residents were beginning to make their homes there.

Between the push factors of racial tension and aging infrastructure in city centers, and the pull factor of new infrastructures being constructed outside the city, the 1950s and '60s saw white residents leave for the suburbs in large numbers (Jakle & Wilson, 1992). Poverty and racist housing covenants enacted by white suburban governing boards prevented most African American families from moving to suburbia (Sugrue, 1998). With the housing stock deteriorating, cities began widespread demolition projects shrinking the size of African American neighborhoods and increasing occupancy in the remaining old homes (Hackworth, 2018). By the end of the 1960s, cities and African American residents were caught in a cycle of poverty and infrastructural decay. As white residents left and African American residents were paid less than their white peers, the tax base for many cities became insufficient to pay for large-scale infrastructure maintenance (Hackworth, 2019; Sugrue, 1998). This was exacerbated by the financial crisis of the 1970s, when manufacturers left the region for less expensive job markets. Neoliberal political and economic strategies rose to popularity in the 1980s in response to that crisis, slashing redistributive social welfare programs and public sector jobs, further impoverishing African Americans who relied heavily on the public sector for employment and welfare programs for food and housing (Konzelmann et al., 2016; Peck, 2012). Inner cities adopted strict austerity policies, leading them to defer maintenance on critical infrastructure, and allowing it to fall further into disrepair (Donald et al., 2014; Peck, 2012).

This racialized decline that began in the post-war era is a kind of slow infrastructural violence. Here we draw on Rodgers and O'Neill's (2012) concept of infrastructural violence in the sense of the state committing violence against its citizens through the intentional neglect of infrastructure and combine it with Nixon's (2011) theory of slow violence to highlight that this process of decay was gradual, occurring over the course of decades. Although not driven by overt, legal forms of discrimination, the purposeful urban disinvestment that led to today's context of infrastructural decay has had significant racial consequences. As part of an intersectional narrative of water insecurity, it should be recognized that African American residents of deindustrialized cities tend to bear the brunt of environmental contamination associated with living in the late industrial period (Hackworth, 2018; Silver, 2021).

"Our children are being poisoned": experiencing Milwaukee's lead epidemic

Milwaukee's pre-war African American population was tiny and fairly affluent, being mostly comprised of middle-class professionals, but it tripled in size during the post-war boom (Jones, 2010) and today accounts for 38.8 percent of the population (US Census Bureau, 2021). The impact of slow, racialized infrastructural violence in the city is evident in the fact that much of Milwaukee's oldest housing stock is in the predominantly African American North Side area (Schmidt, 2011), and that the children who live there exhibit disproportionately high elevated blood lead levels (EBLLs). In 2016, reports on childhood blood lead levels showed that the city's African American children had EBLLs at rates two to four times higher than in Flint, MI during their lead-tainted water crisis (Lynch & Meier, 2020). Milwaukee's health department emphasized that the majority of lead poisoning comes from lead paint in old homes. But with Flint fresh in their minds, many locals began to view the city's 70,000 lead service lines (LSLs) with concern (Mendez, 2019): aging LSLs, in addition to the interior lead plumbing of older homes, can leach lead into drinking water (US EPA, 2021; Wisconsin DNR, n.d.).

Fears were exacerbated in 2018, when an internal audit revealed that the health department had mismanaged the city's lead poisoning response program, failing to ensure that the families of 8,000 children with EBLLs were properly notified and falling short of state-mandated support for those children (Hagy, 2018; Mendez, 2019).[5] These failures led to ongoing mistrust of the water among the residents of the North Side. This concern is not unfounded – the EPA estimates that up to 20 percent of children's lead exposure may be from drinking water, and for infants drinking formula that number jumps to 40–60 percent (US EPA, 2016). While this may be a point of minimal concern for people in Milwaukee who are not exposed to lead from other sources, in areas of the city with deteriorating housing, it is the *cumulative* effect of lead from multiple sources that drives concern over lead in their drinking water.

Grassroot narratives of water insecurity and toxic homes

The in-depth interviews and photovoice methods of this study produced rich narrative data on the responsibilities taken on by participants in response to HWI and living in an environment contaminated by lead. While the guiding question of "How has the lead epidemic affected your daily life?" did not require participants to consider water insecurity, all participants took photos of items relating to water. As participants described changes they made to their lives in response to either a diagnosis of lead poisoning in their family or increased knowledge of lead risks, it became apparent that while men *and* women are involved in responding to this issue, the types and timing of these responsibilities have gendered components. Broadly speaking, the responsibilities described by participants can be divided into, a) one-time or immediate responsibilities pertaining to environmental modifications that will reduce or eliminate the sources of lead in their home, or b) long-term, ongoing care-work responsibilities that ensure lead exposure does not recur or mitigate the existing effects of lead exposure in children.

Immediate environmental modifications

Most of the solutions to water insecurity did not (and could not) involve immediate environmental adjustments. The most effective environmental change to address the concerns regarding water would be to replace the interior plumbing and the LSLs connecting the home to the city water system. For home-owning participants this would not be financially feasible. Voluntarily replacing the LSL costs a homeowner around $5000.00. The city only subsidizes this cost if there is a leak in the line or if the city is replacing the public side.[6] Even in this case, the homeowner must pay $1700.00, which makes it cost-prohibitive for most. For renters, this solution is not even an option – it is the landlord who must pay for a replacement, and most will not consider it.

Subsequently, the environmental changes that can be made for water safety are limited. The main option is installing a filter that mounts directly onto the sink. For example, Shy, a mother of four whose oldest child was severely affected by lead as an infant, rents the house where she lives. For her the faucet filter was a clear instant solution. She explained:

> I definitely use [faucet] filter water. They don't use filtered water for bathing, but they do use filtered water for brushing their teeth, eating, drinking, washing their hair. I try my best to get as much filtered water circulating through the house as possible.
>
> *(February 2022)*

Filters are a "solution" heralded by utilities as simple and accessible. However, this option is not available to everyone as old faucets are often incompatible with mounted filters, and even those who use them described them as "getting in the way" or "inconvenient," due to the filters' need for constant management and affordability. For example, Ivy, a mother of three boys, explained that faucet filters are more expensive to replace and there are fewer options for free replacements compared to pitcher filters, creating a financial burden.

Laresha, another participant, also highlighted the challenges of constant management. She noted that although the faucet filters seem like the obvious first solution, having to constantly replace sink filters is yet another task and another expense. As a result, water filtration in many homes is inconsistent. She explained:

> It was almost instant, like, 'okay, let's put filters in all the houses.' And so we started off there. But even then, at times, every family can't keep up with that. We'll have a filter for a year and then my mom takes it off and we don't have a filter for two years. Then she randomly buys a new one. . . . It's very off and on.
>
> *(March 2022)*

Zipporah faced similar frustrations, initially trying to set her mother up with a faucet filter, but ultimately, she explained, "She doesn't like to keep it up, she doesn't like to replace the filters . . . so she gave up and just uses bottled water now" (March 2022).

Limited options for larger, environmental solutions to water insecurity and the unpopularity of the few available options meant that participants spoke of coping strategies for water insecurity more frequently in the context of ongoing solutions. The strategies involving environmental modifications were mostly associated with lead paint or dust. Participants described dealing with these sources by testing for lead paint, placing duct tape over wall cracks, sealing off windowsills, painting over lead-painted walls to seal them, scraping wooden trim before repainting, taking up carpets, bleaching down walls, and replacing old metal blinds. For example, Ruby, whose family purchased a home only a block from where she grew up, explained that her husband went into the house before she and the children moved in to test the paint and do what he could to eliminate possible sources including repainting and replacing blinds (see Figure 9.1). She recalled:

> He had these little lead [test] strips, from work . . . so he wanted to check the house out. It didn't reach high in none of the rooms, maybe went up a little in some. So, we knew the house wasn't filled with lead, but we still knew it was a possibility. So, before we moved in . . . he was in the house about a good two weeks before he even let me back in. He did all the cleaning. Him and his friend from work. You know, on the weekend they came in here they took all the carpets up out of all the rooms, they bleached everything down, and they put paint up, all around the house.
>
> *(March 2022)*

Once she moved into the house, they installed water filters. She recalls having the filter first in the kitchen sink, but they moved it to the basement after she broke several while washing dishes. Her husband now fills bottles of water and brings them upstairs.

Figure 9.1 "Change for the better."

Source: Ruby, March 2022.

Even when it is possible for participants to make a home safer through various environmental solutions, there is always the concern that they might be evicted if they fall behind on housing payments or if the city notifies their landlord of required lead abatement. As Shy explained, "I don't want to leave my house because I've made a safe environment. . . . It's hard because the history of Milwaukee and renting is, you move to one house, it's got lead, horrible conditions, move to another house, horrible air conditions and so on, so forth." Every move represents the potential to restart the laborious process of making their home lead-safe again – if they can.

Continuous care-work

The bulk of the work for rendering a home lead-safe, especially for ensuring access to safe water, comes after those initial changes. Participants described the continual work to prevent repeated exposures to irremovable sources, like lead pipes, as becoming part of their everyday life. Water was most often rendered safe either through filtration pitchers or the purchase and consumption of bottled water (Figure 9.2), depending on convenience and financial ability. But the biggest challenge wasn't obtaining bottled water and filters, or

Figure 9.2 "Clean water."
Source: Maria, February 2022.

even maintaining them. It was keeping children from using the tap water despite the availability of safer alternatives. As explained by Deanna,

> I still make sure I tell [my son] don't drink the tap water when he's brushing his teeth because, you know, kids love to play in the faucet. I have to let him be more aware not to do that and make sure he's drinking the lead-free water, the pitcher, not at the tap. . . . Even when we're making like, a Kool-Aid or something, 'Be sure you're using the bottled water, filtered water for this and for that.
>
> *(March 2022)*

Some version of the phrase "because, they're, you know, *kids*," was uttered by every single participant at least once as they described why constant watchfulness is necessary to keep kids away from sources of lead. The subtext being that kids are driven by curiosity, entertainment, and convenience – so it is up to their guardians (usually their mothers or grandmothers) to monitor daily interactions with water and prevent accidental exposure. Participants described the struggle to keep children from drinking from unfiltered taps or from catching shower water in their mouths while bathing. Ms. Ledora is a mother and grandmother many times over, and a foster parent[7] who is especially worried about the presence of lead pipes inside her home. Her home was nearly 100 years old when she bought it, 30 years ago, and has never been updated. So, when her newest round of foster kids moved in and she heard them running the tap water in the bathroom after everyone went to bed, she was concerned:

> I start hearing them getting up, going to the bathroom at night. And so, one day I asked them, I said, 'Why y'all keep going in that bathroom so much at night?' They said, 'We be going for water.' I said, 'Oh, no. Don't do that . . . don't get the water out the bathroom no more. Please always come down, look in the refrigerator, get you a bottled water. Drink it from the bottle, not from the tap. Brush your teeth, take a bath, or shower. That's all you do with that water. Nothing else.
>
> *(April 2022) (Figure 9.3)*

Figure 9.3 "Don't drink from that faucet!"
Source: Ms. Ledora, March 2022.

The general feeling about this need to monitor was one of slightly amused, slightly resigned exasperation. The constancy of the observation normalized it to a point that it became second nature, for most. While they found it frustrating, often rolling their eyes when explaining this responsibility, they saw humor in what they viewed as "kids being kids." That is not to say they accept this as normal. As Derek commented, they "shouldn't have to constantly think about . . . oh, if I tell my daughter, go get some water, [what if] she gets it out the sink?" (March 2022). To reduce the chances that their children and grandchildren would drink tap water if left unmonitored, participants try to educate their children about lead and employ a variety of tactics to incentivize safer water sources: allowing children to go to the store and pick out personal water bottles to fill up with filtered water at night or tempting them with bottles of flavored waters. Ruby provides her foster children and grandchildren with bottles of water from the refrigerator because,

> Kids love to grab a cold bottle of water versus water out the sink, and it's cool with me. I mean, I know there's a lot of waste . . . but I'd rather see them drinking bottled water than thinking they gonna go to somebody's sink, and you don't know what kind of water coming out.
>
> *(March 2022)*

The monitoring, and the invention of tactics designed to reduce the need for it, seemed to primarily be the responsibility of women. Laresha framed it this way: "I think moms are just a little bit more attentive to things because they're mom, they're just there . . . dads are more like 'My wife will handle it'. Or, you know, the mom will handle it, it's fine" (March 2022). Her comment was echoed by others and speaks to the perception that mothers are most closely involved in their children's day-to-day care. However, this monitoring also included grown daughters, setting their parents up with filters or bottled water to reduce exposure.

Conclusions – intersectional issues of housing precarity and household water insecurity in the Rust belt

The history of racism and infrastructural violence in the American Rust Belt shaped the present conditions of racialized infrastructural decay which render participants' homes, and the water in them, unsafe from lead exposure. This case is illustrative of the housing-water nexus (Meehan et al., 2020b), which connects housing precarity with increased risk of water insecurity. Strategies developed to mitigate and cope with lead in the home vary by homeownership status, as well as gender. Homeowners are more able to make significant environmental modifications and worry less about having to make those changes again. Renters face a twofold situation of increased housing precarity. Those unable to alter their environment because of lease restrictions must move, and the history of infrastructural violence throughout the North Side (Jones, 2010; Schmidt, 2011) has limited the options for safe homes. Landlords are not always honest about the presence of lead in a property, and participants sometimes had to move multiple times before finding a lead-free location. Additionally, lease terms constrain their ability to move – renters described the difficult choice between hurting their credit history by breaking a lease or staying and hurting their families. Even renters who *can* modify their homes have to worry about financial instability that may result in an eviction, leaving them to restart the mitigation process; this is consistent with the literature that identifies renters as being at increased risk for water insecurity (Deitz & Meehan, 2019; Meehan et al., 2020b).

Within these households, the gendered differences in coping strategies were differences in type, rather than participation. This resonates with findings on gender and water insecurity among First Nations communities in Canada and the United States (Duignan et al., 2022; Eichelberger, 2018; Hanrahan & Mercer, 2019) that show how mechanical interventions (e.g., transporting water by vehicle) were the responsibility of male family members, while women were responsible for the daily task of economizing water use. In our study, men were more likely to be involved in one-time environmental modifications, while women shouldered the burden of ongoing care-work. In carrying out constant care work, women shoulder a dual burden of physical *and* emotional labor. There is significant anxiety connected to always keeping an eye on which water their children are drinking, or in monitoring the filters. That vigilance expends mental energy that the men do not have to expend in their more short-term role. The responsibility for physical labor articulates with the literature on African American gender roles, which situates the responsibility for house-care and child-care on women, despite a more egalitarian ideology in regards to household decision-making and employment outside the home (Dill & Johnson, 2002; Dow, 2019). It is also consistent with the global literature on gender and HWI indicating that women are often disproportionately burdened by water insecurity because of the constant and water-intensive nature of care-work: cleaning, cooking, and bathing are all ceaseless and require water to complete (Cairns, 2021; Crow & Sultana, 2002; Radonic & Jacob, 2021). This places a greater burden on women than on men, even when the responses to living in a toxic environment are shared. Thus, we see that even in high income countries like the United States, where gender roles seem to be more flexible, women are still responsible for the bulk of care-work associated with household water management.

Notes

1 This is the parent-run education and advocacy arm of the Coalition on Lead Emergency (COLE), a coalition of organizations and individuals tackling the lead epidemic in Milwaukee.

2 Taken from Fortun (2014), late industrialism refers to a period characterized by an environment chemically transformed by industries, and where societies must manage the conditions of an environment characterized by outdated infrastructure, inefficient policies, and ongoing contamination.

3 A key component of feminist methodologies, positionality acknowledges that the subjectivities of the researcher(s) may influence every part of the research process from design to analysis. We disclose our positionalities here in recognition of that fact. Jacob is a white, Arab-American woman who has not experienced the same axes of oppression as her participants. Radonic is a long-term Peruvian immigrant woman living in the United States, and Jayakodi is an international female graduate student from Sri Lanka. All are researchers affiliated with Michigan State University and none are residents of Milwaukee.

4 The Rust Belt is sometimes expanded to include deindustrialized cities of the Northeast.

5 Following the 2018 audit, new administrators were selected, and the department is now making a concerted effort to work closely with community groups to develop and implement interventions that go above and beyond the state mandates. While some federal funding from the Federal American Rescue Plan (ARPA) has been set aside for LSL replacement, the health department has not received any increase in funding to support more comprehensive medical or social services for children with EBLLs, making the implementation of these plans difficult.

6 LSLs are comprised of a "public" portion which runs from the water main to the curb stop and a "private" portion which runs from the curb stop to the home.

7 Foster parents are individuals licensed by the state to provide short-term "parental" care to children when their biological parents are unable to care for them or when the state has determined that living with the biological parents is unsafe.

References

Butler, J. (2009). Performativity, precarity, and sexual politics. *AIRB. Revista de Antropologia Iberoamericana*, *4*(3), i–xiii.

Cairns, K. (2021). Caring about water in Camden, New Jersey: Social reproduction against slow violence. *Gender, Place & Culture*, *29*(10), 1423–1445. https://doi.org/10.1080/0966369X.2021.1997940

Carr, J., Truesdell, N. D. Catharine, M., Orr, C. M., & Anderson-Levy, L. (2021). Future thinking and freedom making: Antidiversity as an intervention to plantation politics of higher education. In C. Bianca, B. C. Williams, D. Squire & F. Tuitt (Eds.), *Plantation politics and campus rebellions: Power, diversity, and the emancipatory struggle in higher education* (pp. 141–170). SUNY Press.

Cho, S., Crenshaw, K. W., & McCall, L. (2013). Toward a field of intersectionality studies: Theory, applications, and praxis. *Signs*, *38*(4), 785–810. https://doi.org/10.1086/669608

Clair, A., Reeves, A., McKee, M., & Stuckler, D. (2019). Constructing a housing precariousness measure for Europe. *Journal of European Social Policy*, *29*(1), 13–28. https://doi.org/10.1177/0958928718768334

Crenshaw, K. (1989). Demarginalizing the intersection of race and sex: A Black feminist critique of antidiscrimination doctrine, feminist theory and antiracist politics. *University of Chicago Legal Forum*, *1989*(1), 139–167.

Crow, B., & Sultana, F. (2002). Gender, class, and access to water: Three cases in a poor and crowded delta. *Society & Natural Resources*, *15*(8), 709–724. https://doi.org/10.1080/08941920290069308

Deitz, S., & Meehan, K. (2019). Plumbing poverty: Mapping hot spots of racial and geographic inequality in U.S. household water insecurity. *Annals of the American Association of Geographers*, *109*(4), 1092–1109. https://doi.org/10.1080/24694452.2018.1530587

DeMyers, C., Warpinski, C., & Wutich, A. (2017). Urban water insecurity: A case study of homelessness in Phoenix, Arizona. *Environmental Justice*, *10*(3), 72–80. https://doi.org/10.1089/env.2016.0043

Department of City Development. (2000). *Milwaukee urban atlas*. https://city.milwaukee.gov/DCD/Planning/PlansStudies/Census-data

Dill, B. T., & Johnson, T. (2002). Between a rock and a hard place. In S. Harley & Black Women and Work Collective (Eds.), *Sister circle: Black women and work* (pp. 67–83). Rutgers University Press.

Donald, B., Glasmeier, A., Gray, M., & Lobao, L. (2014). Austerity in the city: Economic crisis and urban service decline? *Cambridge Journal of Regions, Economy and Society*, *7*(1), 3–15. https://doi.org/10.1093/cjres/rst040

Dow, D. M. (2019). *Mothering while black: Boundaries and burdens of middle-class parenthood*. University of California Press.

Duignan, S., Moffat, T., & Martin-Hill, D. (2022). Be like the running water: Assessing gendered and age-based water insecurity experiences with Six Nations First Nation. *Social Science & Medicine (1982)*, *298*, 114864. https://doi.org/10.1016/j.socscimed.2022.114864

Eichelberger, L. (2018). Household water insecurity and its cultural dimensions: Preliminary results from Newtok, Alaska. *Environmental Science and Pollution Research*, *25*(33), 32938–32951. https://doi.org/10.1007/s11356-017-9432-4

Fortun, K. (2014). From Latour to late industrialism. *HAU: Journal of Ethnographic Theory*, *4*(1), 309–329. https://doi.org/10.14318/hau4.1.017

Gaber, N., Silva, A., Lewis-Patrick, M., Kutil, E., Taylor, D., & Bouier, R. (2021). Water insecurity and psychosocial distress: Case study of the Detroit water shutoffs. *Journal of Public Health*, *43*(4), 839–845. https://doi.org/10.1093/pubmed/fdaa157

Hackworth, J. (2018). Race and the production of extreme land abandonment in the American Rust Belt. *International Journal of Urban and Regional Research*, *42*(1), 51–73. https://doi.org/10.1111/1468-2427.12588

Hackworth, J. (2019). *Manufacturing decline: Race, the conservative movement, and the American rust belt*. Columbia University Press.

Hagy, A. (2018). *Childhood lead poisoning prevention program: Assessment of operations and recommendations for corrective action*. City of Milwaukee Health Department.

Hanrahan, M., & Mercer, N. (2019). Gender and water insecurity in a subarctic Indigenous community. *The Canadian Geographer/Le Géographe Canadien*, *63*(2), 211–224. https://doi.org/10.1111/cag.12508

Harley, S., Wilson, F. R., & Logan, S. W. (2002). Introduction: Historical overview of black women and work. In S. Harley & Black Women and Work Collective (Eds.), *Sister circle: Black women and work* (pp. 1–10). Rutgers University Press.

Harris, L., Kleiber, D., Goldin, J., Darkwah, A., & Morinville, C. (2017). Intersections of gender and water: Comparative approaches to everyday gendered negotiations of water access in underserved areas of Accra, Ghana and Cape Town, South Africa. *Journal of Gender Studies*, *26*(5), 561–582. https://doi.org/10.1080/09589236.2016.1150819

Heynen, N. (2018). Urban political ecology III: The feminist and queer century. *Progress in Human Geography*, *42*(3), 446–452.

Hill Collins, P. (2006). *Black sexual politics: African Americans, gender and the new racism*. Routledge.

Hill Collins, P., & Bilge, S. (2016). *Intersectionality*. Wiley.

Jakle, J. A., & Wilson, D. (1992). *Derelict landscapes: The wasting of America's built environment*. Rowman & Littlefield.

Jepson, W., & Vandewalle, E. (2016). Household water insecurity in the Global North: A study of rural and periurban settlements on the texas – Mexico Border. *The Professional Geographer*, *68*(1), 66–81. https://doi.org/10.1080/00330124.2015.1028324

Jones, J. (1985). *Labor of love, labor of sorrow: Black women, work, and the family from slavery to the present*. Basic Books.

Jones, P. D. (2010). *The Selma of the North civil rights insurgency in Milwaukee*. Harvard University Press.

Konzelmann, S., Gray, M., & Donald, B. (2016). Assessing austerity: Introduction to the virtual special issue. *Cambridge Journal of Regions, Economy and Society*, *7*(1), 1–12.

Kornberg, D. (2016). The structural origins of territorial stigma: Water and racial politics in metropolitan Detroit, 1950s–2010s. *International Journal of Urban and Regional Research*, *40*(2), 263–283. https://doi.org/10.1111/1468-2427.12343

London, J. K., Fencl, A. L., Watterson, S., Choueiri, Y., Seaton, P., Jarin, J., Dawson, M., Aranda, A., King, A., Nguyen, P., Pannu, C., Firestone, L., & Bailey, C. (2021). Disadvantaged unincorporated communities and the struggle for water justice in California. *Water Alternatives*, *14*(2), 520–545.

Lynch, E. E., & Meier, H. C. S. (2020). The intersectional effect of poverty, home ownership, and racial/ethnic composition on mean childhood blood lead levels in Milwaukee County neighborhoods. *PLoS ONE*, *15*(6), e0234995. https://doi.org/10.1371/journal.pone.0234995

Meehan, K., Jepson, W., Harris, L. M., Wutich, A., Beresford, M., Fencl, A., London, J., Pierce, G., Radonic, L., Wells, C., Wilson, N. J., Adams, E. A., Arsenault, R., Brewis, A., Harrington, V., Lambrinidou, Y., McGregor, D., Patrick, R., Pauli, B., & Young, S. (2020a). Exposing the myths of household water insecurity in the global north: A critical review. *WIREs Water*, *7*(6), e1486. https://doi.org/10.1002/wat2.1486

Meehan, K., Jurjevich, J. R., Chun, N. M. J. W., & Sherrill, J. (2020b). Geographies of insecure water access and the housing – water nexus in US cities. *Proceedings of the National Academy of Sciences*, *117*(46), 28700–28707. https://doi.org/10.1073/pnas.2007361117

Mendez, E. (2019, March 3). Is Milwaukee safe? What you need to know about the city's lead crisis. *OnMilwaukee*. https://onmilwaukee.com/articles/nns-milwaukee-lead-crisis-what-to-know

Mollett, S., & Faria, C. (2013). Messing with gender in feminist political ecology. *Geoforum*, *45*, 116–125. https://doi.org/10.1016/j.geoforum.2012.10.009

Nixon, R. (2011). *Slow violence and the environmentalism of the poor*. Harvard University Press. https://doi.org/10.4159/harvard.9780674061194

O'Reilly, K., Halvorson, S., Sultana, F., & Laurie, N. (2009). Introduction: Global perspectives on gender – water geographies. *Gender, Place & Culture*, *16*(4), 381–385. https://doi.org/10.1080/09663690903003868

Peck, J. (2012). Austerity urbanism: American cities under extreme economy. *City*, *16*(6), 626–655. https://doi.org/10.1080/13604813.2012.734071

Pierce, G., & Jimenez, S. (2015). Unreliable water access in U.S. mobile homes: Evidence from the American housing survey. *Housing Policy Debate*, *25*(4), 739–753. https://doi.org/10.1080/10511482.2014.999815

Radonic, L., & Jacob, C. (2021). Examining the cracks in universal water coverage: Women document the burdens of household water insecurity. *Water Alternatives*, *14*(1), 60–78.

Radonic, L., Jacob, C., Kalman, R., & Lewis, E. Y. (2021). It's a sprint, not a marathon: A case for building short-term partnerships for community-based participatory research. *Qualitative Research*, *23*(2), 146879412110294. https://doi.org/10.1177/14687941211029477

Rocheleau, D., Thomas-Slayter, B., & Wangari, E. (Eds.). (1996). *Feminist political ecology: Global issues and local experience* (1st ed.). Routledge.

Rodgers, D., & O'Neill, B. (2012). Infrastructural violence: Introduction to the special issue. *Ethnography*, *13*(4), 401–412. https://doi.org/10.1177/1466138111435738

Rubado, M. (2019). The role of local government collaboration in legacy cities. In J. R. Tighe & S. Ryberg-Webster (Eds.), *Legacy cities: Continuity and change amid decline and revival* (pp. 21–36). University of Pittsburgh Press.

Schmidt, D. H. (2011). Urban triage: Saving the savable neighbourhoods in Milwaukee. *Planning Perspectives*, *26*(4), 569–589. https://doi.org/10.1080/02665433.2011.601609

Shaw, S. J. (1996). *What a woman ought to be and to do: Black professional women workers during the Jim Crow Era*. University of Chicago Press.

Silva, B. B., Sales, B., Lanza, A. C., Heller, L., & Rezende, S. (2020). Water and sanitation are not gender-neutral: Human rights in rural Brazilian communities. *Water Policy*, *22*(1), 102–120. https://doi.org/10.2166/wp.2020.126

Silver, J. (2021). Decaying infrastructures in the post-industrial city: An urban political ecology of the US pipeline crisis. *Environment and Planning E: Nature and Space*, *4*(3), 756–777. https://doi.org/10.1177/2514848619890513

Sugrue, T. J. (1998). *The origins of the urban crisis: Race and inequality in postwar Detroit* (1st paperback ed.). Princeton University Press.

Truelove, Y. (2011). (Re-)Conceptualizing water inequality in Delhi, India through a feminist political ecology framework. *Geoforum*, *42*(2), 143–152. https://doi.org/10.1016/j.geoforum.2011.01.004

UN Department of Economic and Social Affairs. (2022). *Goal 6: Ensure availability and sustainable management of water and sanitation for all*. Sdgs.Un.Org. https://sdgs.un.org/goals/goal6

US Census Bureau. (2021). U.S. Census Bureau QuickFacts: Milwaukee city, Wisconsin. *Census.Gov*. www.census.gov/quickfacts/milwaukeecitywisconsin

US EPA. (2016, February 2). *Basic information about lead in drinking water* [Overviews and Factsheets]. www.epa.gov/ground-water-and-drinking-water/basic-information-about-lead-drinking-water

US EPA. (2021, July 15). *EPA, Milwaukee leaders discuss lead in drinking water* [News Release]. www.epa.gov/newsreleases/epa-milwaukee-leaders-discuss-lead-drinking-water

Wanamaker, B. (n.d.). *Lake Michigan*. Great Lakes Commission. Retrieved July 7, 2022, from www.glc.org/lakes/lake-michigan/

Wang, C., & Burris, M. A. (1997). Photovoice: Concept, methodology, and use for participatory needs assessment. *Health Education & Behavior*, *24*(3), 369–387. https://doi.org/10.1177/109019819702400309

Wisconsin DNR. (n.d.). *Drinking water and lead*. Retrieved July 20, 2022, from https://dnr.wisconsin.gov/topic/DrinkingWater/lead.html

Wutich, A. (2009). Intrahousehold disparities in women and men's experiences of water insecurity and emotional distress in Urban Bolivia. *Medical Anthropology Quarterly*, *23*(4), 436–454. www.jstor.org/stable/40541932

Wutich, A., & Ragsdale, K. (2008). Water insecurity and emotional distress: Coping with supply, access, and seasonal variability of water in a Bolivian squatter settlement. *Social Science & Medicine*, *67*(12), 2116–2125. https://doi.org/10.1016/j.socscimed.2008.09.042

10

POVERTY, WATER SECURITY, AND WOMEN'S ACTIVISM IN LIBERIA

Chantal Victoria Bright

Introduction

Wars adversely impact water availability, not only during periods of conflict but also during moments of peacebuilding, and post-conflict recovery. Throughout armed conflict, water is weaponised by armed actors; services are disrupted, infrastructure deteriorates, internally displaced persons search for reliable water sources, groundwater depletion occurs, and waterborne diseases are rampant (Chellaney, 2013; King, 2015; Schillinger et al., 2020). In the aftermath of conflict, water infrastructure challenges remain and can multiply the risk of conflict relapse. Poor water access during conflict significantly impacts women, who tend to be responsible for household reproduction (Lossow, 2015; USAID, 2014). Thus, women in situations of conflict and instability are primarily responsible for collecting and managing water and are disproportionately affected by its scarcity (Fröhlich & Gioli, 2015; Green, 2017). Still, the gender-water-war nexus is an area of underexplored scholarship, and consequently, the inclusion of water on the women, peace, and security (WPS) agenda is not an action priority (Centre for WPS at LSE et al., 2021; Jensen & Kron, 2018). This chapter examines the connections between water security, war and its aftermath, and women's activism in Liberia. It argues that since women are mainly affected by conflict, as well as by the scarcity of water that tends to follow its aftermath, they should participate in initiatives and discussions on water security in the post-conflict context.

Liberia is a conflict-affected West African country still facing immense water challenges from the ruins of the 1989–1997 and 1999–2003 civil wars, compounded by the 2014 Ebola epidemic (WaterAid, 2021). The prevalence of extreme poverty remains, with two of the five million total population estimated to be living under the international poverty line of USD 1.90 per day (World Bank, 2021a). Poverty in Liberia impedes water affordability since many households do not have enough money to invest in improved water services and purchase sachet water or bottled mineral water (GoL, 2009). With rising female-headed households in Liberia (IMF, 2021), women play a central role in dealing with poverty and managing water, balancing water management for domestic use and agriculture – a vital income source for women. While water is not a conflict driver in Liberia, water insecurity can exacerbate socio-economic inequalities and potentially compromise peace in

DOI: 10.4324/9781003100379-13

crisis-affected situations (Sadoff et al., 2017). This is especially alarming in Liberia, where class disparities and income inequalities ultimately contributed to war (Hegre et al., 2009).

Inequality as a driver of conflict in Liberia is best understood by tracing the historical roots of the country's path to sovereign statehood. Therefore, the chapter begins by taking a social-historical approach to the ideals of the 'civilising mission' of Liberia's early to mid-18th century (Brown, 1982; Dorsey, 2000; Liebersohn, 2016). This broad ideology divided and exposed inequalities within the country, one of the main catalysts that led to war. Second, the chapter highlights critical events of women's activism and political participation from the 1920s to give background to women's interventions in the peace process. Describing historic moments in women's history, it illustrates how women's dismissal of the notion of the civilising mission allowed them to organise for peace in Liberia. Third, the chapter describes how water distribution reflects poverty and considers the gender gaps in political representation and water governance in the midst of post-conflict recovery. Finally, the chapter summarises the gender-water-war nexus in the case study of Liberia. It concludes with some ideas on the role of women's activism in striving for water security and peace.

The 'civilised' cause of the Liberian conflict

The origins of Liberia's civil conflict are intrinsically tied to the formation of the country (Brown, 1982; Dalton, 1965). The story of Liberia's pathway to modern statehood begins with the American Colonialization Society (ACS). The private American organisation was formed in 1816 to repatriate free African Americans to the African continent. The group – primarily made up of an alliance of enslavers from southern American states and religious liberals – shared the consensus that the resettlement to Africa would solve the problem of the growing Black population in America (Gavins & Baker, 2016; Library of Congress, 2020; Whyte, 2016). Despite initial resistance to repatriation,[1] the idea became more appealing as Black people had become disillusioned at the prospect of liberation and fair treatment due to increasingly harsher conditions in the United States. They started to buy into the romanticism of migration (Leone et al., 2004), which included associated ideas of masculinity. The institution of slavery had been seen to have deeply emasculated Black men (Orelus, 2010), and colonisation had emphasised connections between the ideologies of the right of conquest and the occupation of land and people as validation of manhood and power. Black men began to view emigration to Africa as restoring the Black masculinity lost with the master-slave relationship (Arvin et al., 2013; Orelus, 2010; Whyte, 2016). Women were part of the migration movement to Africa, but their participation was less acknowledged and recorded (Dorsey, 2000; Leone et al., 2004).

In 1821, the ACS received financial backing from the United States government to facilitate the first US Black resettlement to West Africa (Dalton, 1965; Gavins & Baker, 2016; Library of Congress, 2020). Between 1822 and 1867, under the ACS umbrella, 19,858 free African Americans were resettled by different independent state-led settlements (Fraenkel, 1964; Spicer, 2016). Integrating the settlers and Indigenous Africans was challenging, and violent fights between the groups developed (Spicer, 2016). The unification of all state-led settlement colonies declared independence on 26 July 1847 as a single nation known as Liberia, making Liberia the first established independent republic in Africa and only the second in the world following the Republic of Haiti. Liberia's first president, Joseph Jenkins (J.J) Roberts, described in a speech in 1846 that the settlers 'left their native land to seek on these shores a residence for civil and political freedom . . . and established a government, with executive, legislative, and judicial powers, in the distant and inhospitable wilds of Africa' (Spicer, 2016, p. 47). African Americans embraced settler colonialism and the promise of the civilising mission, 'here (in America) we are ignorant, idle, a nuisance, and a drawback on the resources of the country. But as abandoned as we are, in Africa, we shall civilise and Christianise all that heathen country' (Dorsey, 2000, p. 85). Though Americo-Liberian elites only made up a small percentage of the population, this minority group dictated the standard of civilisation and determined who was to be considered 'civilised' – the prerequisite to joining the ruling class (Brown, 1982; Dalton, 1965; Spicer, 2016).

The 1980 coup d'état led by military leader Samuel K. Doe, who would later become president, was the first time in the country's 133 years that someone of Indigenous African background in Liberia ruled. Despite his violent coming to power, Doe's ascension to leading the nation was a significant moment in the anti-settler movement in Liberia. The 1989 violent capture and overthrow of President Doe by rebel groups was the start of Liberia's intermittent 14-year civil unrest (Library of Congress, 2020; Whyte, 2016). Conflict activities tended to be more frequent in relatively well-off regions, and rebel group recruitments were also prevalent in the same affluent areas of the country (Hegre et al., 2009). Research supports the theory that the war's roots lie in historical socio-economic and political inequalities (Hegre et al., 2009). Liberia's social stratifications and hierarchies destabilised Africa's first republic. Liberian female activists understood that this concept of the 'civilised class' divided the country and that reshaping the division and imaginaries around social classes would create the conditions for them to organise for peace successfully.

Women's participation and intervention for peace

Women were political and social leaders during Liberia's traditional governance – part of the exclusive secret societies known as *Sande* – centuries before women's activism of the 1920s began to take shape in the country's modern statehood (Fuest, 2008). Women's rights pioneers Sarah Simpson George and Maude A. Morris led movements which included the participation of the Women's League and the Liberian Women's Social and Political Movement for women's suffrage in the 1930s before the 1946 Liberian constitutional referendum guaranteed women 'possessing real estate' the right to vote (Collins & UN, 1969; GoL, 1946). The extension of voting rights to women was not exclusive to women; provisions included for the first-time Indigenous ethnic groups who owned property.

In 1959, Ellen Mills-Scarborough became the first woman elected legislature to Liberia's House of Representatives (NFLWO, 1962). By the 1960s, women-led movements advanced women's political leadership to hold positions in the executive, legislative, and judiciary branches of government and the Foreign Service (Collins & UN, 1969; NFLWO, 1962) (see Tables 10.1 and 10.2). To increase the effectiveness of women's movements activism, Mills-Scarborough conceived the idea to streamline the efforts by forming the National Federation of Liberian Women in 1962, later renamed in 1975 as the Liberian Federation of Women Organization (NFLWO) (NFLWO, 1962).

Women's organisations in Liberia worked together across class, ethnicity, and religious lines to formulate peace movements (see Table 10.3). This has been described as the 'distinctive feature' of Liberian women's organising methods (Moran, 2007). Nobel Peace Laureate and peace activist Leymah Gbowee recounted how women dismissed social inequalities that fuelled the war in her memoir. Gbowee writes, 'We are not lawyers, activists, or wives here. We are not Christian or Muslims, we are not Kpelle, Lorma, Krahn

Table 10.1 Government Offices Filled by Women (1948–2006)

Administration/Year	*Minister*	*Secretary of State*	*Deputy Minister*	*Assistant Minister*
Tubman (1948–1970)	-	2*	-	-
Tolbert (1971–1980)	8	1	1	-
Doe (1980–1989)	3	-	2	-
Civil War Interim Regimes/ Taylor (1990–2006)	15	1	4	2

*One of these women was the Assistant Secretary of State

Source: Fuest, 2008

Table 10.2 Notable Liberian Women in History

Name	*Contribution*
Ellen Mills Scarbrough (1900–1983)	First woman in the Legislature in 1959. Mills-Scarborough conceived the idea to streamline the efforts with the formation of the National Federation of Liberian Women in 1962, later renamed in 1975 as the Liberian Federation of Women Organization (NFLWO).
Emma Shannon Walser (1929–2021)	First female judge in 1971.
Angie Brooks-Rudolph (1928–2007)	Only African female president of the United Nations General Assembly and the second woman from any country to hold the position. 1970–77; The first woman to serve on the Supreme Court of Liberia.
Mary Antoinette Brown-Sherman (1926–2004)	First Female President of the University of Liberia.
Dr Florence Chenoweth (1945–2023)	First female Minister of Agriculture in Africa; and at the time was the only female Minister of Agriculture anywhere. Tolbert (1977–1980) and Johnson Sirleaf (2009–2015).
Ruth Sando Perry (1939–2017)	First female African to serve as head of state as interim-president between 1996–1997.
Ellen Johnson-Sirleaf (1938–)	Deputy Minister of Finance from 1971–74 and received a cabinet appointment as Minister of Finance in 1980 before becoming the first democratically twice-elected female Head of State in Africa in 2005 and 2011.

Source: Badri & Tripp, 2017; Collins & UN, 1969; Dunn & Holsoe, 1985.

or Mandingo. We are not Indigenous or elite. We are only women' (Gbowee & Mithers, 2013, p. 124). Before the Women of Liberia Mass Action for Peace, women's involvement in humanitarian and peace activism existed in silos nationwide. In what concerns water, it is important to mention how, during the war, women facilitated secret channels to secure food and water (Johnson, 2011; Ndongo, 2020). They developed safe times and routes to travel, and places to source/hide water. The search for clean water during the conflict added to women's vulnerability to sexual and gender-based violence – a dark legacy in Liberia

Table 10.3 Instrumental Women's Organisations

Association of Female Lawyers in Liberia (AFELL)
Christian Women's Initiative
Coalition of Women of Political Parties in Liberia (CWPPL)
Concerned Women for Liberia (CWO)
Liberian Women's Initiative (LWI)
Muslim Women's Association for Peace and Social Justice
Women's Federation
Women Development Association of Liberia (WODAL)
Women in Liberian Liberty (WILL)
Women of Liberia Peace Network (WOLPNET)
Women of Lutheran Church in Liberia
Women's Peace Initiative (WPI)
United Muslim Women's Advocacy and Empowerment Organization (UMWAEO)

Source: Fuest, 2009; Tripp, 2015.

today (Meger, 2021). Mary Brownell, founding member of the Mano River Women's Peace Network (MARWOPNET) and former president of the Liberian Women's Initiative (LWI), stated in an interview in 2000 'that the level of violence that women experienced or witnessed during the war compelled them to come to centre stage of peace efforts' (Pan African News Agency, 2000). MARWOPNET was awarded the United Nations Prize in the field of human rights in 2003. The groups' peacebuilding efforts in the region continue today (UN Women, 2019). In 2001, the Women in Peacebuilding Network (WIPNET) was launched as a West African regional network to support capacity-building tools to enhance women's role in peacebuilding and post-conflict reconstruction. Leymah Gbowee became the WIPNET chapter leader for Liberia and worked with Muslim anti-war activist Asatu Bah- Kenneth and others to rally women for peace. The coalition of women peacebuilders and organisations birthed the Women of Liberia Mass Action for Peace.

Liberia and its water resources in the context of climate change: impacts on women

Liberia is one of the wettest places in the world, with water resources spread across 15 primary river basins and six main rivers and water availability per capita is the third highest in Sub-Saharan Africa (USAID & SWP, 2021; WaterAid, 2021). The ratio of water extractions to renewable supply is less than 1% (USAID & SWP, 2021). For this reason, Liberia is not often seen as having a 'water problem'. However, the devastation of the war affected the overall water infrastructure, which has compromised potable water. Before the conflict, 24 urban areas across the country had piped water, and most households in the capital Monrovia had coverage (USAID and &, 2014). Groundwater – the country's primary drinking water source – is vulnerable to contamination due to poor sanitation systems, water pollution, and watershed degradation (Kaba & Madan, 2014; USAID & SWP, 2021; WaterAid, 2021; World Bank Group et al., 2021). In a 2017 analysis of the water sector, USAID reported water provision in Liberia to be 'very poor', where most communities' primary water sources are contaminated wells 'not fit for human consumption'.

Sixty-two percent of the population has limited-standard access to sanitation (USAID et al., 2017). There is limited data on surface and groundwater quality, availability, and balance in urban areas (USAID & SWP, 2021). Additionally, little technical expertise and funding impede the development of the water sector. Nevertheless, climate change will definitely impact water security (USAID & SWP, 2021). Liberia's rainy season occurs between May and November and receives an average rainfall of 200 mm inland and 4500 mm along the coast (Liberia Hydrological Services, NVE, 2022). Rain is expected to intensify with rising sea levels and increase flood risks along the coast, where Monrovia's is located (USAID & SWP, 2021).

Climate change also presents immense challenges for poverty reduction. World Bank data collected in 2016 shows how 44% of the population of Liberia lives on the international poverty line at USD 1.90 per person (World Bank, 2021a). Increased rainfall may reduce crop production, affecting food quality, access, and availability. It is worth mentioning that currently, women make up 60% of Liberia's vegetable farmers and have the conscious challenge of using water wisely for food utilisation in the household and agricultural commodities (Visser & Wangu, 2021). Approximately 34% of households are female-headed, and 47% of the employed population below the international poverty line are female (UN Women, 2022; World Bank, 2020a, 2020b). In an environment affected by climate change, unemployment will rise, poverty will increase and both will significantly impact women (Quiggin et al., 2021).

Moreover, climate-induced migration of coastal populations will is predicted to affect women particularly because it will likely increase the time spent on water collection and the risk of sexual and gender-based violence – a practice still prevalent today stemming from wartime rape (Meger, 2021). For example, West Point is one of Monrovia's most densely populated informal settlements and has reported high numbers of sexual assaults. The settlement, located on a 0.53 km^2 peninsula into the Atlantic Ocean and built between the Mersurado and Saint Paul rivers, has a population of approximately 35,000 and poor access to clean water and sanitation. The residents, who have some of the lowest incomes in the country, cannot afford to use paid toilets, and therefore, recur to open defecation. This sanitation practices have created sea and riverine pollution, leading to health hazards since river-water is used for drinking, and there is also fishing in the area (UNEP, 2010).

Poverty and water security in post-conflict recovery

In 2005 Ellen Johnson-Sirleaf became Liberia's first president in the post-conflict period and the first democratically elected female head of state in Africa. Her administration was challenged with post-conflict reconstruction and the 2014 Ebola epidemic. Additionally, Liberia acquired important debts under previous administrations, and Johnson-Sirleaf was left to seek debt relief from the World Bank Group and International Monetary Fund (IMF) to allow opportunities for economic growth. As a result, Liberia is required to produce Poverty Reduction Strategy Papers (PRSPs) every three years as a condition for debt relief through the Heavily Indebted Poor Countries (HIPC) initiative. PRSPs are a tool used mainly in post-conflict recovery settings to show approaches to achieving economic growth and poverty reduction (Obwona & Guloba, 2009). In the first PRSP, Johnson-Sirleaf's argued that clean water was one of the priority themes throughout the country (International Monetary Fund, 2008). The devastation of the war affected the overall water infrastructure of the country (Kaba & Madan, 2014; WaterAid, 2021; World Bank Group et al., 2021).

Under Johnson-Sirleaf's leadership, the water sector was restructured, and three sub-sections were created: (1) Monrovia, (2) other urban areas, and (3) the rural country. Liberia's water, sanitation, and hygiene (WASH) sector is spread across eight ministries and agencies with fragmented responsibilities (USAID et al., 2017). The government regulatory authority on water and sanitation was enacted only as of late 2017 (Water Governance Facility, 2019). Her successor, George M. Weah – who took office in 2017 – developed a Pro-Poor Agenda for Prosperity and Development (PAPD), which had limited results in context of high inflation and slow economic growth (GoL, 2018; World Bank, 2021b).

Despite the steps taken by the Liberian government with the support from foreign aid (GoL, WASH Liberia Commission, 2013; USAID et al., 2017), in 2021 Water Aid reported that in a country with a total population of five million, 1.2 million people are still without clean water (WaterAid, 2021). Poverty is the primary factor associated with barriers to accessing clean water in Liberia. Liberia's Water Supply and Sanitation Policy states:

> Poverty is a principal impediment to increasing access to services, from the household to the national level. Within communities, some households can only afford the costs of improved services with assistance from other families or the state. Many poor households pay much more of their incomes towards their daily needs for water supply and sanitation services from informal private providers.
>
> *(GoL, 2009)*

An affordable alternative source of drinking water in Liberia is locally produced sachet water – 500 millilitres of water packaged in single-use plastic mini bags extracted from privately owned wells or boreholes. A study conducted in 2018 found that the daily consumption rate is at least six bags of sachet water (3,000 millilitres or three litres of water) per individual in Monrovia. The study participants reported sachet water as affordable, available, and safe (Apeh, 2018). Sachet water is affordable for most Liberians at an average of £5.00 (USD 0.033) per bag. The proportion of income on the water source selection is unknown. However, since 2018, the government established the Water, Sanitation, and Hygiene Commission (WSSC) – a new regulatory agency responsible for tariffs, licenses, Public-Private Partnerships (PPP), service standards, and compliance with water laws (see Table 10.4). WSSC has required sachet water producers to register with the Liberia Water and Sewer Corporation (LWSC) to pay tariffs. As a result, the cost of a water sachet has increased to £25.00 (USD 0.16) per bag (Koinyeneh, 2019).

While sachet water is widely available and has high consumption, the Liberian Environmental Protection Agency assessed the industry in 2020 and found '17 of the 18 sachet water brands to be unsafe for human consumption' (Genoway, 2020). The sachets and bottled water volume raise serious environmental sanitation issues and public health risks (Wardrop et al., 2017). People not affected by water poverty rely on private water delivery services and imported mineral bottled water (USAID et al., 2017). Plastic contributes to 14.2% of Monrovia and surrounding suburbs waste (David et al., 2019). Liberia is challenged with implementing sustainable waste management practices when there is 10% open burning, 25% landfilling, and 65% indiscriminate dumping (David et al., 2020).

Table 10.4 Critical Events in Water Sector Reform in Post-Conflict Recovery

Year	*Event*
2008	Liberia PRS (2008–11) includes water and sanitation as part of Pillar IV, infrastructure and service delivery and sets targets for water and sanitation. Approval of Integrated Water Resources Management Policy.
2009	Approval of National Water and Sanitation Policy.
2010	National Environmental and Occupational Health Policy – To strengthen existing standards to public health (GoL, 2002).
2011	Water, Sanitation, and Hygiene, (WASH) sector Strategic Plan 2011–17.
2016	Approval of Urban Water Supply and Sanitation Project – The project seeks to (1) improve access to adequate, safe and reliable water supply and public sanitation services to the people in the towns of Monrovia, Buchanan, Kakata, and Zwedru on a sustainable basis, (2) enhance the institutional, operational, and management capacity, and (3) Ensure the long-term institutional and financial viability of Liberia Water and Sewer Corporation (World Bank, 2016).
2017	Government of Liberia passed an act to establish a national regulatory agency, Liberia Water, Sanitation, and Hygiene Commission (WSSC) (Water Governance Facility, 2019).
2019	Approval of Additional Financing for the Urban Water Supply Project is to increase access to piped water supply services in the project area in Monrovia and improve the operational efficiency of LWSC (World Bank, 2019).

*Adapted from the AMCOW report (GoL, AMCOW, 2011).

Table 10.5 Drinking Water Sources in a Study Conducted in Monrovia in 2018 (Apeh, 2018).

Variable	*Mean Score (x)*	*Standard Deviation*
Pipe or pump	3.65	1.23
Water vendor (clean water)	2.91	1.28
Well	2.69	1.31
River, lake, creek	1.93	0.98
Sachet water or "pure water"	4.37*	0.81
Bottled water	2.44	1.12
Other Sources	2.01	1.00

Source: Field Survey 2018: * Response equal to or above cut-off point of 3.00.

Women's political participation and water governance

In Liberia's recent past, women's associations worked by challenging the differences across class, ethnicity, and religion, to establish certain agreements for the end of the conflict. Thus, in a context where income differences hinder equal access to water, these associations could play an important role again. However, women are underrepresented in governance at different levels and water management in the country is still predominately male-dominated (Lamptey, 2023). Studies have shown that when women are represented in politics, confidence in democracy improves, there is more excellent responsiveness to all citizens, and policies are more likely to address quality of life, such as education and health, including basic water services (UN Women, 2021).

On paper, the Liberian state recognises 'gender equality as a means of maintaining peace, reducing poverty, enhancing justice, and promoting development in the country' (USAID, 2017; p. 5). Nevertheless, since the 1946 Liberian constitutional referendum, only 47 women have been elected to office, and only 27 women have been elected to the Liberian legislature since 2005. There has been a decline in women in the national legislature from 16% in 2005 when Johnson-Sirleaf was elected to 11% in 2021, post-by-election. Factors explaining women's underrepresentation include economic, institutional, psychological, and socio-cultural barriers and violence against women in elections (UN Women, 2021). As of 2022, women comprise 50% of Liberia's population, but only 11 out of 103 legislators are women (World Bank, 2021c). When writing in 2023, Jewel Taylor Howard is the current and first Vice President, previously serving as a senator. Howard is the ex-wife of former President Charles Taylor, who was convicted by an international tribunal for war crimes in Sierra Leone in 2012 (Ensign, 2013; GoL, 2022). In 2022, out of 19 current government cabinet positions, five women were appointed to head the following ministries: Ministry of Public Works, Ministry of Agriculture, Ministry of Health and Social Welfare, Ministry of Commerce and Industry, Ministry of Gender, Children and Social Protection – all ministries which interconnect with Liberia's water supply and sanitation sector (GoL, 2022; GoL, AMCOW, 2011). Liberia will hold its fourth post-war election in October 2023. In May 2023, UN Women and the National Election Commission facilitated the signing of a Memorandum of Understanding (MoU) with 25 political parties to promise a minimum of 30% representation of women to be included in each party's candidate nomination lists for the 2023 elections (Hussein & Ganyani, 2023). The MoU responded to the failed attempts by women's rights advocacy for legislative reforms. While this is a significant milestone in expanding women's inclusion in politics, unfortunately, the political parties fell short of the commitment. The National Elections Commission (NEC) reported women comprised only 15% of the candidate list of political parties (Hussein & Ganyani, 2023).

The prominence of the women's movement in water governance, advocacy and leadership is not prioritised due to the prevalence of gender-based violence (GBV) in Liberia. Keeping women safe from all forms of violence is the most pressing issue in Liberia (World Bank Group et al., 2021). One of the four key pillars of UN Women Liberia is ending violence against women and girls. In the 2023 election year, women's mobilisation efforts are predominately focused on the prevention of and response to violence against women in elections (UN Women, 2021). The Afrobarometer Round 9 study conducted in 2022 found that Liberians rank gender-based violence as the most critical women's rights issue. In the same study, 85% of respondents reported no access to water inside the house, compound, or outside the compound for household use (Afrobarometer, 2023). Without safely managed water, women and girls are more vulnerable to abuse. However, not enough activity from women's organisations is happening on exploring the links and steps for the prevention of water insecurity and GBV, even though there has been some research on tackling GBV in the context of climate change (Aguilar et al., 2012; CSW66 et al., 2022). There needs to be more evidence of women mobilising to address gender-specific vulnerability to water insecurity in Liberia. This does not suggest that engagement and advocacy for improved water governance are not happening, but water governance initiatives have yet to be shown on a larger scale. Thus, water governance must tackle challenges posed by contextual complexities in Liberia.

Conclusion

Much work has been done in the context of women, peace, and security since the adoption of the United Nations Security Council Resolution 1325 in 2000, which officially recognised the gender differential impact of war and the importance of women's involvement in peace and security issues (Mader et al., 2020). However, there is a gap in the response to gender, water security, and peace. In many regions, people flee due to rising sea levels or droughts. Experts know that water insecurity disproportionately affects women in conflict and conflict-affected situations, and the lack of safe water and poverty are jointly reinforcing significantly in these settings. Poverty and conflict are closely interconnected (Hegre et al., 2009). Conflict-affected countries are ten times more likely to be prone to poverty. These countries face more significant consequences of political instability where there is water insecurity. Conflict makes water harder to access (Sadoff et al., 2017; World Bank, 2020b).

Liberia may be overlooked as a country without significant water challenges because of its abundant water resources. However, water availability in Liberia is compromised by pollution, poor sanitation, degraded wetlands and watersheds, climate change, and a lack of technical expertise or proper water governance. Additionally, post-conflict recovery compounded with the 2014 Ebola crisis contributes to the slow recovery of the water infrastructure. While it is unlikely for Liberia to experience conflict relapse directly linked to water, water insecurity is a risk multiplier. Inequality is being exacerbated by water insecurity – one of the leading causes of Liberia's conflict (Sadoff et al., 2017). Conflict-affected countries are ten times more prone to poverty (Sadoff et al., 2017; World Bank, 2020b). In Liberia, poverty-induced vulnerability reveals that failing to achieve water security and sustainable development will not reduce the underlying causes of risk, political instability, and the socio-cultural and political perpetuation of inequality.

This chapter advocates equitable water distribution to mitigate conflict and envisions women at the forefront of water management in conflict-affected situations because of the potential risk pathways and more significant impact on women. Liberian women's activism throughout history underscores how women's meaningful participation can lead to effective change. Women played an essential role in the peace processes by rejecting the civilising mission and disrupting the inequalities in the country. For this reason, women's organisations and local forms of collective action in Liberia can also play an essential role in developing strategies toward sustainable water management solutions and addressing overlapping drivers that threaten peace and security.

Acknowledgements

The UK government's Turing Scheme partly provided support for this chapter. Research on women's social and political movement in Liberia was made possible by the Liberian Archives Collections at Indiana University Bloomington.

Note

1 Many African Americans were critical of the ACS and exposed the organisation as merely an unconventional model to uphold and protect slavery in America (Dorsey, 2000; Leone et al., 2004). African Americans saw emigration as a plot designed to deny them their full rights to citizenship in America.

References

Afrobarometer. (2023). *Summary of results Afrobarometer Round 9 survey in Liberia, 2022*. https://www.afrobarometer.org/publication/liberia-round-9-summary-of-results/

Aguilar, L., Rogers, F., & Bobtoya, S. (2012). *Climate change gender action plan for the Government of Liberia*. https://portals.iucn.org

Apeh, C. C. (2018). Survey of sachet water waste disposal in Liberia. *Journal of Health and Pollution, 8*. https://doi.org/10.5696/2156-9614-8.20.181211

Arvin, M., Tuck, E., & Morrill, A. (2013). Decolonizing feminism: Challenging connections between settler colonialism and heteropatriarchy. *Source: Feminist Formations, 25*, 8–34.

Badri, B., & Tripp, A. M. (2017). *Women's activism in Africa: Struggles for rights and representation*. Zed Books Ltd.

Brown, D. (1982). On the category "civilised" in Liberia and Elsewhere. *Journal of Modern African Studies, 20*, 287–303.

Centre for WPS at LSE, Women's International Peace, Gender Action for Peace and Security. (2021). *Defending the future*. www.lse.ac.uk/women-peace-security/assets/documents/2021/Defending-the-Future-Policy-Brief-03.pdf

Chellaney, B. (2013). *Water, peace, and war: Confronting the global water crisis*. Rowman & Littlefield Publishers.

Collins, E. K., & UN. (1969). *Country report – Liberia: Paper presented by Mrs. Elizabeth K. Collins, National president Liberian women's social and political movement, Sinkor, Liberia*. Addis Ababa.

CSW66, IUCN, Generation Equality, & UN Women. (2022). *Tackling violence against women and girls in the context of climate change*. https://www.unwomen.org/en/digital-library/publications/2022/03/tackling-violence-against-women-and-girls-in-the-context-of-climate-change

Dalton, G. (1965). History, politics, and economic development in Liberia. *Journal of Economic History, 25, 569–591*.

David, V. E., John, Y., & Hussain, S. (2020). Rethinking sustainability: A review of Liberia's municipal solid waste management systems, status, and challenges. *Journal of Material Cycles and Waste Management*, 22, 1299–1317. https://doi.org/10.1007/s10163-020-01046-x

David, V. E., Wenchao, J., John, Y., & Mmereki, D. (2019). Solid waste management in Monrovia, Liberia: Implications for sustainable development. *Journal of Solid Waste Technology and Management, 45*, 102–110. https://doi.org/10.5276/JSWTM.2019.102

Dorsey, B. (2000). A gendered history of African colonization in the Antebellum United States. *Journal of Social History, 34*, 77–103.

Dunn, D. E., & Holsoe, S. E. (1985). *Historical dictionary of Liberia*. Scarecrow Press.

Ensign, R. L. (2013). OFAC removes 12 people from sanctions lists [WWW Document]. *The Wall Street Journal*. Retrieved June 10, 2022, from www.wsj.com/articles/BL-252B-89

Fraenkel, M. (1964). *Tribe and class in Monrovia*. Routledge.

Fröhlich, C., & Gioli, G. (2015). Gender, conflict, and global environmental change. Peace review. *Journal of Social Justice Gender*, *27*, 137–146. https://doi.org/10.1080/10402659.2015.1037609

Fuest, V. (2008). "This is the time to get in front": Changing roles and opportunities for women in Liberia. *Source: African Affairs, 107*, 201–224.

Fuest, V. (2009). Liberia's women acting for peace: Collective action in a war-affected country. In Stephen Ellis and Ineke van Kessel (Eds.), *Movers and shakers: Social movements in Africa* (pp. 114–137). Brill.

Gavins, R., & Baker, E. J. (2016). Back-to-Africa movement. In *The Cambridge guide to African American history* (pp. 25–25). Cambridge University Press. https://doi.org/10.1017/CBO9781316216453.026

Gbowee, L., & Mithers, C. (2013). *Mighty be our powers: How sisterhood, prayer, and sex changed a nation at war – A memoir*. Beast Books.

Genoway, E. Jr. (2020). *Liberia: Local company producing sachet drinking water in filth*. Front Page Africa.

Government of Liberia GoL. (1946). *1946 Liberian constitutional referendum*. Retrieved June 11, 2022, from www.emansion.gov.lr/

Government of Liberia GoL. (2002). *Act adopting the environmental protection and management law*. Retrieved June 11, 2022, from www.emansion.gov.lr/

Government of Liberia GoL. (2009). *Water supply and sanitation policy*. Retrieved June 11, 2022, from www.emansion.gov.lr/
Government of Liberia GoL. (2018). *Pro Poor Agenda for Prosperity and Development (PAPD)*. Retrieved June 11, 2022, from www.emansion.gov.lr/
Government of Liberia GoL. (2022). *Executive mansion of Liberia*. Retrieved June 11, 2022, from www.emansion.gov.lr/
Government of Liberia GoL, AMCOW. (2011). *Water supply and sanitation in Liberia: Turning finance into services for 2015 and beyond*. Retrieved June 11, 2022, from from www.emansion.gov.lr/
Government of Liberia GoL, WASH Liberia Commission. (2013). *WASH sector investment plan summary report (2012–2017)*. Retrieved June 11, 2022, from from www.emansion.gov.lr/
Green, A. H. (2017). "Mind the gap": Measuring and understanding gendered conflict experiences. In *The Oxford handbook of gender and conflict* (pp. 316–327). Oxford University Press. https://doi.org/10.1093/oxfordhb/9780199300983.013.25
Hegre, H., Østby, G., & Raleigh, C. (2009). Poverty and civil war events: A disaggregated study of Liberia. *Journal of Conflict Resolution*, *53*, 598–623. https://doi.org/10.1177/0022002709336459
Hussein, G. W. B., & Ganyani, G. (2023). *The United Nations in Liberia commends NEC on the signing of an MOU on Section 4.5 of the Elections Law*. https://liberia.un.org/en/231571-united-nations-liberia-commends-nec-signing-mou-section-45-elections-law
International Monetary Fund. (2008). *Liberia: Poverty reduction strategy paper*. https://www.imf.org/external/pubs/ft/scr/2008/cr08219.pdf
International Monetary Fund. (2021). *Liberia, IMF staff country reports, 2021(010), A001*. Retrieved January 11, 2024, from https://doi.org/10.5089/9781513566313.002.A001
Jensen, D., & Kron, A. (2018). Environmental peacebuilding and the United Nations. In A. Swain & J. Öjendal (Eds.), *Routledge handbook of environmental conflict and peacebuilding*. Routledge.
Johnson, S. A. (2011). Women, shared leadership, and policy: The Mano river women's peace network case study. *Journal of Pan African Studies*, *4*, 59–73.
Kaba, A., & Madan, G. (2014). *Walking with Villagers: How Liberia's land rights policy was shaped from the grassroots*. iied Sustainable Development Institute & NAMATI Innovation in Legal Empowerment. www.iied.org/sites/default/files/pdfs/migrate/G03832.pdf
King, M. D. B. (2015). The weaponization of water in Syria and Iraq. *Washington Quarterly*, *38*, 153–169. https://doi.org/10.1080/0163660X.2015.1125835
Koinyeneh, G. C. (2019). *Liberia: Water producers threaten to increase sachet price if Gov't levy tax on boreholes*. Front Page Africa.
Lamptey, C. (2023). *Liberia: A final push to make the 2023 candidate registration count for Liberian women*. Daily Observer.
Leone, S., Author, L., & Blyden, N. A. (2004, April). "Back to Africa:" The migration of new world blacks to, the Atlantic world. *OAH Magazine of History*, *18*(3), 23–25.
Liberia Hydrological Services, NVE. (2022). Liberia hydrological services [WWW Document]. *Rainfall*. Retrieved August 29, 2022, from https://lhsliberia.com/hydrology-for-the-curious/rainfall/#:~:text=The%20annual%20amount%20of%20rainfall,2000%20mm%20in%20the%20interior
Library of Congress. (2020). History of Liberia: A timeline [WWW Document]. *Library of Congress: Maps of Liberia, 1815 to 1997*. Retrieved December 23, 2020, from www.loc.gov/collections/maps-of-liberia-1830-to-1870/articles-and-essays/
Liebersohn, H. (2016). Introduction: The civilizing mission. *Source: Journal of World History*, *27*, 383–387.
Mader, K., Porcelli, D., Birki, N., Bracco, N., Roele, J., & Udell, G. (2020). *Mapping women, peace and security in the UN security council: 2019*. https://www.womenpeacesecurity.org/resource/mapping-women-peace-and-security-in-the-un-security-council-2019/
Meger, S. (2021). *Sexual violence in times of war and peace*. In Tarja Väyrynen, Swati Parashar, Élise Féron, Catia Cecilia Confortini (Eds.), *Routledge handbook of feminist peace research*. Routledge.
Moran, M. (2007). Liberian women peacemakers: Fighting for the right to be seen, heard and counted by African women and peace support group. *AFR Today*, *53*, 112–113.
Ndongo, O. (2020). Women's wartime struggle for peace and security in the Mano River Union. In *Preventive diplomacy, security, and human rights in West Africa* (pp. 153–174). Palgrave Macmillan. https://doi.org/10.1007/978-3-030-25354-7_5

NFLWO. (1962). *Historical background. E. Reginald Townsend and Evelyn Diggs Townsend papers, 1905–2002*. Liberian Collections, Indiana University Bloomington.
Obwona, M., & Guloba, M. (2009). Poverty reduction strategies during post-conflict recovery in africa. *Journal of African Economies, 18*. https://doi.org/10.1093/jae/ejp008
Orelus, P. W. (2010). Chapter 2: Black masculinity under white supremacy: Exploring the intersection between black masculinity, slavery, racism, heterosexism, and social class. *Counterpoints*, 351, 63–111. http://www.jstor.org/stable/42980552
Pan African News Agency. (2000). *War prepared women's role in solving conflicts*. Pan African News Agency.
Quiggin, D., De Meyer, K., Hubble-Rose, L., & Froggatt, A. (2021). *Climate change risk assessment 2021 The risks are compounding, and without immediate action the impacts will be devastating*. https://www.chathamhouse.org/2021/09/climate-change-risk-assessment-2021
Sadoff, C. W., Borgomeo, E., & de Waal, D. (2017). *Turbulent waters: Pursuing water security in fragile contexts*. World Bank Group. http://documents.worldbank.org/curated/en/885171489432062054/Turbulent-waters-pursuing-water-security-in-fragile-contexts
Schillinger, J., Özerol, G., Güven-Griemert, S., & Heldeweg, M. (2020). Water in war: Understanding the impacts of armed conflict on water resources and their management. *WIREs Water*, 7(6), e1480. https://doi.org/10.1002/wat2.1480
Spicer, C. (2016). The perpetual paradox: A look into Liberian colonization. *The Ascendant Historian*, *3*, 36–52. https://doi.org/10.2307/2715813
Tripp, A. M. (2015). *Women and power in Postconflict Africa*. Cambridge University Press. https://doi.org/10.1017/CBO9781316336014
UNEP. (2010). *Africa water Atlas*. https://wedocs.unep.org/20.500.11822/7919
UN Women. (2019). *Statement: Peace hut women of Liberia wins 2019 UN population award* [WWW Document]. Retrieved June 30, 2022, from www.unwomen.org/en/news/stories/2019/6/statement-ed-phumzile-peace-hut-women-of-liberia-wins-2019-un-population-award
UN Women. (2021). *2021 annual report Liberia*. https://africa.unwomen.org/en/digital-library/publications/2022/08/un-women-liberia-annual-report-2021
UN Women. (2022). *Women count: Liberia data* [WWW Document]. Retrieved July 14, 2022, from https://data.unwomen.org/country/liberia
USAID. (2014). *Water & conflict: A toolkit for programming*. https://pdf.usaid.gov/pdf_docs/PNAEC948R.pdf
USAID. (2017). *Liberia: Water for world country plan*. https://www.globalwaters.org/sites/default/files/wfw_liberia_country_plan.pdf
USAID, Pattle, S., Popkin, B., & Lazarte, E. (2017). *Options Analysis (OA) for A PPP in the water sector in Liberia*. https://pdf.usaid.gov/pdf_docs/PA00TCVZ.pdf
USAID, & SWP. (2021). *Liberia water resources profile overview*. https://winrock.org/resources/liberia-water-resources-profile/
USAID, & Tetra Tech. (2014). *Liberia WASH sector assessment desk review: WASH sector status and trends*. https://pdf.usaid.gov/pdf_docs/PA00KMCM.pdf
Visser, J., & Wangu, J. (2021). Women's dual centrality in food security solutions: The need for a stronger gender lens in food systems' transformation. *Current Research in Environmental Sustainability*, *3*. https://doi.org/10.1016/j.crsust.2021.100094
Von Lossow, T. (2015). Gender in inter-state water conflicts. *Peace Review*, *27*, 196–201. https://doi.org/10.1080/10402659.2015.1037628
Wardrop, N. A., Dzodzomenyo, M., Aryeetey, G., Hill, A. G., Bain, R. E. S., & Wright, J. (2017). Estimation of packaged water consumption and associated plastic waste production from household budget surveys. *Environmental Research Letters*, *12*, 1–12. https://doi.org/10.1088/1748-9326/aa751f
WaterAid. (2021). *Liberia WaterAid country profile* [WWW Document]. WaterAid. Retrieved February 28, 2019, from www.wateraid.org/where-we-work/liberia
Water Governance Facility. (2019). *UNDP-SIWI water governance facility* [WWW Document]. Retrieved June 9, 2022, from https://siwi.org/goal-wash/liberia?accordion-strategy
Whyte, C. (2016). Between empire and colony: American imperialism and pan-African colonialism in Liberia, 1810–2003. *National Identities*, *18*, 71–88. https://doi.org/10.1080/14608944.2016.1095493

World Bank. (2016). *Liberia urban water supply project* [WWW Document]. Retrieved June 8, 2022, from https://projects.worldbank.org/en/projects-operations/project-detail/P155947

World Bank. (2019). *Liberia urban water supply project additional financing* [WWW Document]. Retrieved June 8, 2022, from https://projects.worldbank.org/en/projects-operations/project-detail/P169031?lang=en

World Bank. (2020a). *Female headed households (% of households with a female head) – Liberia* [WWW Document]. Retrieved July 13, 2022, from https://data.worldbank.org/indicator/SP.HOU.FEMA.ZS?locations=LR

World Bank. (2020b). *Fragility and conflict: On the front lines of the fight against poverty* [WWW Document]. Retrieved June 11, 2022, from www.worldbank.org/en/topic/poverty/publication/fragility-conflict-on-the-front-lines-fight-against-poverty

World Bank. (2021a). *Poverty & equity brief (Africa western & central): Liberia*. https://databankfiles.worldbank.org/public/ddpext_download/poverty/987B9C90-CB9F-4D93-AE8C-750588BF00QA/AM2020/Global_POVEQ_LBR.pdf

World Bank. (2021b). *The World Bank in Liberia* [WWW Document]. Retrieved June 10, 2022, from www.worldbank.org/en/country/liberia/overview#1

World Bank. (2021c). *Proportion of seats held by women in national parliaments (%) – Liberia* [WWW Document]. Retrieved June 10, 2022, from https://data.worldbank.org/indicator/SG.GEN.PARL.ZS?locations=LR

World Bank Group, Kelly, J. T., Rubin, A., Ekhator-Mobayode, U., & Arango, D. J. (2021). *The risk that travels with you links between forced displacement, conflict and intimate partner violence in Colombia and Liberia*. https://documents1.worldbank.org/curated/en/449471635478676087/pdf/The-Risk-That-Travels-with-You-Links-between-Forced-Displacement-Conflict-and-Intimate-Partner-Violence-in-Colombia-and-Liberia.pdf

11

GENDER, HUMAN RIGHTS AND WATER GOVERNANCE IN INDONESIA

Stroma Cole, Paula Skye Tallman, Gabriela Salmón-Mulanovich, Binahayati Rusyidi and Yesaya Sandang

Introduction

This chapter explores gender and water governance in Indonesia from a human rights lens. As elsewhere in the world, women in Indonesia are responsible for water management at the household level but are largely excluded from its governance at larger scales. The chapter explores the causes and consequences of this paradox.

Following a background focused on gender, water and human rights and the importance of the UN guiding principles, the paper outlines Indonesia's water governance and how gender mainstreaming has been incorporated at the national framework level. We then identify reasons why these national initiatives are not translating into increased gender equality in water governance on local levels. These reasons include patriarchy, legal gaps, overlapping mandates, lack of law enforcement, competition with other industries, unequal urbanisation and climate change, which all lead to increased water insecurity for women.

The paper then uses examples from across the archipelago to illustrate the consequences of poor water governance for women in Indonesia, including reduced opportunities, increased socio-economic inequalities, unheard voices and socio-cultural conflict. Through case-studies, we demonstrate that women have experienced little improvement in having their voices heard in water governance, but that this does not need to be the case. We end our chapter with three examples of how women can be better included in water governance to improve their access to water.

Background

Water, gender and human rights

For close to 50 years, international human rights organizations have recognised that all people are entitled to access to clean drinking water (UN, 1977) and that women in particular require support to ensure this right is materialised (Singh et al., 2008; UN, 1979). In 2015, the United Nations General Assembly added the Sustainable Development Goal (SDG-6) of "clean water and sanitation for all", which included six "outcome-oriented targets" such

DOI: 10.4324/9781003100379-14

as ensuring safe and affordable drinking water, ending open defecation, providing access to sanitation and hygiene and improving water quality, among others.[1] Yet, in 2016, the report of the Special Rapporteur consulting on gender, human rights, and water and sanitation for the United Nations Office of the High Commissioner found that gender inequalities in accessing water and sanitation permeate the life cycle of women (Heller, 2016).

The 2021 Summary Progress Update on SDG-6 was equally dismal. This update stated that:

> [B]illions of people worldwide still live without safely managed drinking water, safely managed sanitation and basic hygiene services, especially in rural areas and least developed countries; the current rates of progress need to quadruple in order to reach the global target of universal access by 2030.
>
> *(UN Water, 2021)*

Furthermore, the one "outcome-oriented target" for SDG-6 that specifically mentions women (SDG-6.2 – adequate and equitable sanitation for all – with special attention to the needs of women and girls) showed worse outcomes between 2015 and 2017 (UN Water, 2021).

Despite the need for more attention to gender in achieving the human right to water, progress is slow in incorporating women's participation in water governance. Indeed, the UN-WATER Progress Update (2021) found that fewer than half of the reporting countries have laws or policies that mention women's participation in sanitation and/or water management. In this context, achieving SDG-6, specifically for women, will require large investments and accountability, particularly including changes in governance to address gender, water and human rights.

Gender and the United Nations Guiding Principles on Human Rights

The United Nations Guiding Principles on Business and Human Rights (UNGP) are a non-binding international legal instrument that provides a global standard for preventing and addressing violations of human rights associated with business activities (OHCHR, 2011). According to UNGP principles 11 and 14, businesses have a responsibility to respect the human right to water and these concerns must consider gender. Women and girls are affected differently and disproportionately by business operations. For example, if a water source is contaminated as a result of a business activity, affected women in rural areas may have to travel longer distances every day to obtain potable water (Hall et al., 2014).

There have been efforts to incorporate gendered dimensions in business activities and their human rights impacts with clear landmarks. This includes 2017 and 2019 UNGPs Working Groups (hereafter Working Group), which brought a "gender lens to the UNGPs", by developing practical guidance incorporating a three-step cycle: (1) gender-responsive assessment, (2) gender-transformative measures and (3) gender-transformative remedies. These steps ask companies to evaluate gendered risks and impacts (e.g., gender disaggregated statistics), develop gender-sensitive and gender-responsive policies and plans, identify overlapping vulnerabilities, support women's equal participation in consultations and negotiations, and provide equal access to compensation and other forms of reparation (Deva, 2019).

According to Deva (2019), following the three-step cycle, impact assessments for businesses should (1) assess how biased norms and patriarchal power relationships impede women's human rights to water, (2) develop subsequent actions and remedies that attend to these dynamics and ultimately (3) transform the patriarchal norms and uneven power relations, which are at the root of discrimination, gender-based violence, and gender stereotyping. These efforts are important because in many contexts, women lack access to decision-making and men have legal authority over natural resources, such as land titles, ensuring secure access to water.

Gender and water governance

Water governance "refers to the political, social, economic, and administrative systems that influence water's use and management".[2] It is essentially how we get water, how much we can access, and how frequently we use it. It has been stated that issues regarding water problems are governance crises (UNDP, 2006) and they do not occur in a "social vacuum" (Cleaver & Hamada, 2010). Rather, "the social attributes, roles, opportunities, status and the relationships between women, men, girls and boys in access to water (resources) and participation in water governance are socially constructed and context-specific" (Bagonza Asaba, 2013).

Across a variety of geographic areas, women are predominantly responsible for household water management while facing barriers to participating in water governance elsewhere (Assefa et al., 2021; Cairns et al., 2017; Das, 2014; Eaton et al., 2021; Imburgia, 2019). Barriers include legal complications in participating in water associations, gendered roles and power relations that reinforce the notion of masculine work related to irrigation and other water-related projects (Imburgia, 2019). Das (2014) also found that entrenched social norms along with socioeconomic status prevented women's participation in water governance (Das, 2014). Despite these barriers, women tend to express their willingness to participate in water governance and decision-making, likely because of their motivation to ensure adequate supply (Das, 2014; Imburgia, 2019; Naiga et al., 2017). Moreover, Naiga et al. (2017) found that when women were relegated to minor responsibilities in water associations, they were more heavily invested in water managing activities compared to men (i.e., participating in water users meetings, cleaning water infrastructure, etc.).

Suggestions for including more women in decision-making roles for water governance include establishing a gender quota for water associations that comprise a proportion of women trained in technical knowledges, raising awareness of the importance of including women in all levels of the water association and facilitating their participation (Imburgia et al., 2021). Naiga and colleagues (2017) also showed that conducting an analysis of the cost-benefits of women's involvement in water governance can help to push back on gender stereotypes. They also suggested the implementation of simple technologies that could alleviate time for water provision at home, freeing time for women to participate in water governance activities and proposed school programming that included discussions of gender roles among boys and girls in the classroom. These actions can exert an influence on non-material resources such as policy and knowledge, or rules to access water, which can have differential impacts on men and women facilitating or precluding women from assuming leadership roles (Hannah et al., 2021).

Human rights and water governance in Indonesia

In Indonesia, the SDG's have been translated into four national priorities, or "outcomes", developed through the UN Partnership for Development Framework. "Outcome 2" in

Indonesia stated that the goal for 2020 was to have, "the poor and most vulnerable have better and more equitable access to quality basic social services, and to comprehensive social protection, and better access to water supply and sanitation" (UNPDF, 2017). Additionally, Article 6 of Indonesian Water Resource Law of 2019, stipulates that the state is responsible for fulfilling people's rights to access water. To address these aims, the Indonesian government set targets for providing access to clean water and sanitation for all people in the National Development Plan of 2015–2019. While progress has been made, national statistics indicate there are still gaps. In 2021, 90.78% of households had access to a proper water source, with greater access in urban areas (96%) compared to rural regions (83.91%). In 2021, 80.29% of Indonesian households had access to proper sanitation (Badan Pusat Statistik, 2022a, 2022b).

Indonesia has implemented political reforms to try and make its water development process more inclusive, incorporating gender equality in water governance. According to Al'Afghani et al. (2019), water management in Indonesia can be categorised into two approaches: top-down and bottom-up. Top-down water policies are developed by the central government and mandated to provincial and municipal governments to fulfil the needs of all citizens, for example through local tap water agencies (Al'Afghani et al., 2019). Since 2009, Indonesia has implemented the Water Supply and Sanitation for Low Income Community Programme (known as PAMSIMAS) that explicitly requires the engagement of women, men and persons with disabilities in the whole process. Within the periods from 2006 to 2018, 17.2 and 15.4 million Indonesians in 33 provinces have benefited from improved water and sanitation respectively, making PAMSIMAS the biggest community-based water-related provision programme in Indonesia (World Bank, 2019). At the same time, water-associated programmes are encouraged to design bottom-up initiatives in lower government levels through *Musyawarah Perencanaan Pembangunan* (*Musrenbang*). *Musrenbang* is a consultative meeting that involves government, non-government and community sectors to discuss and decide annual development priorities at the village, sub-district and district levels.

Available evidence suggests that, in general, Indonesian women are not regarded as key actors and beneficiaries in water management programmes that are top-down or bottom-up. At higher levels of government, there are a very limited number of women who have strategic positions in decision making, including in sectors related to water development (Akbar et al., 2020; Arum et al., 2006; Prasetyo et al., 2019). At the local level, women representatives who attended the *Musrenbang* are selected by the local elites and their attendance is merely symbolic and only done to meet formal administrative requirements. While there is evidence indicating that women's voices and needs are considered in discussions at the household level, particularly with regard to water-facilities installation (e.g., the location and height of the water tap) and that they have participated in the implementation of community-based water and sanitation provision programmes (Yuerlita & Saptomo, 2008), often they are discouraged to voice their perspectives or suggestions about water projects because water projects are considered "men's business" (Prasetyo et al., 2019). Thus, despite the government's efforts to advance gender equality, a discrepancy remains between policies and local implementation.

Causes of poor water governance in Indonesia

In this next section, we use examples from across Indonesia based on studies by the authors to identify the causes and consequences of poor water governance for women in Indonesia.

These studies include research in Bali, a well-established tourism island, Labuan Bajo, an emerging destination on the Island of Flores, Yogyakarta, a city on the island of Java and Sumba, a remote island in the East Nusa Tenggara Province, where the authors have recently conducted a study exploring the links between household water insecurity and gender-based violence.

Patriarchal formations

Indonesia is deeply patriarchal (Wahdiniwaty & Rustam, 2019) and this is one of the root causes of the inequitable water governance. In Yogyakarta, Prihatiningtyastuti et al. (2020) found that traditional values governing access and distribution of clean drinking water in rural areas failed to recognise the role of women in the household economy. Even when women raised concerns, their perspectives and experiences were frequently dismissed (Prihatiningtyastuti et al., 2020).

Similarly, Cole's (2017) work in Labuan Bajo, Flores, East Indonesia demonstrated that the paying of bridewealth by the groom's family to the bride's was a patriarchal practice that lay at the root of women's water struggles. Both men and women expressed that wives have been bought and therefore all domestic (including water-related) work was their duty. Husbands who participated in domestic duties or queued for water tended to be socially criticised and would feel embarrassed to do "women's work". A Sumba health worker recounted, "the culture here obliges women to serve their husband. At home, all the work is women's responsibility". Women's experiences, however, are not monolithic but are rather influenced by intersectional oppression, based on gender, ethnicity, class, caste and other forms of discrimination. Poverty compounds gender inequality and this is frequently related to ethnicity.

In Labuan Bajo for example, the original settler families – historically peripheral and impoverished fisher folk – cannot use wells due to saltwater intrusion and do not have the physical space or financial resources for a water tank (Cole, 2017). In Sumba, a caste system operates, impacting women's burdens in relation to water access. The authors interviewed local women and community leaders in East Sumba in 2020. A religious leader in one of the study sites stated, "if the water source is located on land owned by the aristocrats (maramba), people from the lower classes cannot take water easily". Women who are from the lowest "slave" (*hamba*) caste suffer the most. They have total responsibility for finding water and are the target of their owner's irritation whenever problems arise. As one respondent in a focus group held in Sumba in 2021 explained "Whatever is the wish of the master, even if the slave is tired, she has to carry on" or as another put it: "As long as she is still breathing, she must still fetch water". Furthermore, due to patrilocal residence daughters in law become prime water collectors.

Gaps in the laws concerning the right to water

Foundational for all Indonesian laws, the Constitution (set out in 1945), articulates the notion of human rights (Marzuki, 2011). However, it was not until 2019 that the Indonesian Water Resource Law made the human right to water explicit. The phrase used is "the people's right to water", following the definition of the human right to water in the UN General Assembly's Resolution 64/292 (A/RES/64/292 of 28 July 2010). Despite this, the group Indonesian Women Solidarity highlighted the absence of a specific provision for gender equality within water regulations in Indonesia. Furthermore, the law allows water to be considered a commodity that can be privatised, monopolised and industrialised in the name of investment and development. Water privatization, as seen in Jakarta, causes price spikes, increasing the burden on women as household finance managers (Indonesian Women Solidarity, 2020).

The Water Resource Law 2019 places the government and public in the principal roles to ensure that corporate use of water does not negate the human right to water. Nonetheless, full community involvement, necessary to safeguard this right by law, is limited to the environmental impact assessment mechanism. Adherence to rigorous water impact assessments is incredibly troublesome, especially for women (Sandang, 2021). Meanwhile, water regulations at the provincial level give greater emphasis to the technical aspects of groundwater utilization, which is the most common source of water for the community. Thus, the Water Resource Law 2019 does not require a thorough impact assessment of business water use that includes women's participation. As a result, the gendered impact of business water use is absent (Sandang, 2021).

Multiple levels of governance

Indonesia is divided into 34 provinces, each of which is divided into several regencies, which are subdivided into several districts which are further subdivided into villages. At the national, provincial and regency level, each is vertically divided into several departments, eleven of which have mandates for water (Cole, 2012). This complex picture results in overlaps and gaps.

The problems of overlapping mandates by the multiple government departments involved in water and its management have been discussed widely. The lack of coordination between different levels of local government was observed in Sumba. For example, a *Musrenbang* at the village level might agree on water infrastructure needs, only for those needs to be ignored or over-ruled at the district level. Compounding the overlapping mandates, a study by Sandang (2021) in Yogyakarta documented how the provincial government struggled to protect the human right to water amidst the rise in population and the conflicting interest of the national government to promote business investment, which further pressures scarce water supplies.

Problems also arise from the fact that authorities from different districts manage different aspects of water flows, such as sites of water collection versus water extraction. In Bali, the provincial level government grants groundwater extraction permits while the districts collect groundwater extraction taxes. These taxes are an important source of district revenue, thus businesses are encouraged to use underground water supplies (Cole et al., 2021). Additionally, there is the issue of sustainable management of water sources. Specifically, the tourism industries located in Badung primarily use groundwater from the Denpasar-Tabanan aquifer. The recharge areas for this aquifer are situated in the neighboring districts of Tabanan, Gianyar or Bangli. However, the districts that are responsible for protecting and preserving the water recharge areas receive nothing from the taxes paid to Badung for water extraction (Cole et al., 2021). These examples echo literature on water which dominantly see water crises as a reflection of failures due to governance gaps in Indonesia (Mulyana & Prasojo, 2020; Mulyana & Suganda, 2017).

Lack of law enforcement

As Cole (2012) reported in Bali, there are laws in place, but they lack enforcement. Respondents throughout the research in Bali were keen to point out that the problem is "weak law enforcement", "lack of control over law breakers" and that "regulations do not function", "supervision is weak", and that "laws are not implemented". By 2019 there was some improvement, and the registering of wells was being taken more seriously. However, as Cole et al. (2021) argued, this has more to do with income generation than water conservation.

Meanwhile, Sandang's (2021) study in Yogyakarta argued that a lack of law enforcement correlates with the government's limited ability to control and monitor commercial water use. Yogyakarta's rapid hotel development has not been matched by the government's capacity to assess and supervise existing developments.

Prioritization of industrial use of water

In the cases included here, the competition for water came from the tourism industry and agriculture, which are the main economic activities in the study sites. In Labuan Bajo, the rapid development of tourism has made women the victims and bearers of the greatest burden due to the strain to manage increasingly complex water needs because of increased demand for water in the tourism sector (Cole, 2017). Similarly, in Bali, the tourism industry's excessive use of groundwater has negatively impacted groundwater levels, causing saltwater intrusion, land subsidence and poor water quality. According to the International Tourism Partnership-Destination Water Risk Index, Bali is currently ranked first in terms of extreme vulnerability to water stress (ITP, 2018). Meanwhile, three Indonesian destinations (Surabaya, Jakarta and Bandung) are classified as highly vulnerable to future water stress.

In Sumba, private sugarcane plantations compete for water supplies. One plantation currently covers 12,000 hectares and will be expanded to 50,000 over the next few years (Daniel et al., 2021; Susilowati et al., 2019). However, there is resistance from the local community who believe it violates the rights of traditional tribes in the area. A study by Daniel et al. (2021) suggested arguments voiced by a local non-government organization centered on the substantial use of water needed to support the plantation which will reduce water availability for local people. Furthermore, these companies monopolise groundwater wells and exclude local people. This competition for water risks abusing the human rights of the local people to access water by a private company and is likely to lead to social conflict.

Urbanization and climate change

The unequal distribution of water is amplified by urbanization and climate change. Indonesia is the fourth most populous country in the word, with a current average population of 278,823,441 people.[3] Much of this population is moving to urban centers: Indonesia has had the highest rate of urbanization in recent decades among Asian countries (Marta et al., 2020). Such high levels of rural-urban migration and increases in usage in centralised areas will continue to stress local water sources.

Climate change in Indonesia will also exacerbate the unequal water distribution, putting more pressure on women. The International Food Policy Research Institute estimated climate change impacts on Indonesia and predicted that the country will experience temperature increases of approximately 0.8°C by 2030. Their models also predict changing rainfall patterns with the rainy season ending earlier (Oktaviani et al., 2011). Shortened rainy seasons are already being experienced in some of the driest regions of the archipelago such as East Nusa Tenggara. In 2020, the island of Sumba reportedly had 249 days without rain.[4]

Consequences of poor water governance in Indonesia for women

Reduced opportunities

Poor water governance in Indonesia reduces women's opportunities to pursue economic opportunities, due to time spent accessing water and the fact that economically productive activities such as agriculture and animal husbandry require water. Working in East Sumba, the authors heard a local woman lament "too much time is spent to carry water, so it hampers other activities". She continued, "because water is still an issue, I don't think I can do other productive activities while I am burdened with getting water and taking care of domestic chores".

These dynamics were found across the archipelago. In Labuan Bajo, Cole (2017) found that younger respondents commonly cited water as a reason that women struggled to work. As one marine dive equipment controller explained: "Someone must stay at home to wait for the water to flow. I think I would let my wife work if the water flowed 24 hours a day, but now she needs to be at home to wait for it". More simply stated by another study participant in Labuan Bajo, "Modern women, they want to work, but because of dealing with the lack of water there's no time for working, waiting for water for hours every day" (Cole, 2017). Prihatiningtyastuti et al. (2020) quantified the association between water scarcity and reduced opportunities. In Yogyakarta they found that during the dry season women's working hours are reduced as they spend more time collecting water. The reduction in working hours resulted in a loss of at least 50% of their monthly income.

Increasing socio-economic inequalities

Water scarcity and privatization increased socio-economic inequalities across the field sites. For example, on Sumba Island, a local authority stated, "In Prailangina, some water comes from springs, other from water tankers that cost 350 thousand rupiah each". Despite the provincial minimum wage being IDR 1,975,000 a month, most rural families earn less than IDR 1,250,000. This means that buying a tank of water costs about 25% of their income, far exceeding the recommended guidelines in the human right to water. The expense of water from tankers is unfeasible for many and further reinforces socio-economic inequalities between those who can and those who cannot purchase from these privatised sources. Further, in Labuan Bajo, Cole (2017) observed that the more water that is purchased the cheaper it is, and the more regularly you order the more consistent your supply. Those who cannot afford to buy and store a full tanker end up paying up to three times as much as those that can (Cole, 2017), aggravating socioeconomic disparities.

Increased social conflict

In Yogyakarta, since 2014, social conflict between community and hotels has increased. As depicted in a documentary entitled *Behind the Hotels* (Laksono, 2015), women from Miliran district seek water from neighbors or nearby markets for daily household needs, such as cooking and bathing, because their wells dried up after a hotel began operating near their community. Escalating the tensions, in December 2019, the regional water authority instructed households to start storing their own water due to high intensity use by hotels and shopping malls. The measure was widely criticised, claiming that water services for hotels were being prioritised over water services for residents.

Intense criticism has also been levied against mineral water companies operating in Indonesia. These multi-nationals and national establishments extract the groundwater based on the volume approved by the government using advanced technology. Tensions and conflicts between local people and companies are growing. For instance, conflict between groups of farmers and local people against a large water company in Central Java lasted a decade. The majority of local communities who work in agriculture accused the company's operation of reducing water availability needed to irrigate their fields. The farmers were forced to lease water pumps to keep watering their fields and many households had to buy expensive water from water trucks (Cara, 2014).

In response to local protests, this mineral water company implemented a corporate social responsibility (CSR) initiative (Nashrullah, 2021), including providing WASH facilities and support for community-based water management. While CSR has been used to compensate local people for corporate abuse of community water rights, women are under-represented in the negotiations, advocacy and bargaining process; therefore, their concerns and interests may not be properly heard or addressed.

Women's unheard voices

Throughout the cases we found that women's voices were excluded in governance and decisions which affect water access and supply. For example, in the district government in Labuan Bajo there was only one woman out of 30 representatives (Cole, 2017). This severely restricts the chances of women's voices being heard in the policy development or planning of water services which is likely to further entrench policies that fail to provide for the needs of women.

In Yogyakarta, women have voiced concerns about water struggles and competition for water with hotels. They said water from the municipal water provider is still not accessible and is deemed unaffordable. At the same time, the groundwater quality is inconsistent and extra efforts such as filtering, or boiling are needed to make the water safe. However, when it came to deliberations about water governance and conflicts with hotels attended by government agencies, hotel representatives and NGOs, none of the women from the community participated. Furthermore, in a separate focus group discussion with numerous government agencies, only one woman participated (Sandang, 2021).

Thinking through alternatives

The United Nations Convention on the Elimination of All Forms of Discrimination Against Women highlights the state's responsibility to ensure that no public authority discriminates against women. As a result, the government should be more active in involving women in water management decisions.

Women's participation can be encouraged by inviting them to meetings to discuss water-related issues. Women then can actively participate by sharing their experiences of water use and management. Furthermore, gender equality necessitates greater efforts and collaboration between the government and non-governmental organizations to provide training and workshops. Leadership development and empowerment programmes can encourage women to attend and participate in public meetings. Training and skill development programmes are also critical for increasing women's confidence in participating in public decision-making.

The Srikandi Sungai (River Heroine) is an example of a community initiative to encourage women's active participation. Facilitated by Gajah Mada University in Yogyakarta, it aims to increase women's leadership in water management by organizing and developing their leadership and activism skills. Women who participate in this initiative come from different professional backgrounds, education and ages. Through the River Heroine programme women show that they are empowered individuals with qualities to lead, participate and act for the betterment of water management in their communities (Rohmatin & Habsari, 2021). Consequently, women are becoming more involved in the decision-making process, determining water and sanitation policies in their environment (Gusti, 2021).

The authors found another example of women taking initiative in decision making processes in the village of East Sumba, where women held positions in village government. Village residents took their voices and concerns into account, including with regard to water access. In this case, they strongly supported the development of a village piped water system, believing it would benefit women. When water distribution was disrupted due to broken pipes, these women would push the village leaders to make immediate repairs to keep water distribution, reducing their burden to collect water elsewhere. These women also pressured the village government to provide water torrents to maintain the availability of water when the water supply was disturbed. In a specific instance, a woman held the key to the water source and filling of jerry cans could only take place when she unlocked it. She ensured that all households could only fill five-litre jerry cans per day, irrespective of caste, addressing existing socioeconomic disparities in access. She also ensured that extraneous activities, such as washing motorcycles and watering livestock could not take place at this protected source, prioritizing water for drinking and household use. Many women believe access to water is crucial for women not only for the domestic needs but also because some women in their villages depend on water to earn money (e.g., growing vegetables, farming chicken/goats or weaving traditional fabric). Thus, their experiences have translated into advocacy for effective water management on a local level. This local finding is in line with existing evidence demonstrating the benefits of involving women in water management. Importantly, participation in these roles must be compensated if women are to sustainably enter water governance and management.

On a larger scale, another solution is embedding a gender lens into a Human Rights to Water Impact Assessments (HRWIA). As an instrument for examining policies, legislation, programmes, projects and identifying (as well as measuring) their effects on the human right to water, a HRWIA needs to meaningfully address gender relations and the rights of women. To increase gender responsive HRWIA, three points need to be considered (Götzmann & Bainton, 2021). First, gender-specific context analysis should be implemented. This includes basic steps such as increasing the collection of gender-specific data. This sex-disaggregated data should be used to fully understand how proposed projects affect gender relations and women's rights. Second, women's participation in decision making around water should increase. This allows impact assessment practitioners to understand and positively impact how gender stereotypes, cultural protocols, women's workload, logistics and other factors limit women's participation. Third, tools, methods and skills should be adapted for enhanced gender responsiveness, due to the inadequacies of a gender-neutral approach to conventional environmental and social impact assessment. In other words, there is an urgent need to enhance gender analysis within the current set of impact assessment tools and methodologies set out by the UNGP that can support the human right to water.

Conclusion

This paper identified several reasons for the lack of women's involvement in water governance in Indonesia. Women's integral role in water provisioning combined with this lack of involvement in planning and decision-making means that the right to water is increasingly under threat. This has consequences: the reinforcement of gender and socio-economic inequalities, the reduction of opportunities for women's economic empowerment, and increasing conflict around water resources and access.

Women's human rights are an inalienable, integral and indivisible part of universal human rights. Both governments and businesses should take concrete steps to identify, prevent and address gender-based discrimination and inequalities in all aspects of life including the right to water (Deva, 2019). The Guiding Principles provide a framework for human rights based on a tripartite division between the government, business and community. However, what we observe is weak governance due to unclear legal frameworks, lack of law enforcement, overlapping mandates and a lack of coordination between departments. Meanwhile businesses either ignore the law or deploy paternalistic CSR to placate communities so they can operate "business as usual". Finally, communities, and particularly women, have limited opportunities to recourse when their human right to water is abused. As we have identified, patriarchal systems in Indonesia burden women with the worst effects of water insecurity but offer them the least opportunity to remedy these issues.

Alternatives must be found as we face increasing climate change, unequal and unplanned urbanization in Indonesia. Such alternatives include the proliferation of women's leadership development and empowerment programmes, human rights impact assessments with a gendered lens, and women's control over local distribution of water.

Acknowledgements

We would like to thank the British Academy for their support for the research in Sumba made possible by grant KF400136 and their support for the Labuan Bajo research SG143314. We would like to acknowledge the support of the Indonesian Endowment Fund for Education for their support for the research in Yogyakarta. We also would like to thank the NGO SOPAN for their field support and assistance in East Sumba.

Notes

1 United Nations (2017) Resolution adopted by the General Assembly on 6 July 2017, Work of the Statistical Commission pertaining to the 2030 Agenda for Sustainable Development (A/RES/71/313)

2 https://siwi.org/undp-siwi-water-governance-facility/what-is-water-governance?iproject=undp-siwi-water-governance-facility

3 www.worldometers.info/world-population/indonesia-population/

4 www.reuters.com/article/us-climate-change-indonesia-sumba-wideri/as-crops-fail-indonesias-sumba-seeks-lifeline-in-weaving-fishing-idUSKBN23936P

References

Akbar, A., Flacke, J., Martinez, J., & van Maarseveen, M. F. A. M. (2020). Participatory planning practice in rural Indonesia: A sustainable development goals-based evaluation. *Community Development, 51*(3), 243–260. https://doi.org/10.1080/15575330.2020.1765822

Al'Afghani, M. M., Kohlitz, J., & Willetts, J. (2019). Not built to last: Improving legal and institutional arrangements for community-based water and sanitation service delivery in Indonesia. *Water Alternatives, 12*(1), 285–303.

Arum, H. R. T., Sastrosasmita, S., & Isk, D. A. (2006). *Clean water governance in a small community: A case study of community development approach in the water provision in Yogyakarta.* Proceedings Regional Conference on Urban Water and Sanitation in Southeast Asian Cities, Vientiane, Lao. https://citeseerx.ist.psu.edu/viewdoc/summary?doi=10.1.1.125.4999&rank=1&q=Clean%20water%20governance%20in%20small%20community:%20A%20Case%20study%20of%20community%20development%20approach%20in%20the%20water%20provision%20in%20Yogyakarta,%20Indonesia&osm=&ossid=

Assefa, G. M., Sherif, S., Sluijs, J., Kuijpers, M., Chaka, T., Solomon, A., Hailu, Y., & Muluneh, M. D. (2021). Gender equality and social inclusion in relation to water, sanitation and hygiene in the oromia region of ethiopia. *International Journal of Environmental Research and Public Health, 18*(8), 4281. https://doi.org/10.3390/ijerph18084281

Badan Pusat Statistik. (2022a). *Proporsi Rumah Tangga yang Memiliki Akses Terhadap Layanan Sanitasi Layak Menurut Daerah Tempat Tinggal (Persen), 2020–2022.* www.bps.go.id/indicator/29/1270/2/proporsi-populasi-pendudukyang-memiliki-akses-terhadap-layanan-sanitasi-layak-dan-berkelanjutan-menurutdaerah-tempat-tinggal.html

Badan Pusat Statistik. (2022b). *Persentase Rumah Tangga Menurut Provinsi, Tipe Daerah dan Sumber Air Minum Layak (Persen), 2020–2022.* www.bps.go.id/indicator/29/854/1/persentase-rumah-tangga-menurut-provinsi-tipe-daerah-dan-sumber-air-minum-layak.html

Bagonza Asaba, R. (2013). *Gender, power and local water governance in Rural Uganda* [Doctor of Philosophy, National University of Ireland Maynooth]. https://mural.maynoothuniversity.ie/6752/

Cairns, M. R., Workman, C. L., & Tandon, I. (2017). Gender mainstreaming and water development projects: Analyzing unexpected enviro-social impacts in Bolivia, India, and Lesotho. *Gender, Place & Culture: A Journal of Feminist Geography, 24*(3), 325–342. https://doi.org/10.1080/0966369X.2017.1314945

Cara, C. C. (2014, September 14). *KONFLIK AQUA KLATEN: Spanduk Penolakan, Gesekan Warga Tulung dengan Wangen Terjadi.* Solopos.com. www.solopos.com/konflik-aqua-klaten-spanduk-penolakan-gesekan-warga-tulung-dengan-wangen-terjadi-535950

Cleaver, F., & Hamada, K. (2010). "Good" water governance and gender equity: A troubled relationship. *Gender & Development, 18*(1), 27–41. https://doi.org/10.1080/13552071003599996

Cole, S. (2012). A political ecology of water equity and tourism. *Annals of Tourism Research, 39*(2), 1221–1241. https://doi.org/10.1016/j.annals.2012.01.003

Cole, S. (2017). Water worries: An intersectional feminist political ecology of tourism and water in Labuan Bajo, Indonesia. *Annals of Tourism Research, 67*, 14–24. https://doi.org/10.1016/j.annals.2017.07.018

Cole, S., Wardana, A., & Dharmiasih, W. (2021). Making an impact on Bali's water crisis: Research to mobilize NGOs, the tourism industry and policy makers. *Annals of Tourism Research, 87*, 103119. https://doi.org/10.1016/j.annals.2020.103119

Daniel, D., Djohan, D., Machairas, I., Pande, S., Arifin, A., Al Djono, T. P., & Rietveld, L. (2021). Financial, institutional, environmental, technical, and social (FIETS) aspects of water, sanitation, and hygiene conditions in indigenous – Rural Indonesia. *BMC Public Health, 21*(1), 1723. https://doi.org/10.1186/s12889-021-11800-x

Das, P. (2014). Women's participation in community-level water governance in Urban India: The gap between motivation and ability. *World Development, 64*, 206–218. https://doi.org/10.1016/j.worlddev.2014.05.025

Deva, S. (2019). *Gender dimensions of the guiding principles on business and human rights* (SSRN Scholarly Paper N.º 3496398). Report of the UN Working Group on Business and Human Rights, A/HRC/41/43. https://papers.ssrn.com/abstract=3496398

Eaton, J., Krishna, A., Sudi, C., George, J., Magomba, C., Eckman, A., Houck, F., & Taukobong, H. (2021). Gendered social norms change in water governance structures through community facilitation: Evaluation of the UPWARD intervention in Tanzania. *Frontiers in Sociology*, *6*, 672989. https://doi.org/10.3389/fsoc.2021.672989

Götzmann, N., & Bainton, N. (2021). Embedding gender-responsive approaches in impact assessment and management. *Impact Assessment and Project Appraisal*, *39*(3), 171–182. https://doi.org/10.1080/14615517.2021.1904721

Gusti, G. (2021, August 10). *Srikandi Sungai Indonesia Dorong Pendirian Taman Sungai*. Universitas Gadjah Mada. https://ugm.ac.id/id/berita/19859-srikandi-sungai-indonesia-dorong-pendirian-taman-sungai

Hall, R. P., Van Koppen, B., & Van Houweling, E. (2014). The human right to water: The importance of domestic and productive water rights. *Science and Engineering Ethics*, *20*(4), 849–868. https://doi.org/10.1007/s11948-013-9499-3

Hannah, C., Giroux, S., Krell, N., Lopus, S., McCann, L. E., Zimmer, A., Caylor, K. K., & Evans, T. P. (2021). Has the vision of a gender quota rule been realized for community-based water management committees in Kenya? *World Development*, *137*. https://doi.org/10.1016/j.worlddev.2020.105154

Heller, L. (2016). *Report by the Special Rapporteur on the human right to safe drinking water and sanitation on his mission to El Salvador*. United Nations Digital Library. https://digitallibrary.un.org/record/847924

Imburgia, L. (2019). *Livelihood processes of gender roles and relations in water governance: A comparative study in Ethiopia and Argentina* [PhD Thesis, University of Reading]. https://centaur.reading.ac.uk/94859/

Imburgia, L., Osbahr, H., Cardey, S., & Momsen, J. (2021). Inclusive participation, self-governance, and sustainability: Current challenges and opportunities for women in leadership of communal irrigation systems. *Environment & Planning E: Nature & Space*, *4*(3), 886–914. https://doi.org/10.1177/2514848620934717

Indonesian Women Solidarity. (2020). *Equality & inclusion*. Solidarity Center. www.solidaritycenter.org/category/what-we-do/equality-inclusion/

ITP. (2018). *ITP destination water risk index*. UN Global Compact CEO Water Mandate. https://ceowatermandate.org/resources/itp-destination-water-risk-index-2018/

Laksono, D. D. (2015, February 6). *Belakang hotel*. Watchdoc Documentary.

Marta, J., Fauzi, A., Juanda, B., & Rustiadi, E. (2020). Understanding migration motives and its impact on household welfare: Evidence from rural – urban migration in Indonesia. *Regional Studies, Regional Science*, *7*(1), 118–132. https://doi.org/10.1080/21681376.2020.1746194

Marzuki, P. M. (2011). *An introduction to Indonesian law*. Setara Press.

Mulyana, W., & Prasojo, E. (2020). Indonesia urban water governance: The interaction between the policy domain of urban water sector and actors network. *Planning*, *15*(2), 211–218.

Mulyana, W., & Suganda, E. (2017). Water governance for urban resilience analysis of key factors and the role of stakeholders in Metropolitan Area. *Indonesian Journal of Planning and Development*, *2*(1), 11–18.

Naiga, R., Penker, M., & Hogl, K. (2017). Women's crucial role in collective operation and maintenance of drinking water infrastructure in Rural Uganda. *Society & Natural Resources*, *30*(4), 506–520. https://doi.org/10.1080/08941920.2016.1274460

Nashrullah, N. (2021, July 9). *Belajar dari Komitmen Danone-Aqua Terhadap Sosial Lingkungan*. Republika Online. www.republika.co.id/berita/qvz6b8320/belajar-dari-komitmen-danoneaqua-terhadap-sosial-lingkungan

OHCHR. (2011). *Guiding principles on business and human rights: Implementing the United Nations "protect, respect and remedy" framework*. United Nations Human Rights Office of the High Commissioner. www.ohchr.org/documents/publications/GuidingprinciplesBusinesshr_eN.pdf

Oktaviani, R., Amaliah, S., Ringler, C., Rosegrant, M. W., & Sulser, T. B. (2011). *The impact of global climate change on the Indonesian economy* (Discussion Paper N.o 01148; Environment and Production Technology Division). International Food Policy Research Institute. www.ifpri.org/publication/impact-global-climate-change-indonesian-economy

Prasetyo, D. D., Astini, F. N., Fillaili, R., & Widjanarko, H. (2019). *Gender and Social Inclusion (GESI) contextual analysis/formative research for the plan's water for women project in Indonesia*. The SMERU Research Institute. https://smeru.or.id/en/research/gender-and-social-inclusion-gesi-contextual-analysisformative-research-plan%E2%80%99s-water-women

Prihatiningtyastuti, E., Dayaram, K., & Burgess, J. (2020). Women's role in water management: A tale of two villages. In K. Dayaram, L. Lambey, J. Burgess, & T. Wulida Afrianty (Eds.), *Developing the workforce in an emerging economy* (pp. 68–82). Routledge.

Rohmatin, F., & Habsari, S. K. (2021, March 26). *Srikandi Sungai Indonesia: The role of women in water conservation*. OSF Preprints. https://doi.org/10.31219/osf.io/cdp7n

Sandang, Y. (2021). *Hotels and the human right to water: Prospects and challenges in Yogyakarta, Indonesia* [PhD Thesis, Faculty of Environment and Technology, University of the West of England]. https://uwe-repository.worktribe.com/output/7562679/hotels-and-the-human-right-to-water-prospect-and-challenges-in-yogyakarta-indonesia

Singh, N., Åström, K., Hydén, H., & Wickenberg, P. (2008). Gender and water from a human rights perspective: The role of context in translating international norms into local action. *Rural Society*, *18*(3), 185–193. https://doi.org/10.5172/rsj.351.18.3.185

Susilowati, M. H. D., Wibowo, A., & Susiloningtyas, D. (2019). Spatial pattern of drought disaster area and types of agriculture plant in Lebak Regency, Banten Province. *IOP Conference Series: Earth and Environmental Science*, *311*(1), 012022.

UN. (1977). *United Nations water conference: Mar del Plata action plan*. United Nations Digital Library. https://digitallibrary.un.org/record/136713?ln=en

UN. (1979). *Convention on the elimination of all forms of discrimination against women*. United Nations Human Rights Office of the High Commissioner. www.ohchr.org/en/instruments-mechanisms/instruments/convention-elimination-all-forms-discrimination-against-women

UNDP. (2006). Human development report 2006. *UNDP (United Nations development programme)*. https://hdr.undp.org/content/human-development-report-2006

UNPDF. (2017). *Government – United Nations partnership for development framework (UNPDF) 2016–2020*. Republic of Indonesia and United Nations System in Indonesia. https://indonesia.un.org/sites/default/files/2020-09/unpdf%202016-2020.pdf

UN Water. (2021). *Summary progress update 2021: SDG 6 – water and sanitation for all* (p. 58). www.unwater.org/publications/summary-progress-update-2021-sdg-6-water-and-sanitation-for-all/

Wahdiniwaty, R., & Rustam, D. A. (2019). Patriarchy as a barrier to women entrepreneurs in Indonesia. *IOP Conference Series: Materials Science and Engineering*, *662*(3), 032042.

World Bank. (2019, July 29). *Indonesia: Expanding access to clean water for the rural poor*. World Bank. www.worldbank.org/en/results/2019/07/29/indonesia-expanding-access-to-clean-water-for-the-rural-poor

Yuerlita, R. F., & Saptomo, A. (2008). *People's participation in rural water supply and sanitation project: A case study in Jorong Kampung Baru, Solok, West Sumatra, Indonesia*. 3rd WEPA International Forum on Water Environmental Governance in Asia, Bangkok Thailand. www.wepa-db.net/pdf/0810forum/paper09.pdf

12
PEACE, POWER, PARTICIPATION
Transboundary water cooperation through a gender lens

Rozemarijn ter Horst

Introduction

The topic of water diplomacy has gained attention over the past years, as the transboundary impacts of environmental issues are increasingly acknowledged at the international level. This has also increased the role and visibility of diplomats in environmental discussions (Pohl et al., 2014; Sehring et al., 2022). The way diplomats work influences policies (Neumann, 2012), and thus practices of water diplomacy shape water governance and management at different levels. In addition, gender influences water diplomacy and related negotiations, determining who is selected and allowed at the negotiation table as well as who is heard.

There is a small, yet growing body of literature that discusses gender, diplomacy, international negotiations and water. In this chapter, I take the space to share a brief overview of the literature on the topic of gender and international negotiations in general, as it provides an understanding of how diplomacy is gendered. I then discuss the literature that is specific to water diplomacy and international negotiations over water. Based on previous research, I discuss this separately as it brings together the fields of diplomacy as well as engineering. These are two distinct yet equally masculinized fields in which norms, values and histories are mainly shaped by men and male experiences (Sehring et al., 2023). I end this chapter by reflecting on questions that have been left unexplored, specifically on the relation between gender and the securitization of water and its influence on participation.

Gender, diplomacy and international negotiation

Feminist studies are increasingly attending to the participation of women and the genderedness of diplomacy and international negotiations. Genderedness entails how the:

> [A]dvantage and disadvantage, exploitation and control, action and emotion, meaning and identity, are patterned through and in terms of a distinction between male and female, masculine and feminine. Gender is not an addition to ongoing processes,

DOI: 10.4324/9781003100379-15

> conceived as gender neutral. Rather, it is an integral part of those processes, which cannot be properly understood without an analysis of gender.
>
> *(Acker, 1990, p. 146)*

The research shows that it is both valuable to "count heads" as well as to go deeper by unpacking the construction and influence of gendered dynamics on individuals, interactions and organizations. I will continue by summarizing insights provided by different researchers who made notable contributions to understanding power and participation of women in diplomacy.

Cynthia Enloe's *Bananas, beaches and bases: Making Feminist Sense of International Politics* (2014) brought the issue of women and international relations to the forefront. First published in 1990, the text asks, "where are the women?" and explores the meaning of masculinity and femininity in different international contexts and the ways in which these meanings shape power relations. The presence and experiences of women are often obscured from accounts on how international relations are or were practiced. While pointing out that feminist studies are centered around investigations of power, Enloe specifically asks: "Who gains what from wielding a particular form of gender-infused power? What do challenges to those wielding of that form of power look like? When do those challenges succeed? When are they stymied" (Ibid, p. 9)? She investigates these questions through stories of women affected by international tourism, wives living at army bases, domestic workers and wives of diplomats. While doing so, she makes those women and their experiences and struggles visible.

Gendered diplomacy and the mainly masculinized norms and practices of diplomacy and international negotiations influence existing stereotypes about how one should behave as a man or as a woman. Ifat Maoz (2009) studies the effects of the "women and peace hypothesis," which is a persistent idea that women are more cooperative. She researches this gendered evaluation effect in Palestinian-Israeli negotiations, finding that people expect women to look for joint solutions and thus value a compromise proposal offered by women as more realistic than such an offer made by men. She reaches a similar conclusion in a later study (David & Maoz, 2015). Sarai Aharoni (2017) confirms the influence strong stereotypes may have, including the power for women to use these stereotypes as strategic choice, as well as how it may limit them. She shows how women are not inherently more peaceful. It thus becomes much more interesting as to how these "incoherent" stereotypes come into existence, as well as what really influences practices. She points towards intersectionality to unpack how collective identity, class, unequal gender regimes, as well as patterns of political violence influence practices. By doing so, she shows the added value of feminist theorization in the discussion on the women and peace hypothesis.

Another notable contribution comes from Ann Towns and Birgita Niklasson (2017), who research the position of men and women in diplomacy through analyzing ambassador appointments. They study the positions of over 7000 ambassadors from the 50 highest ranked countries in terms of GDP in 2014. The research shows that there is an increase in the number of women who join as ambassadors, but that the number of women is still significantly less than the number of men; 85 percent are male. In addition, economic and military stations of posting seem to relate often with gender. In many cases, it appears that the higher the status of the postings in terms of military and economic power, the less women are appointed in these positions. The authors note that in many of the lower positions connected to these postings, the number of women is not higher than men. Women do not gradually disappear once positions increase in status, but generally have less access to positions with higher economic and military status in the cases they studied. The authors warn against generalizations though, as policies and culture that influence appointments can differ greatly between countries. Moreover, they point out that policies and culture are not static, especially as the research shows that the number of women seems to be on the rise in diplomacy.

Karin Aggestam and Ann Towns (2018) also engage with the questions posed by Enloe, starting with "where are the women?" in contemporary diplomacy and international negotiation. During the nineteenth century, concurrent with the professionalization of diplomacy, women generally have been barred from official diplomatic positions (Ibid, pp. 279–280), limiting the oftentimes important formal roles played by women (see also Sluga & James, 2016). This is currently changing with more women entering the field, and through programs such as the special attention of the United Nations for participation of women in conflict resolution, peacebuilding, peacekeeping, humanitarian response and in post-conflict reconstruction through the UN Security Council Resolution 1325 in 2000. Yet, the authors importantly add that there is a difference between counting heads, or descriptive representation, and understanding how women have influence or subjective representation. Subjective representation is highly influenced by norms and practices in diplomacy and negotiations. They show that these masculinized norms and practices exclude women, as they make it "difficult to visualize women in leading diplomatic positions" (Aggestam & Towns, 2018, p. 278), and make the case that it is necessary to include power, position and gendered hierarchies in studies on women's participation in diplomacy and negotiation.

In addition to counting heads and analyzing corresponding positions, Niklasson and Towns further unpacked gendered power dynamics in the edited special issue *Understanding the Gender of Ministries of Foreign Affairs* (Niklasson & Towns, 2022). This special issue in *The Hague Journal of Diplomacy* shows that the topic has indeed gained more ground over the past years. Towns and Niklasson focus on Ministries of Foreign Affairs (MFAs) as these ministries hire and select ambassadors and can be seen as the connection between the local and the international level. The special issue analyzes these ministries in three ways: first, as gendered institutions, in which the institutional roles, rules, practices, and power relations influence who is selected for certain positions. Secondly, it invites people to rethink what Ministries of Foreign Affairs (MFAs) are and how they work, moving away from a solely Western understanding, as most research on gender and diplomacy is done by Western researchers based on Western case studies. In a comparison, authors note how similar diplomatic femininities and masculinities are in the cases of the Bulgarian, Czech and Turkish MFAs, pointing towards an international institution, or circuit, of diplomacy. Lastly, the special issue also confirms how a division based on gender, combined with a hierarchy that ranks men and masculinity higher than women and femininity, diminishes opportunities for women to advance in the organization (see McGlen & Sarkees, 1993 on gendered hierarchies).

The literature reviewed in this section points towards how institutions, norms and practices of diplomacy are gendered, as well as how this gendered nature cannot be understood without including questions related to power. By looking at power relations, the literature shows that the differentiation between men and women, combined with expectations and hierarchies based on masculinities and femininities, creates different and gendered possibilities for advancement. For this reason, identifying who is in what position, or counting heads, is a good first step to see one of the ways the power of gendered diplomacy works. They show that there are globalized norms, values and stereotypes that shape the diplomatic circuit but also stress the importance of case studies that use an intersectional approach to avoid generalizations and recognize the interlocked workings of different forms of oppression based on gender, race, ethnicity, class, sexuality, caste, ability and other forms of discrimination (see Ojeda et al., 2022).

Gender in transboundary water governance

In the past section, I have shown how diplomacy and negotiations are gendered, both in terms of the number of men and women that participate, as well as the gendered norms and practices that influence who participates and how. In this section, I turn to gender and water diplomacy and international negotiations over water. In this chapter I define water diplomacy as:

> [T]he deliberative political processes and practices of preventing, mitigating, and resolving disputes over transboundary water resources and developing joint water governance arrangements by applying foreign policy means, embedded in bi- and/or multilateral relations beyond the water sector and taking place at different tracks and levels.
>
> *(Sehring et al., 2022, p. 36)*

Different tracks include a diplomatic track, a track that connects diplomacy with science, religion or civil society, and a track that connects people with people. Negotiations are an example of a deliberative process that is part of water diplomacy, through which two or more parties aim to come to a joint agreement on a certain issue. These negotiations can take place through ministries of foreign affairs, related to an ad hoc crisis including the construction of dams. Oftentimes they take place within the context of river basin organizations, platforms that are set-up to facilitate interactions of countries that share joint rivers and aquifers.

In the previous section I discussed how central the Ministry of Foreign Affairs is to diplomacy. When it comes to water, it is not only diplomats that take part in negotiations. Oftentimes it is a mix between professional diplomats or representatives of the ministry that is designated to manage water, such as a ministry of water or a ministry of the environment, as well as people with specific relevant experience related to water. Most probably, the gendered norms and practices of such a ministry or organization are very different from those of a ministry of foreign affairs. Essentially, water management remains a field that is dominated by engineering, and similarly to diplomacy, water management is also historically and currently a highly masculinized field. This influences who has access to certain education, positions, field work and rituals, and who is seen or envisioned as a "real" water manager (Liebrand & Udas, 2017; Rap & Oré, 2017; Shrestha et al., 2019; Zwarteveen, 2008, 2017). As with diplomacy, these norms and practices are also subject to change over time.

When it comes to descriptive representation, a survey done between 2018 and 2019 showed that women represent less than one-fifth of the staff in the highest leadership positions. This shows similar gender disparities found in the studies done by Towns and Niklasson (2017; Niklasson & Towns, 2022). Despite these numbers, women in decision making at the transboundary level, as well as representation in diplomacy and international negotiations is rarely researched. Anton Earle and Susan Bazilli (2013) are among the first to point out that gender is seemingly absent in discussions over, and analysis of, transboundary water management. They point towards the "hydraulic mission," the idea that nature can be dominated and that thus every drop of water can be used for human development, mainly through dams and other infrastructural projects, as one potential reason for this absence (see also Molle et al., 2009). Furthermore, women are portrayed mainly as users of water at the household and village level in the context of this "hydraulic mission," and not as agents that participate in its governance (De Silva et al., 2018). Natasha Carmi et al. (2019) showcase that women are part of water governance and make women and their agency in water diplomacy visible. They advocate for more involvement of women by pointing to obstacles women face to both access and participate in water diplomacy in Jordan, Lebanon and the State of Palestine.

The studies described above formed the main body of scientific literature on women, water diplomacy and international negotiations. While sparking interest and bringing important issues to the table, many questions are left open. These are similar to the questions that are brought up by Enloe (2014), Towns and Niklasson (2017) and Aharoni (2017) described in the section above. Where are the women? Why is their participation stymied? How are water diplomacy and international negotiations gendered, and how does this influence those who are involved in and impacted by international decision-making? The reason why such questions are asked is not value-free. They are meant to make imbalances visible with the aim of diminishing these imbalances. Thus, the final aim of asking these questions is to create more just water governance, based on an understanding that more equal representation is one of the main keys to achieving this.

Bringing together scientific insights on gender and water diplomacy

For the past three years, I was part of a research team that worked towards contributing to knowledge creation and drawing attention to the genderedness of water diplomacy. We conducted a workshop in 2021 on "(En)Gendering Transboundary Water Governance: Feminist Perspectives on Water Conflict and Cooperation" (Offutt, 2020) that brought together scientists and practitioners working on gender and transboundary waters. Based on the workshop, Jenniver Sehring, Margreet Zwarteveen and I worked on an edited volume, *Gender Dynamics in Transboundary Water Governance: Feminist Perspectives on Water Conflict and Cooperation* that aimed to bring together research on gender, water, diplomacy and negotiations and to create a starting point for conversations and research on the topic. We decided to collect contributions on this topic as there is still very little written on how gender plays out in negotiations and interactions over transboundary waters. With contributions of 20 authors, the book includes case studies from the Brahmaputra, the Chu-Talas, the Danube, the Indus, the Jordan, the Nile, the Rhine and the Zambezi basins. Jointly, they show that gender influences who has access to the negotiation table. I will briefly share the main insights of the book, as the chapters contribute to an understanding as to how water diplomacy and international negotiations are deeply gendered.

First, the framing of gender's role in transboundary water governance shapes our understanding. Is the focus merely on states, and state relations? Or do we acknowledge the entanglement of the personal and the professional to play a role as well? Ritu Priya and Tania Debnath (2023) show that these boundaries and frames influence literature on international relations and theoretical choices. "Seeing" gender requires critically re-thinking ontological categories and definitions. In practice, this is oftentimes challenging, as people working on water diplomacy and international negotiations often do not acknowledge the deep genderedness of their work and working environment. Those who work for the adoption of gender policies and gender mainstreaming may not be taken seriously, as they bring in ideas that counter the general norms, ideas that may even be resisted (Kunz et al., 2019).

Further research on gender dynamics in negotiations show that there are strong perceptions as to how women and men should behave, with aggressive and confrontational behavior accepted and expected of men, and cooperative behavior, including attentively listening, accepted and expected of women. These expectations influence self-governance as well as interactions between individuals. For instance, studies have shown how women's participation leads to more cooperative behavior in mixed negotiation teams in the Nile, Chu-Talas and Rhine basins (Mattur & ter Horst, 2023; Said, 2023; Sehring, 2023). However, gender is not a stand-alone issue, as it intersects with other determinants that influence power in a certain context, including caste, class, ethnicity and income. People navigate these dynamics in very different ways, from adhering to gendered norms to adopting hegemonic norms of the other gender, or by disengaging and protesting.

Many of the insights are similar to what is found in the research on diplomacy and international negotiations. Insights from the volume also show that context matters in order to understand the impact of how water diplomacy finds itself at the intersection of both the masculinized norms of water engineering and diplomacy. Context is also critical in order to acknowledge the diversity in cases and experiences of individuals, all based on different interactions between collective identity, class and unequal gender regimes (Aharoni, 2017). In this sense, the edited volume also confirms the point made by Ann Towns and Birgita Niklasson (2017) who warn against generalizations. Many of the interviewees in the case studies on the Chu Talas and Rhine basis indicated that they had not thought about gender often and were curious to learn why we brought up the topic, a situation echoed in research on diplomacy. Others indicated that gender was not the most important issue, or that being a woman benefited rather than limited them. Towns and Niklasson shared a similar experience when they conducted interviews with diplomats in Stockholm in 2014. They were told that "women face few limits *as women* in diplomacy" (Towns & Niklasson, 2017, p. 522). However, theirs and our data tell that oftentimes the story is different. In most cases, once the interviews progressed, examples were given of situations in which gender, gendered stereotypes and norms and values have influenced access and participation. Interestingly, for several interviewees this meant that contradictions could be found in their narratives: some expressed that gender did not play a (big) role while sharing experiences and practices in which it did matter.

Reflection: water securitization, participation and gender

There are many questions left open on how water diplomacy and international negotiations are gendered, how these dynamics change over time and how they shape water governance. One element that has come up based on the presentation of the book *Gender Dynamics in Transboundary Water Governance* is the impact of water securitization. Securitization of water is the idea that states ensure water availability and are tasked to protect and secure these resources. This happens through discourse and infrastructure, or institutionally through exclusion of certain groups from decision making processes.

Water securitization is oftentimes triggered by sudden events, including droughts or infrastructure development (Fischendler, 2015). Outside of the water sector, and based on the case of Israel, Aharoni (2018) shows that securitization restricts women's access to diplomacy. She bases her analysis on the secret meetings of Golda Meir and King Abdullah in 1947–48, the Oslo peace process 1993–2000 and the 2007 Annapolis Peace Summit. Her analysis shows how the extreme conditions of militarization, secrecy and structurelessness work to create an environment of exclusion, and that based on gendered hierarchies, mostly women are excluded.

How does securitization of water influence gender and participation? In case there is little conflict over the shared resources, is there more space for women to be included in its governance? And in case conflict flares up and water becomes an issue of high politics, will the men move (back) in, for instance because other ministries with other gendered norms and practices become involved? Securitization and conflict are part of the entanglement between cooperation and conflict, yet in research on transboundary water and gender, interactions between water, gender, conflict and securitization and participation have not been made explicit (Offutt, 2020).

Interestingly, Ann Towns (2020) shows that diplomacy is oftentimes associated with a feminine approach to transboundary interactions, as opposed to military power and violence. Listening and negotiating is seen as a feminine activity and as an art. This is also often mentioned in case studies on the Chu-Talas, Nile and Rhine basin. Collaborative skills, including listening and creatively identifying joint solution is seen as related to being a woman, and antagonistic behavior, including shouting and non-cooperation is associated with being a man (Sehring et al., forthcoming). Yet, as shown before, seeing diplomacy as a feminine approach does not translate into a larger number of women involved. We have shown that this relates to many factors, including education, hiring practices, job structures, as well as ideas of what a water expert looks like. Another factor may be the level of securitization of water.

Aharoni also brings up the issue of *tokenism*, that is, how the participation of one or few women can be used to show a symbolic effort, but not a real and transformative change. She expresses the expectation and hope that with participation of non-elites in negotiation, there will be space for "transformative politics that involves re-allocation of resources and a deeper commitment to non-violent ways of conflict resolution" (Aharoni, 2018, p. 209). This hope is shared by Eric Blanchard (2003) who writes that participation of women in masculine high politics will lead to a more comprehensive security, as with a feminist critique, new approaches and questions are being introduced. This includes questioning war and its influence on civilians and the focus on borders and securing them.

Based on a case of environmental security in Colombia, Keina Yoshida and Lina Céspedes-Báez (2021) also confirm that securitization works to promote exclusion. They criticize the Women, Peace and Security agenda of the United Nations, based on the UN Security Council Resolution 1325 in 2000, which treats women mainly as victims of war and conflict. This is similar to the dynamics of the water sector, as described in the previous section on "Gender in transboundary water governance." They argue that to "add women and stir" is not enough. This superficial approach can easily disregard gendered identities and power dynamics, as well as the high diversity amongst individual experiences. They point out that an intersectional approach can help to both make women visible, as well as draw attention to the differences in experiences and power. Yet, as shown in the cases of Israel and Colombia, a securitization agenda limits the space to do so.

Diplomacy and international negotiation are changing, both with the participation of women, and the participation of broader sectors of civil society, press, religious groups and science (Sending et al., 2011). But adding different groups "and stirring," as Yoshida and Céspedes-Báez (2021) point out, may not be enough to lead to structural changes, which are changes that may last when securitization comes into play. For this to happen, it is necessary for the norms and practices, or "the rules of the game" to fundamentally transform.

The relation between gender, participation and securitization in water diplomacy and international negotiation has not been deeply explored. I will broadly reflect on this issue based on two different case studies. The first case study is on the International Commission for the Protection of the Rhine (ICPR), composed of France, Germany, Luxembourg, the Netherlands, Switzerland and the European Union. The Commission was set up directly after the Second World War to facilitate international exchanges and agreements on the management of the waters of the Rhine. Initially, the activities were limited to monitoring and data exchange (Dieperink, 2000; Ruchay, 1995). Cooperation remained limited until the 1986 chemical spill in Sandoz, Switzerland, which greatly affected the Rhine, killing most salmon in the river. This led to civil society campaigns and joint political support in the basin to develop activities to bring back the salmon and protect water quality. The common policy project facilitated cooperation in the basin, which is currently increasingly supported through European cooperation, for instance through the EU Water Framework Directive. The ICPR sees itself as an example of cooperation, receiving visitors from all over the world to share their experiences. The ICPR history shows a story of decreasing securitization. Although water quantity is becoming a sensitive issue, the ICPR continues to provide a framework in which its members can explore options to engage.

The ICPR is also an example of a river basin organization that strives for a gender balance, or a gender imbalance in the favor of women. Interestingly, women are in large majority at the level of minister. At the level in the secretariat, more space for women was only created after the implicit rule was abandoned that a Dutch representative, as a downstream country, was to be selected as secretary. Before this, the pool of people to select from consisted mainly of Dutch male engineers. According to interviewees, women's participation at the country level started to increase 15 years ago, first in the Netherlands, then Germany, then France, yet in Switzerland little has changed over the years (Mattur & ter Horst, 2023). Differences between countries delegations thus do exist, highly influenced by the organizational cultures of the ministries and organizations that select delegates. There are also differences in gender representation in various working groups, in which interviewees could identify topics that attracted more men, such as dikes, and those that attracted women, such as ecology, in which ideas about what are suitable studies for men and women play a large role.

Through interviews on the interactions in negotiations, it was noted that these changed from being cold, distant, sometimes aggressive and focused on maintaining a country's position, to warm, familiar, explorative and cooperative. Interviewees at the ICPR indicated that they thought the change came as the major transboundary issues were addressed and issues related to the Rhine are not securitized. Many also indicated that this change happened due to the participation of women in negotiations, and perceived norms of how to behave appropriately towards "the other gender" (Sehring et al., forthcoming). Interestingly, the interviews also showed that there is indeed a hegemonic masculine norm. One example is how an interviewee learned how to negotiate as a man, which entailed her taking more space, talking louder and compromising less, as she felt that this approach would lead to better results (Mattur & ter Horst, 2023). Although more research is needed to explore the link between securitization and gender participation in the case of the Rhine, the absence of securitization allowed for an open and cooperative atmosphere, in which participants could challenge the hegemonic status quo.

The second case study is on the work of Nadia Gefoun, a diplomat with an impressive track record in foreign service. She was a former Ambassador of Sudan to Norway and Denmark, Deputy Head of Mission of Sudan to Sweden, Chargé D'Affaires of Sudan to Malaysia, member of the Sudanese Mission to the United Nations in Geneva and Press Attaché for the Sudanese Embassy in Cairo. She shares that "traditionally, women were not part of the top leadership in anything related to natural resources, for example roads, mining, oil, and also water" (ter Horst et al., 2023, p. 188). She shared that she thought that the main reason why women were not present in negotiations was due to education, as few women trained as engineers.

A link with securitization and her position comes through the announcement of the Grand Ethiopian Renaissance Dam. The announcement and construction of the dam in Ethiopia securitized the waters of the Nile, including the participation of army officials in interactions between Egypt, Sudan and Ethiopia. Although she has worked on water in terms of reviewing agreements and being part of bilateral committees, she was not part of the negotiations that happened after the announcement of the Grand Ethiopian Renaissance Dam. While no conclusions can be drawn based on this limited information, it is interesting to look further at what happened to Sudan in the past years. Nadia Gefoun shared that she had hoped that the Sudanese revolution, and the prominent role women had played in this revolution, could lead to more equality in representation (Ibid). The Minister of Water who was part of the government that took over after the revolution in 2019 explicitly chose to include young people and women in the team that helped to prepare negotiations. The military coup in 2021 had a negative effect on the participation of women and young people. Two years is too short to change underlying norms, practices and policies, but they show the influence of civilian movements and determination of powerful and visionary individuals.

Conclusion

In this chapter I summarized insights on gendered participation in diplomacy and international negotiation in general, and for the water sector specifically, noting that dynamics present in diplomacy and international negotiation hold true for the water sector. In terms of descriptive representation, or counting heads, women are generally underrepresented, especially in positions with high economic and political status. Research shows that it is critical to look beyond descriptive representation and to understand how women and men have agency, what space they have to contribute to content, decisions and what power they have to ultimately shape water governance. When it comes to this subjective representation, we have discussed several elements related to both agency and structure.

Specifically, when it concerns diplomacy and international negotiation related to water, two different fields come together, each with its own norms and values. Both are historically shaped by male experiences. One of the ways that this shapes water diplomacy and negotiations is through education. Some topics are considered more for men, including engineering. Anton Earle and Susan Bazilli (2013) show how water management at the international level is largely dictated by the "hydraulic mission," which relies on engineers, while including less input from lawyers, geographers or biologists.

Different case studies show how securitization can limit the space for the true participation of women or those who do not adhere to the status quo, leading to tokenism and symbolic participation. On the other hand, two case studies show that increased cooperation over the course of decades creates space for the increased participation of women, both through different policies of sending institutions and through changing norms. The case study on the ICPR shows that hegemonic masculinized norms remain pervasive, and that learning how to behave according to those norms can help someone in her and his job. The case study on the work of Nadia Gefoun in Sudan shows that sudden change can happen, through large supported civil society demands as well through visionary and powerful individuals.

Lastly, we discussed the centrality of contextualization and situatedness in feminist scientific literature. The experience of one is not the same as the experience of another; acknowledging and embracing difference is critical to better understand how water diplomacy and international negotiation are gendered, and with what effect. We highlighted intersectionality as a concept to unpack and understand these differences, and how elements such as class, caste and education impact representation and participation. Our work showed how intersectionality can help to understand changing practices in diplomacy and international negotiations. We also call for more case studies to acknowledge situatedness, contribute to the small body of literature on gender and water diplomacy and international negotiation, as well as to keep track of potentially transformational changes through increased participation of the non-elite.

References

Acker, J. (1990). Hierarchies, jobs, bodies: A theory of gendered organizations. *Gender & Society*, *4*(2), 139–158. https://doi.org/10.1177/089124390004002002

Aggestam, K., & Towns, A. (Eds.). (2018). *Gendering diplomacy and international negotiation*. Palgrave Macmillan.

Aharoni, S. B. (2017). Who needs the women and peace hypothesis? Rethinking modes of inquiry on gender and conflict in Israel/Palestine. *International Feminist Journal of Politics*, *19*(3), 311–326. https://doi.org/10.1080/14616742.2016.1237457

Aharoni, S. B. (2018). Diplomacy as crisis: An institutional analysis of gender and the failure to negotiate peace in Israel. In K. Aggestam & A. Towns (Eds.), *Gendering diplomacy and international negotiation* (pp. 193–211). Palgrave MacMillan.

Blanchard, E. M. (2003). Gender, international relations, and the development of feminist security theory. *Signs*, *28*(4), 1289–1312. https://doi.org/10.1086/368328

Carmi, N., Alsayegh, M., & Zoubi, M. (2019). Empowering women in water diplomacy: A basic mapping of the challenges in Palestine, Lebanon and Jordan. *Journal of Hydrology*, *569*, 330–346. https://doi.org/10.1016/j.jhydrol.2018.12.011

David, Y., & Maoz, I. (2015). Gender perceptions and support for compromise in the Israeli-Palestinian conflict. *Peace and Conflict: Journal of Peace Psychology*, *21*(2), 295–298. http://doi.org/10.1037/pac0000092

De Silva, L., Veilleux, J. C., & Neal, M. J. (2018). The role of women in transboundary water dispute resolution. In C. Fröhlich, G. Gioli, R. Cremades & H. Myrttinen (Eds.), *Water security across the gender divide. Water security in a new world* (pp. 211–230). Springer.

Dieperink, C. (2000). Successful international cooperation in the Rhine catchment area. *Water International*, *25*(3), 347–355. https://doi.org/10.1080/02508060008686842

Earle, A., & Bazilli, S. (2013). A gendered critique of transboundary water management. *Feminist Review*, *103*, 99–119.

Enloe, C. (2014). *Bananas, beaches and bases: Making feminist sense of international politics* (2nd ed.). University of California Press.

Fischendler, I. (2015). The securitization of water discourse: Theoretical foundations, research gaps and objectives of the special issue. *International Environmental Agreements: Politics, Law and Economics*, *15*, 245–255. http://doi.org/10.1007/s10784-015-9277-6

Kunz, R., Prügl, E., & Thompson, H. (2019). Gender expertise in global governance: Contesting the boundaries of a field. *European Journal of Politics and Gender*, *2*(1), 23–40. https://doi.org/10.1332/251510819X15471289106112

Liebrand, J., & Udas, P. B. (2017). Becoming an engineer or a lady engineer: Exploring professional performance and masculinity in Nepal's Department of Irrigation. *Engineering Studies*, *9*(2), 120–139. https://doi.org/10.1080/19378629.2017.1345915

Maoz, I. (2009). The women and peace hypothesis? The effect of opponent negotiators' gender on the evaluation of compromise solutions in the Israeli-Palestinian conflict. *International Negotiation*, *14*(3), 519–536. http://doi.org/10.1163/138234009X12481782336267

Mattur, R., & ter Horst, R. (2023). Governance in the Rhine basin through a Gender Lens: The international commission for the protection of the Rhine. In J. Sehring, R. ter Horst & M. Zwarteveen (Eds.), *Gender dynamics in transboundary water governance* (pp. 146–162). Routledge.

McGlen, N. E., & Sarkees, M. R. (1993). *Women in foreign policy: The insiders*. Routledge.

Molle, F., Mollinga, P. P., & Wester, P. (2009). Hydraulic bureaucracies and the hydraulic mission: Flows of water, flows of power. *Water Alternatives*, *2*(3), 328–349.

Neumann, I. B. (2012). *At home with the diplomats: Inside a European foreign ministry*. Cornell University Press.

Niklasson, B., & Towns, A. E. (2022). Introduction: Approaching gender and Ministries of Foreign Affairs. *The Hague Journal of Diplomacy*, *17*(3), 339–369. http://doi.org/10.1163/1871191x-bja10123

Offutt, A. (2020). Making gender visible in transboundary waters. *FLOWs: The Water Governance Blog at IHE Delft Institute for Water Education*. https://flows.hypotheses.org/5728.

Ojeda, D., Nirmal, P., Rocheleau, D., & Emel, J. (2022). Feminist Ecologies. *Annual Review of Environment and Resources*, *47*, 149–171. https://doi.org/10.1146/annurev-environ-112320-092246

Pohl, B., Carius, A., Conca, K., Dabelko, G., Kramer, A., Michel, D., Schmeier, S., Swain, A., & Wolf, A. (2014). The rise of hydro-diplomacy: Strengthening foreign policy for transboundary waters. *Adelphi*. www.adelphi.de/en/publication/rise-hydro-diplomacy

Priya, R., & Debnath, T. (2023). Is academic knowledge production on transboundary waters in South Asia gender neutral? In J. Sehring, R. ter Horst & M. Zwarteveen (Eds.), *Gender dynamics in transboundary water governance* (pp. 39–57). Routledge.

Rap, E., & Oré, M. T. (2017). Engineering masculinities: How higher education genders the water profession in Peru. *Engineering Studies*, *9*(2), 95–119. http://doi.org/10.1080/19378629.2017.1342255

Ruchay, D. (1995). Living with water: Rhine River basin management. *Water Science & Technology*, *31*, 27–32. https://doi.org/10.1016/0273-1223(95)00353-O

Said, A. (2023). The role of gender in transboundary water governance of the Nile Basin. In J. Sehring, R. ter Horst & M. Zwarteveen (Eds.), *Gender dynamics in transboundary water governance* (pp. 163–178). Routledge.

Sehring, J. (2023). Gendered river basin institutions: The Chu-Talas commission in Central Asia. In J. Sehring, R. ter Horst & M. Zwarteveen (Eds.), *Gender dynamics in transboundary water governance* (pp. 129–145). Routledge.

Sehring, J., Schmeier, S., ter Horst, R., Offutt, A., & Sharipova, B. (2022). Diving into Water diplomacy – Exploring the emergence of a concept. *Diplomatica*, *4*(2), 200–221. https://doi.org/10.1163/25891774-bja10082

Sehring, J., ter Horst, R., & Said, A. (forthcoming). Gender perceptions and women's participation in transboundary water governance. Insights from the Nile, Chu-Talas, and Rhine basins. *Political Geography*.

Sehring, J., ter Horst, R., & Zwarteveen, M. (2023). *Gender dynamics in transboundary water governance*. Routledge.

Sending, O. J., Pouliot, V., & Neumann, I. B. (2011). The future of diplomacy: Changing practices, evolving relationships. *International Journal*, *66*(3), 527–542.

Shrestha, G., Joshi, D., & Clement, F. (2019). Masculinities and hydropower in India: A feminist political ecology perspective. *International Journal of the Commons*, *13*(1), 130–152. https://doi.org/10.18352/ijc.920

Sluga, G., & James, C. (Eds.). (2016). *Women, diplomacy and international politics since 1500*. Routledge.
ter Horst, R., Amakali, M., Barua, A., Gefoun, N., Jekel, H., & Villar, P. C. (2023). Negotiating water: Lived experiences of female practitioner. In J. Sehring, R. ter Horst & M. Zwarteveen (Eds.), *Gender dynamics in transboundary water governance* (pp. 179–192). Routledge.
Towns, A. E. (2020). 'Diplomacy is a feminine art': Feminised figurations of the diplomat. *Review of International Studies*, *46*(5), 573–593. http://doi.org/10.1017/S0260210520000315
Towns, A. E., & Niklasson, B. (2017). Gender, international status, and ambassador appointments. *Foreign Policy Analysis*, *13*(3), 521–540. https://doi.org/10.1093/fpa/orw039
Yoshida, K., & Céspedes-Báez, L. M. (2021). The nature of women, peace and security: A Colombian perspective. *International Affairs*, *97*(1), 17–34. https://doi.org/10.1093/ia/iiaa173
Zwarteveen, M. (2008). Men, masculinities and water powers in irrigation. *Water Alternatives*, *1*(1), 111–130.
Zwarteveen, M. (2017). Hydrocracies, engineers and power: Questioning masculinities in water. *Engineering Studies*, *9*(2), 78–94. https://doi.org/10.1080/19378629.2017.1358730

13

NEW SPACES FOR WATER JUSTICE? GROUNDWATER EXTRACTION AND CHANGING GENDERED SUBJECTIVITIES IN MOROCCO'S SAÏSS REGION

Lisa Bossenbroek and Margreet Zwarteveen

Introduction

For the last couple of decades, groundwater tables in the Saïss agricultural plain, mid-north Morocco, have been dropping. This is partly due to intensified groundwater use in agriculture and changing cropping patterns. Specifically, since the 1980s, farmers have increasingly dug wells to engage in commercial agriculture. From growing rain-fed crops, they have shifted to higher-value irrigated crops, notably onions and other vegetables. Changing tenure relations (land privatization, most notably) and new agricultural policies that actively promoted modernization have further intensified groundwater use. Today, water-intensive crops produced in the region are sold on regional, national and even international markets. They produce new material and cultural linkages between hitherto unconnected life worlds, interactively re-shaping socio-natural relations. They open up new possibilities of living and being but also close off old ones.

In this chapter, we describe and analyze these changes to interrogate them from a feminist-justice perspective. We use material from extensive ethnographic fieldwork, conducted over a one-year period from 2011 to 2013, complemented by various recurring several-week visits over 2014 and 2015. We focus on anecdotes regarding changing groundwater and irrigation dynamics that illustrate how they are intimately linked to changing gender relations. Rather than merely assessing or mapping the gendered impacts of intensifying groundwater use, our purpose is to tease out how prevailing gendered ways of being, working and relating shape and in turn are shaped by new agricultural and groundwater dynamics, in ways that are not always straightforward or easy to predict. Hence, our argument is that groundwater abstraction and gender relations dynamics co-constitute each other, meaning one cannot be comprehended without the other. We show this by focusing on how the gendered organization of labor and space shape and make possible intensifying groundwater use and the larger changes it provokes.

DOI: 10.4324/9781003100379-16

Contemporary agrarian policies in Morocco intended to promote a new kind of Green Revolution are premised on promoting a particular gendered organization of farming (and consequently of families), attributing specific roles, responsibilities and rights to men and to women. The two main protagonists of agricultural modernization that the Moroccan government focuses on and wants to support are new agricultural investors and those existing farmers who have the spirit and willingness to modernize. In public policies, these protagonists are clearly cast as men – with farms figuring as enterprises that operate with a logic of profit maximization. Through agricultural support programs, extension services and subsidies, the farming policy model and farmers co-shape what happens on agricultural fields both materially and discursively. Yet, a closer look at how different people engage with changing farming realities and groundwater dynamics reveals a much larger diversity of practices and experiences, many of them hybrids combining peasant traditions and entrepreneurial logic. Complex intersections of class, generation and gender mingle with existing ways of doing agriculture, interacting with new policy directions and market opportunities to contingently produce a range of possible farming and water-use configurations. Each comes with different gender arrangements – different options for relating and performing as men or as women. The chapter presents some of these configurations and arrangements, based on ethnographic documentation of people's everyday experiences and engagements with changing groundwater dynamics. We do this to suggest that there is merit in theorizing water dynamics and gender dynamics as intimately linked: materially (through labor and property relations), and symbolically and discursively (through norms, meanings and symbols), with "gender" and "water" continuously defining and re-defining each other – moving together, as in a perpetual dance.

We first briefly review our theoretical sources of inspiration for understanding interactions between gender relations, gendered subjectivities and groundwater dynamics. We then provide a brief description of the Saïss agricultural plain and discuss the various processes that have contributed to intensifying groundwater use. In the third section we show how the Moroccan government's current agricultural modernization plans are founded on a particular farming model and farming identities – a model that both co-shapes contemporary agrarian dynamics and serves as the mainstream frame of reference that many researchers use to make sense of it. We then continue by using our empirical material to interrogate this model, showing how gendered meanings and identities evolve and are actively (re-)negotiated and performed through changing ways of dealing with groundwater. We conclude our paper with a brief feminist reflection on what these observations mean for the feminist analysis of, or feminist struggles for, water justice.

The interlinkages between gender and environmental change

Linkages between gender and the environment are a recurrent topic in feminist scholarship and debate. Much of this work starts from the original definition of gender itself, a definition that, by separating someone's biological sex from someone's social gender, is premised on the analytical possibility to distinguish Nature from society. Early feminist work on the environment, much of which was done under the broad umbrella of eco-feminism, attempted to demonstrate how, in modern societies, this separation (between Nature and society) justifies and produces social-valuational hierarchies, between people and animals, between people of different ethnicities, but importantly also between men and women. This work showed how modern societies organize activities (work), identities and much else through binary distinctions in which (some) men, culture, public life, rationality, productive work are considered as separate, opposed and often superior to women, Nature, private life, emotion and domestic work. In a feminist attempt to revalue the feminine side of this binary, much eco-feminist effort was invested in showing the importance of all that was associated with women. This often happened as part of a more encompassing critique of modernity, in which feminists formed productive alliances with a rapidly emerging environmental movement. Eco-feminists thus convincingly argued and showed how modern progress presupposes exploiting both women and Nature. Maria Mies' seminal work *Patriarchy and Accumulation on World Scale* (1986) demonstrates how capital accumulation is premised on subordinating and exploiting women, Nature and colonies[1] (Mies, 1986, p. 2). For Mies, just as for other eco-feminist scholars, the logical implication was that women were the protagonists *par excellence* of social movements for resistance and change (see Mies & Shiva, 1993). This was founded on the idea that women, because of how societies relegate them to the "Nature" side of the binary, have a closer relationship to Nature than men.

While ecofeminist analyses provide a powerful feminist critique of modernity and science, they are less useful for understanding everyday gendered experiences of living with the environment. Further, the idea that womanhood is a good basis for environmental activism proved questionable in practice: people's everyday relationships and dealings with the environment are only partly determined by their gender. In this chapter, we therefore propose complementing eco-feminist thinking with two other strands of feminist thought.

The first is a body of work that questions a-priori assumptions about Nature-society distinctions that inform both some eco-feminist as well as early feminist political-ecology analyses. The argument is that these distinctions did not exist prior to the analysis, history or human experience, but are instead themselves the product of ways of being, relating and conceptualizing that are always context specific. The boundary between Nature and society, in other words, is not dictated by a world that exists irrespective of how human beings make sense of it: this boundary comes into being through the very conceptual categories and definitions used to describe it. For gender, this line of thinking has most famously been articulated by Judith Butler. She took issue with how the definition of gender continues to assume some unchangeable biological core, which gets its context-specific and changeable feminine or masculine shape and meaning through processes of socialization. Butler instead posited that being or behaving as a man or a woman is always a performance, something that is not necessarily linked to biology. Gender is not something one "*is*" but something one "*does*". The performance of particular gender identities is based on "the tacit collective agreement to perform, produce and sustain discrete and polar gender as cultural fictions" (Butler, 1990, p. 138). Conceptualizing gender as performative implies that gender identities are never complete(d); they are continuously becoming. Identities, therefore, are necessarily fragmented and provisional; they are wrought through the interplay among various fields of power and regulatory frameworks (see also Resurreccion & Elmhirst, 2008, p. 15). As Butler (1990) argues, the coexistence (clashing or convergence) of various discursive injunctions itself enables re-configuration or re-deployment. It leads to theorizing the everyday social practices that constitute gender relations and environmental change as a spiral in which the two constantly (re-)define each other. Rather than theorizing gender subjectivities or relations as pre-existing or separate from the environment, this theorization posits that changes in the environment – in our case, changes in groundwater use, access and control – always happen and are defined through changing gender subjectivities and relations: gender and the environment move together.

A second strand of feminist work that we build on for our analysis entails a plea for taking seriously women's and men's own creativity, experiences, actions and self-perceptions in their dealings with others and with environmental (or more generally capitalist) transformations. Dissatisfied with analyses that reduce these experiences to Marxist false consciousness or interpret them as the result of Foucauldian disciplining and normalization, this work suggests that there may not be one larger structure or totality that holds reality together. Rather than helping uncover or expose this larger structure or totality (to either further strengthen it or to more effectively contest and resist it), feminist scholars in this tradition use ethnographic methods to trace the multiple ways in which realities or worlds are constituted and enacted in practice, accepting that there may be clashes, convergences or overlaps between these worlds, making them more or less durable or mobile. Methodologically, this work starts by emphatically acknowledging and learning from people's own ways of navigating and making sense of the multiple force fields that they belong to. Rather than using this as empirical material to either provide evidence for or dismiss larger theories, these scholars propose mobilizing it to *interfere*, *think with* or *destabilize* them in search of alternative ways of articulating and creating realities (Domínguez-Guzmán, forthcoming). The study by Katherine Gibson et al. (2001), although focusing more generally on global capitalist economic developments, provides an interesting example here. It focuses on the actions and life trajectories of female migrant workers in the Philippines, illustrating these women's multi-faceted experiences. To make sense of these experiences, the authors move beyond existing neo-liberal or Marxist frames of analysis, which tend to respectively frame female migrant workers as either "heroes" of national development, emancipation and modernization, or as "victims" of capitalist growth and exploitation. Their attempts to do more justice to migrant workers' own stories reveal the paradoxes and ambiguities in their experiences. Female migrant workers face hardship and exploitation, but also consciously strive to purchase particular assets and capacities that allow them to move upward socially. This careful ethnographic account of capitalist economic expansion experiences illustrates how gendered subjectivities and relations become redefined through new economic possibilities, with female workers assuming new aspirations and roles by identifying economic and cultural possibilities to support their home communities and realizing their own dreams in their host community.

Similarly, the work of Priti Ramamurthy attempts to grasp the ambiguous experiences and subjectivities of female rural wage workers (2003) and smallholders (2011) in South India within capitalist growth. She mobilizes the perplexity concept to mark tension among overlapping, opposing and asymmetric force-fields, expressing people's puzzlement as they experience both the joys and aches of global everyday life, often simultaneously (2003, p. 525). She argues that their perplexity captures their experiences of capitalism's contradictions through a structure of feeling, combining on-going struggles against existing socio-cultural norms and existing socio-economic structures (2011, p. 1054) with pleasures in the benefits and luxuries that come with new opportunities for wage work or more purchasing power. Taking inspiration from these bodies of thought, we place the ambiguous experiences and aspirations of men and women who experience groundwater intensification at the heart of our attempt to make a feminist contribution to thinking about water justice.

Taken together, one clear implication of these theorizations for thinking about feminist water justice is that it becomes more difficult to identify winners and losers from environmental and capitalist change. Questions of (in-)justice, examined closely, often lose the neatness they may have when observed from the comfortable teleological distance of more structural frames of analysis. When situated within larger power structures, analytical/methodological insistence on taking seriously people's own sometimes contradictory or indeed perplexing feelings and experiences usefully directs attention to how they themselves subtly maneuver, adapt or navigate situations of injustice, transforming and re-positioning themselves in the process. While such everyday transformations may not immediately add up to fuel larger change movements, they may provide pragmatic, more modest inspirations for alternative ways of engaging and intervening in water – ways that are more friendly, caring, convivial and companiable, or indeed more just.

Intensifying groundwater abstraction in the Saïss agricultural plain

Over the last century, groundwater abstraction in the Saïss agricultural plain has gradually intensified. In particular since the 1980s, farmers have increasingly accessed and used water through wells and deep-tube wells, partly prompted by liberalization policies, which provoked changes in tenure relations, facilitated access to new irrigation technologies (such as deep tube-wells and drip irrigation), and integrated the Saïss rural economy into national and globalizing markets. Groundwater extraction further intensified during the last two decades, partly as result of proactive government policies to restructure the agricultural sector. Most notably, in 2006, the government decided to privatize the lands of 93 state cooperatives (created in the early 1970s as part of a national land reform program). These reforms redistributed land that had been owned by foreign settlers during the French protectorate (1912–1956). Some peasant families received an individual land plot with land-use rights, organized into state cooperatives, while others became laborers working in kolkhoz-inspired state cooperatives.[2] The 1970s land reforms aimed to create a class of peasant farmers who would contribute to agricultural development. These plots could not be sold until 2006, when former members could acquire formal land title from the state and become landowners.[3] Many families were subsequently forced to sell their land as they were either in debt or had inheritance problems, attracting urban investors and new "entrepreneurs" to the region. Today, land fetches ever-higher prices (45,000 euro–55,000 euros a hectare) with brokers who help connect willing buyers and sellers. In the same vein, the recently established (2004) private-public partnership arrangements for formerly state-owned land favor mostly those investors who are able to set up "productive", "modern" farm projects.

New possibilities to obtain land combine with availability of water and (female) labor at relatively low prices to attract many urban investors and "entrepreneurs" to the region. As they can afford to pay much more, they make it ever more difficult for those born and raised in the region to access land and water. Newcomers' farm projects tend to be highly water intensive. As soon as they have purchased the land, they fence it, dig new tube wells, install drip irrigation and usually plant high-value crops such as grapes or fruit trees. In contrast to the resident peasant families, who irrigate only between 30% to 40% of their plots, new investors usually irrigate up to 80% to 90% of their newly purchased land. Altogether, these transformations have drastic consequences for the groundwater tables. Over the last three decades, groundwater levels have dropped ten meters (Kuper et al., 2016). This was even more noticeable in the confined aquifer, where the groundwater table dropped up to 2.6 meters a year (Ibid.).

New farm projects not only require water, they also strongly rely on relatively cheap female labor, with women often working without contracts and under poor labor conditions. They are hired because of their nimble fingers and other supposedly feminine skills that make them particularly suitable for the tasks they need to do. This is one way in which, and as we show in more detail below, intensifying groundwater use is premised on changing the gendered organization of labor and space.

The intimate relationship between gendered norms/practices and groundwater dynamics

Intensifying groundwater use, as described previously, not only impacts prevailing gendered ways of being but is also itself shaped by gendered practices and norms. The new agrarian policies, intending to promote a new kind of Green Revolution, are significantly premised on a particular gendered organization of and distinction between farming and families. The Moroccan Government identifies two main protagonists of the modernization of the agricultural sector: new farm "entrepreneurs" and peasant farmers who are willing and able to modernize. Both are clearly imagined as men, while their ideal performance assumes a particular gendered distinction between farms and families. While co-shaping actual farm realities, for instance, through targeted subsidy programs, our analysis of contemporary rural actors' experiences and farming practices show that groundwater dynamics provide a much wider range of new opportunities for performing farming identities and constructing social relations than imagined or foreseen by policy frameworks. New possibilities for being and relating are forged, while old ones are closed off. This happens along intersecting axes of gender, class and generation.

A new green revolution founded on a particular farming model and farming identities

These new farm projects epitomize the Moroccan Government's ambition, articulated in its 2008 Green Morocco Plan (PMV or Plan Maroc Vert): to modernize the agricultural sector, turning it into a profitable part of the national economy. As the plan is based on a particular capital- and resource-intensive form of entrepreneurial farming, it should come as no surprise that it favors the more well-to-do, larger farm enterprises and private investors, to peasant farms neglect (for a critical analysis, see Kadiri & El Farah, 2013; Akesbi, 2011). Private investors are therefore the PMV subsidy system's main beneficiaries. In contrast, peasant/family farming – which in Morocco is referred to as "the second pillar of agriculture" – receives much less state support. The little that it does receive falls under solidarity, rather than productivity: it is meant to assist small- and medium-scale farmers in marginal regions to survive.

Most government officials that we conversed with align with the ideas and ideals of the PMV: they proudly refer to the unfolding agrarian transformations as modernity and progress. For many of them, the PMV offers much needed institutional and technological support to speed up Morocco's development. It is clear that the main protagonists of this vision of modernity are the new farming "entrepreneurs" who buy and invest in land in the region. Government officials refer to these new "entrepreneurs" as individuals "who contribute to the real development and who bring investments to the agricultural sector". Yet, existing farmers are also allowed to play a potentially important role in helping bring about modernity, provided they adopt new technologies and increase their productivity. These two main protagonists of the current transformations – farming entrepreneurs and modernizing farmers – are clearly (although largely implicitly) cast as men. We conducted an interview with a government ministry of agriculture official in which we asked whether there were also female farmers, or female new entrepreneurs. His answer is telling: "Yes, I know one, but she is more man than a man. She has about 2000 hectares and everything is equipped with drip irrigation". Promoting current agrarian developments is based on distinct gendered ideas about the preferred way to organize farming: the agricultural sector is expected to largely modernize through some audacious men's work and initiative, who are pictured as heading modern farm enterprises, enterprises that are separate and distinct from families and homes. This representation foregrounds the work and achievements of some – the male farmers who supposedly direct the farm enterprise – while rendering the work, efforts and strivings of all others – including women – invisible. It is also an account that, by focusing on marketable profits, leaves much social and environmental farming costs – and particularly water – out of the equation.

New possibilities for groundwater access, use and control and performing new masculinities

New groundwater access and use modes are key to realizing the new "modern" farm projects and becoming new farm "entrepreneurs". As previously mentioned, these new farm projects produce high-value crops, are fully equipped with drip irrigation, and rely on several deep tube-wells for their water access. Their owners usually do not live on the farms, nor do they themselves engage in physical farm work. They only come every once in a while, driving to and through the area in fancy four-wheel drive cars and dressed in distinctly urban outfits. They visit occasionally just to inspect their projects or to spend a weekend of leisure in the countryside. Hassan (aged 46) typifies this new farm entrepreneur. He is an architect who lives in the city of Fes. Two and half years ago (in 2014), he bought a plot because, according to him, "the agricultural sector offered interesting business opportunities and I like to spend my weekends here". He planted one hectare of peaches, one hectare of nectarines, one hectare of apples and four hectares of prunes. On his newly acquired land, he also built a two-storied house with a terrace overlooking a swimming pool.

Where these farm projects are a hobby or investment for people like Hassan, for people who are originally from the region, farming is the backbone of their livelihoods. A number of residential families started their own farms after having benefited from the early 1970s land reforms. They, or their parents, became members of the Socialist-inspired state cooperatives as laborers, and started working collectively on the land. In the early 1990s, these large farms were divided into smaller state cooperative farms with members receiving an individual plot with land-use rights. From that moment onwards, families started to work on their own plot and became responsible for their own farms. Some, in particular the wealthier ones, began to dig wells and to diversify their production by planting a couple of irrigated crops such as onions, peppers, carrots and potatoes. These were the people who were relatively better positioned to benefit from land privatization since 2006. There were many others who failed to hold onto their plots. In particular, those families who were already quite heavily indebted were forced to sell their land. Some decided to move to the city, often using the money earned by selling the land to repay their loans. Others remained living in the countryside and became laborers on the land that they used to own. Families who obtained their private land titles usually continue to farm as they used to, irrigating 30%–40% of their plot. Some have installed drip irrigation and have the aspiration, one day, to dig a second tube-well.

To the young men in these farming families, newcomers' agricultural projects are a source of inspiration. The new farmer-entrepreneurs are exciting role models for them, embodying new, more modern ways of doing agriculture and being a farmer. Young men may refer to these newcomers when articulating dreams of starting a fruit tree farm, renting extra land to realize their own farm project, or engaging in bio-farming. They are fully aware of how important groundwater access is to realize these aspirations. Young male farmers, for instance, state: "access to groundwater opens up new horizons"; "now we can engage in real agriculture"; "now we can install a drip system and plant fruit trees". On the face of it, this vision of farming, built on intensive water use, contrasts with older generations farming vision. These older farmers believe in mixed-cropping patterns in which water is used with care, and with consideration for longer term futures. Some elderly men even reject drilling deeper tube wells, along with the fruit trees and grapes, altogether. This is not because they cannot afford them, but because they wish to continue with their existing farming practices; it is part of who they are and what works for them (e.g., Bossenbroek et al., 2017). For younger male farmers, (potential) groundwater access opens up attractive new farming horizons, fueling dreams of planting more lucrative crops, and of using new technologies such as drip irrigation. The difference between the generations may nevertheless be less stark than it appears at first sight. Many younger men who aspire to modernize also wish to respect their parents farming practices and values (Bossenbroek & Kadiri, 2016). They wish to make changes, but not to the detriment of their key values of sharing and caring for each other and their environments. Where they imitate new entrepreneurial farmers' dress, behavior or advanced technologies to appear modern, they also continue to express strong adherence to their territory and family. Farming is part of who they are and sustains their livelihood. They see themselves as *fellah* (farmer) and *rajal âamal* (businessman) at the same time (see Bossenbroek et al., 2015).

New ways to access and use groundwater play an essential role in performing new masculine farming identities. Groundwater access gives young men the opportunity to work with drip irrigation, which they hope to actively take up to reinvent more modern versions of themselves. In these young men's eyes, drip irrigation makes the job of irrigating cleaner, more technical and less tiring, as the following quote of Mohammed (aged 23) illustrates:

> Before, when my father irrigated with the sequia [small earthen channels through which water is conducted to crops in the fields], irrigating took a lot of time, sometimes four to five hours. His clothes were muddy and dirty. Today, I only have to open the valves and, in the meantime, I can do something else. I can call my clients, for example, or oversee the laborers working on the field.

Mohammed considers drip irrigation not just as a technology that facilitates irrigating but also as a way he can positively distinguish himself from the older generation of peasant farmers. Khalid (aged 26) likewise considers drip irrigation as offering new possibilities to farm, while performing a new farmer identity. Drip irrigation is "a new way of farming", he explains: "with drip irrigation I can incorporate the fertilizer into the irrigation water. Previously, this was done manually. Drip irrigation also gives the possibility to plant other crops". When we asked if there were other benefits to drip irrigation, he replied, "Yes, I don't get dirty anymore and can start to irrigate from my house without going to the field, while engaging in other activities". One day, we passed by his land when he had just opened the drip-system valves. He was heading home, where he was working on his future farm project: a small tree farm. Like some other young male farmers we interviewed, Khalid was neatly shaven and proudly featured a fashionable hair style to match his sporty jacket and boat shoes, complemented by a ring and a silver chain. This costume is not very different from young people in the city; it marks Khalid as cool and in vogue, thereby also distinguishing him from the older generation of male farmers.

New farming possibilities based on accessing groundwater thus provide exciting new opportunities for becoming a modern rural man. "The farm" plays an important role in performing these new professional masculine identities. As a newly designated, distinct space for engaging in agricultural activities, it enables performing particular gender identities (Pratt, 1998). When compared to the old peasant farm, the new farm is increasingly becoming defined as a masculine space, with farming re-defined as a masculine activity. As we explain in more detail in the following section, this re-definition of gendered organization of work and space happens by more strictly separating the private-home-women's domain and the public-work-men's domain.

Changing gendered meaning of space, new labor practices and performing new femininities

Alongside and partly because of how current groundwater dynamics re-define farms and the activity of farming in more strictly masculine terms, definitions of domestic and "feminine" work are also changing. During the period of the state cooperative, women used to engage in agriculture almost matter of fact and casually, considering farm activities as a more or less logical extension of their domestic duties. Today, the domestic or private sphere seems to become ever more narrowly defined and spatially separated, with farm enterprises that symbolize the productive and public domain, contrasting ever more starkly with the protected, bounded family home space. This forces women into a homebound traditionalism that is newly invented, even when justified by supposedly traditional norms of female virtue and chastity. We use this section to further illustrate how different rural women actively take up and navigate this newly bounded "private" home domain to re-create their femininity, among others by taking up new "bourgeois" female identities. Whereas many belonging to the older generation proudly engage in those farming activities that can be done within the safe homestead space, such as husbandry activities, many younger women prefer to abandon farm work altogether. Instead, they prefer to engage in what they perceive as "clean" work, such as craft skills. There is also a large group of women, who belong to households with little or no land, for whom it is much more difficult to perform clean, modern forms of rural femininity. Struggling to make ends meet, they are forced to accept low-prestige, low-paying wage work. For them, it is nearly impossible to find public recognition or personal pride in what they do. They therefore hesitate to talk about it, and literally try to prevent others from seeing them doing it, by covering their faces behind veils that reveal only their eyes.

Women's work used to be embedded in collective family-based farm organization, in which "private" and "public" were deeply entangled (e.g., Bossenbroek & Zwarteveen, 2015). Farmers' wives and their daughters recall how, when they cultivated rainfed crops, they would work in the field alongside their husbands, fathers and other male and female family members. They would be in charge of sowing, weeding, husbandry and help with harvesting, activities that they fluidly combined with housework such as cooking, cleaning and educating children. Although there was some gender division of labor, all collaborated in and were responsible for ensuring family and farm prosperity and wellbeing. Women's identity was significantly co-formed, enacted and re-affirmed through their active contributions to farm work. Through their everyday domestic work, as well as by helping with collective work on the land, farm women continuously re-affirmed their commitment to their marriage, while also asserting their productive value by helping assure family farm reproduction (Bossenbroek & Zwarteveen, 2015). Their identities as a wife, mother and farmer were thus closely interwoven. Likewise, for young unmarried women, work on the land was often regarded as "simply part of our responsibilities".

With intensified farming and irrigation, and with men actively occupying a newly emerging public sphere to perform their masculinity and professionalism, women's involvement in farm work is gradually changing. Ilham (aged 32) for example, explains that when her family was still engaging in rainfed crops "everyone helped during the harvest period and women would be in charge of harvesting and weeding. Today we have laborers and I hardly help in the fields and my mother's biggest concern is the stables". In a similar vein, Aziza (aged 42) explained how she used to walk "*far*" to work in the fields. Today, however, she is relieved of those duties as she is now mainly looking after what have come to be seen as strictly domestic duties; "my husband is in charge of the irrigation. For harvesting or for weeding he hires laborers, usually women". The experience of Ilham and Aziza is typical of women belonging to land-owning families, who have obtained their private land titles during the privatization process. These women appreciate the fact that they do not have to work in the fields anymore as something positive. It is a source of pride to them, something that reflects their new social standing; from being laborers they now have become co-owners of an irrigated land plot of seven to 13 hectares, with husbands who can afford to hire laborers to do the work.

By actively taking up more narrowly defined domestic activities, they not only show off their new social standing but also re-invent what it means to be a good (rural) woman, defining it through emphasizing what used to be typical more traditional urban bourgeois virtues of femininity: smoothly running the "private" domain of the household while remaining relatively invisible and unseen by the public eye. Although their physical presence on the farm has become much less, they continuously emphasize the importance of their household activities for the reproduction of the farm. Aziza (aged 43) for instance explains: "I am in charge of cooking for the laborers and for my family, raising the children, washing the clothes. Of course it's work, it's the motor of the farm"! In a similar way, Halima (aged 48) is proud to talk about her full responsibility for all the husbandry activities, consisting of feeding and milking the cows, and cleaning the stable. All this, according to her, is "a key activity of the farm". With a twinkle in her eyes, she tells us how she assisted a cow to give birth; "it was in the early morning at 4 o'clock that the contractions started. Around six in the morning I helped her. It was very difficult and you have to be careful". Proudly she adds, "*I did it all by myself*". She aspires to start a milk cooperative, together with her husband. In their emphasis on the continued dependencies and linkages between farm and family, and between their own work and that of their husbands, the stories and experiences of Halima and Aziza challenge the masculinist accounts of farming – with farms being likened to continuously innovating enterprises – that dominate policy documents as well as much research. Their professional pride does not reside in identifying with what increasingly comes to be seen as the masculine farm domain, with mastery of technology and networking skills as the defining characteristics. Instead, they emphasize their hard work, and express the importance of this work for the prosperity of both farm and family. This is similar to what Haugen (1998) observed when studying changing rural femininities in Northern Europe (p. 143).

Many belonging to the younger generation, while often being full of admiration and respect for the hard work of their mothers, seem less invested in portraying themselves as part of the farm. They instead articulate a clear desire to distance their future feminine selves from physical farm work. They refer to their mother as being "strong", "hardworking" and "caring" but also note that their mothers' work is very exhausting and demanding, while also being "dirty". Many therefore opt for expanding and re-defining domestic work, as an avenue for performing new professional rural female identities. Hence, Saïda (aged 28) started a business baking sweets that she sells at her elder brother's café in town. Woman-owned home-based enterprises are a known phenomenon in the city, but so far remain a rarity in the Saïss countryside. Saïda explains that it was her own idea; she had to persuade her brother to support it, by arguing that she could save money so they could buy a car together. Although she did just obtain her driving license, she admits that "he will probably use it more often, but at least it helped to persuade him". Also, Hind (aged 23) is eager to learn new things and take up activities that may generate some income: "I love to cook and to learn to cook new dishes. I also would like to learn how to use a sewing machine; I am currently taking classes in the city". When we asked her what kind of clothes and models she intended to sew, she replied "*modern clothes, like pants and modern djellabas*". Other young women we interviewed also aspired to sew, embroider or weave carpets. For them, such activities more closely align with what they have come to consider as appropriately feminine work.

While seemingly traditional, young women also express subtle forms of rebellion when engaging in virtuous feminine crafts. When weaving carpets, for instance, young women do not necessarily reproduce the previous generation's same motifs or patterns. Fatima's last creation (aged 28) consisted of vast bright pink, bright yellow and blue rectangles; quite a contrast to the dark red, white and dark blue carpets that decorate most houses. For her, the motifs represented "my thoughts" and "dreams"; the carpet mirrors her creativity and emotions. The different patterns and colors can also reflect a dialogue between generations: by reproducing particular patterns taught by their mothers or grandmothers, young women identify with their roots. On the other hand, innovating or including other patterns and colors is a way of self expression and re-negotiating one's femininity (cf. Merini, 2007). Some young women want to sell their carpets through a yet-to-be-formed cooperative, indicating their sense of adventure and entrepreneurialism. Although such cooperatives exist in various other regions in Morocco, in the Saïss such cooperatives are rare.

All this illustrates that women, particularly those belonging to landowning families, consciously attempt to give a positive interpretation to what it means to be a modern rural farm woman. The younger generation's activities seem to further reinforce an emerging division between the private-home-women's domain and the public-work-men's domain, also re-inscribing a gendered distinction between different types of work. Their wish to engage in home-based "clean" skills and crafts expresses their desire to distance themselves from dirtier, outdoor, physical labor. They re-invent what it means to be a virtuous woman, also setting standards that are difficult to achieve for those who cannot afford to refrain from physical, "dirty", public labor. It is here where class, gender, and generation intersect to create new hierarchies and exclusions: emerging ideals of womanhood are unattainable for a growing group of female agricultural wageworkers, many of whom belong to landless families. It is telling that women from landowning families would often depreciatingly talk about these women, particularly those who find their jobs through *mouquéfs*,[4] referring to them as illicit women. Malika (aged 42), whose husband's family has eight hectares of land, says; "these women work hard and it is not rewarding. Sewing or weaving is much better. It is clean, you don't have to work under the burning sun. These women even have to hide their faces, as they don't want to be recognized."

Indeed, most women working for wages wear their scarves so they reveal only their eyes. This has earned them the nickname of *ninjas*. This all-covering outfit marks their paradoxical situation: although newly set up farm projects rely heavily on their relatively cheap labor, their contributions tend to go unnoticed and they even want to be invisible. While explaining that they dress that way to protect themselves from the sun, dust and pesticides, when prompted many also admitted that they use their scarves to remain invisible and anonymous. Their work and income are a source of shame and embarrassment, of inner struggle and frustration, rather than something they can take professional pride in. Indeed, many prefer to use other sources of identification when presenting themselves (Bossenbroek, 2019). Women wage workers would often complain about poor work conditions (low wages, not being paid, no work contract, no insurance, long working days, etc.) and common harassments. Nevertheless, even for these women, the new work opportunities may offer new possibilities for being and becoming, as the following example of Donja (aged 27) illustrates. For her, agricultural work offered a much sought after chance to earn an independent income, while also enabling her to "*meet other women*" while going to work, and exchange experiences and ideas with them. Especially for younger female wageworkers, agricultural wage work often provides one of the few opportunities to leave the confined spaces of their homes. Hence, female wageworkers experiences combine feelings of happiness and satisfaction, along with feelings of distress and confusion.

Conclusion

In this chapter, we have illustrated how changing groundwater-use practices, and the larger changes it provokes and forms a part of, happen through and are accompanied by re-negotiating gendered identities, social relations and spaces. Groundwater dynamics alter farms gendered labor organization, as well as the gendered meanings of labor and space. These changes open up new possibilities of living and being but also close off others.

New water-intensive farming futures and identities are actively promoted by the Moroccan government; clearly favoring particular actors and very specific forms of farming. New farm "entrepreneurs" and peasant farmers who have the willingness and ability to modernize, both clearly cast as men, mostly benefit from the government's various subsidies. Yet, representing contemporary processes of agrarian change in Moroccan policy terms (also the language preferred by many researchers who favor modernization) reproduces the idea that it is mainly through the work and achievements of the male farmers and entrepreneurs who supervise the farm enterprise that change happens. It makes the work, efforts and strivings of all others – including women – invisible. Additionally, it is an account that, by focusing on marketable profits, leaves much of farming's social and environmental costs – and in particular water – out of the equation. We have reconstructed the story of contemporary groundwater intensification in Morocco from a feminist angle, based on an ethnographic study of people's own engagements with and interventions in groundwater.

Our analysis shows that only a privileged few young men want and can pursue those modern farming identities, mimicking entrepreneurial stereotypes, actively seizing new technological opportunities to invent new ways of combining manhood with rurality. New investors emerging farms also modify existing gendered organization and meanings of farm work, as they heavily rely on (cheap) wage labor for the labor-intensive activities of planting, weeding and harvesting, much done by women. Female engagement in casual wage work for others ("strangers") provokes and happens through a re-articulation of the divide between "private" and "public" spaces. The overall effect is that farming increasingly becomes or is seen as the exclusive professional domain of men, with "the farm" being reinvented as an exclusive space to perform a new type of farming masculinities. This space becomes increasingly distinct and separated from the private family sphere of the home, which in the process is reinvented as women's place to perform a new kind of femininity. Some women, particularly those belonging to landowning families and of the older generation, proudly assume a more "bourgeois" feminine identity by taking up and performing supposedly more domestic roles. Younger women instead creatively re-invent the domestic domain to carve out new, and perhaps more appropriately feminine, professions for themselves. By engaging in what they call "clean" work, such as craft skills or small businesses, they create new, more "graceful" rural feminine ways of being and behaving. For women belonging to landless families, who rely on agricultural wage work to make ends meet, it is difficult to comply with these newly emerging normative definitions of womanhood. They enjoy little personal pride in what they do, and many even try hiding their wage work involvement from the public eye.

In line with Mies (1986), we could use this analysis to show that economic progress and modernity are enabled by exploiting Nature – water resources – and women's labor. The conclusion from such an analysis could be that female wage workers are both the victims of new capitalist processes of agrarian change, and protagonists of movements for more social and environmental justice. This is an interesting analysis, but one that is difficult to translate or explain to the men and women we studied and therefore also of little direct use to identify entry-points for change. This is why we instead tried to emphatically acknowledge and learn from people's own ways of navigating the multiple force fields that they are part of. This yields an account in which winners and losers are less easily identifiable, one in which perplexities and paradoxical experiences abound. Categories and differences – including those based on gender – are re-made and re-invented in production and trade, in ways that are often contingent and not always clear cut or consistent.

Without denying the existence of larger power structures and without denying that some actors are relatively better off than others, our study shows how all rural actors actively navigate existing spaces or create new ones to fuel, co-shape, and fulfill emerging desires. This happens, for instance, through negotiations about labor practices and the meaning of labor, negotiations that are always deeply entwined with re-articulations of class, gender or generational differences. Paraphrasing an older article by Katz, simultaneous existence of diverse, articulated structures of dominance, exploitation and oppression means that there are multiple, connected positionings from which to confront and potentially transform various forms of domination (1992). Documenting these diverse grounded positionings and spaces may pave the way for pragmatic and perhaps modest strategies of interrogating challenging injustices; strategies that risk being overlooked by more structural analyses.

Acknowledgement

This chapter has been reproduced with permission of Cambridge University Press. The chapter was originally published as: Bossenbroek, L. and M. Zwarteveen (2018) *New spaces for water justice? Groundwater extraction and changing gendered subjectivities in Morocco's Saïss region* in Rutgerd Boelens, Tom Perrault and Jeroen Vos (eds.) *Water Justice*. Cambridge University Press; pp. 330–345.

Notes

1 Inspired by Rosa Luxemburg's work (1923) Mies uses the term "colonies" to refer to "non-capitalist" milieus and areas to appropriate more labor, more raw materials and more markets.

2 Only three of such state cooperatives were nationally created, and in the 1990s these lands were redistributed: members received an individual plot with land use rights.

3 On the condition that former state-cooperative members pay off their debts, they can now own the land they used to work on, in return for a fee of approximately 70,000 Dirham (approximately €6,250), depending on land size and soil quality. Former cooperative members could thus become private landowners, with the land becoming private property.

4 Places where wageworkers gather in order to find a job for the day, which are usually situated at the outskirts of the Saïss' various small agricultural centers.

References

Akesbi, N. (2011). Le Plan Maroc Vert: Une analyse critique. In A. Akesbi, N. Akesbi, K. Askour, W. Benaabdelaali, N. El Aoufi, A. El Houmaidi, S. Hanchane, M. Hollard, A. Kamal, M. Lahlou, M. Mzaiz & H. Touzani (Eds.), *Questions d'économie marocaine* (pp. 9–48). Presses universitaires du Maroc.

Bossenbroek, L. (2019). Les ouvrières agricoles dans le Saïss au Maroc, actrices de changements sociaux? *Alternatives Rurales*, *7*. https://doi.org/10.60569/7-a7

Bossenbroek, L., & Kadiri, Z. (2016). Quête identitaire des jeunes et avenir du monde rural. *Economia*, *27*, 46–50. www.economia.ma/fr/content/qu%C3%AAte-identitaire-des-jeunes-et-avenir-du-monde-rural

Bossenbroek, L., Kuper, M., & Zwarteveen, M. (2017). Sour grapes: Multiple enclosures of ground water in the region of the Sais in Morocco. In J. P. Venot, M. Kuper & M. Zwarteveen (Eds.), *Drip irrigation: Untold stories of efficiency, innovation & development*. Earthscan.

Bossenbroek, L., van der Ploeg, J. D., & Zwarteveen, M. (2015). Broken dreams? Youth experiences of agrarian change in Morocco's Saïss region. *Cahiers Agricultures*, *24*(6), 342–348. http://doi.org/10.1684/agr.2015.0776

Bossenbroek, L., & Zwarteveen, M. (2015). One doesn't sell one's parents: Gendered experiences of land privatization in the Saïss, Morocco. In C. Archambault & A. Zoomers (Eds.), *Global trends in land tenure reform: Gender impacts* (pp. 152–168). Routledge.

Butler, J. (1990). *Gender trouble: Feminism and the subversion of identity*. Routledge.

Domínguez Guzmán, C. (forthcoming). Grandes Narrativas, Pequeños Agricultores: Explorando Desbordes Académicos en el Caso de los Excedentes de Agua en el Valle de Motupe, Perú. *Estudios Atacameños*.

Gibson, K., Law, L., & McKay, D. (2001). Beyond heroes and victims: Filipina contract migrants, economic activism and class transformations. *International Feminist Journal of Politics*, *3*(3), 365–386. https://doi.org/10.1080/14616740110078185

Haugen, M. (1998). The gendering of farming: The case of Norway. *The European Journal of Women's Studies*, *5*, 133–153. https://doi.org/10.1177/135050689800500202

Kadiri, Z., & El Farah, F. Z. (2013). L'agriculture et le rural au Maroc, entre inégalités territoriales et sociales. *Rabat: Blog Scientifique Farzyates/Inégalités du Centre Jacques Berques*. www.umifre.fr/c/1522

Katz, C. (1992). All the world is staged: Intellectuals and the projects of ethnography. *Environment and Planning D: Society and Space*, *10*, 494–510. https://doi.org/10.1068/d100495

Kuper, M., Faysse, N., Hammani, A., Hartani, T., Hamamouche, M. F., & Ameur, F. (2016). Liberation or anarchy? The Janus nature of groundwater on North Africa's new irrigation frontier. In T. Jakeman, O. Barreteau, R. Hunt, J. D. Rinaudo & A. Ross (Eds.), *Integrated groundwater management* (pp. 583–615). Springer.

Merini, A. F. (2007). Rêve et transmission: De la tradition à la créativité. In F. Mernissi (Ed.), *A quoi rêvent les jeunes* (pp. 49–53). Marsam.

Mies, M. (1986). *Patriarchy and accumulation on a world scale*. Zed Books.

Mies, M., & Shiva, V. (1993). *Ecofeminism*. Zed Books.

Pratt, G. (1998). Grids of difference: Place and identity formation. In R. Fincher & J. M. Jacobs (Eds.), *Cities of difference* (p. 26). Guilford.

Ramamurthy, P. (2003). Material consumers, fabricating subjects: perplexity, global connectivity discourses, and transnational feminist research. *Cultural Anthropology*, *18*(4), 524–550. https://doi.org/10.1525/can.2003.18.4.524

Ramamurthy, P. (2011). Rearticulating caste: The global cottonseed commodity chain and the paradox of smallholder capitalism in south India. *Environment and Planning A*, *43*(5), 1035–1056. https://doi.org/10.1068/a43215

Resurreccion, B., & Elmhirst, R. (2008). *Gender and natural resource management*. Routledge.

14

LIQUID HERITAGE

Can water museums facilitate a new gendered water ethics?

Sara Ahmed

Introduction

The water challenges that we face today suggest the need for a new paradigm of water management and governance, which requires re-thinking what water 'is' and what water 'means' for all users. A value-based framework to guide water policy is not new: the Dublin Statement on Water and Sustainable Development (1990) called for participatory water decision-making at the lowest appropriate level with a commitment to women's participation at all levels in water resources programs (Ahmed, 2005). In 2010, the United Nations voted to recognize access to safe water and sanitation as a fundamental human right despite the opposition from large water companies and contentious debates at various global water forums on the profit-making motives underlying public-private partnerships couched in the language of social contracts. Parallel to these value-driven debates is the hegemonic paradigm of Integrated Water Resources Management (IWRM) applied by many international and national organizations intent on finding win-win solutions through the integration of competing demands. While this is not the place to critique the rubric of IWRM, suffice it to say that it ignores many aspects of our relationship with water, including 'cultural diversity and corresponding water ontologies or the theories and beliefs that different cultures hold about water's fundamental nature,' (Groenfeldt, 2019, p. 271).

It is in this context that the search for a new water ethics, a moral compass that can guide how we use, conserve, manage and 'listen' to water has become significant. It is in this context too, that I would like to explore the role of water museums in facilitating social change and raising awareness on water conservation from an inclusive and gendered lens. First, a short caveat, as this is also a personal journey: the how and why of starting this museum as a digital space is as important as the museum itself.

My journey with water began in the late 1970s at a boarding school, aptly named Woodstock, in the Indian Himalayas in the then small town of Mussoorie – a hill station prized by the British as a wellness destination because it had an abundance of water. Like so many other hill-stations across India, Mussoorie is currently facing a water crisis every summer, as rampant urbanization and unsustainable tourism has affected water resources such as mountain springs. As part of our advanced geography class (I was the only girl in the class)

DOI: 10.4324/9781003100379-17

we were taken inside the under construction and rather controversial high dam, the Tehri Dam. We also interacted with activists protesting the displacement of people and the submergence of entire villages – a frightening proposition for a 17-year-old. The Living Waters Museum is now back at Woodstock, to start the Himalayan Water Repository, beginning with the Mussoorie Water Narratives, inspiring students to work on local water issues and supporting water conservation, recycling and sustainability efforts at the school.

In contrast to rural India, my doctoral research in the late 1980s took me to the crowded holy city of Varanasi, to analyze the politics of participation and notions of pollution in the cleaning of the mighty Ganga. Why would people want to participate in cleaning a river they believed was sacred and would wash away all their sins? According to the American anthropologist, Mary Douglas (1966), 'dirt is matter out of place' – culture defines what is clean and dirty as does caste in India, in terms of who has access to what water. Many of the people who were being affected by the technology of the Ganga Action Plan,[1] which was based on imported knowledge systems (developed by the Thames Water Authority, United Kingdom), were from communities belonging to less-privileged castes. Specifically, they were identified as *Doms*, those that cremated the dead and *Dhobis*, those that washed dirty clothes along the *ghats* (riverbank steps). However, their voices were ignored by the upper caste leaders behind the Ganga Action Plan, led by the Sankat Mochan Temple Foundation, with which I was also associated at the time. It then became evident that some knowledges were socially and politically marked as different and insignificant due to intersectional caste and gender power relations and based on the judgment of dominant groups.

Over the past 25–30 years I have found myself caught in these polarizing debates between the top-down approaches to water management which impact the environment, the poor and vulnerable, especially women, and the bottom-up alternatives suggested by NGOs. In my search for a middle path, an alternative discourse, I met the artist Basia Irland at a conference in Duluth, Minnesota, in 2000 aptly titled, 'The Bonds between Women and Water.' Here too, I heard the voices of women from Indigenous or First Nation communities from North America, using poetry, songs, performance and music to talk about their diverse water challenges. A few years later, at a conference on 'Water: Ecologies, Cultures, History' at the University of Perth in Western Australia, I saw the art of Aboriginal women on their water worlds, and heard engineers and educators talk about the need to bring storytelling and oral histories into education systems. In the winter of 2004–05, Basia and I travelled to parts of Gujarat and Rajasthan, visiting stepwells where she would take samples and images and make her silk scrolls, exploring the different waterborne diseases affecting communities in the area.[2] We had numerous conversations about forming a museum, even romantically, in a stepwell, but then I realized that would require considerable funding. Moreover, what would you put in a water museum: art and artefacts (no), water utensils (whose collection), models of water technologies (again funds required)? And there were many practical questions: who would manage such a water museum and so on and on.

Hence, I decided to start virtually, by creating a digital 'museum' that would be a repository for stories about our rich water heritage, to work with young people and to use my learning on the impacts of our unsustainable development. I also intended to question the paradox implicit in the values our culture was reinforcing, revering our collective waters on the one hand, but equally excluding the 'other.' That is, excluding those who face oppression based on gender, caste, ethnicity, class, sexuality, ability, and other forms of discrimination. I then aimed to build stories that were rooted in interdisciplinary and inclusive frameworks. In early 2017, the Living Waters Museum was born. A few months later,

I was privileged to be a part of the founding of the Global Network of Water Museums (www.watermuseums.net) launched in Venice in May 2017, and endorsed a year later by UNESCO's Intergovernmental Hydrology Programme (IHP) as a special initiative towards SDG 6 on clean water and sanitation through knowledge management, dissemination and public outreach.

Building a digital museum?

According to UNESCO (2015), museums have a tremendous role to play in protecting and promoting both our cultural and natural diversity. A recent and exhaustive 'how-to' resource guide asks, 'How can museums help put the world on a path to a sustainable future, through working to support the Sustainable Development Goals' (McGhie, 2019)? No doubt this is a challenging proposition as museums tend to be comfortable remembering and recording what it was like in the past, but they are on less firm ground when dealing with a dynamic present, let alone an unknown and fluid future compounded by other complex uncertainties such as climate change.

I believe that museums can address this conundrum through the power of storytelling. In general, cultures comprise various components – values, beliefs, practices, rituals, traditions – but what strings them together are our stories. 'Stories give meaning to human existence. . . . They keep history alive, give geography meaning and the physical world a human face,' (Ashwin & Ishan, 2019). The Talanoa Dialogue, which forms part of the UN Framework of Countries on Climate Change (UNFCCC, 2018), is a testimony of the power of storytelling to facilitate conversation. *Talanoa* is a traditional word used in Fiji and across the Pacific to reflect a process of inclusive, participatory and transparent dialogue around the sharing of stories to build empathy and make wise decisions for the collective good. However, in the hustle and bustle of our modern worlds, stories are not just what you read your children at bedtime, or to yourself, but have the potential to come alive through new media and technological innovation that can engage communities, especially our children and youth, in reflecting on pathways to a more sustainable water future. But stories also raise complex questions: how stories are told, by whom and to whom is as important as understanding questions of gender and participation, the 'who participates and why.'

As stated on our platform, the Living Waters Museum (www.livingwatersmuseum.org/) seeks 'to collect, curate and communicate our rich and diverse water heritage, both tangible and intangible, through a digital repository of visual narratives, which can celebrate the past, inspire the present and be a source of learning for the future.' The process of developing this digital museum is collaborative and interdisciplinary. In our team[3] we engage young people in 'storytelling' around water and its intersection with natural and built environments, water wisdoms, social justice and livelihoods, using multi-media and the creative and performing arts. Even though we are a virtual museum, we are more than just a website: whether we have been in Ahmedabad, or now in Pune, and even during the Covid pandemic lockdowns in India, we have always found ways to engage with people, physically or online.

At Ahmedabad University, we provided internships to exhibition design students to understand water and culture, held critical conversations questioning our growth paradigm in cafes and at stepwells, worked with children with special needs, facilitated heritage and water walks, and contributed to the development of a foundation studio on water from a multidisciplinary lens for all new undergraduate students. Throughout the Covid pandemic, when we went into online teaching, students were connected to civil society organizations across the country so that they could understand how gender and equity were being mainstreamed in water and resource management projects. Speaking to leaders and community actors on video, students looked at innovative ways to produce their projects, working in online groups with differential internet access or space at home. In Pune, we used the content from Punyache Paani, an exhibition on the stories of Pune's waters launched in March 2022, (https://punyachepaani.livingwatersmuseum.org) and the lessons from our foundation studio, to develop and test our interactive Water Classrooms (www.waterclassrooms.in) for middle school children. We talked about purity and pollution in relation to the caste system and access to water in India through the 'privilege game,' looked at women's burden of water work from the stories of their own homes or the utensils they had seen visiting their grandparents in villages (see Figure 14.1).

Museum experts increasingly talk about the 'soft power' of museums (Nye, 1990): their ability to engage the audience as co-creators and re-invent or re-imagine our lived reality to create new opportunities or address complex societal problems, for example, fostering new water values through active citizenship. The use of digital technologies to rediscover our tangible and intangible water heritages through sensory experiences can not only tell a powerful story but engage citizens in new narratives that challenge our perceptions of

Figure 14.1 Women and water rituals in Pune.

Source: https://punyachepaani.livingwatersmuseum.org.

Figure 14.2 Water classrooms, Pune.
Source: www.waterclassrooms.in.

museums as dank, musty spaces housing relics from the past (Kenderdine, 2014). Additionally, academics and water experts in India and globally are looking at the creative arts as a means of offering 'playful' solutions which can provide a powerful mechanism for social change, bringing science to people while acknowledging their water wisdom and agency (Moench, 2015) (see Figure 14.2).

In the next few sections, I will explore some of the stories that have emerged from the projects implemented by the Living Water Museum, which reflect on the dynamic relationship between gender, women and water.

Hydro-feminism: stepwells in Ahmedabad

There are more than 3,000 stepwells built between the seventh and mid-nineteenth centuries dotting the semi-arid landscape of Gujarat and Rajasthan in Western India, extending along the trade routes that carry into Central Asia. These elaborate architectural wonders, usually from three to nine levels deep, or 60 to 80 feet, mark the invisible landscape of underground water, providing cool shade, life and sustenance to weary travelers, villages and communities (see Figure 14.3).

The gift of water in India is considered a pious act; consequently, many stepwells were funded by, and named after, women and men of wealth – kings, queens and merchants – often to honor a deceased relative or a deity – typically a female goddess. Indeed, water is largely associated with the feminine in India and stepwells were also special spaces for women: to fetch water, meet friends, and spend time away from the confines of home and domestic work. According to Mehta-Bhatt (2014), of the 3000 stepwells in Gujarat, at least one fifth or more were commissioned by women patrons, and named after them. However,

Figure 14.3 Toorji Ka Jhalra (stepwell), Jodhpur.
Source: www.jaljharokha.livingwatersmuseum.org.

despite it being a female space (*'her space'*), women could only access the stepwells in the mornings or evenings when they were not visited by menfolk – and these were valuable moments for sharing stories, ideas or gossip.

As Jain-Neubauer remarks in her seminal work, stepwells reinforced patterns of social exclusion: 'Architecture negotiates space, it designs space. When it engages with water, it *designs* water,' (2016, p. 9). Often well-diggers and water diviners came from marginalized socio-economic groups (e.g., they belong to Muslim communities or to non-privileged castes) that were denied access to the community water assets that they had helped build. Today, not only are these stepwells largely neglected, poorly maintained by official authorities, but early examples of hydro-feminism celebrating the feminine spirit are masked by everyday realities and vulnerabilities characterizing women's relationship with water (Zwarteveen et al., 2012). Across the same arid landscapes, poor women belonging to the *Dalit*, a non-privileged caste, often have to walk further to fetch water as they are denied access to the village well or the community standpost. This happens because of gender and caste hierarchies and culturally embedded notions of purity and pollution prevailing in contemporary rural India (Ibid).

Many stepwells have been renovated and re-purposed to serve as community halls for the celebration of marriages and festivals. Several coffee table picture books have been written on their amazing architecture and design (Lautman, 2017; Livingston, 2002), water music festivals and dance performances are hosted in their corridors. However, much of the architectural knowledge or these events tend to ignore local communities. In this context, in early 2019, the Living Water Museums launched the 'Steps to Hope' initiative, aiming to (i) visualize narratives of stepwell as connectors with the past when water was revered as sacred; (ii) use digital media in designing an interdisciplinary, social interface with water which can encourage youth and children to re-value water; (iii) facilitate greater public awareness and involvement in addressing challenges around water and (iv) develop strategies to impart education in schools on water heritage and equity.

Work on an interactive stepwell map is underway. It was designed through two layers – the first one is informational and the second delves into storytelling. The latter will engage in narration, music and visualization experiences, such as virtual tours, and photogrammetry.[4] A pilot is almost ready for demonstrating the interactive and immersive power of storytelling and digital media which is accessible to local communities (content is written in Gujarati and Hindi), and those who are visually or hearing impaired (through story narration and 360-degree virtual tours).

Britto N., a graduate student from the National Institute of Design (NID), Ahmedabad, joined the team in April 2019 to develop a digital narrative on the Ahmedabad stepwells as part of his graduation project to design interactive map fiction. He aimed to accomplish storytelling through gamified experiences (a fluid category), to excite young people about their water heritage. Emerging discourse is increasingly concerned with the design of socially relevant games to address contemporary challenges and in the process, empower the gamer to problem-solve or find solutions which can then be translated to the real world. Britto (2019) describes that the intent of his project is 'to instigate reflection on the present water scenario and initiate thoughts about probable future water crisis through the fictionalized past of the stepwells.'

This exercise produced the interactive narrative called *'Muna. The journey of elixirs'* which stretches across two timelines and three worlds. World 1 is the establishment of the theme 'Water crisis (urban-rural),' World 2 is fictional, wherein the stepwell is an important element. World 3 is more fantastical and is imagined to be the most interactive portion of the experience, which was born out of the ambitious and bold exploration of the narrative building process.

Muna revolves around a curious and courageous ten-year-old girl, named Muna, who has grown up listening to her grandma's stories of the stepwells. One day, in her search for water for her family, Muna falls down a borewell, much like Alice in Wonderland. From the desolate landscape of her city, Ahmedabad, Muna enters a fictional village set a millennium back to discover how urban growth has exploited their water resources. In her quest to help the villagers and find her way home again, Muna finds a beautiful stepwell and learns of its role, only to see it being destroyed by an earthquake. She then comes across its source, the Holy Ocean (groundwater), which is being depleted by inverse (household) taps drawing water out and pushing it upwards. With the help of female guardians, spirits with different roles, Muna finds her way across the ocean to close the taps and restore water. This contextualized digital story reflects on the agency and everyday struggles of women and young girls living in a patriarchal social context, that is, a context where a masculine form of domination based on the figure of the providing father takes shape (Ojeda et al., 2022) and where women have limited access to resources, including education.

Women, water and work

In late 2018 we launched a call for images, stories or poems on women's relationship with water and the triple burden of work they face: (i) the daily grind of fetching water for household use as part of the care economy; (ii) women's roles and rights in water for agriculture and other livelihoods and (iii) women's participation in community water governance and management. The exhibition was called 'Women, water and work' (www.livingwatersmuseum.org/women-water-work).

While we were not too successful in the global reach of the call we received some interesting photographs from Isaac Kojo Biney Aggrey, a Ghanaian artist. The series, called 'Women's rights' registers different performances where the artist dresses in women clothes and is surrounded by water pots, thus inverting the social order around women's water work through creative representation. This reminded me of a young girl who had come with her grandmother to a public talk I gave in August 2022 at the Catskills Water Museum in upstate New York. I had shown images of women walking long distances to fetch water, and she asked me, 'Why do they do this in India, who decides,' and so on – thus began an explanation of patriarchy for an 11-year-old, further deepening the meaning and power of a virtual museum in communicating social inequalities around water. This also reminded me of the 2018 meeting of the Global Network of Water Museum in Den Bosch, the Netherlands. I was surprised by archival images of women washing clothes in public squares in Ecuador as part of a presentation being made by the Yaku Museum of Water in Quito. I have since seen similar historical images from Porto, Portugal as part of a global exhibition, hosted by WAMU-NET and supported by the Living Waters Museum entitled "I Remember Water" (https://irememberwater.watermuseums.net/).

Talking pots: water stories

Established in 1981, the Veechar Museum of Utensils is a unique private collection of water pots and other popular artefacts, developed passionately by Surendra Patel and housed in the sprawling campus of his rustic Vishalla restaurant, which provides traditional Gujarati food as an 'experience.' However, like many museums in India, it is a static space with little to excite the young visitor, so, in collaboration with Surendrabhai (brother, as he's fondly called), his staff and creative professionals, we decided to co-facilitate a value perspective on water by celebrating water wisdom and exploring the tangible and intangible heritage around water and food in Gujarat through various media. While the research, design and implementation of the exhibition, games and educational activities were largely led by Swarnika Nimje, another exhibition design student intern from NID, we worked collectively to develop short stories with selected pots and water serving utensils talking in the first person on their form and function, supported by 'did you know' water facts (Nimje, 2018). These were translated into Gujarati for exhibition panels and take-away postcards which were used to generate a discussion with students as part of the foundation studio on water at Ahmedabad University and our Water Classrooms in Pune. Students reflected on seeing similar pots in the ancestral homes of their grandparents, in villages across the state and at large community gatherings in the past.

But the pots also told a gendered story – of the time and labor spent by women and girls in water collection, storage and use, and of women's roles in promoting hygiene (handwashing) and managing the WASH needs of their children (see Figure 14.4). As the *ghada*, an earthen pot that women in Gujarat typically use for fetching water speaks:

Figure 14.4 The ghada.
Source: www.livingwatersmuseum.org.

> Do you know who brings water home for hundreds of rural families across Gujarat? Made of bronze, my long neck and wide collar help Jigna to hold me gracefully in the curve of her hip, as she walks back from the village well, careful not to spill water she has spent several hours collecting. On her head, she confidently balances another broad-based, brass pot, while sharing local news and singing along with her friends. I know it's hard work for Jigna, particularly in the dry summer months, but she doesn't complain even when she has to miss going to school on some days.

In contrast, the water utensils associated with men, such as the leather Mashq (see Figure 14.5), speak of travel, mobility and markets:

> Aman steps down from his trusted horse, and walks towards the small step well in the distance. As we descend the cool, intricately carved interior levels, Aman opens my wooden stopper and stoops low to fill me with water. He quenches his thirst and fills me again before continuing his journey under the hot sun. Made of cast brass parts, I have been hammered together by artisans in Kutch and Jaisalmer for centuries. My contemporary, the Mashaq, is made from leather and used for supplying water to small establishments and individuals by the Bhistis, a fast-disappearing community of traditional male water suppliers.

Figure 14.5 The mashaq.
Source: www.livingwatersmuseum.org.

The story of the Bhaturo, a vessel for serving water, highlights the complementary roles and responsibilities of women and men:

> I spend most of my days dangling under a bullock cart (typically driven by men) watching shadows go by. Amina always fills the water pot up to the brim and then fits me in the mouth of the pot. I make sure not a drop of water spills out as we move along the dusty roads. Some days we don't find a water source for many hours. Every hour, Amina pours water from the pot into me and lets her children sip. I am quite multi-functional, which is a good quality for a frequent traveler like me.

In the evening, after a long day of work, many of the empty water pots, the *ghadas and ghatams*, become drums, providing entertaining music for tired souls. Sukrit Sen, our arts and outreach coordinator, has performed several Jal Tarang events, highlighting how water is a medium for sound. Not only do the different materials in which water is contained (a glass, pot, steel tumbler) create different sounds, but whether a vessel is half full or empty also affects sound. Through simple workshops with children in schools or water heritage sites, we have used sound and our everyday objects to talk about the larger values of water, and the physics of water.

The river within: re-visiting our roots

From the dry and dusty waterscapes of Ahmedabad we move to Northeast India, where in late 2020, during the Covid lockdown, we partnered with Minket Lepcha, a storyteller, artist and award-winning filmmaker, to bid for a small British Council digital heritage grant to work with young girls from Indigenous communities on exploring social and environmental change through the arts.

> Discussions on rivers are dominated by trans-boundary issues, hence, diluting conversations on rivers to borderline issues of national interest. The most that local conversations are recognized is in terms of floods, disasters and upstream-downstream water disputes. The danger of declining local rivers is reflected in the poor portrayal of local communities that safeguard the river on micro-scales.
>
> People's beliefs, fears, ambitions, motivations and morals have considerable effect on environmental attitudes. These stories play a major role in ecological conservation practices and remain an important part of collective environmental history. The custodians of this essential intangible heritage are the future generation. Dominant education, media and history has not been inclusive of oral storytelling tradition of marginalized communities.
>
> *(Lepcha, 2021)*

Across Northeast India, rivers and water have been an integral part of cultural practices, from weaving, to cooking, music, dance and storytelling, especially for women who are respected for having a special, sacred bond with life, children and nature (www.livingwatersmuseum.org/remeeting-the-roots). However, when Covid-19 hit remote rural communities, the vulnerability of elders and the knowledge and oral traditions they represent became a growing concern. In this project, Minket and her team of young women professionals from design, water practice and academia worked with 30 young women, aged between 14 and 24, from Indigenous communities across seven diverse states of Northeast India. Drawing on local ecologies through the creative and performing arts, online workshop sessions provided a safe and trusting environment for the participants to start expressing themselves and their relationship to water through an intersectional lens, using different media.

In the context of tribal patriarchy (most village elders are male), and the quintessential exoticness with which women from the Northeast are represented, isolated from mainstream identities as 'foreign,' the project gave the young women an opportunity to voice an alternative, fresh and empowering narrative building on their agency, skills, capacities and imagination. Northeast India also has several women-led grassroots movements, such as the Borok's Women's Group, and many gave their time to this endeavor, helping the participants and facilitators to forge bonds of sisterhood that continue today in the many regional, national and global spaces where the project team has shared their work. The co-learning process helped the participants to travel virtually beyond their villages, re-visit their roots and the river/water 'within,' and use their own language, artistic renditions and even food to re-imagine a more sustainable and just world. They depicted waterfalls, hot springs and rivers that they remembered visiting with their family and friends, shared memories of local food, customs and folktales, made short films, recorded music and songs, wove, painted and embroidered pieces for a collective exhibition, all drawing on the intergenerational transfer of knowledge.

Sara Ahmed

Flowing forward: water, the arts and educations

> In these confronting times, we must find strategies not only to preserve our heritage, but to let its stories be rediscovered and reinvented.
>
> *(Kenderdine, 2014)*

The logo of the Living Water Museum, steps to water, was inspired by the stepwells of Gujarat but also by the *ghats* of Benares along the river Ganga where I started my water journey back in 1987. It has always been my goal that art and academia engage with communities and that whatever we produce must not only have aesthetic value, or academic rigor, but must address issues of inequality and social exclusion. Typically, a visitor to a museum is a silent receptor of knowledge, but through our digital platform we as a team have not only been able to co-create innovative content *with* and *for* young people, but we have also embraced new ideas and responded to emerging opportunities. Collaboration and partnerships have been key to our approach, as we have worked with limited financial and human resources. While digital tools may be the best way in the twenty-first century 'to learn about our past, capture our present and imagine our future' (Ibid.), we are also conscious that as a water museum we have the potential to be an agent of social change.

Going forward, we are clear that our exhibition content needs to be bilingual (current exhibitions are in Hindi, Marathi, Gujarati) and accessible for the visually impaired (using audio, podcasts, music) and that it needs to be taken beyond the virtual 'museum' to schools, water practitioners and decision-makers. Our Water Classrooms initiative, with the support of the *Transforming Education for Sustainable Futures* project, was built off our Pune urban waterscapes digital exhibition, Punyache Paani. Together with educators, academics, researchers, artists and NGOs, we developed interactive sessions for middle school children to learn about water from an interdisciplinary and inclusive lens, in contrast to the physical and natural sciences focus on water in schools today. The sessions were grouped in four modules: water and our self, shared waters, systems and water, and sustainability/the planet. Over a two-month period, we had 25–30 students from sixth to eighth grades across six schools in Pune, with at least one teacher, come for Sunday morning workshops with fun, activity-based hands-on learning. Students initiated actions that they could take in their homes, schools and peer communities and reflected on their learning through art and media. Plans are afoot to take the exhibition of learning to different places and to translate and contextualize content for other schools. At the global level, there are water museums that want to collaborate, but finding appropriate resources, both financial and human, to lead the process is a challenge.

Equally encouraging are the many conversations across the country about the role of public art in facilitating critical engagement on the environment, including the development of smart cities with a culturally sensitive ethos. The ongoing civil society agitation against the riverfront development project in Pune is using media and the arts to raise awareness of this potentially destructive endeavor. In our first collective urban online exhibition on Mumbai's waterscape, (www.confluence.livingwatersmuseum.org) we invited the Koli artist, Parag Tandel, to share traditional dry fish recipes passed on across generations from mothers to daughters. Parag illustrated the book with his quirky art about our dying rivers and oceans. Later a beautiful physical version of the book was produced by the TARQ Gallery in Mumbai which represents him, and funds from the sale are being used to support the

Tandel Fund of Archives being created by Parag and his partner to document the rich history, customs and practices of the Koli community, the first citizens of Mumbai, currently being displaced from their *koliwadas* by the massive Coastal Road project.

This year (2023) we took part in the Mumbai Urban Arts Festival with a panel on Water as a Muse and dance and music performances around the theme of access to water in the city. I was also asked to speak and moderate a panel as part of a series called 'Align and Disrupt' at the international India Art Fair, Delhi. With three other women curators, we looked at the challenge of curating for the public, broadly defined, from Indigenous communities in Australia to a virtual audience, examining issues of access, language, knowledge and interpretation. It is so encouraging to see the growing number of water festivals across the country with specially curated events for children on water and the arts.

There is no blueprint approach to building a water museum, as the diversity of museums in the global network illustrates. Amongst the many lessons that I have learned over the past six years is that investing in and believing in people is as important as the partnerships that sustain them, and that good design takes time. The young are restless, and their staying power is going to be limited in any initiative, so there is effort required to mentor them before they move on. However, it is important to recognize that their aspirations are not necessarily yours. Still, we are slowly building a network of young people passionate about water and as they find their space in the professional world, we know that they will remain connected to their liquid heritage. It is up to us to find ways of harnessing this potential towards wise water management principles and practices beyond the creative storyboard. This may be a tall order for a museum, and some would argue that it is not a priority, but as museums seek to reinvent themselves as more inclusive, democratic spaces, we have the moral responsibility to talk to our waters and ask, 'what do you want?'

Next steps

> Art has the ability to change our minds, inspiring us to take on different perspectives, and to reimagine our (water) worlds.
>
> *(Nossel, 2016, p. 103)*

While development practice is beginning to recognize the importance of 'new' ways of encountering and experiencing'"development' (Lewis et al., 2022) beyond academic papers and evidence-based policy briefs, art alone cannot change the world. As we initiate new chapters in Goa, Mussoorie and Pondicherry, we are also designing small fellowship grants with local institutional partners to mentor youth, women and men, especially those from marginalized communities to document and visualize their own water stories within the larger framework of water, social equity and sustainability. We recognize that stories can be retold in many formats including music, songs, performance and art, and we respect the many inter-generational, multi-lingual and cultural traditions of storytelling we share. Moving forward, we envisage the Living Waters Museum as a 'hub' around which we can develop a network of contextual and locally embedded water museums, physical or digital chapters, which will use the power of knowledge generation and communication to advocate for water justice through education, public outreach and new creative learning alliances across geographic borders and disciplinary boundaries. In so doing, this collective, bottom-up approach can also facilitate a more inclusive dialogue on gendered roles, rights and responsibilities around water use, control and management. By building local knowledge, skills and capacities, especially for and with young girls and women, we can better equip and empower them to participate in decision-making on community water governance and question long-standing patriarchal norms.

Notes

1 The Ganga Action Plan was launched in 1986 to improve water quality by intercepting, diverting and treating domestic sewage and prevent toxic and industrial chemical wastes from entering the river (see Ahmed, 1994, 1995).
2 See Irland (2007).
3 For more on our team, see www.livingwatersmuseum.org/about.
4 Photogrammetry has been defined as the 'art, science, and technology of obtaining reliable information about physical objects and the environment through processes of recording, measuring, and interpreting photographic images and patterns of recorded radiant electromagnetic energy and other phenomena' (Aber et al., 2010, p. 23).

References

Aber, J. S., Marzolff, I., & Ries, J. B. (2010). *Small format aerial photography: Principles, techniques and geoscience applications*. Elsevier.

Ahmed, S. (1994). The rhetoric of participation re-examined: The state, NGOs and water users at Varanasi, Uttar Pradesh, India. *The Environmentalist, 14*(1), 3–16.

Ahmed, S. (1995). Whose concept of participation? State-society dynamics in the cleaning of the Ganga at Varanasi. In G. P. Chapman & M. Thompson (Eds.), *The quest for sustainable development in the Ganges Basin*. Mansell.

Ahmed, S. (Ed.). (2005). *Flowing upstream: Empowering women through water management initiatives in India*. Centre for Environment Education and New Delhi: Foundation India Books.

Ashwin, C., & Ishan, P. (2019, September 19–20). *Museum experience through stories*. [Conference presentation abstract]. International workshop on proposed national water museum, Ministry of Jal Shakti, New Delhi.

Britto, N. (2019). *Muna: A digital interactive narrative on water in the context of stepwells*. [Unpublished doctoral dissertation, National Institute of Design, Ahmedabad].

Douglas, M. (1966). *Purity and danger – An analysis of the concepts of pollution and taboo*. Routledge.

Groenfeldt, D. (2019). Why we need water ethics. *American Scientist, 107*, 270–273.

Irland, B. (2007). *Water library*. www.basiairland.com/projects/book/waterlibrary.html

Jain-Neubauer, J. (Ed.). (2016). *Water design: Environment and histories*. Marg Foundation.

Kenderdine, S. (2014). *How will museums of the future look?* TEDxGateway [Video]. www.youtube.com/watch?v=VXhtwFCA_Kc

Lautman, V. (2017). *The vanishing stepwells of India*. Princeton Architectural Press.

Lepcha, M. (2021). (Re)meeting the roots: Visualising water heritage through storytelling. *Living Waters Museum*. www.livingwatersmuseum.org/remeeting-the-roots

Lewis, D., Rodgers, D., & Woolcock, M. (Eds.). (2022). *New mediums, better messages? How innovations in translation, engagement and advocacy are changing international development*. Oxford University Press.

Livingston, M. (2002). *Steps to water: The ancient stepwells of India*. Merrill Publishers.

McGhie, H. A. (2019). *Museums and the sustainable development goals: A how-to guide for museums, galleries, the cultural sector and their partners*. Curating Tomorrow.

Mehta-Bhatt, P. (2014). *Her space, her story: Exploring the stepwells of Gujarat*. Zubaan.

Moench, M. (2015). *Re-envisioning climate change through the arts*. TEDxGateway [Video]. www.youtube.com/watch?v=iZRGUADAu2Q

Nimje, S. (2018). *Living waters museum: A digital museum of water* [Unpublished doctoral dissertation, National Institute of Design, Ahmedabad].

Nossel, S. (2016). Introduction: On "artivism," or art's utility in activism. *Social Research: An International Quarterly, 83*(1), 103–105.

Nye, J. S. (1990). *Bound to lead: The changing nature of American power*. Basic Books.
Ojeda, D., Nirmal, P., Rocheleau, D., & Emel, J. (2022). Feminist ecologies. *Annual Review of Environment and Resources*, 47(1), 149–171. https://doi.org/10.1146/annurev-environ-112320-092246
UNESCO. (2015). *Recommendation concerning the protection and promotion of museums and their collections, their diversity and their role in society*. https://unesdoc.unesco.org/ark:/48223/pf0000246331
UNFCCC. (2018). *The Talanoa Dialogue platform*. https://unfccc.int/process-and-meetings/the-paris-agreement/the-paris-agreement/2018-talanoa-dialogue-platform
Zwarteveen, M., Ahmed, S., & Gautam, S. R. (Eds.). (2012). *Diverting the flow: Gender equity and water in South Asia*. Zubaan and SaciWATERs.

PART 3

Sanitation stories

15

THE CONTENTIOUS PATH OF MENSTRUAL HEALTH

Reflections on the past and provocations for the future of the water sanitation and hygiene sector

Jacqueline Gaybor Tobar

Introduction

Menstrual stigma is powerful. It has not only made us keep menstruation out of sight by hiding the bleeding, or feel ashamed of inhabiting a menstruating body, but also by keeping it out of our mind and thinking about it critically as a feminist issue. Menstrual stigma is a complex issue deeply rooted in social, cultural, and historical norms and beliefs surrounding gender and sexuality. Such norms and beliefs result in negative attitudes and discriminatory behaviors toward menstruating individuals, and even their exclusion from specific social and cultural activities (Chrisler et al., 2016; Johnston-Robledo & Chrisler, 2020). The stigma around menstruation, or the "menstruation closet" that Iris Marion Young (2005, p. 106) refers to, has made it imperative to hide our menstrual processes: conceal the menstrual blood and the physical sensations (such as pain, discomfort, or relief), the emotions and related thoughts, blocking the possibility of discussing this topic openly.

Sarah Ahmed (2012, p. 2) says, "every research project has a story, which is the story of an arrival." My interest in menstrual health was ignited in 2010 when three Ecuadorian menstrual activists gave a workshop on menstrual health at a bio-health retreat in a small town in the Ecuadorian highlands. Until that encounter, menstruation for me was solely a biological event. There was a limit to what I could and had to know: menstruation is part of a cycle and arrives approximately every 28 days, indicating no pregnancy. Pads and tampons are needed to manage the menstrual flow and pass as a non-bleeder (Vostral, 2008). Some people experience pain. Try not to stain. That was about it. The subtlety with which the three activists expressed themselves in front of an audience of 80 people raised powerful questions in me and triggered a slow process of building a profoundly different relationship with my own body and, some years later, prompted me to conduct research on menstrual health in an academic space. My desire to dedicate my research to this topic was born from that first encounter where I experienced how unexplored important social, economic, cultural, or environmental aspects filled the silences around menstruation. The stigma had been such that I had never questioned the prejudices about our (my) menstrual bod(y)ies. Nor had I considered the relevance of looking into my embodied knowledge and experience of this event as central to understanding what constitutes knowledge about the body.

DOI: 10.4324/9781003100379-19

Bringing menstruation into an academic space has allowed me to delve into the complex value systems in which it is rooted and explore its social, environmental, and economic dimensions. Nevertheless, it took me some time to decide to investigate this topic. Along the way, there was no shortage of comments that disqualified the topic of menstruation as relevant in the field of development studies. Maybe it is hard to imagine this now in 2023, as menstruation has rapidly become a subject of public debate. This has been partly due to the relevance it acquired in the development field. Menstruation has slowly transitioned from a neglected topic to a central issue on the agenda for achieving gender equality. While significant progress has been made in thinking about menstruation, this path has been contentious. Within the Water Sanitation and Hygiene (WASH) development sector, menstruation was initially framed as an overly simplified biological problem to solve through practical solutions. This approach, known as Menstrual Hygiene Management (MHM) showed the dominance of a gender-blind thinking that disregarded the complex intersections of gender, race, class, and other social identities in development. More recently, different forms of knowledge and voices that had been in the background for decades have become more prominent and are shaping the conversation and action, situating menstruation as a category of analysis that holds significant potential for inquiry on "how systems of power and knowledge are built upon its understanding" (Bobel, 2020, p. 4). Currently, menstruation is perceived as a public Sexual and Reproductive Health and Rights (SRHR) concern. The Menstrual Health (MH) or Menstrual Health and Hygiene (MHH) framings offer a more comprehensive view of the issue by acknowledging that menstruation is a complex subject that goes beyond the biological but instead is affected by context-specific gender norms and power relations.

In this chapter, I explore this contentious path by drawing on a critical analysis of existing scientific literature on menstrual health compiled between 2015 and 2023. I situate this chapter in the field of critical development studies and explore the limitations of the conceptualization and development interventions around menstruation, in particular, its emphasis on addressing symptoms without looking at the underlying causes of gender inequalities that contribute to stigma and discrimination against menstruation and menstrual bodies.

I structure this chapter as follows: in the next section, I build upon race and feminist scholars' critique of modernization theory and look at development as a system informed by a predominantly white and Western perspective. Then, I provide an overview of the different frameworks through which menstruation has become a public concern. In this section, I examine some of the universal assumptions and question their simplified approach and limitations. Here, I delve into and question the different debates on menstruation management and education, as well as menstrual pain and the workplace. Finally, I argue for the need to question how restrictive sexual and gender norms promote unequal gender relations that adversely affect perceptions and people's relations with menstruation and the menstrual body.

A color-blind call for modernization: unpacking the development discourse of menstrual hygiene management

Two important interconnected concepts underlie the MHM agenda and are visible in its scientific production, policy, and social interventions: development and modernization (e.g., of practices to manage menstruation). The concept of development has been at the center of a long interdisciplinary debate among scholars. Post-structuralism, for example, sees development as a social construction that seeks to exercise power through dominant Western ideas and representations (Peet & Hartwick, 2015). From a post-colonial perspective, development does not confine itself to the sole focus of economic growth. It also extends a Western project of social transformation, through which potentially the whole fabric of social life is subject to control (Young, 2005). From a post-development viewpoint, development as a Western control method encompasses a set of parameters to judge other societies (Walsh, 2010, pp. 15–16). Despite this and other powerful critiques, modernization theory has been one of the most influential ideas by which societal progress and improvement are evaluated in terms of economic growth (Brohman, 1995). From a modernization theory perspective, the concept of development in a world ideologically divided into developed and developing countries is connected with "the construction of a single model of modernity based on the experience of a few industrialized countries. If this model is followed, it is assumed that all countries may reach the goal of a similar type of 'modern society'" (Brohman, 1995, p. 122). As development is considered a linear project, development aid is thought of as a way to bring economic and political progress to developing societies in the same manner that more developed societies have followed.

Modernization theory has been extensively criticized by different approaches that question its Eurocentric bias, assuming that Western societies represent the ideal path to progress and suggesting that all societies should aim to emulate them. Criticism has also emerged around modernization theory's underestimation of local agency and the exclusion of minorities from decision-making and, therefore, its limited ability to influence development programs and policies. Connected to this is the assumption that the same development model can be successfully applied to all societies disregarding their unique circumstances and historical legacies (Bretón & Palenzuela, 2016). Feminist scholars have provided strong critiques of modernization theory in development studies, pointing to its tendency to ignore gender inequalities and women's experiences in development. Kabeer (1994) has critiqued modernization theory's limited understanding of the complex ways by which multiple social categories such as race, gender, and class interact to shape individuals' experiences and opportunities in development. Another feminist critique of modernization theory relates to its reinforcement of patriarchal power structures, which negatively affect gender minorities access to education, employment, and political participation (Chant, 2006). Nevertheless, despite the emergence and dissemination of powerful critiques, modernization theory continues to influence academic production, development assistance, and policy-making at the national and international levels.

Development has also been criticized for being color-blind. For Sarah White (2002), race has rarely been directly addressed, even in the influential critiques of "Eurocentrism" or "neo-colonialism," making it a taboo subject. The failure to see race within development is paradoxical, especially considering that its historical foundation, scientific production, social interventions and influence in public policy are themselves racial projects that create and continue to reproduce structures of power based on racial hierarchies and identities (Gaybor, 2020). From a modernization theory perspective, development often focuses on economic growth and material progress. Following White, "the secret of development's power lies in its capacity to enlist others to its own agenda, so that they want what it claims to offer" (2002, p. 410). For Grosfoguel (2013), development policies and interventions often prioritize the interests of dominant racial groups while ignoring the needs and experiences of marginalized racial and ethnic groups. Moreover, there is a generalized assumption that there is a homogeneous "Developing World," "Third world," or "African girls" population, ignoring the differences between and within racial and ethnic groups and overlooking the cultural, linguistic, and historical factors that shape their experiences of development. These forms of racial difference and hierarchy shape the development field at various levels. Often, the traditional culture of brown and black bodies found in developing countries symbolizes ignorance and backwardness and accounts for the slow progress in embracing modernity, while there is a direct association of expertise and knowledge with whiteness and the West, comprising "a (racialized) distinction not only between who is developed and underdeveloped but who gives assistance and who must be grateful for it" (Kothari, 2006, p. 14).

Assumptions and challenges of MHM and education

While the relationship between the right to education and menstruation might not be immediately clear, the MHM framework, an approach that emerged from the WASH development sector, consolidated the link between the two in the development agenda since the early 2000s. The study of Sowmyaa Bharadwaj and Archana Patkar (2004), one of the first publications on MHM in WASH, points to the masculine and Western bias of this sector that for many years blocked the consideration of menstruation as a relevant topic in the WASH agenda. Bharadwaj and Patkar (2004) discuss the overall absence of menstruation in WASH and therefore the lack of investment and action to address menstruation-related issues. The authors found that very few WASH professionals "actively engaged with the issue although it has crossed many a mind in passing" but also that they "were astonished at the absence of this issue from both technical and rights-based discourses," and finally, that the literature on gender mainstreaming in the WASH sector, "[was] silent on menstrual management" (Bharadwaj & Patkar, 2004, p. 4).

Nevertheless, between 2008–2018, MHM became the second main research focus area within WASH (MacArthur et al., 2020). The fast-growing interest in MHM in WASH and its formulation as a problem needing practical solutions relates to closing the gender gap in education in countries in the Global South (Sommer et al., 2015). By linking MHM with the right to education, MHM was constructed as an issue of societal importance to achieve gender equality. This relates to what Sommer et al. (2015, p. 8) refer to as "the moral perspective given to menstruation" in the context of a notable trend in the gender and development landscape wherein a strong emphasis had been placed on educating girls as a key strategy to end poverty for themselves, their families and the world (Hayhurst, 2013).

MHM interventions and research have primarily focused on young adolescent girls and pointed to the lack of access to menstrual management products and (separate) bathroom facilities as factors that lead to their school absenteeism (Banerji & Thomson Reuters Foundation, 2018; Crofts & Fisher, 2012). Yet, it is important to emphasize that school absenteeism is difficult to track and even more challenging to attribute to the lack of menstrual products. As noted by Benshaul-Tolonen et al. (2020), most existing studies on menstrual-related school absenteeism have three main methodological limitations. Firstly, they focus only on menstrual-related absence and not on the broader impact of menstruation on school participation and the psychosocial experiences of schoolgirls. Secondly, they rely on self-reporting and recall bias without alternative validation methods. Thirdly, they do not consider the influence of stigma and taboos on behavior and experiences about menstruation. Therefore, instead of approaching menstruation as a systemic issue, it is framed as a problem to be addressed with technical solutions, or as Geertz et al. (2016) put it, the focus was primarily on the "hardware." This imaginary spread fast, covering up structural causes proven to be decisive in affecting girls participation in schools. Many of these point to social norms that place women and girls in unequal power relations, assigning them specific family responsibilities and household chores. For example, a study conducted by Grant et al. (2013) stressed that girls are also expected or forced to marry early and bear children. Likewise, if a family lacks economic resources and decisions need to be made regarding who can go to school, boys have often been given priority over girls. Furthermore, menstrual pain has also been identified as a substantial obstacle to girls access to education and their overall academic engagement (Hennegan et al., 2017).

The narrative about the precarious situation of schoolgirls in Sub-Saharan African and South Asia and the need for immediate product-based solutions rapidly positioned itself as an uncontested truth. (I)NGOs, social entrepreneurs, and Fem Care companies quickly jumped into designing and implementing interventions targeted at these girls. With WASH's traditional focus on hygiene and tangible solutions, interventions that involved the distribution of menstrual products were seen as the appropriate responses to these challenges (Gaybor & Harcourt, 2022). In this model, menstruation was "a nuisance at best, pathological at worst, and void of any cultural meaning, further alienating women from their own bodies" (Vostral, 2008, p. 171). Furthermore, girls school absenteeism has also been attributed to the lack of hygiene facilities and overall cleanliness of schools. The WHO and UNICEF's definition of *menstrual hygiene* helps illustrate the emphasis hygiene has been given:

> Women and adolescent girls are using clean material to absorb or collect menstrual blood, that can be changed in privacy as often as necessary for the duration of the menstrual period, using soap and water for washing the body as required, and having access to facilities to dispose of used menstrual management materials.
>
> *(WHO & UNICEF, 2013, p. 16)*

The quote lists material requirements and preconditions that must be met for adequate menstrual management. This definition (WHO & UNICEF, 2013) spread widely, serving as the basis for advancing research, programming and policy making on MHM (Sommer & Sahin, 2013), simultaneously reinforcing an imaginary about the precarious menstruating girls and women in the Global South and their need to be rescued by modern hygiene practices. MHM literature often signaled that girls and women in the Global South manage their menstruation in an *unhygienic, incorrect, poor,* or *deficient* manner. For example, Adinma and Adinma's (2008, p. 82) study undertaken amongst adolescent secondary school girls in Onitsha, Southeastern Nigeria, concludes that "perceptions on menstruation amongst adolescent secondary school girls are poor, and practices often incorrect." According to Crofts and Fisher's (2012) study conducted in Uganda, the lack of access to menstrual products and sanitation facilities renders the experience of menstruation unsanitary, inappropriate and deficient. As indicated above, there is inconclusive evidence, as other studies show that school attendance by girls is not affected by characteristics of the school environment such as cleanliness, the privacy of toilet facilities, or soap (Grant et al., 2013).

MHM has also relied on human rights institutions and discourses to gain ground in the development agenda. A large part of human rights arguments associated with menstruation reinforced the urgency to act and reproduced the modernizing figure of the West as the savior of the backwardness of girls and women from the Global South (UNICEF, 2018).

The broad and unsubstantiated statement that "one in ten girls in Sub-Saharan Africa misses school during their menstruation," disseminated by the WHO and UNICEF (2013) and other human rights organizations (UNESCO, 2014), traveled uncritically around the world to the point of acquiring the status of a fact. The claim was widely mobilized to show the moral foundation of the urgency which prompted paternalistic action.

Menstrual health and SRHR: more than pads and hygiene mandates

The historical neglect of menstrual health within important international SRHR guidance and resources highlights a significant gap in addressing a critical aspect of individuals SRHR (Wilson et al., 2021). This is still apparent, for instance, when examining the list of SRHR topics on the WHO (2023) website, where menstrual health is absent. This oversight could be attributed to the historical stigma and taboo surrounding menstruation, resulting in insufficient consideration of this matter in policies, discussions, and research. Nonetheless, ongoing global efforts are fostering positive transformation. The work of Catarina de Albuquerque, the first Special Rapporteur on the rights to safe drinking water and sanitation, was instrumental in creating synergies between WASH-MHM and SRHR and in gendering the conversation (Roaf & de Albuquerque, 2020). For context, it is important to highlight that the recognition of the human right to clean water and adequate sanitation by the United Nations General Assembly in resolutions adopted by the Human Rights Council (HRC) (2010, 2011) and the Millennium and Sustainable Development Goals[1] did not engage with the topic of menstrual health initially. De Albuquerque embedded it in her mandate, making menstruation a central focus. She also problematized menstrual stigma, highlighting how it was "an element of the structural and social dynamics which (re)produce unequal power relations" (HRC, 2012, p. 5). Furthermore, in 2016, the HRC cautioned against framing issues or adopting practices that confirm or worsen stereotypes and shame about menstruation. This HRC resolution nuances the relevance given to addressing women's material needs but emphasizes that this is not the ultimate goal, taking distance from the product-focus approach of MHM. The resolution calls for an end to stigma and the combat of practices based on harmful stereotypes (HRC, 2016). This evolution of the framework shows that there is a growing appeal to bridge practical gender needs with strategic gender interests as a critical pathway to achieving transformational changes in gender equality (MacArthur et al., 2020). More recently, the interdisciplinary diversity of critical menstruation studies has opened up a less action-oriented pathway, revealing a growing entanglement of the topic of menstruation with historical concerns regarding SRHR (Gaybor & Harcourt, 2022) while contributing with evidence-based studies and a more nuanced approach to this broad theme.

Provocations for the future: menstrual pain and the workplace

Menstrual leave has gained global media and public attention in recent years, with some countries and companies in the Global South and North offering it as a means to address the menstrual-related challenges of pain faced by menstruating people (although currently it is mainly directed towards women) in reproductive age, in the workplace (King, 2021). Recent development of legislation and public policies across the globe demonstrate this growing concern: Taiwan, China, South Korea, Indonesia, Zambia, Mexico, and, more recently Spain, have established policies around menstrual leave. In other countries, these policies were introduced earlier. Japan's national policy was introduced in 1947 (Barnack-Tavlaris et al., 2019). In Russia, menstrual leave was introduced in 1880, specifically in some industrial sectors such as textile, where female workers experienced physically demanding working conditions. This policy was not developed to protect women workers' rights or to accommodate the workspaces to their embodied needs: rather, it aimed to protect the reproductive function of women workers and safeguard national-level fertility (King, 2021). The policy was eventually retracted in 1927 as it led to increased discrimination against women in the workplace (Ilic, 1994). Additionally, private corporations, like Culture Machine and Gozoop in India (Singh, 2021) or Carrefour in France (Lemaitre, 2023), have also recently implemented menstrual leave policies. These policies and regulations differ from each other: some offer paid or unpaid days off work, others emphasize flexibility in the workplace, meaning that the employee does not need to be in the office but can work remotely instead of taking actual time off. Determining how these policies are implemented and the frequency of their use by women proves challenging due to the scarcity of publicly available data and the restricted accessibility of human resources policies and procedures, which companies do not commonly disclose (Levitt & Barnack-Tavlaris, 2020).

Menstrual leave has been positioned as a highly progressive step toward acknowledging the validity of menstrual pain – a form of pain that has been belittled, not only in the field of development but also in the fields of medicine and psychology (Hudson, 2022). This aligns with a broader pattern of gendered omissions and assumptions concerning women's bodies (Culley et al., 2013). It is one among other manifestations of the systemic neglect and lack of understanding surrounding women's health concerns and experiences. Therefore, menstrual pain cannot be discussed without also considering the role of menstrual stigma. Menstrual stigma's power lies in its ability to quietly establish societal standards of normality and acceptability, while simultaneously delegitimizing experiences that fall outside of these norms, which "relate back to power relationships in a society characterized by sexism and misogyny" (Olson et al., 2022, p. 3). While I consider the recognition of menstrual pain a critical step in advancing the conversation on de-stigmatizing menstruation and "making menstruation matter" in the policy agenda, it still needs to systematically integrate a gender lens in order to question if these policies attain the societal benefits for which they are enacted. As previously mentioned, there are still limited studies that evaluate the effects of menstrual leave policies on health and well-being or on improved conditions in the workplace (Barnack-Tavlaris et al., 2019). Nevertheless, important ongoing debates question their (possible) positive impacts and raise valid concerns on the current over enthusiasm. Sally King (2021) points out that these policies risk reframing the problem as one of the female body, reinforcing inaccurate stereotypes that position menstruating people as weaker or less reliable. As a gendered experience, menstrual leave cannot be examined in isolation, as it is entrenched within the framework of patriarchal societies, where menstruation has historically been mobilized to marginalize and oppress women (Johnston-Robledo & Chrisler, 2013). Menstrual leave, therefore, potentially creates another point of stigma and risks perpetuating gender inequality, as it frames the female body as frail, inferior or disproportionately prone to illness. Disclosing menstrual pain can result in disapproval and social censure, leading to misguided notions concerning the bodily processes of menstruating individuals and casting doubt upon their resilience, vitality, and fertility. For instance, McCurry and Leavenworth (2016) reported that in Japan, the cultural beliefs surrounding menstruation contribute to women's reluctance to use menstrual leave days. Many women fear disclosing their menstrual status and taking leave, which could reinforce harmful gender stereotypes and lead to another form of gender discrimination.

Furthermore, a few studies to date have documented the implications of these policies on gender relations in the workplace. It is important to note that comprehensive evidence of their impact and effectiveness remains elusive. For example, Min-ho et al. (2012) documented the risk of resentment among fellow employees despite only a small number of women opting for it. Ilic (1994) detailed how, decades ago, these policies made employers reluctant to hire female employees. On the flip side, the implementation of menstrual leave policies has been found to hold significant potential in normalizing or *neutralizing* the discussion about menstruation among all employees (Barnack-Tavlaris et al., 2019). However, this same study found that implementing menstrual leave could be perceived as unfair towards men, as it might lead to women receiving additional days off (Ibid).

The idea of menstrual leave as unfair largely remains due to a lack of widespread awareness and limited empirical evidence regarding the nature, prevalence, and effective management of pain and other symptoms associated with menstruation. Menstrual leave policies exemplify the impatience of jumping into policy development without first understanding the complexity of the problem. King (2021, p. 9) refers to this as a global menstrual ignorance which "serves to maintain gender inequality, by preventing established myths from being challenged by empirical data." Nonetheless, the conversation on menstrual pain has the potential to evolve into a more comprehensive framework – one that transcends misconceptions about menstruation, is rooted in empirical data and considers the fundamental needs of workers and the possibility of reasonable workplace adjustments (King, 2021). Policies should carefully consider chronic illnesses associated with the menstrual cycle, which are distinct from the normal menstrual pain experienced during healthy periods. Conditions such as endometriosis, ovarian cysts, fibroids, and autoimmune disorders can cause severe symptoms, including pain in menstruating people. Recognizing these as health conditions means acknowledging that they deserve effective and timely medical diagnosis, treatment, and the same consideration as other accepted chronic health conditions that qualify for sick leave.

Concluding reflections

MHM is a key ground to observe the power relations and global structures of inequality that operate in the complex field of international development and how the dynamics of aid become racially symbolized. The concepts of development and modernization undergirded the WASH agenda in MHM, which emerged as a band-aid approach to address the link between menstruation and education and menstruation and health in an effort to achieve gender equality. The emphasis on hygiene in WASH created and reinforced an imaginary about the precariousness of menstruating girls and women in the Global South and their need to be modernized by the West. The magnified focus on menstrual management as an urgent problem and modern products and facilities as the ideal solutions has overshadowed the systemic roots of menstruation-related challenges. Although WASH's MHM agenda aimed for social change, it did not (paradoxically) systematically integrate a gender lens into its analysis. In its early stage, WASH worked around gender inequality, moving the focus away from the root causes of menstrual stigma. This has had an impact by, for example, failing to problematize the context-specific underlying assumptions that lead, in the first place, to the denigration of menstrual bodies. By focusing solely on the provision of products, there was a risk of perpetuating the notion that menstruation is bothersome and the menstrual body a nuisance. Moreover, the emphasis on products and managing bleeding contributed to framing menstruation as a simplistic and narrow topic, perpetuating racist, misogynistic, and colonial ideologies surrounding menstruating bodies in the Global South.

While the conversation on menstruation has expanded its scope from a focus on products and has evolved into a more comprehensive framework under the broader SRHR umbrellas of Menstrual Health or Menstrual Health and Hygiene, the current developments and global debates on menstrual leave raise essential questions that invite us to look back and reflect. On the surface, policies on menstrual leave appear to be a progressive step towards addressing the challenges menstruating individuals face in the workplace, particularly concerning pain. They hold the promise of providing an opportunity for menstruating people to exercise their rights. Moreover, menstrual leave policies offer a chance to initiate conversations about menstruation and, in a few cases, normalize it. However, drawing lessons from the MHM framework, I argue for caution and urge against blindly following a path of "acting first and asking questions later." Menstrual leave policies underscore a concerning disconnection between the commendable aim of upholding the right to work for individuals enduring menstrual pain and the broader socio-cultural milieu. This phenomenon mirrors the simplified approach observed in MHM initiatives since the early 2010s, where menstrual products were presented as isolated remedies to ensure gender parity in education, inadvertently overlooking the significance of and the myriad social constraints that intersect with girls school attendance (Chandra-Mouli & Patel, 2020). Therefore, we must dare to engage and question how sexual and gender norms promote unequal gender relations that adversely affect perceptions and people's relations with menstruation and the menstrual body in the school, the household, the workplace and beyond. There must be more evidence and investigation that addresses these concerns comprehensively. Before the wholesale design and implementation of these policies, we need to understand the potential impact of policies in reinforcing or challenging stigma, or shaping the perception of menstruation as an illness requiring special consideration, or fostering a more inclusive learning and working culture.

Finally, before jumping straight into policy development, it might be helpful to question the assumptions and biases that underpin current approaches to menstrual health. MHM has shown the risks of acting without evidence, and, relatedly, the importance of having evidence that captures the diverse voices and experiences of menstruating individuals and their connections to the broader social, economic, and political factors that shape these menstrual experiences.

The conversation around menstruation and education and menstruation and work opens another door to reflect on the disembodied nature of the school and the workplace, particularly the systematic disregard for those with bodies that deviate from the normative standards of masculinity and able-bodiedness and thus omit the physical and bodily experiences of queer, menstruating, or disabled individuals, among others. So, a question to ask in parallel is: for whom are these spaces designed? Or better yet, who is the ideal student or the ideal worker that fits them?

It is clear that action is needed to address the challenges menstruators face in the workplace, school, and beyond. However, a deeper understanding of menstrual stigma's systemic and root causes must inform these actions. We must engage in a process of *questioning, listening, pausing, and learning* (Gaybor & Harcourt, 2022). Previous experiences from the WASH sector have shown that a simplistic approach to menstrual health is insufficient. Menstrual health is a complex ecosystem that intersects with gender norms and power dynamics. By adopting a more thoughtful, nuanced, and feminist approach to menstrual health, we can develop policies and programs that are better aligned with the needs and realities of menstruating people and that can support gender equality and social justice.

Note

1 While the SDGs do not have a specific goal or indicator on MHM, menstruation has been linked to Goals 3, 4, 5, 6, 8, and 12.

References

Adinma, E., & Adinma, J. (2008). Perceptions and practices on menstruation amongst Nigerian secondary school girls. *African Journal of Reproductive Health*, *12*(1), 74–83.

Ahmed, S. (2012). *On being included: Racism and diversity in institutional life*. Duke University Press.

Banerji, A., & Thomson Reuters Foundation. (2018, May 22). Third of girls in South Asia miss school during periods – study. *Reuters*. www.reuters.com/article/us-india-women-health-idUSKCN1IN00F

Barnack-Tavlaris, J. L., Hansen, K., Levitt, R. B., & Reno, M. (2019). Taking leave to bleed: Perceptions and attitudes toward menstrual leave policy. *Health Care for Women International*, *40*(12), 1355–1373.

Benshaul-Tolonen, A., Zulaika, G., Sommer, M., & Phillips-Howard, P. A. (2020). Measuring menstruation-related absenteeism among adolescents in low-income countries. In C. Bobel, I. T. Winkler, B. Fahs, K. A. Hasson, E. A. Kissling & T. Roberts (Eds.), *The Palgrave handbook of critical menstruation studies* (pp. 705–723). Palgrave Macmillan.

Bharadwaj, S., & Patkar, A. (2004). Menstrual hygiene and management in developing countries: Taking stock. *Junction Social, Social Development Consultants*. www.ircwash.org/resources/menstrual-hygiene-and-management-developing-countries-taking-stock

Bobel, C. (2020). Introduction: Menstruation as lens – menstruation as opportunity. In C. Bobel, I. T. Winkler, B. Fahs, K. A. Hasson, E. A. Kissling, & T. Roberts (Eds.), *The Palgrave handbook of critical menstruation studies* (pp. 1–6). Palgrave Macmillan.

Bretón, V., & Palenzuela, P. (2016). Desarrollo y Colonialidad: Una Epistemología para el Análisis Crítico del Desarrollismo (Development and coloniality: An epistemology for critical analysis of developmentalism). *Revista Andaluza de Antropología*, *10*, 119–142.

Brohman, J. (1995). Universalism, Eurocentrism, and ideological bias in development studies: From modernisation to neoliberalism. *Third World Quarterly*, *16*(1), 121–140.

Chandra-Mouli, V., & Patel, S. V. (2020). Mapping the knowledge and understanding of menarche, menstrual hygiene and menstrual health among adolescent girls in low-and middle-income countries. In C. Bobel, I. T. Winkler, B. Fahs, K. A. Hasson, E. A. Kissling & T. Roberts (Eds.), *The Palgrave handbook of critical menstruation studies* (pp. 609–636). Palgrave Macmillan.

Chant, S. (2006). *Gender, generation and poverty: Exploring the 'feminisation of poverty' in Africa, Asia and Latin America*. Edward Elgar Publishing.

Chrisler, J. C., Gorman, J. A., Manion, J., Murgo, M., Barney, A., Adams-Clark, A., Newton, J. R., & McGrath, M. (2016). Queer periods: Attitudes toward and experiences with menstruation in the masculine of centre and transgender community. *Culture, Health & Sexuality*, *18*(11), 1238–1250.

Crofts, T., & Fisher, J. (2012). Menstrual hygiene in Ugandan schools: An investigation of low-cost sanitary pads. *Journal of Water, Sanitation and Hygiene for Development*, *2*(1), 50–58.

Culley, L., Law, C., Hudson, N., Denny, E., Mitchell, H., Baumgarten, M., & Raine-Fenning, N. (2013). The social and psychological impact of endometriosis on women's lives: A critical narrative review. *Human Reproduction Update*, *19*(6), 625–639.

Gaybor, J. (2020). *The body politics of menstruation: Technologies, sustainability and destigmatization* [PhD Thesis, ISS-Erasmus University Rotterdam].

Gaybor, J., & Harcourt, W. (2022). Seeing the colour red: Menstruation in global body politics. *Global Public Health*, *17*(10), 2388–2400.

Geertz, A., Iyer, L., Kasen, P., Mazzola, F., & Peterson, K. (2016). *An opportunity to address menstrual health and gender equity*. FSG.

Grant, M., Lloyd, C., & Mensch, B. (2013). Menstruation and school absenteeism: Evidence from rural Malawi. *Comparative Education Review*, *57*(2), 260–284.

Grosfoguel, R. (2013). The structure of knowledge in Westernized universities: Epistemic racism/sexism and the four genocides/epistemicides of the long 16th century. *Human Architecture: Journal of the Sociology of Self-Knowledge*, *11*(1), 73–90.

Hayhurst, L. M. C. (2013). Girls as the "new" agents of social change? Exploring the "girl effect" through sport, gender and development programs in Uganda. *Sociological Research Online (Special Issue: Modern Girlhoods)*, *18*(2).

Hennegan, J., Dolan, C., Steinfield, L., & Montgomery, P. (2017). A qualitative understanding of the effects of reusable sanitary pads and puberty education: Implications for future research and practice. *Reproductive Health*, *14*, 1–12.

Hudson, N. (2022). The missed disease? Endometriosis as an example of "undone science". *Reproductive Biomedicine & Society Online*, *14*, 20–27.

Human Rights Council. (2010). *A/HRC/RES/15/9*. https://www.right-docs.org/doc/a-hrc-res-15-9/

Human Rights Council. (2011). *A/HRC/RES/16/2*. https://www.right-docs.org/doc/a-hrc-res-16-2/

Human Rights Council. (2012). *A/HRC/21/42*. https://www.right-docs.org/doc/a-hrc-res-21-2/

Human Rights Council. (2016). *A/HRC/33/49*. https://www.right-docs.org/doc/a-hrc-33-49/

Ilic, M. (1994). Soviet women workers and menstruation: A research note on labour protection in the 1920s and 1930s. *Europe-Asia Studies*, *46*(8), 1409–1415.

Johnston-Robledo, I., & Chrisler, J. C. (2013). The menstrual mark: Menstruation as social stigma. *Sex Roles*, *68*(1–2), 9–18.

Johnston-Robledo, I., & Chrisler, J. C. (2020). The menstrual mark: Menstruation as social stigma. In C. Bobel, I. T. Winkler, B. Fahs, K. A. Hasson, E. A. Kissling & T. Roberts (Eds.), *The Palgrave handbook of critical menstruation studies* (pp. 181–199). Palgrave Macmillan.

Kabeer, N. (1994). *Reversed realities: Gender hierarchies in development thought*. Verso.

King, S. (2021). Menstrual leave: Good intention, poor solution. In J. Hassard & L. D. Torres (Eds.), *Aligning perspectives in gender mainstreaming: Gender, health, safety, and wellbeing* (pp. 151–176). Springer.

Kothari, U. (2006). An agenda for thinking about "race" in development. *Progress in Development Studies*, *6*(1), 9–23.

Lemaitre, C. (2023, April 27). Après l'Espagne, l'idée d'un "congé menstruel" fait son chemin en France pour les femmes souffrant d'endométriose. *L'Independent*. www.lindependant.fr/2023/04/27/apres-lespagne-lidee-dun-conge-menstruel-fait-son-chemin-en-france-pour-les-femmes-souffrant-dendometriose-11161266.php

Levitt, R. B., & Barnack-Tavlaris, J. L. (2020). Addressing menstruation in the workplace: The menstrual leave debate. In C. Bobel, I. T. Winkler, B. Fahs, K. A. Hasson, E. A. Kissling & T. Roberts (Eds.), *The Palgrave handbook of critical menstruation studies* (pp. 561–575). Palgrave Macmillan.

MacArthur, J., Carrard, N., & Willetts, J. (2020). WASH and gender: A critical review of the literature and implications for gender-transformative WASH research. *Journal of Water, Sanitation and Hygiene for Development*, *10*(4), 818–827.

McCurry, J., & Leavenworth, S. (2016, March 4). Period policy in Asia: Time off "may be seen as a sign of weakness". *The Guardian*. www.theguardian.com/lifeandstyle/2016/mar/04/period-policy-asia-menstrual-leave-japan-women-work

Min-ho, J., Bo-eun, K., & Eun-ji, B. (2012, October 30). Menstrual leave – an entitlement men reject. *The Korea Times*. www.koreatimes.co.kr/www/news/culture/2014/04/399_123535.html

Olson, M. M., Alhelou, N., Kavattur, P. S., Rountree, L., & Winkler, I. T. (2022). The persistent power of stigma: A critical review of policy initiatives to break the menstrual silence and advance menstrual literacy. *PLOS Global Public Health*, *2*(7), e0000070.

Peet, R., & Hartwick, E. (2015). *Theories of development: Contentions, arguments, alternatives*. Guilford Publications.

Roaf, V., & de Albuquerque, C. (2020). Practice note: Why we started talking about menstruation – looking back (and looking forward) with the UN special rapporteur on the human rights to water and sanitation. In Chris Bobel, Inga T. Winkler, Breanne Fahs, Katie Ann Hasson, Elizabeth Arveda Kissling, Tomi-Ann Roberts (Eds.), *The Palgrave handbook of critical menstruation studies* (pp. 475–483). Palgrave.

Singh, B. (2021, October 26). 12 companies in India that offer period leave to their female employees. *India Times*. www.indiatimes.com/trending/social-relevance/indian-companies-that-offer-period-leave-to-female-employees-552433.html

Sommer, M., Hirsh, J., Nathanson, C., & Parker, R. (2015). Comfortably, safely and without shame: Defining menstrual hygiene management as a public health issue. *American Journal of Public Health*, *105*(7), 1302–1311.

Sommer, M., & Sahin, M. (2013). Overcoming the taboo: Advancing the global agenda for menstrual hygiene management for schoolgirls. *American Journal of Public Health*, *103*(9), 1556–1559.

United Nations Children's Fund (UNICEF). (2018). *Fast facts: Nine things you didn't know about menstruation*. Retrieved March 16, 2021, from www.unicef.org/press-releases/fast-facts-nine-things-you-didnt-know-about-menstruation

United Nations Educational, Scientific and Cultural Organization (UNESCO). (2014). *Puberty education & menstrual hygiene management: Good policy and practice in health education*. UNESCO.

Vostral, S. (2008). *Under wraps: A history of menstrual hygiene technology*. Lexington Books.

Walsh, C. (2010). Development as Buen Vivir: Institutional arrangements and (de)colonial entanglements. *Development*, *53*(1), 15–21.

White, S. (2002). Thinking race, thinking development. *Third World Quarterly*, *23*(3), 407–419.

Wilson, L. C., Rademacher, K. H., Rosenbaum, J., Callahan, R. L., Nanda, G., Fry, S., & Mackenzie, A. C. (2021). Seeking synergies: Understanding the evidence that links menstrual health and sexual and reproductive health and rights. *Sexual and Reproductive Health Matters*, *29*(1), 44–56.

World Health Organization (WHO). *Topics: Sexual and reproductive health*. Retrieved August 1, 2023, from www.who.int/teams/sexual-and-reproductive-health-and-research-(srh)/overview

World Health Organization (WHO) and United Nations Children's Fund (UNICEF). (2013). *Progress on sanitation and drinking-water – 2013 update*. UNICEF Data. https://data.unicef.org/resources/progress-on-sanitation-and-drinking-water-2013-update/

Young, I. M. (2005). *On female body experience: 'Throwing like a girl' and other essays*. Oxford University Press.

16

THE MANY MEANINGS OF MENSTRUATION

Practices, imaginaries and access to water and sanitation infrastructure in Lusaka, Zambia

Amie Jammeh and Tatiana Acevedo-Guerrero

Introduction

The Sustainable Development Goals highlight the importance of menstrual hygiene management and sanitation services to protect the wellbeing of women and girls. Within development discussions, menstruation is often associated with stress and stigma and is related to the loss of educational opportunities for women and young girls (Sapkota et al., 2013, UNICEF, 2019). In this context, WASH United, a global non-profit organization that works "to end the global sanitation and hygiene crisis" has made May 28 "Menstrual Hygiene Day."[1] Since 2014, when it was first celebrated, "Menstrual Hygiene Day" has created a platform for menstrual related information to be shared publicly and has also given an opportunity for advocates and funders to engage policy makers (Bobel, 2019). Menstruation, considered previously as a taboo topic within the WASH sector and social sciences, has surfaced as a topic of interest (Joshi et al., 2015). Moreover, Menstrual Hygiene Management (MHM) has emerged as a growing domain of development interventions. In 2018, the World Bank reported that 500 million women and girls have inadequate access to facilities to manage their menstrual cycles, most prominently in public spaces such as schools (World Bank, 2018). MHM interventions in schools are said to contribute to the success of girls in education and help them reach their full potential in life (Bharadwaj & Patkar, 2004; Sapkota et al., 2013, Jewitt & Ryley, 2014; Redman-MacLaren et al., 2018).

Development interventions conducting MHM have been met with critiques from feminist approaches. These critiques have highlighted how MHM can reinforce imaginaries wherein women and girls from the Global South are to be educated on how to manage their own menstruation in a proper and hygienic way (Bobel, 2019; Joshi et al., 2015). These critiques also highlight that inequities exacerbated by a lack of access to water and sanitation cannot be solved with technological fixes, such as the distribution of sanitary pads. Instead, efforts must be made to address the roots of gender inequalities and disparities (Joshi et al., 2015). The problems and challenges girls and women face in their menstrual lives are deeply rooted in specific power relations and on the socio-cultural norms and beliefs of a particular context. These critiques also point out that the research conducted to quantify the impact of MHM programmes on the schooling of adolescent girls has been empirically unable to ascertain

DOI: 10.4324/9781003100379-20

the relationship between sanitary pad distribution and an increase in school attendance, and that there are other leading factors contributing to girls absenteeism from schools (Bobel, 2019; Joshi et al., 2015). Moreover, feminist scholars point out that although considerable research has been undertaken on menstruation, most of it has concentrated on the practices and handling of menstruation among women and adolescent girls and findings are confined to indicating what is considered "good" or "hygienic" practices of menstruation. Most of these studies give little room to the socio-economic and national contexts and to the fact that women and girls experience and handle menstruation in different ways among different cultures and at different points of their lives (Lahiri-Dutt, 2014).

This chapter documents expectations and experiences of menarche and menstruation among adolescent girls in school from Lusaka, Zambia. With a population of three million people, Lusaka is the most densely populated city in Southern Africa (Zambia Statistics Agency, 2022). The study is based on photovoice elicitation,[2] a community-based research methodology that puts cameras in the hands of participants, who then discuss and interpret images to reveal their ideas, values and beliefs (Fantini, 2017; Harley, 2012; Shaw, 2020). Pre-menarche and post-menarche girls in two public schools were invited to take photos and also share narratives relating to their expectations, dreams or fears surrounding their (first) periods and monthly menstruation. In order to explore sanitation from gendered perspectives, the chapter situates these groups of girls in their socio-economic and cultural contexts, while at the same time listening to their own experiences. It contributes to critical studies on menstrual health by talking about schoolgirls in their own terms, and *seeing* menarche, or the first menstrual period, and menstruation through their own contexts, eyes and views.

Research participants came principally from two informal settlements, George and Chawama. This research collaborated with the National Water Supply and Sanitation Council (NWASCO) that has been conducting MHM programmes in the city. Two "basic cycle" schools in peri-urban Lusaka were selected[3] and agreed to collaborate in the research. Parents, teachers and students participated in ethics and introductory meetings before giving informed consent. Students from grades seventh and nineth were targeted, and 22 girls aged 12–17 expressed their willingness to participate. Participants were provided with and trained in the use of cameras. Each camera had a capacity of 27 pictures. Participants were allowed to take pictures for five to eight days and the cameras were then collected and the film developed into pictures. Afterwards, different discussions about the pictures were organised with all participants.[4] Three focus group discussions were also held with schoolboys, teachers and community members from George and Chawama.

Sanitation and the birth of menstrual hygiene management

UNICEF's handbook on menstrual health and hygiene specifies that "gender inequality, discriminatory social norms, cultural taboos, poverty and lack of basic services often cause girls' and women's menstrual health and hygiene needs to go unmet" (UNICEF, 2019, p. 13). According to reports by the consultancy firm, Reimagining Social Change (FSG), in countries of Sub-Saharan Africa and South Asia, women and girls lack access to menstrual and sanitation facilities and receive limited menstrual-related information prior to menarche. FSG, a consultancy whose reports inform development organizations, has also documented the different kinds of absorbents used for controlling menstrual blood by women and girls around the globe, ranging from homemade materials to disposable and reusable sanitary pads (Geertz et al., 2016). In this vein, different United Nation-led programmes (Water, WatSan, WASH, WinS) have been implemented by the WASH sector in the last decade in an effort to put women and girls at the center of development interventions. UNICEF states that *good* Menstrual Hygiene Management entails:

> Women and adolescent girls using a clean menstrual management material to absorb or collect menstrual blood, that can be changed in privacy as often as necessary for the duration of a menstrual period, using soap and water for washing the body as required, and having access to safe and convenient facilities to dispose of used menstrual management materials. They understand the basic facts linked to the menstrual cycle and how to manage it with dignity and without discomfort or fear.
>
> (UNICEF, 2019, p. 8)

Emergent interest in menstrual hygiene intensified among governments, international organizations, researchers and charity institutions after the publication of research called the "Girl Effect" which indicated that "one in ten girls in Africa miss class or drop out of school entirely because of their periods" (Thomson, 2015, para. 4). This research argued that in African countries, school environments are incapable of providing for the needs of adolescent girls to privately manage their monthly flows or render emotional support and adequate menstrual health awareness before menarche.

It was in this context that UNICEF started promoting WASH in Schools. "WinS" policies specifically tackle the needs of menstruating girls: ensuring a satisfactory number of 1) separate (female and male) toilets with locks inside the doors, 2) trash cans for disposing menstrual pads and materials and 3) access to water for washing the blood off stained clothes and hands (Sommer et al., 2017). In this context, the governments of Kenya, Uganda, Niger, Senegal, India, South Africa, Philippines and Zambia, among others, have ratified national development policies to cater to the menstrual hygiene needs of adolescent girls in schools (Bobel, 2019; Sommer et al., 2017). These governments made a commitment to 1) supply free menstrual pads to schoolgirls, 2) improve sanitation facilities, 3) incorporate MHM in schools curricula, 4) preserve the dignity of girls and 5) train teachers for an effective deliverance of MHM and for rendering emotional support or counseling to girls (Sommer et al., 2017).

Zambia has been included in UNICEF's project "WinS4Girls: Advocacy and Capacity Building for Menstrual Hygiene Management through Water, Sanitation and Hygiene in Schools Programmes" (WinS4Girls), implemented in 14 countries. The project, which promoted "evidence-based policies and interventions that could be taken to scale through education systems," was implemented in collaboration with the national government. It entailed the adoption of "MHM national guidelines to promote effective MHM programmes in schools." The state committed to providing drinking water and single-sex toilets (at least two toilets per school) and incorporating MHM in the curricula. According to the guidelines, kits were to be delivered to all schools. These toolkits included "washable and disposable menstrual materials and ready to stitch materials for schools to locally produce washable pads" (see www.wins4girls.org/countries/zambia.html).

Looking at menstruation from a gender lens

Paechter (2003) refers to gendered roles as temporal acts of performance which are socially constructed through time. Similarly, Bina Agarwal explains how gender relations are power-laden and revealed in a "range of practices, ideas and representations." These include, for example, the division of roles and resources and "the ascribing to them of different abilities, attitudes, desires, personality traits, behavioral patterns and so on" (1997, p. 51). Intersectional feminism, in turn, recognises the interlocked workings of different forms of oppression (and emancipation) based on gender, race, ethnicity, class, sexuality, caste and ability (Ojeda et al., 2022). This chapter understands gender as the socially constructed roles ascribed to males and females. These roles are always time and place specific and interact with other social relations. In this vein, the perceptions and imaginaries around menstruation vary among cultures and change with time.[5] In this regard, Lahiri-Dutt (2014) has pointed out that we can learn about gender power relations in a particular society by looking at how menstruation is perceived and understood.

Power dynamics in which women and girls from different contexts are immersed go far beyond the availability of menstrual pads. Technical solutions (giving out disposable and reusable pads, menstrual cups and/or tampons) to the so called "menstrual crisis" are a way to manage the female body, and do not do much to erase the different stigmas associated with menstruation (Bobel, 2019). The body of feminist critiques to "Menstrual Hygiene Management" can be grouped around three arguments. Firstly, authors argue that MHM focuses on technical "fixes," leaving behind the structural roots of gender inequalities. Lahiri-Dutt (2014) contends that the WASH sector continues to be a predominantly masculine sector, and therefore its designs and services are geared towards technical and engineering solutions to complex social issues with limited involvement and awareness of women and girls needs. Gaybor (2020) in turn argues that the WASH sector tends to construct "menstrual hygiene" as a public health problem to be remedied with the promotion of "modern" hygiene measures in combination with infrastructural and technical solutions. What appears important in MHM initiatives is then to *hygienically* hide and get rid of the blood in discreet ways. A predominantly male WASH sector privileges technical solutions and hardware service provision to water related problems; therefore, it is not surprising that MHM interventions are focused on sanitary pad distribution (Lahiri-Dutt, 2014). Policy approaches thus tend to relate the promotion of gender equality to the provision of accessible water and sanitation services while neglecting the everyday dynamics of gender relations and power dynamics that were reflected and/or reinforced (Zwarteveen & Ahmed, 2012).

MHM targets and strategies are frequently translated into checklists measuring achievements and promoting gender equality without the targeted population's (menstruating women and girls) active participation. At the end of the day, the burden and responsibility for managing the provided new devices, facilities or services (toilets, water kiosks, menstrual pads and their disposal) is placed on women and girls without considerable wages, as it is considered normal and appropriate due to socially constructed gender roles (Joshi et al., 2016; McCarthy & Lahiri-Dutt, 2020). In their study of Northern Ghana, Joshi et al. (2015) document the intervention of international donors working in partnerships with local NGOs to promote MHM in schools. In the setting of rural Ghana, where male teachers tend to be in charge of children's education, it is these teachers who are given the responsibility of distributing sanitary pads to adolescent girls. Because of the gender and age power differences between teachers and their students in this particular context, this dynamic can potentially create awkward situations, putting adolescent girls in a state of increased vulnerability.

Secondly, these authors warn about the consequences of making natural female bodily functions universal, pathological and in need of hygienic measures – in other words, subject to *medicalization* (in need of the attention of medical practitioners and prescriptions).

They draw attention to the word "management" in MHM, which suggests girls menstruation is something that needs urgent attention, without which disaster might occur. Gaybor (2020) draws attention to MHM literature concentrating on the Global South, specifically on Sub-Saharan Africa and South Asia, as a product of racially laden colonial imaginaries and assumptions about health discourses and practices. Similarly, McCarthy and Lahiri-Dutt (2020, p. 16) raise a critique on the checklists and indicators used to assess or *manage* menstrual hygiene:

> The benevolence and utilitarianism of this language obscures the fact that these indicators are embedded in specific contexts, and that a variety of structural, religious, cultural, and gendered practices – that both construct and obstruct the 'management' of menstruation – are involved in determining an individual's menstrual management practices.

Joshi et al. (2015) have also pointed to the existence of another type of discourse around menstruation, beyond those of management and medicalization. The *sexualised* notion of menstruation entails girls reaching maturity and taking up the path of womanhood, and perhaps pregnancy. Their research found that among teachers, local NGOs and pupils in Ghana, anxieties were built more around the sexualised discourse of menstruation. School teachers and principals would use the medicalised discourses around menstruation mostly while dealing with donors.

Thirdly, this literature highlights that the commercialization and industrialization of menstruation has yielded valuable economic gains for corporations and industries. The economic opportunities in making sanitary products available to the huge potential market in the Global South, especially India, have had an influence in the propaganda surrounding MHM initiatives by corporations (Lahiri-Dutt, 2014). Women and young girls in different areas of the Global South are thus bombarded with advertisements on how the use of sanitary pads will empower them, making their lives easier, more modern and comfortable. These advertisements stand in contrast to their daily lives, where many face inadequate access to water, sanitation facilities and sanitary products (Bobel, 2019; Joshi et al., 2015). Scorgie et al. (2016) highlighted in their research in South Africa: the difficulties, struggles and anxieties women faced in not only concealing their menstruation but also in disposing of used sanitary pads in secret and out of sight. The secrecy attached to menstruation and the inadequate provision of waste bins have impacted the functions of sanitation facilities; women put waste sanitary pads in pit latrines, leading them to fill up quickly (Scorgie et al., 2016). In the context of limited access to urban infrastructures and services and restrictive cultural norms, managing menstrual waste has important sustainability implications and impacts difficulties of women's menstrual experiences.

Gender relations in urban Zambia

George and Chawama are informal neighborhoods with self-built infrastructure including outdoor toilet blocks, unpaved roads and poor drainage and solid waste collection. Water supply is intermittent, with water kiosks locked at certain times of the day. Women and girls also walk to nearby wells and fetch extra water for bathing, laundry and other domestic chores (VisionRI, 2016).

In general, communities living on informal peri-urban settlements reached the secondary level of education (Milambo, 2019). Metropolitan Lusaka has a total of 758 primary schools and a 79.4% enrolment rate of girls (MoGE, 2018). More than 50% of schools in Lusaka are operated by the government with free education from grade one to seven (Milambo, 2019). According to the interviewed teachers, only about 30% of students in the compounds pursue studies after high school. Community members and teachers believe that some of the main causes for school dropout are economic hardship within the households (sons and daughters abandon school to work or help around the house). In focus groups and interviews, community members claimed that girls are taught to be wives and mothers, while boys are taught to be husbands, fathers and leaders. Girls and women are more burdened by household chores and responsibilities than boys and men. In general, young women and girls explained that there is a lot of respect in what concerns parents and the elderly in general: "*We naturally follow the rules and what our parents decide, without questioning*" (Bertha,[6] 46 years).

Schlyter (1999) has pointed out that increasing urbanization in Zambia has catalyzed changes in gender relations as a result of new working and living conditions. These changes disrupt power hierarchies, sometimes leading to women and men bargaining for new gender relations. According to Evans (2014), in a context of economic instability, women have been entering the workforce and contributing more and more to the community's economy. This became evident in interviews and focus group discussions: in both George and Chawama women are frequently the main economic providers for the family. They are involved in small businesses, either selling food or vegetables along the streets of the city or in the local markets, while men are mostly engaged in services and the informal labour market.

Despite women's participation in paid labor, this research found that men are predominately considered as masters of the household. In both compounds, it is considered taboo for men to engage in household chores that are associated with females. These chores are considered to devalue the charisma and status of men. Dinwali (42), a married woman, explained: "*It is not appropriate according to tradition for someone other than me to cook for my husband, I always do the cooking for him; sometimes if I am travelling, I cook in advance and store the food for him in the refrigerator.*" Elder daughters also take responsibility for taking care of the house. They act as "second in command" after their mothers: "*Boys are valued more. After school we do domestic work which gives us little time to study our books while the boys do not, so they have more time to study*" (Nancy, 15). Elder sisters have become the main economic providers of their households if the parents are not working.

Menarche, moye and womanhood

Although it is expected, menarche cannot be forecasted with precision. It is an important event in the life of females, as in many cultures it marks the entrance to womanhood (Yeung et al., 2016). A first group of pictures analyzed in photovoice workshops relates to the acquisition of menstrual knowledge from their mothers, sisters and friends. Information on menstruation is commonly given by mothers, aunties, grandmothers, sisters and friends. Most of the details are only given after the girls have seen their first period.

As in many other communities around the world, menstruation is seen beyond hygiene and is a moment to celebrate and rejoice among girls, mothers and the community (Joshi et al., 2015; Lahiri-Dutt, 2014). Some of the pictures taken by the schoolgirls talk about this moment when mothers make gifts and impart knowledge and advice on menstruation:

> *The day I saw my period, I was shocked and cried and told my mother because I did not know what it was. She covered my face with a new chitenge and took me to the room. My father, brothers and other men were not allowed to see me. My mother was so happy, she called her friends, they came and a party was made for me. She bought chitenges, pads and underpants for me. . . . I was given 200 kwacha* (14 USD).
>
> *(Chusima, 16)*

Like Chusima, other girls shared pictures of the *chitenges* and underpants that mothers offered them after their first menstruations. *Chitenges*, which are pieces of traditional fabric, are to be used in different forms: firstly, pieces of the fabric are cut into menstrual clothes to absorb menstrual blood. As such they are placed inside lycra shorts or regular underwear (see Figure 16.1). Secondly, traditional colorful *chitenges* are rolled around the hips as skirts (see Figure 16.2). Before menarche girls dress like "children" but afterwards, wearing *chitenges* symbolises the start of womanhood.

The first period is an important event in the life of girls in peri-urban Lusaka. All participants narrated how, although they were afraid or scared when they saw blood in their underpants, they were also overwhelmed with gifts, blessings and teachings (see Figure 16.3). Upon seeing their first period, girls are taught new practices to live with their periods. After these lessons, girls are celebrated for achieving a milestone in their lives and are ready to face the world as "*matured and knowledgeable women.*" During the week of their

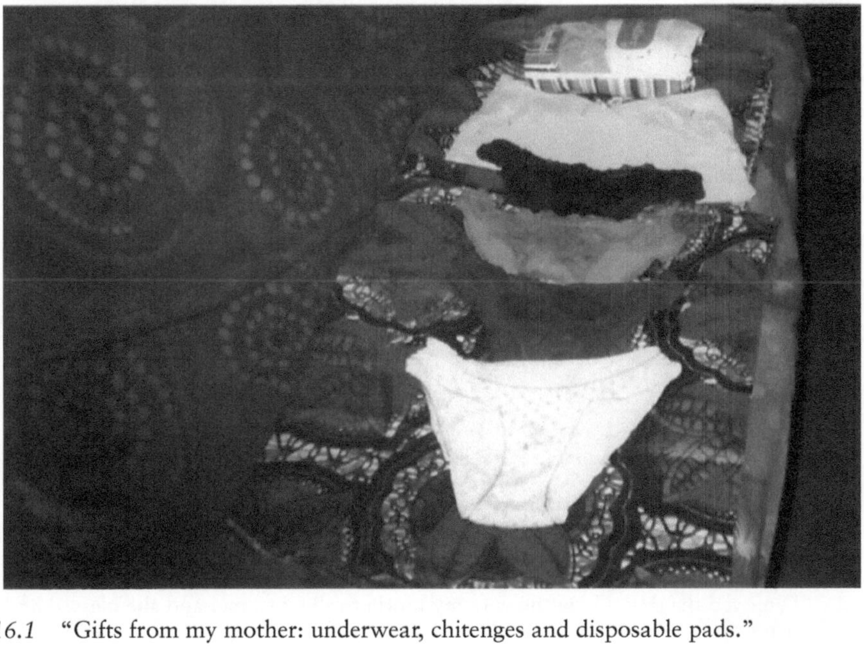

Figure 16.1 "Gifts from my mother: underwear, chitenges and disposable pads."
Source: Chusima.

Figure 16.2 "Our new chitenges."
Source: Tasha.

Figure 16.3 "I enjoyed the day, I love the way my mother talked to me, and she blessed and prayed for me."

Source: Amy.

first menstruation some girls were given herbs to prevent menstrual cramps. Many girls are celebrated by their mothers with a "nice meal" (see Figure 16.4):

> *When I saw my first period. My mother went to the market and bought three panties and two chitenges, one was for me to wear and the other one she cut it into pieces for me to use and control the blood. My mum also bought a live chicken which she cooked for me without adding salt to eat, this is the tradition to welcome me to womanhood.*
>
> (*Tasha, 16*)

It is common among girls reaching menarche to undergo a traditional initiation ceremony, commonly called *Moye* in the local language. Traditionally during *Moye*, girls were confined indoors for between one to three months, after which a ceremony was organised for them. During this month they were taught how to cook, respect their in-laws and look after their husbands. They are also told how to shave their pubic hair and to sexually satisfy their husbands, with different sex positions showcased through dance moves. People brought gifts (money, cloth and chickens) and in most cases after a few months of the *Moye* celebration, the girl was promised in marriage. The initiation processes of *Moye* depends largely on the ethnic group to which a family belongs.[7] In the context of Lusaka's compounds however, only some families still practice it, as many within the community have converted to Christianity and do not observe traditional rituals. The girls whose families do practice *Moye* explained that initiation rituals have transformed deeply with time. They now spend only one week home after their first period and a party with food and music is organised after the seventh day. Community

Figure 16.4 "We ate chicken with rice."

Source: Tasha.

members explained that "*nowadays, people are aware of the advantage of educating a girl, so Moye is done only for one week and during the summer holidays.*" Furthermore, sexuality lessons are no longer included, and the girls are not sent off to marriage after the party. In both compounds, "*educating a girl child is an important achievement for a family.*" Unlike compound girls, boys do not undergo any process of initiation to learn maturity, adulthood or to celebrate their manhood.[8] There is a saying in the Nyanja Language "*when a boy grows up, there is no process of beating drums to show that he is an adult.*"

Some girls explained that during *Moye*, they are treated "*like queens*" and asked to rest. They have discussions with their aunties, grandmothers or a medicine woman called *alangizi* about the entrance to womanhood and the expectation that lies ahead of them. During the conversations about the meanings of "adulthood" or "womanhood" that the girls had with elder family women or from the *alangizis*, they are taught about how to clean themselves and wear *chitenges* or pads. The girls are told how to respect their parents and elders and how to behave in their presence: "*We have to behave like adult women and therefore cannot hug or sit on the lap of our fathers like when we were children*" (Sophia 15). Good behaviour also entails sitting properly as a woman, holding their skirts while sitting, not playing with young children, and becoming friends with other girls that have seen their period (reached womanhood). Girls also explained that they have to take care of themselves to look beautiful. One of them took a picture of herself and explained she felt insecure about her beauty because she now had pimples. Another one took a picture of herself brushing her teeth, explaining that because she is an adult she has to take "*care of herself.*"

It is worth mentioning that the girls told different stories on the occurrence of their first period. Although most of them consulted their mothers or female relatives for assistance, some hid and did not tell anyone during the first days of their periods while traditional ceremonies or initiations were performed for others. Some girls complemented their menstrual knowledge with information from their teachers or through programmes organised by local NGOs.

Besides stories about *Moye*, an important group of pictures referred to the expectations of womanhood related to marriage, families and economic security. Pregnancy is associated with menstruation and the girls mothers fear the "*dangers of*" pregnancy after menarche. It is worth mentioning that the focus is not so much on the loss of virginity as it is on teen pregnancy: "*After my period my mother told me if you are having sex, you should stop because you can get pregnant now*" (Mulonda, 14). Despite the warnings against teen pregnancy, many of the girls took pictures of babies when asked about their imaginaries and expectations about menarche. Amy (aged 16), for example, said "*I am happy anytime I see my period because it reminds me that I can have children and that in the future I will find someone I love, marry and have kids.*"

Girls explained how, at school, boyfriends help them with lunch money, snacks and gifts. As explained by Lungowe (aged 16): "*Sometimes I feel envy of my classmates with boyfriends because the boy shares with them the money they have, it is not much, sometimes they just offer two kwacha but it shows that someone cares about you.*" Munzi (aged 16) added: "*With someone you love you feel protected, you feel you are not alone, someone is always there for you and to treat you with sweet words like honey, baby and darling.*" Some girls explained how boyfriends can give them money for uniforms and to continue with their education (see Figure 16.5).

Figure 16.5 "I want to have someone on my side to protect me and comfort me when I am in trouble and provide for me, maybe help me pay for school."

Source: Chimunya.

Experiences of menstruation

The first theme regarding menstruation that came up in photovoice workshops had to do with discomfort and uncomfortable physical symptoms: feeling tired in the classroom and having headaches. Five girls photographed (re-enactments of) menstrual cramps, for which they said they can take painkillers or lay down to rest (see Figure 16.6). Some girls keep themselves from sport activities for the fear of staining or because of the uncomfortable nature of their periods. Girls also disclosed that during their periods they sometimes do not feel comfortable around boys because they always suspect themselves of staining their uniforms and continuously look behind their backs and always stay alert. During the discussions, as many girls pointed to the physical pains related to menstruation, some others showed pictures with different stories. While one stated that she did not feel any menstrual cramps, the other explained how she did feel some pain, but upon arrival at school she was so distracted by being with her friends that she "*forgot about it*" (see Figure 16.7). This diversity in experiences is important as it highlights the fact that women physically experience menstruation in very different ways.

Another one of the main themes that came up in photovoice workshops was water and sanitation infrastructure. Many participants took pictures of outdoor water taps in both their compounds and at their schools. They pointed out how, at schools, there are toilets that "*haven't worked for years*" and how it is difficult to wash their hands or take a shower after sports, due to broken sinks and broken showers. Decaying school infrastructure is portrayed in the photographs; similar pictures were taken in the compounds, where the water supply is intermittent (see Figure 16.8). Some taps are broken and the supply is expensive. Some of the girls explained having to fetch water from wells when there is none in the taps (see Figure 16.9). Intermittency contributes to anxieties, as after menarche girls are in charge of discretely washing their menstrual clothes and underwear.

Figure 16.6 "I feel sleepy and lazy in class and I feel tired and lazy during my period."
Source: Mehai and Nbanji.

Figure 16.7 "When I arrive to school I forget about the pain because I am with my friends."
Source: Mulonda.

Figure 16.8 "The toilets are broken and there is no water. . . . Since grade fourth (four years ago) this shower has been broken."

Source: Betty and Mulonda.

Figure 16.9 "Today there is no water in the compound."

Source: Josephine.

The third theme which came up in photovoice workshops concerned MHM initiatives championed by NGOs, in partnership with the government. Female students are invited to reusable menstrual-pad distribution days (see Figure 16.10). These days are done on specific occasions, such as Menstrual Hygiene Day and World Women's Day. While some girls explained that they had switched to the *new* reusable pads, some others mix the new pads with the regular *chitenge* cloth, and others are not using the pads yet, but plan to use them in the future. As with *chitenge* cloth, girls have difficulty accessing water to wash disposable and reusable pads.

Although the government plans foresee "incorporating MHM in the curricula," there is limited biological knowledge of menstruation and the processes it entails. According to Funke (aged 13), "*period is when blood comes from down here.*" Despite the fact that some teachers have attended trainings or workshops on MHM, there have not been major changes in the curricula. Topics such as menstruation and the details of human reproduction are left in the hands of the science teachers. One of the girls explained that she did not feel comfortable asking any questions about sexuality or female reproduction in science class, since it is taught by a young male teacher from the community. Some boys who participated in the focus groups heard about menstruation from their teachers or female friends, while others had no idea about it. According to John (aged 17), "*periods are a natural thing for girls and women, so I cannot tease or mock them about it.*" During the discussions some of the boys felt shy and embarrassed to talk about menstruation.

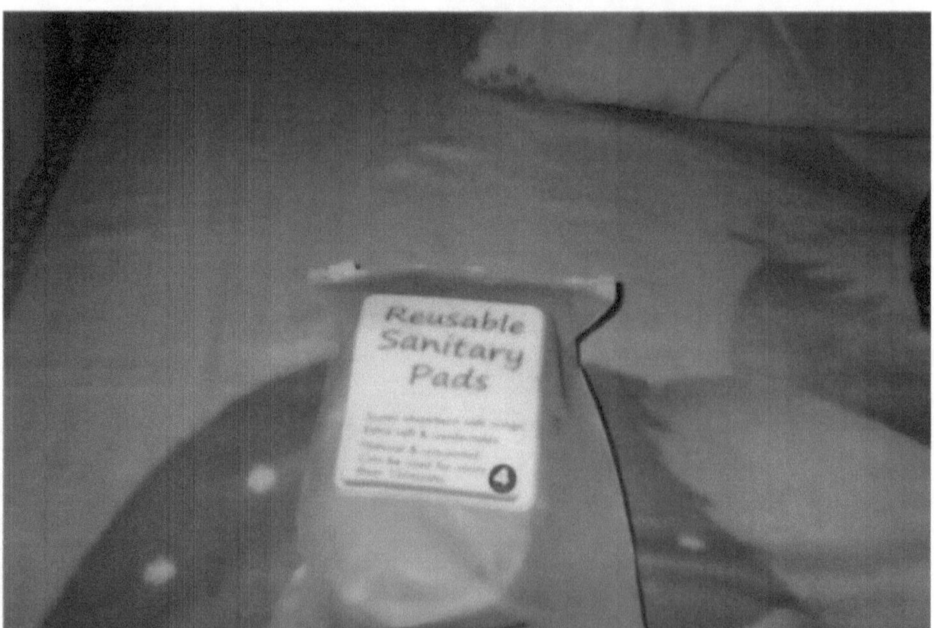

Figure 16.10 Reusable pads delivered to girls from the George and Chawama communities.

Source: Josephine.

The fourth theme was related to the special care and support they receive from family during period days. Jenny (aged 14) explained that her mother pampers her during her period days by liberating her from all house chores, even washing her clothes, so that she can rest and sleep more (see Figure 16.11). Petri (aged 14), in turn explained that during her period days her mother prepares her favorite dinner of okra and fried fish stew. Estonia (aged 15) took a picture of a leather bag her mother offered her to keep her menstrual pads and *chitenges*. Maurice (aged 16) explained how she likes to go and take walks in the bushes around her home to be in nature and look at flowers. Nyambe (aged 15) explained how she feels comfortable spending time with her siblings and her dog.

The fifth theme which came up in photovoice workshops has to do with *modesty*. Young girls are discouraged from talking about their periods with boys and men. During an NGO activity to distribute reusable pads in one of the schools, the girls were told to hide the reusable pads before leaving the class, and not to display them around school. Most adolescent girls are also expected to follow certain standards in what concerns modesty: *chitenges* cloth pads, underpants and underwear are not to be hung outside for drying, these items should be air dried inside the home or hidden from boys and men. Special places or bags are provided by mothers for their daughters to keep their menstrual products safe and out of sight. Some girls who wear disposable pads explain that their mothers help them with money to buy the pads (or buy the pads for them). They argue that they prefer disposable pads due to the lack of good water services to wash *chitenges*, and they prefer burning

Figure 16.11 "When I am menstruating my mother does the home chores for me to rest and recover from my period."

Source: Jenny.

the disposable pads because it is difficult to dispose of them with discretion: "*I use pads because I do not like washing cloth pads. After, I burn the pad and then collect the ashes and flush them in the toilet*" (Amy, 16).

Conclusion

As mentioned, Zambia was included in UNICEF's project "WinS4Girls: Advocacy and Capacity Building for Menstrual Hygiene Management through Water, Sanitation and Hygiene in Schools Programmes" (WinS4Girls), implemented in 14 countries. However, in both visited schools the programme has only translated into the delivery of reusable pads whenever there are international funds available. Moreover, although the infrastructure (sinks, toilets and showers) exists in the schools, students involved in the photovoice initiative reported the intermittence of water, the lack of maintenance in toilets and drains and the constant state of breakdown of the showers.

In both George and Chawama, girls are gifted with materials and products to control their monthly flow at the commencement of their first period. Although they face multiple uncertainties regarding economic stability, infrastructural malfunctions in their compounds and socio-cultural changes due to widespread unemployment among men, girls were not particularly challenged about clothes, pads and stains.

According to Wang (1999, 2006) one of the main objectives of photovoice is to advocate for change through policy or decision makers. Following this call, this chapter concludes with three points for discussion vis-à-vis MHM projects in peri-urban settlements from the Global South:

The reality of weak infrastructure in both the compounds and the schools.

Despite efforts to extend infrastructure, toilets and taps seldom work and students complained about the widespread lack of maintenance. This has to do with the fact that some of the development projects do not tackle maintenance issues in the long run. Public schools thus lack the funding to hire maintenance personnel. Thus, it is the financing of primary and secondary education in general which needs restructuring. Beyond infrastructure, it would be necessary to think about redistributive solutions that challenge the inequality and lack of opportunities that affect women and girls in Zambia.

The importance of the bonds between mothers and daughters.

Mothers show love, affection, care and guidance. Mothers also gift materials (*chitenges*, pads, shorts), cook their daughters' favorite dishes and also do home chores in order to make the period days more comfortable. Mothers are the ones concerned about the possibility of early pregnancy. One of the important findings is that some mothers warn against pregnancy but are more lenient concerning virginity. This link of care and affection between mothers and daughters has not been explored by MHM initiatives. Parents from the Global South are portrayed usually as ignorant of their daughters' menstrual histories and sexuality or as intransigent and conservative.

Analyzing bonds between mothers and daughters, studies can acknowledge that menstruation is not necessarily a source of problems or discomfort but is also a moment of celebration, of being *cared for*, of receiving presents. By focusing exclusively on technical and medical issues, MHM risks missing the culturally embedded meanings of menstruation.

The complexities surrounding sexuality, romantic love and marriage.

Households face economic instability as a result of urban inequality and the lack of employment among men. Women have been going out of the house to work in markets or as domestic workers. This situation coexists with persisting imaginaries of men as providers and heads of the family. Even though economic roles have shifted, ideals of masculinity have not. As a result, expectations around sexuality and romantic love are complex. School girls see romantic love as a vehicle to obtain economic stability (having someone to "have your back") and relationships are not always incompatible with school. As one of the girls explained, boyfriends can help with education costs. Any project aiming to study or intervene in the lives of women in the urban South should first aim at understanding these complexities in order to contribute to context-specific strategies that go beyond technical fixes and challenge gender-based asymmetries.

Acknowledgments

We would like to first of all thank the students. Young women and girls who shared their time and ideas, giving meaning to this work. We are grateful to the National Water Supply and Sanitation Council (NWASCO) of Lusaka, most especially Chola Mbilima and Chrispin Lukwanda for their generosity and advice.

Notes

1 See https://wash-united.org/.
2 Photovoice has been used to analyse diverse health, environmental and societal issues. This method promotes critical dialogue and enables underrepresented and/or disadvantaged communities to make their opinions heard. It can also promote the contribution of young people in a genuine way when done in school contexts (Warne et al., 2013). By following the everyday lives of participants, photovoice offers insight to both significant "objective" aspects (such as activities, facts and dates) and "subjective" factors (such as emotions, opinions, beliefs, ideas) therefore contributing to greater comprehension of human behavior (Ciolan and Manasia, 2017).
3 This research also received approval from the District Board of Education (DEBS) of the Lusaka District.
4 The participatory process of analysis was conducted through a process of selecting, contextualising and analysing (Wang, 1999). Participants went through their photos and selected some for discussion. The SHOWeD technique was followed to structure discussions around the questions: what do you see here? How does this relate to your lives? Why does this problem/concern, expectation or imaginary exist? What can we do about it (Catalani and Minkler, 2010)? Participants were also invited to build narratives about the photos they took as some felt shy and nervous with the structured questions of SHOWeD. They were encouraged to categorise and cluster the photos to discuss recurrent themes and topics (Spencer et al., 2019).
5 It is important to mention that transgender menstruators also deal with social expectations (according to expectations of femininity), stigma and shame generated by stereotypes around menstruation (see Rydström, 2020).
6 All names have been changed.
7 Different ethnic groups in Zambia have different names for the initiation rituals: Nyanga people call it *MOYE*, Tonga people call it *NKOLOLA*, Chewa people call it *CHINAMWALI* and Bemba people call it *MBUSA*.

8 The Luvale people in North-Western Zambia have an initiation ceremony for boys called *Mukanda*, where 13-year-old boys are recruited to undergo circumcision and the teachings of adulthood.

References

Agarwal, B. (1997). "Bargaining" and gender relations: Within and beyond the household. *Feminist Economics*, *3*(1), 1–51. https://doi.org/10.1080/135457097338799

Bharadwaj, S., & Patkar, A. (2004). Menstrual hygiene and management in developing countries: Taking stock. *Junction Social Development Consultants*. www.ircwash.org/resources/menstrual-hygiene-and-management-developing-countries-taking-stock

Bobel, C. (2019). *The managed body: Developing girls and menstrual health in the global South* (pp. 69–109). Palgrave Macmillan.

Catalani, C., & Minkler, M. (2010). Photovoice: A review of the literature in health and public health. *Health Education Behavior*, *37*(3), 424–451. https://doi.org/10.1177/1090198109342084

Ciolan, L., & Manasia, L. (2017). Reframing photovoice to boost its potential for learning research. *International Journal of Qualitative Methods*, *16*(1), 160940691770290. https://doi.org/10.1177/1609406917702909

Evans, A. (2014). "Women can do what men can do": The causes and consequences of growing flexibility in gender divisions of labour in Kitwe, Zambia. *Journal of Southern African Studies*, *40*(5), 981–998.

Fantini, E. (2017). Picturing waters: A review of Photovoice and similar participatory visual research on water governance. *WIREs Water*, *4*(5), e1226. https://doi.org/10.1002/wat2.1226

Gaybor, J. (2020). *The body politics of menstruation: Technologies, sustainability and destigmatization* [Doctoral thesis, Erasmus University Rotterdam].

Geertz, A., Iyer, L., Kasen, P., Mazzola, F., & Peterson, K. (2016). An opportunity to address menstrual health and gender equity. *FSG Boston*. www.fsg.org/resource/opportunity-address-menstrual-health-and-gender-equity/

Harley, A. (2012). Power, ethics and politics in using photovoice. *The International Journal of Qualitative Methods*, *11*(4), 320–339. https://doi.org/10.1177/160940691201100402

Jewitt, S., & Ryley, H. (2014). It's a girl thing: Menstruation, school attendance, spatial mobility and wider gender inequalities in Kenya. *Geoforum*, *56*, 137–147. https://doi.org/10.1016/j.geoforum.2014.07.006

Joshi, D., Kooy, M., & van den Ouden, V. (2016). Development for children, or children for development? Examining children's participation in school-led total sanitation programmes. *Development and Change*, *47*(5), 1125–1145. https://doi.org/10.1111/dech.12258

Joshi, D., Sahin, M., Buit, G., & González-Botero, D. (2015). Menstrual hygiene management: Education and empowerment for girls? *Waterlines*, *34*(1), 51–67. https://doi.org/10.3362/1756-3488.2015.006

Lahiri-Dutt, K. (2014). Medicalising menstruation: A feminist critique of the political economy of menstrual hygiene management in South Asia. *Gender, Place & Culture*, *22*(8), 1158–1176. https://doi.org/10.1080/0966369x.2014.939156

McCarthy, A., & Lahiri-Dutt, K. (2020). Bleeding in public? Rethinking narratives of menstrual management from Delhi's slums. In C. Bobel, I. T. Winkler, B. Fahs, K. A. Hasson, E. A. Kissling & T. A. Roberts (Eds.), *The Palgrave handbook of critical menstruation studies* (pp. 15–30). Palgrave Macmillan.

Milambo, H. (2019). *Determining the socio-economic effects of lack of water security in Lusaka district* [Bachelor Undergraduate Thesis, University of Zambia].

MoGE. (2018). *Educational statistical bulletin 2017*. Zambia Ministry of General Education, Directorate of Planning and Information.

Ojeda, D., Nirmal, P., Rocheleau, D., & Emel, J. (2022). Feminist ecologies. *Annual Review of Environment and Resources*, *47*(1), 149–171. https://doi.org/10.1146/annurev-environ-112320-092246

Paechter, C. (2003). Masculinities and femininities as communities of practice. *Women's Studies International Forum*, *26*(1), 69–77. https://doi.org/10.1016/s0277-5395(02)00356-4

Redman-MacLaren, M., Barrington, D. J., Harrington, H., Cram, D., Selep, J., & MacLaren, D. (2018). Water, sanitation and hygiene systems in Pacific Island schools to promote the health and education of girls and children with disability: A systematic scoping review. *Journal of Water Sanitation and Hygiene for Development*, *8*(3), 386–401. https://doi.org/10.2166/washdev.2018.274

Rydström, K. (2020). Degendering menstruation: Making trans menstruators matter. In C. Bobel, I. T. Winkler, B. Fahs, K. A. Hasson, E. A. Kissling & T. A. Roberts (Eds.), *The Palgrave handbook of critical menstruation studies* (pp. 945–959). Palgrave Macmillan.

Sapkota, D., Sharma, D., Budhathoki, S. S., Khanal, V. K., & Pokharel, H. P. (2013). Knowledge and practices regarding menstruation among school going adolescents of rural Nepal. *Journal of Kathmandu Medical College*, *2*(3), 122–128.

Schlyter, A. (1999). *Recycled inequalities: Youth and gender in George compound, Zambia* (Report No. 114). Nordic Africa Institute.

Scorgie, F., Foster, J., Stadler, J., Phiri, T., Hoppenjans, L., Rees, H., & Muller, N. (2016). "Bitten by shyness": Menstrual hygiene management, sanitation, and the quest for privacy in South Africa. *Medical Anthropology*, *35*(2), 161–176. https://doi.org/10.1080/01459740.2015.1094067

Shaw, P. A. (2021). Photo-elicitation and photo-voice: Using visual methodological tools to engage with younger children's voices about inclusion in education. *International Journal of Research & Method in Education*, *44*(4), 337–351. https://doi.org/10.1080/1743727X.2020.1755248

Sommer, M., Figueroa, C., Kwauk, C., Jones, M., & Fyles, N. (2017). Attention to menstrual hygiene management in schools: An analysis of education policy documents in low- and middle-income countries. *International Journal of Educational Development*, *57*, 73–82. https://doi.org/10.1016/j.ijedudev.2017.09.008

Spencer, R. A., McIsaac, J. D., Stewart, M., Brushett, S., & Kirk, S. F. L. (2019). Food in focus: Youth exploring food in schools using photovoice. *Journal of Nutrition Education and Behavior*, *51*(8), 1011–1019. https://doi.org/10.1016/j.jneb.2019.05.599

Thomson, S. (2015, December 9). 1 in 10 girls in Africa will drop out of school for this reason. *World Economic Forum*. www.weforum.org/agenda/2015/12/1-in-10-girls-in-africa-will-drop-out-of-school-for-this-reason/

UNICEF. (2019). *Guidance on menstrual health and hygiene*. www.unicef.org/documents/guidance-menstrual-health-and-hygiene

VisionRI. (2016). *Baseline, socio-economic and health assessment, and willingness to pay study*. www.visionri.com/Home/ProjectDetails/3dc21aa4-7b9c-44f1-80ab-8b67e4a8c277

Wang, C. C. (1999). Photovoice: A participatory action research strategy applied to women's health. *Women's Health*, *8*(2), 185–192. https://doi.org/10.1089/jwh.1999.8.185

Wang, C. C. (2006). Youth participation in photovoice as a strategy for community change. *Journal of Community Practice*, *14*(1–2), 147–161. https://doi.org/10.1300/J125v14n01_09

Warne, M., Snyder, K., & Gillander Gadin, K. (2013). Photovoice: An opportunity and challenge for students' genuine participation. *Health Promotion International*, *28*(3), 299–310. https://doi.org/10.1093/heapro/das011

World Bank. (2018). *Menstrual hygiene management enables women and girls to reach their full potential*. www.worldbank.org/en/news/feature/2018/05/25/menstrual-hygiene-management

Yeung, D. Y. L., Tang, C. S.-K., & Lee, A. (2016). Psychosocial and cultural factors influencing expectations of menarche. *Journal of Adolescent Research*, *20*(1), 118–135. https://doi.org/10.1177/0743558404271134

Zambia Statistics Agency. (2022). *Zambia statistics agency – Quality statistics for development*. www.zamstats.gov.zm/

Zwarteveen, M., & Ahmed, S. (2012). Gender, water laws and policies: An introduction. In M. Zwarteveen, S. Ahmed & S. R. Gautam (Eds.), *Diverting the flow: Gender equity and water in South Asia* (pp. 67–85). Zubaan.

17

ACCESS TO WATER, SANITATION, AND HYGIENE FOR ALL

Focusing on transgender experiences in India

Durba Biswas

Introduction

Sanitation's differential health and wellbeing impacts on various groups of people is well recognized. Research spanning multiple nations and time periods has shown that poor WASH services impact health and wellbeing at all ages (Million Death Study Collaborators, 2010; Spears, 2013; Hirve et al., 2015; Bisung & Elliott, 2016). The focus of much of the research and policy making has been on WASH related vulnerabilities for women and children, especially those from the Global South (GS) (Nallari, 2015). Sanitation services across many countries in the GS have improved over the past decades, although many of the challenges persist (WHO, 2020; World Bank, n.d.; Beard et al., 2022). In the meantime, populations that tend to be ignored, such as people with disability, the elderly, migrants, and the LGBTQIA+ community, have entered the sanitation discourse (Slater et al., 2018; Slater, 2023). Socially, a just sanitation framework and its implementation necessitate acknowledging the limitations, and even harm, of an assumption of a universal and homogeneous gender binary.

This chapter explores the barriers for trans individuals in accessing public bathrooms, based on literature from various areas of social science such as history, youth and sexuality studies, sociology, and law. In South Asian countries like India, for example, there is an underlying assumption of evenness in transgender identity and that public bathrooms earmarked for them address their total sanitation needs (see Boyce et al., 2018). Presenting India as a case study, this chapter argues that to meet the overall goal of sustainable and safe sanitation for all – and achieve various national and international standards – requires inclusive sanitation that also encompasses the identities and experiences of transgender and gender non-conforming individuals throughout the various stages of their lives. The overarching aim of the chapter is to broaden the discussion on what makes public bathrooms transgender-inclusive in the Indian context by engaging more deeply with issues such as the need for privacy, individuality, social marginalization, disability, menstrual health, and parenthood. This chapter recognizes the interlocked workings of different forms of oppression and how transgender identity intersects with oppression based on gender, race, coloniality, ethnicity, class, caste, ability, and other forms of discrimination which can determine bathroom access.

DOI: 10.4324/9781003100379-21

Including the introduction, this chapter is divided into six sections. Section two briefly presents the historical role of public bathrooms in defining appropriate sanitation practices across the Global North (GN) and the GS. Section three presents an account of heterogeneous trans experiences and how these experiences impact sanitation needs. Section four focuses on the status of sanitation facilities for transgender individuals in India. Section five examines the role of caste, age, and the ability to pass as their stated gender in bathroom access for transgender Indians. The chapter concludes with section six, which provides suggestions for future research on transgender-inclusive sanitation, especially in a GS country like India.

Sex-segregated bathrooms: coding gender into bathrooms

Historically, public bathrooms were established exclusively for men and the lack of public bathrooms for women limited their movement in public spaces. The first recorded sex-segregated bathroom for women was in 1739 for a Parisian ball (Cavanagh, 2010, p. 19). Catering to women from elite homes, these bathrooms were initially considered a novelty. As women began stepping into public spaces for their work and leisure activities, women's public bathrooms served to enclose their bodily functions for privacy and to shield their virtue in increasingly dense mixed-gender spaces. By the nineteenth and the early twentieth century, sex segregated bathrooms in workplaces had permeated in Europe and North America. By the 1920s most states in The United States of America (USA) had passed laws to ensure bathrooms for women at workplaces (Cavanagh, 2010). As women's bathrooms became popular in many places in the GN, these ideas of appropriate feminine private spaces were exported to various colonies in the GS (Mbatha et al., 2008). In colonized India, the Royal Commission report of 1863 reported a mortality rate of 69 per 1000 among the British troops due to venereal diseases and poor sanitation. A year later, under the Military Cantonments Act of 1864, sanitary police were established to improve health and hygiene of the military personnel while Sanitary Boards, and subsequently Sanitary Commissioners, were appointed to address the sanitation facilities for civilians (Mushtaq, 2009). Improvements in sanitation infrastructure were introduced for elite members of society and the military and by the time of Indian independence in 1947, the sanitation coverage was less than one percent. Similarly, in South Africa, the earliest recorded western-style bathrooms were constructed in Cape Town for the Dutch East India Company in 1814. With the introduction of western-style bathrooms, local bathrooms, which mimicked the natural squatting position, were regarded as primitive and backwards – and by extension so were the Indigenous people using them (Patel, 2017). The influx of Western bathrooms into the colonies extended the notion of bathroom spaces where bodies were segregated by gender and race. In the post-colonial era, as new states build their own constitutions and legal frameworks, Patel (2017) highlights that the segregation of bathrooms by race and class became outdated while segregation by gender has remained a common feature.

The GS has experienced improvements in sanitation services over the decades with the percentage of population using at least basic sanitation services increasing from 55 percent in 2000 to 78 percent in 2020.[1] However, these improvements are unevenly distributed. For example, 62 percent of the global urban population has access to safely managed sanitation as compared to only 44 percent of the global rural population.[2] Women's bathrooms, unlike open defecation, can safeguard health, privacy, and safety particularly for girls and women who are economically or socially vulnerable (Hirve et al., 2015).

Globally, the disproportionate lack of safe and accessible public spaces for transgender individuals is evident in the volume of media reports on the violence against transgender persons in public spaces. In men's bathrooms, they face the possibility of sexual and physical violence as their presence is suspiciously viewed as a threat to the notions of both masculinity and male sexuality. Conversely, in women's bathrooms, they are treated as predators and are frequently subjected to verbal and physical assault. Both instances ignore the existing threats of violence against transgender individuals. Even without outward displays of hostility, cisgendered people expressed discomfort in sharing public bathrooms with transgender individuals (Callahan & Zukowski, 2019; Cavanagh, 2010). Underlying transphobic anxieties contort gender nonconformity as a threat to patriarchal notions of gender and sexuality. In literature, "gender-critical" feminists have argued against shared use public bathrooms for cisgendered and transwomen to protect cis women from "men who identify as women" (Pearce et al., 2020; Wynn, 2018; Jones & Slater, 2020). However, empirical research on violence in public bathrooms in the United States shows that accommodating the sanitation needs of transgender individuals had no impact on safety or privacy in these spaces (Hasenbush et al., 2019).

Transgender experience of using bathrooms

The GN and the GS

In scholarship from the GN, safe bathroom use has repeatedly emerged as a core aspect of everyday life for transgender individuals (Wernick et al., 2017; McGuire et al., 2022, Francis et al., 2022, Lerner, 2021). Griffin (2008) points out that "finding a safe bathroom is such a profound part of the transgender experience" that information about transgender-friendly bathrooms is available on mobile applications and websites. Denying bathroom access has been used as a tool of oppression for transgender persons (Griffin, 2008). Using a data set of nearly 2800 individuals from the National Transgender Discrimination Survey data set, Seelman (2014) estimates that nearly a quarter of the transgender or gender nonconforming students in the United States were not allowed to use bathroom facilities at their place of education.

In the GN, scholarship on transgender sanitation rights show that during the formative years and into their early adulthood, bathrooms are the most crucial part of safe school environment for transgender students (Wernick et al., 2017) and many avoid using school bathrooms for the fear of being teased or harmed (Jones, 2016) all of which can impact their mental health and wellbeing (Price-Feeney et al., 2021, Clark et al., 2014). They also are less likely to report incidences of harassment (Francis et al., 2022). Some change to schools with individual bathrooms or transgender-friendly bathrooms or regulate their bodies and daily routines around when and where they can use the bathrooms. For example, transgender students may take baths when they expect to find the bathrooms empty rather than when it is convenient (McGuire et al., 2022). When coerced into using bathrooms which correspond to their sex at birth rather than their stated gender, transgender students experience detrimental emotional impacts (see Ingrey, 2018) whereas transgender-friendly bathrooms benefit their emotional wellbeing (Wernick et al., 2017). Moreover, bathrooms are not only spaces for managing their sanitation needs but also spaces to socialize with their peers. Separate bathrooms can socially isolate them and hamper their ability to build a healthy sense of belonging through social interactions (McGuire et al., 2022; Wernick et al., 2017).

In the United States, a range of legal protections against discrimination are instituted at workplaces (Griffin, 2008). However, at workplaces, bathroom use produces anxieties, a sense of isolation, and discomfort for transgender individuals. Navigating the daily use of bathrooms, if they exist, located at a distance from other sex segregated bathrooms can lead to losing out on portions of their school education, hamper their work as adults and minors, and single them out (McGuire et al., 2022). In one instance, the bathroom for trans students was located on the third floor of their school building which required them to walk up multiple flights of stairs every time they needed to use the bathroom (McGuire et al., 2022). Bornstein (2013), a transgender author who was transitioning while working, was provided a bathroom to use on an isolated and under construction floor of the building. This required her to climb down multiple floors from her office space and navigate through construction debris to reach a poorly maintained bathroom in an isolated part of the empty floor. Transgender individuals do not conform to a homogeneous category and their identities intersect with race, ethnicity, class, caste, sexuality, age, and disability (Patel, 2017; Biswas, 2019). In the context of the United States, McGuire et al. note that, during a group discussion on their bathroom use experience, non-white trans students spent most of their time discussing their experiences with housing, safety, employment, and discrimination (2022, p. 48). Similarly, individuals who do not pass as their stated gender are at a high risk of violence (Griffin, 2008).

In the GS, trans experience and bathroom access is an emerging topic of research and policy advocacy. In 2011, Nepal became one of the first countries to legally recognize the third gender[3] and in 2012 established the country's first gender-inclusive toilet (Boyce et al., 2018). The toilet block generated a sense of safety and visibility for the third gender citizens. The country's first openly gay member of parliament expressed that gender-neutral bathrooms, men's bathroom for cis-men and transmen, women's bathrooms for cis-women and transwomen, and third gender bathrooms for individuals identifying as neither man nor woman reaffirm the court ruling that all gender identities are equal (Boyce et al., 2018, p. 107).

In Cape Town, communities protest and petition for toilets which afford them dignity and safety but demands for sex segregated toilets are not common (Patel, 2017). The participants in Patel's study repeatedly allude to the disgust and violence expressed by cisgendered people towards them – signaling that "Cape Town bathrooms are not for us. Not for transgender individuals. Not for non-binary people" (Patel, 2017, p. 56). Transgender individuals of color in South Africa demonstrate awareness about the role of race in transgender lives as is evident from a young transwoman's narration.

> Cape Town bathrooms are violent . . . unless you're like white, then I think life must be really nice for you."
>
> *(Patel, 2017, p. 56)*

Vilane (2018), a transgender author, recounts similar experiences faced by him and other trans advocates as school students in Swaziland, where gender binary is rigidly enforced in public spaces.

Menstrual and reproductive health, parenthood, and disability

Menstrual and reproductive health is a crucial part of transgender lives which is often overlooked in the discussions about safe bathrooms. Much of the discussion on safe bathrooms is focused on the experiences of violence and harassment while research on trans health is focused mainly on gender affirmation surgery, HIV and other sexually transmitted diseases, and hormone treatments (Chrisler et al., 2016; Reisner et al., 2016). Both strands of inquiry neglect the overall health risks and discomfort from poor bathroom access. Trans individuals may withhold going to public bathrooms where they feel unsafe – exposing them to the risk of infections. Transmen and gender non-conforming individuals who menstruate have to practice additional invigilation and regulation of their surroundings and bodies. Transmen in the United States expressed that the process of changing or cleaning up when using men's bathrooms is isolating and daunting (Lane et al., 2022). Menstrual hygiene products, which are available in many of the schools and public spaces for women and girls are not available for transmen in sex segregated bathrooms matching their stated gender or in gender neutral and bathrooms for the third gender (Lane et al., 2022; Vora, 2020).

Transgender pregnancy and parenthood are important life events (Besse et al., 2020). Women's bathrooms in spaces such as offices, hospitals, airports, and movie theaters, are often designed to accommodate childcare by providing diaper changing stations and diaper disposals. Transmen may struggle to find similar facilities to take care of their children in gender-neutral or men's bathrooms. Similarly, disability is rarely discussed when talking about trans experiences, but it can have a profound impact on an individual's identity and access to public spaces, including sanitation and health services (Slater & Jones, 2021). Puar (2014) argues that due to a history of pathologization of trans identity, discourses on disability identity and transgender identity have resisted overlaps – highlighting that neither "person with disability" nor "transgender" fully captures the experiences of trans disabled people (Puar, 2014).

Transgender inclusive sanitation in India

The World Bank estimates that between 2010 and 2020, the percentage of population in India using basic sanitation has increased from 42 percent to 71 percent (World Bank, n.d.). There is an increasing recognition that effective sanitation for all needs to reflect the requirements of not only women and girls but people with disability, the elderly and transgender individuals (WSSCC & FANSA, 2016). The Transgender Individuals (protection of rights bill) of 2016[4] provided visibility to transgender individuals and their legal rights; however, it has also been argued that the bill dilutes the rights-based framework (Jos, 2017). The subsequent Transgender Individuals (protection of rights bill), 2019 prohibits violence and discrimination in the workplace, educational institutions, and public spaces. The Act also made provisions for constituting the National Council for Transgender Persons (NCTP) to advise the Central Government on policies and programs for trans individuals. The Act also emphasizes improving the access to basic services for trans individuals.

The 2019 Act attracted criticism from academics and transgender activists on multiple grounds, including the requirement for transgender persons to apply for gender verification certificate to the District Magistrate who will forward the application to a Screening Committee to evaluate the application – a process which negates self-affirmation (Bhattacharya et al., 2022). Moreover, while the Act prohibits violence and discrimination, it does not specify measures to apprehend those who discriminate or are violent towards trans individuals (Mudraboyina et al., 2019; Bhattacharya et al., 2022). The level of inclusivity in laws, policies, and their application remains low, and incremental increases in the legal recognition of transgender rights impacts access to public infrastructures, education, employment, and housing, among others. As a result, transgender individuals currently live under a legal, policy, and social framework where they are recognized but have not been fully integrated into society, and, due to underlying transphobia, bathrooms remain sites of social exclusion and violence. In a 2016 national level report on inclusive sanitation, the Water Supply and Sanitation Collaborative Council (WSSCC) and the Freshwater Action Network South Asia (FANSA) reported on the daily struggles and often difficult coping mechanisms that trans individuals use to manage their daily sanitation needs. For example, trans individuals are exposed to sexual harassment and violence if they use the men's bathroom and are unwelcome in sex segregated bathroom as "it is widely believed that they are seeking sex work when they visit the bathrooms" (Pebam, 2018; WSSCC & FANSA, 2016, p. 26). This leaves them with few options and many of them continue to practice open defecation or wait to find a safe time to use the bathrooms.

In recent years, multiple states in India have started to address transgender communities sanitation needs and concerns, which are translated into some form of public bathrooms (Pebam, 2018; Roy, 2016; Government of Tamil Nadu, 2022). However, these exercises remain the exceptions. Currently, the transgender community faces health risks, violence, and humiliation in their daily lives as they manage their sanitation needs. Technical solutions as a response to the sanitation needs of transgender individuals may partially meet their needs at best and can have unintended and even harmful consequences.

For the first time, the 2011 census of India survey offered three options to individuals for declaring their sex – female, male, and others. Individuals who identify as neither male nor female could choose the "others" category. This category captures non-binary identities such as transgender, intersex, and eunuchs (*Hijras*) without distinction (Jos, 2017). The unreleased census of 2021 replaces "others" with "transgender person" in the House Listing and Housing Census Schedule,[5] implying that all individuals who do not identify as male or female are transgender people.

As per the census of 2011, there are 487,803 transgender individuals in India (Press Information Bureau of India, 2022). However, the census data has been criticized for underrepresenting transgender population: for example, in 2005, the All-India Hijra Kalyan Sabha (AIHKS) estimated that there were 30,000 Hijras in Delhi alone (Goel, 2016). Moreover, other data-gathering surveys in India for public health, employment, and education such as the National Family Health Survey, continue to use sex-segregated data which excludes the transgender population (Raman, 2021).

In 2015, Sangama, a human rights organization, conducted a state level survey of transgender individuals for the Government of Kerala, India. Nearly all of the 3,619 respondents were transwomen. Up to 58 percent of the transwomen had dropped out of the school system before completing their tenth-grade education due to bullying and gender related negative experiences. Similarly, in the Delhi National Capital region and Uttar Pradesh, between 45 and 49 percent of the respondents from a survey of 900 transgender individuals had dropped out of school before completing their tenth-grade education (Kosciw et al., 2017). Over 20 percent of the transwomen in Kerala stated that they were denied access to public toilets and hospitals and nearly 34 percent reported being sometimes denied access to these facilities (Sangama, 2015). The categorization and counting of transgender individuals, therefore, misrepresents the total transgender population, and providing transgender friendly bathrooms based on these figures can lead to insufficient sanitation services for transgender communities (Roy, 2016).

The Clean India Mission or the Swachh Bharat Abhiyan (SBA) was launched in 2014 to end open defecation in India. By 2019, a total of 104,802 bathroom seats in community bathrooms were built with another 129,809 bathroom seats to be added in future.[6] There is no data on how many of these seats are designed for the transgender community, but an Internet search for public bathrooms in India for transgender individuals reveals that such bathrooms continue to be an exception. For example, the government of Manipur has introduced separate bathrooms for transgender individuals during festivals at Imphal and Lambuikhongnangkhong, a step that was appreciated by the transgender community (Mazumdar, 2017). However, the bathrooms have not addressed the issue of their safety and dignity (Pebam, 2018).

Going beyond gender in India: the role of age, caste, and class

While increasing the number of transgender friendly bathrooms can help transgender individuals in their daily lives, these facilities may not meet all their sanitation needs. Among policy makers in India, the transgender community is frequently assumed to be homogeneous, meaning they have more or less uniform needs from public and community bathrooms. In India, caste, age, and the ability to pass as their stated gender can define their bathroom use experiences. Moreover, public toilets across many parts of India are often unused due to their poor operation and maintenance after they have been constructed (Kulkarni et al., 2017). It is likely that without adequate resources for periodic cleaning and infrastructure maintenance, sanitation facilities for transgender individuals may similarly fall into disuse in the long term.

The third gender is historically part of gender identities in India and other countries in the GS (Boyce et al., 2018). However, the under-sensitization and poor knowledge about gender non-conforming identities can hinder trans children from articulating their authentic selfhood to avoid transphobia. In India, the discussion on transgender-inclusive sanitation focuses primarily on trans adults and the sanitation needs of trans youth and trans children has had a limited discussion in both policy and in academic circles. However, popular media articles in India provide some insights into their lives. In an article in popular media illustrates the challenges young transgender individuals face where a transgender person narrates their school experience in Delhi:

> I started using the school restroom only after seventh grade. I would either hold it in or make sure that I was the only people in there as it was very embarrassing to use the restrooms assigned to females.[7]

Therefore, to design inclusive sanitation, there is a need to also recognize the sanitation needs of transgender adolescent and minors. This requires a rethinking of not only public bathrooms but also bathrooms in schools and colleges (see Watkins & Moreno, 2017; Slater et al., 2018).

Besides age, it is important to re-examine the role of caste within transgender community. In 2014, the Supreme Court of India added transgender individuals to other backward class (OBC) category, giving them a visible but homogeneous caste dimension. The enquiry of the role of caste in policy decisions for the transgender community lacks nuanced discussions (Dutta, 2014). Transgender individuals from disadvantaged caste communities, for example, may find it more challenging to raise funds or resources as compared to their upper caste counterparts (Goel, 2018). The caste hierarchy can also determine who interacts with policymakers and whose interests are represented in sanitation related decisions.[8]

In India, to be identified as a transgender individual can be undesirable and even dangerous, as public bathrooms are often sites of violence (FANSA & WSSCC, 2016). Mitigating everyday transphobia puts a premium on them passing as their stated gender (see Schilt & Westbrook, 2009; Wynn, 2018, 2019). Furthermore, menstrual hygiene, childcare, and disability is rarely bought up while designing sanitation solutions for transgender communities – indicating the scope of the knowledge gap between trans sanitation needs and what is being offered through the sanitation programs.

The way forward

The review of available literature on this topic indicates a need for more policy advocacy, and research in the GS. The countries' traumatic colonial past and persisting cultural and gender norms acting upon intersections of race, caste, ethnicity, class, sexuality, age, and disability produce trans experiences which are unique. In India, gender-neutral and sex-segregated public bathrooms for transgender individuals do not fully remove them from violent and uncomfortable experiences. In the GN, transgender scholarship and activism have highlighted the exclusionary nature of public bathrooms by actively recognizing the diversity of trans voices. Including transgender voices in the larger sanitation discourse in the GS can lead to more just sanitation for all.

In India, important strides are being made in the right direction; however, there is a need to recognize that simply "adding" transgender to sanitation may lead to a similar fate as adding women to sanitation. Underlying transphobia undermines the welfare benefits of transgender friendly sanitation interventions as violence and transgender bathroom use experience are deeply entangled. Technical solutions can meet their sanitation needs partly, but nuanced examination of the barriers to safe sanitation for transgender communities is essential for achieving various sanitation goals. For example, the Central Government of India has allowed for transgender individuals to use the restrooms of their choice (male or female) in public and community bathrooms, but this move does not automatically ensure safety from violence as trans individuals have reported facing verbal and physical violence while using bathrooms for men and women (Mukherjee et al., 2020; WSSCC & FANSA, 2016). Moreover, having a bathroom which explicitly broadcasts a transgender individual's identity to others may not be desirable to all transgender individuals, as underlying transphobia can make them vulnerable to violence. Moreover, while the Transgender Persons Act (Government of India, 2019) recognizes the transgender community, provisions within the Act offer inadequate protection against violence.

The heterogeneous experience within the Indian transgender community requires that the design of public bathrooms need to provide enough scope for privacy and anonymity to those who do not wish to single out their identity. Trans individuals also have varying sanitation needs at different stages of their lives and transgender-inclusive sanitation can be crucial during their formative years. While speaking to children might be challenging in contexts similar to India, it is possible to speak with trans adults to collect retrospect insights into trans childrens' needs. Further, in educational institutes, transgender-sensitive language, awareness, discussion can complement toilet building exercises. Finally, there is a need to expand the meaning of sanitation for transgender individuals to include menstrual and reproductive care along with childcare. In short, transgender-inclusive sanitation in the GS in general, including India, requires a deeper understanding of their social, cultural, biological needs through various life stages.

Therefore, a one-size-fits-all "bathroom for third gender" or sharing bathrooms with women will not automatically lead to transgender-inclusive sanitation. Gender neutral bathroom stalls where they can safely express their stated gender identity, socialize, and learn social skills as they assert their selfhood can improve trans bathroom use experience. However, such measures need simultaneous changes in society to a point where trans rights as equal citizens are upheld and protected. Moreover, transgender friendly bathroom designs at the very least must cater to the basic needs of trans individuals with disability, transgender parents, and those who are menstruating or are pregnant.

–

Note: a shorter version of this chapter was published as a commentary in the *Economic and Political Weekly* magazine and cited as Biswas, D. (2019). Challenges for Transgender-inclusive Sanitation in India. *Economic & Political Weekly*, 54(18), 19.

Acknowledgements

I am thankful to Priya Sangameswaran, for her comments on an early draft of the paper. I am also thankful to Lisa Bossenbroek and Tatiana Acevedo-Guerrero for providing comments and suggestions on the draft manuscript. The usual disclaimers apply.

Notes

1 People using at least basic sanitation services (percentage of population) | Data (worldbank.org).
2 People using safely managed sanitation services, urban (percentage of urban population) | Data (worldbank.org); *People using safely managed sanitation services, rural (percentage of rural population) | Data (worldbank.org).
3 The Third Gender is a South Asian term which encompasses a range of gender and socio-cultural identities, including gender non-conforming, intersex, and transgender identities (Boyce et al., 2018).
4 1989LS p. 65 (Government of India, 2019).
5 Census Schedule | Government of India (censusindia.gov.in).
6 https://data.gov.in/resources/details-total-number-community-bathrooms-and-bathroom-seats-built-under-swachh-bharat-mission (accessed on 2 January 2019).
7 www.firstpost.com/india/transgender-individuals-demand-safer-gender-neutral-bathrooms-in-india-post-section-377-verdict-5414801.html (accessed on 25 December 2018).
8 https://thewire.in/lgbtqia/caste-religion-hijra-community (accessed on 30 January 2019).

References

Beard, V. A., Satterthwaite, D., Mitlin, D., & Du, J. (2022). Out of sight, out of mind: Understanding the sanitation crisis in Global South cities. *Journal of Environmental Management*, *306*, 114285. https://doi.org/10.1016/j.jenvman.2021.114285

Besse, M., Lampe, N. M., & Mann, E. S. (2020). Focus: Sex & reproduction: Experiences with achieving pregnancy and giving birth among transgender men: A narrative literature review. *The Yale Journal of Biology and Medicine*, *93*(4), 517.

Bhattacharya, S., Ghosh, D., & Purkayastha, B. (2022). "Transgender persons (protection of rights) act" of India: An analysis of substantive access to rights of a transgender community. *Journal of Human Rights Practice*, *14*(2), 676–697. https://doi.org/10.1093/jhuman/huac004

Bisung, E., & Elliott, S. J. (2016). Psychosocial impacts of the lack of access to water and sanitation in low- and middle-income countries: A scoping review. *Journal of Water & Health*, *15*(1), 17–30. http://doi.org/10.2166/wh.2016.158

Biswas, D. (2019). Challenges for transgender-inclusive sanitation in India. *Economic & Political Weekly*, *54*(18), 19.

Bornstein, K. (2013). Gender terror, gender rage. In S. Stryker & S. Whittle (Eds.), *The Transgender studies reader* (pp. 236–243). Routledge.

Boyce, P., Brown, S., Cavill, S., Chaukekar, S., Chisenga, B., Dash, M., Dasgupta, R. K., De La Brosse, N., Dhall, P., Fisher, J., & Gutierrez-Patterson, M. (2018). Transgender-inclusive sanitation: Insights from South Asia. *Waterlines*, *37*(2), 102–117. http://doi.org/10.3362/1756-3488.18-00004

Callahan, M. P., & Zukowski, K. T. (2019). Reactions to transgender women and men in public restrooms: Correlates and gender differences. *Journal of Homosexuality*, *66*(1), 117–138. https://doi.org/10.1080/00918369.2017.1395661

Cavanagh, S. L. (2010). *Queering bathrooms: Gender, sexuality, and the hygienic imagination*. University of Toronto Press.

Chrisler, J. C., Gorman, J. A., Manion, J., Murgo, M., Barney, A., Adams-Clark, A., Newton, J. R., & McGrath, M. (2016). Queer periods: Attitudes toward and experiences with menstruation in the masculine of centre and transgender community. *Culture, Health & Sexuality*, *18*(11), 1238–1250. https://doi.org/10.1080/13691058.2016.1182645

Clark, T. C., Lucassen, M. F., Bullen, P., Denny, S. J., Fleming, T. M., Robinson, E. M., & Rossen, F. V. (2014). The health and well-being of transgender high school students: Results from the New Zealand adolescent health survey (Youth'12). *Journal of Adolescent Health*, *55*(1), 93–99. http://doi.org/10.1016/j.jadohealth.2013.11.008

Dutta, A. (2014). Contradictory tendencies: The Supreme Court's NALSA judgment on transgender recognition and rights. *Journal of Indian Law and Society*, *5*, 225–237.

Francis, J., Sachan, P., Waters, Z., Trapp, G., Pearce, N., Burns, S., Lin, A., & Cross, D. (2022). Gender-neutral toilets: A qualitative exploration of inclusive school environments for sexuality and gender diverse youth in Western Australia. *International Journal of Environmental Research and Public Health*, *19*(16), 10089. http://doi.org/10.3390/ijerph191610089

Goel, I. (2016). Hijra communities of Delhi. *Sexualities*, *19*(5–6), 535–546. https://doi.org/10.1177/1363460715616946

Goel, I. (2018, May 18). Caste and religion create barriers within the Hijra community. *TheWire.in*. https://thewire.in/lgbtqia/caste-religion-hijra-community

Government of India. (2019). *The transgender persons (Protection of Rights) Act. Act No. 40 of 2019*. https://prsindia.org/files/bills_acts/bills_parliament/2019/The%20Transgender%20Persons%20(Protection%20of%20Rights)%20Act,%202019.pdf

Government of Tamil Nadu. (2022). *The Tamil Nadu transgender persons (protection of rights) Rules. G.O. Ms. No. 83, Social Welfare and Women Empowerment [SW-3(1)]*. https://images.assettype.com/barandbench/2023-01/1d404a86-e08f-4958-a353-d1d16dd402ec/Tamil_Nadu_government_gazette.pdf

Griffin, C. J. (2008). Workplace restroom policies in light of New Jersey's gender identity protection. *Rutgers Law Review*, *61*, 409–436.

Hasenbush, A., Flores, A. R., & Herman, J. L. (2019). Gender identity nondiscrimination laws in public accommodations: A review of evidence regarding safety and privacy in public restrooms, locker rooms, and changing rooms. *Sexuality Research and Social Policy*, *16*(1), 70–83. https://psycnet.apa.org/doi/10.1007/s13178-018-0335-z

Hirve, S., Lele, P., Sundaram, N., Chavan, U., Weiss, M., Steinmann, P., & Juvekar, S. 2015. Psychosocial stress associated with sanitation practices: Experiences of women in a rural community in India. *Journal of Water Sanitation and Hygiene for Development*, *5*(1), 115–126. http://doi.org/10.2166/washdev.2014.110

Ingrey, J. (2018). Problematizing the cisgendering of school washroom space: Interrogating the politics of recognition of transgender and gender non-conforming youth. *Gender and Education*, *30*(6), 774–789. https://doi.org/10.1080/09540253.2018.1483492

Jones, C., & Slater, J. (2020). The toilet debate: Stalling trans possibilities and defending "women's protected spaces". *The Sociological Review*, *68*(4), 834–851. https://doi.org/10.1177/0038026120934697

Jones, T., Smith, E., Ward, R., Dixon, J., Hillier, L., & Mitchell, A. (2016). School experiences of transgender and gender diverse students in Australia. *Sex Education*, *16*(2), 156–171. https://doi.org/10.1080/14681811.2015.1080678

Jos, J. (2017). Limiting gender variance: Critical reflections on the transgender individuals bill. *Economic and Political Weekly*, *52*(4), 21–22.

Kosciw, J. G., Greytak, E. A., Zongrone, A. D., Clark, C. M., & Truong, N. L. (2017). The national school climate survey: The experiences of lesbian, gay, bisexual, transgender, and queer youth in our nation's schools. *GLSEN*. www.glsen.org/sites/default/files/2019-10/GLSEN-2017-National-School-Climate-Survey-NSCS-Full-Report.pdf

Kulkarni, S., O'Reilly, K., & Bhat, S. (2017). No relief: Lived experiences of inadequate sanitation access of poor urban women in India. *Gender & Development*, *25*(2), 167–183. https://doi.org/10.1080/13552074.2017.1331531

Lane, B., Perez-Brumer, A., Parker, R., Sprong, A., & Sommer, M. (2022). Improving menstrual equity in the USA: Perspectives from trans and non-binary people assigned female at birth and health care providers. *Culture, Health & Sexuality*, *24*(10), 1408–1422. https://doi.org/10.1080/13691058.2021.1957151

Lerner, J. E. (2021). Having to "hold it": Factors that influence the avoidance of using public bathrooms among transgender people. *Health & Social Work*, *46*(4), 260–267. https://doi.org/10.1093/hsw/hlab027

Mazumdar, P. (2017, November). Separate toilets for transgenders at Manipur festival. *The New Indian Express*. www.newindianexpress.com/nation/2017/nov/23/separate-toilets-for-transgenders-at-manipur-festival-1708703.html

Mbatha, S., Wilson, Z., & Buckley, C. (2008). *Zulu Indigenous practices in water and sanitation: Preliminary field research on indigenous practices in water and sanitation conducted at Ulundi [Conference paper presentation]*. Water Institute of Southern Africa 2008 Biennial Conference & Exhibition, Sun City, South Africa. www.ewisa.co.za/misc/WISAConf/default2008.htm

McGuire, J. K., Okrey Anderson, S., & Michaels, C. (2022). "I don't think you belong in here": The impact of gender segregated bathrooms on the safety, health, and equality of transgender individuals. *Journal of Gay & Lesbian Social Services*, *34*(1), 40–62. https://doi.org/10.1080/10538720.2021.1920539

Million Death Study Collaborators. (2010). Causes of neonatal and child mortality in India: A nationally representative mortality survey. *The Lancet*, *376*(9755), 1853–1860. https://doi.org/10.1016/S0140-6736(10)61461-4

Mudraboyina, R., Jagirdar, S., & Philip, P. C. (2019, August 5). A critique of transgender persons (Protection of rights) bill, 2019. *Feminism in India*. https://feminisminindia.com/2019/08/05/critique-transgender-persons-protection-of-rights-bill-2019/

Mukherjee, S., Sundberg, T., & Schütt, B. (2020). Assessment of water security in socially excluded areas in Kolkata, India: An approach focusing on water, sanitation and hygiene. *Water*, *12*(3), 746. https://doi.org/10.3390/w12030746

Mushtaq, M. U. (2009). Public health in British India: A brief account of the history of medical services and disease prevention in colonial India. *Indian Journal of Community Medicine: Official Publication of Indian Association of Preventive & Social Medicine*, *34*(1), 6. http://doi.org/10.4103/0970-0218.45369

Nallari, A. (2015). "All we want are toilets inside our homes!" The critical role of sanitation in the lives of urban poor adolescent girls in Bengaluru, India. *Environment and Urbanization*, *27*(1), 73–88. https://doi.org/10.1177/0956247814563514

Patel, N. (2017). Violent cistems: Trans experiences of bathroom space. *Agenda*, *31*(1), 51–63. https://doi.org/10.1080/10130950.2017.136971

Pearce, R., Erikainen, S., & Vincent, B. (2020). TERF wars: An introduction. *The Sociological Review*, *68*(4), 677–698. https://doi.org/10.1177/0038026120934713

Pebam, B. (2018, July 28). Transgender inclusive public sanitation in Manipur needs a bigger push. *Varta*. www.vartagensex.org/details.php?p=5b5c74f1c49de

Press Information Bureau of India. (2022, June 30). *Welfare of transgender persons in India: Slew of measures by central government in last three years umbrella scheme "smile" launched in 2022*. https://static.pib.gov.in/WriteReadData/specificdocs/documents/2022/jun/doc202263068501.pdf

Price-Feeney, M., Green, A. E., & Dorison, S. H. (2021). Impact of bathroom discrimination on mental health among transgender and nonbinary youth. *Journal of Adolescent Health*, *68*(6), 1142–1147. https://doi.org/10.1016/j.jadohealth.2020.11.001

Puar, J. K. (2014). Disability. *Transgender Studies Quarterly*, *1*(1–2), 77–81. https://doi.org/10.1215/23289252-2399659

Raman, S. (2021, June 16). Denied visibility in official data, millions of transgender Indians can't access social benefits. *Indiaspends.com*. https://scroll.in/article/997580/denied-visibility-in-official-data-millions-of-transgender-indians-cant-access-social-benefits

Reisner, S. L., Poteat, T., Keatley, J., Cabral, M., Mothopeng, T., Dunham, E., Holland, C. E., Max, R., & Baral, S. D. (2016). Global health burden and needs of transgender populations: A review. *The Lancet*, *388*(10042), 412–436. https://doi.org/10.1016/S0140-6736(16)00684-X

Roy, M. (2016, August 8). India's transgender community still waiting for toilets it was promised. *Reuters India Insight*. www.reuters.com/article/india-transgender-toilets-idINKCN10J0VL

Sangama. (2015). *Transgender Survey Kerala, 2014–15*. http://sjd.kerala.gov.in/DOCUMENTS/Report/Survey%20Report/12157.pdf

Schilt, K., & Westbrook, L. (2009). Doing gender, doing heteronormativity: "Gender normals," transgender individuals, and the social maintenance of heterosexuality. *Gender & Society*, *23*(4), 440–464. https://doi.org/10.1177/0891243209340034

Seelman, K. L. (2014). Transgender individuals' access to college housing and bathrooms: Findings from the national transgender discrimination survey. *Journal of Gay & Lesbian Social Services*, *26*(2), 186–206. http://doi.org/10.1080/10538720.2014.891091

Slater, J. (2023). Mundanity, fascination and threat: Interrogating responses to publicly engaged research in toilet, trans and disability studies amid a "culture war". *The Sociological Review*, 00380261231167747. https://doi.org/10.1177/00380261231167747

Slater, J., & Jones, C. (2021). Toilet signs as border markers: Exploring disabled people's access to space. *International Journal of Disability and Social Justice*, *1*(1), 50–72.

Slater, J., Jones, C., & Procter, L. (2018). School toilets: Queer, disabled bodies and gendered lessons of embodiment. *Gender and Education*, *30*(8), 951–965. https://doi.org/10.1080/09540253.2016.1270421

Spears, D. (2013). The nutritional value of toilets: How much international variation in child height can sanitation explain. *World Bank Policy Research Working Paper*. https://doi.org/10.1596/1813-9450-6351

Vilane, V. (2018). Experiences of transgender individuals in Swaziland. In Z. Matebeni, S. Monroe & V. Reddy (Eds.), *Queer in Africa: LGBTQI identities, citizenship, and activism* (pp. 178–195). Routledge.

Vora, S. (2020). The realities of period poverty: How homelessness shapes women's lived experiences of menstruation. In C. Bobel, I. T. Winkler, B. Fahs, K. A. Hasson, E. A. Kissling & T. Roberts (Eds.), *The Palgrave handbook of critical menstruation studies* (pp. 31–47). Palgrave MacMillan.

Water Supply and Sanitation Collaborative Council (WSSCC) & The Freshwater Action Network South Asia (FANSA). (2016). Sanitation and hygiene in South Asia. Leave no one behind: Voices of women, Adolescent girls, elderly, peoples with disabilities and sanitation workforce. *FANSA and WSSCC, Hyderabad*. https://sanitationupdates.files.wordpress.com/2016/01/leave-no-one-behind-report-2016-wsscc-fansa.pdf

Watkins, P. J., & Moreno, E. (2017). Bathrooms without borders: Transgender students argue separate is not equal. *The Clearing House: A Journal of Educational Strategies, Issues and Ideas*, *90*(5–6), 166–171. https://doi.org/10.1080/00098655.2017.1361285

Wernick, L. J., Kulick, A., & Chin, M. (2017). Gender identity disparities in bathroom safety and wellbeing among high school students. *Journal of Youth and Adolescence*, *46*(5), 917–930. https://doi.org/10.1007/s10964-017-0652-1

World Bank. (n.d.). *Improved sanitation facilities (% of population with access)*. https://data.worldbank.org/indicator/SH.STA.BASS.ZS?locations=IN

World Health Organization. (2020). *State of the world's sanitation: An urgent call to transform sanitation for better health, environments, economies and societies*. State of the World's Sanitation: An Urgent Call to Transform Sanitation for Better Health, Environments, Economies and Societies (who.int).
Wynn, N. [ContraPoints]. (2018). The aesthetic [Video]. *YouTube*. www.youtube.com/watch?v=z1afqR5QkDM
Wynn, N. [ContraPoints]. (2019). Are traps gay? [Video]. *YouTube*. www.youtube.com/watch?v=PbBzhqJK3bg

18

HARVEST OF UTERUS

Poor sanitation, water scarcity, and the political economy of sugarcane in Maharashtra, India

Seema Kulkarni and Abhay Shukla

Asha (29 years old) was married into a sugarcane cutting family in Beed district when she was just 13 years old. Since then, Asha has engaged in the tedious work of sugarcane cutting along with her husband, migrating and travelling widely for work. She was doing strenuous cane cutting even during her first pregnancy and fell down while loading cane when she was three months pregnant. Since then, her health has continuously deteriorated with complaints of excessive bleeding, white discharge, and pain in abdomen. After a difficult first pregnancy, Asha went on to have three children and by the age of 19 she underwent tubectomy. However, the continued hard labour involved in cane cutting, and poor living and working conditions, including lack of sanitation at work sites, aggravated her reproductive health problems. Heavy bleeding and uterine prolapse finally culminated in her having to undergo a hysterectomy which cost her entire life's savings. Even worse was the post-operative infection, for which she incurred additional expenses, bringing the total expenditure to over one hundred thousand rupees. Now she has to visit the doctor frequently as she continues to experience pain and weakness, paying around **₹ 1000** (12 USD) per visit. She is weak and frail and has aged prematurely. She feels the necessity to either get admitted for medical care or else lying down all the time.

In India, stories of exploitation of unorganised sector workers are unfortunately common, in the sphere of production. In the realm of social consumption, the commercialised private medical sector remains in the news for overcharging patients and performing unnecessary treatment procedures to maximise profits. When these two spheres of exploitation converge, while also being framed by deeply rooted caste and gender-based oppression, the consequences are devastating.

The ongoing epidemic of unnecessary hysterectomy operations among women sugarcane cutting workers in Beed district of Maharashtra is an example of such "convergence of oppressions" – a widespread crisis with multiple implications. To unravel the tangled skein of causation of this complex problem, it is important to assemble the composite story of a unique system of exploitation which deprives countless rural women of their wombs, their health, and their hard-earned savings.

DOI: 10.4324/9781003100379-22

Villages of wombless women

Responding to questions raised by legislators in the state assembly in June 2019, the health minister of Maharashtra stated that 4,605 hysterectomies had been performed in Beed district during the previous three years (2016–17, 2018–19) (PTI Beed, 2019). The rates of hysterectomy in many villages are extremely high, for example, in a study commissioned by the Maharashtra State Commission for Women in 2018, ***36 percent of women sugarcane cutters had undergone a hysterectomy***(Navachetana Sarvangin Vikas Kendra, 2018). This figure may be compared with an overall hysterectomy rate of 3.0 percent in Maharashtra and 3.3 percent at all India level, concerning women in the reproductive age group, according to NFHS-5 (IIPS, 2022).

The vast majority of these operations in Beed were performed in private hospitals (86 percent), and the causes for which wombs were removed were often trivial or doubtful, such as "swelling of the womb," "womb has gone bad," and white discharge. This has emerged from a number of interviews conducted by MAKAAM[1] members in Beed district. MAKAAM gathered testimonies of women sugarcane harvesters from across 12 different villages in 2019 who had undergone hysterectomies. These testimonies were gathered from women cane cutters who had undergone hysterectomies and had also volunteered to speak in detail with the MAKAAM members. These testimonies were gathered after conducting focussed group discussions in these villages. These testimonies were presented to the lawmakers and to the media to raise the issue.

In Vanjarwadi village, with many households involved in cutting sugarcane, about half of the women have undergone hysterectomies, and reportedly it is the "norm" to remove the uterus after having two or three children (Jadhav, 2019). Many such women, who get operated in their 20s and 30s, subsequently suffer from chronic health problems such as backache and abdominal pain. Paradoxically, while the main "rationale" for undergoing hysterectomy is to improve their ability to perform hard labour of sugarcane cutting without interruption, many such women find that after the operation, their ability to do such work actually gets compromised.

Arduous working and living conditions

Sugarcane cutting is a back-breaking occupation which involves not just harvesting cane but also tying cane bundles, loading, unloading, and transporting these to the factory. A couple, usually husband and wife, perform these tasks together and with pressure to deliver on time, the *mukadam* (contractor or middleman) insists that daily targets of harvesting two tonnes of cane are met by each couple. Typically, the working day is 12–13 hours long, with additional four to five hours devoted by women for unpaid work such as childcare, cooking, cleaning, and fetching fuel and water.

Timings are dictated by the factory owner; since profits are maximised when freshly harvested sugarcane promptly reaches the factory, women report being woken up at unearthly time of 3:00 a.m. to load the trucks, often at the cost of their health. None can afford illness, since missing work incurs a fine which is usually twice that of the wages earned. Menstruating and pregnant women thus continue to work despite the discomfort caused to them; some women work till the last hour of their pregnancy and deliver at the work site itself. Single women workers often suffer sexual harassment at the workplace and have to carry their young children around during work.

Despite these travails, cane cutting becomes a preferred source of livelihood due to persistent drought, agrarian crisis, and non-availability of employment opportunities in their villages (Shiralkar et al., 2019), forcing people – primarily from denotified and nomadic tribes, dalits, and OBCs (Other Backward Classes) – to migrate to the sugar belt in western Maharashtra such as Kolhapur, Sangli, Satara districts or sometimes as far as Karnataka in search of work. Maharashtra has about 1.2 million sugarcane cutters, with Beed district supplying more than half of them. Unfortunately, the cane cutters are registered neither with the factory owners nor with government, making them unable to claim minimum wages or social security benefits. Not only are they the invisible workforce but they are also effectively working as bonded labour.

Calculation of income earned reveals gross denial of minimum wages; the daily earning of a couple is in the range of ₹ 300–450 (3.5–5 USD), while the working day is long and unregulated. Migrant cane cutters must purchase rations at work sites at inflated prices, while losing out on their right to subsidised rations in their villages. The contractor advances an amount of about ₹ 50,000–60,000 (600–700 USD) per season, usually paid to the men. Women state that often this advance is frittered away on alcohol or other frivolous expenditures by the men, leaving little money for the household or her own health.

Most of these migrant families live in the worst possible conditions, often having just a tarpaulin sheet over their heads, no toilets, and limited access to water and health care. These deficits have adverse effects especially on women's reproductive health, since menstrual hygiene becomes difficult. Unhygienic absorbent material used during menstruation leads to rashes and infections, often leading to multiple reproductive health problems, which provide lucrative ground for promoting profit-seeking irrational care by many private healthcare providers.

A large-scale study done in eight districts of the Marathwada region from where a large number of people migrate for sugarcane harvesting, highlighted the plight of sugarcane cutters, especially the women who remained invisible workers (MAKAAM Mahila KIsan Adhikar Manch, 2020). The study also highlighted the poor working conditions, lack of basic amenities like water, sanitation, and housing that further aggravate the health situation. Lack of water and sanitation facilities, also means that women are unable to maintain menstrual hygiene leading to several illnesses related to that. Health morbidity and more specifically reproductive health morbidity is reported to increase as a result of poor hygiene combined with excessive workloads.

Recent site visits done by the SOPPECOM team in the period between November 2022 and January 2023 to different sugarcane fields show that these worksites and the living spaces lack basic amenities such as drinking water and sanitation. The site visit reports point to water borne illnesses due to water that has been provided by the contractors in contaminated containers.

Unregulated commercialisation promotes rampant profiteering

The private medical sector in Maharashtra (as in the rest of India), is completely unregulated with regards to the content of care that it provides – having no standardisation related to rates, treatment protocols, or quality of care. This situation has left the field wide open for perpetuation of various irrational practices and unnecessary procedures, acknowledged as being widespread even by sections of the medical profession (Gadre & Shukla, 2016). However, within this problematic wider setting, certain private hospitals in Beed district have "distinguished" themselves with a track record of gross medical malpractices such as illegal sex-selective abortions. No wonder Beed has the worst child sex ratio of all the districts in Maharashtra, and one of the most adverse in the country, that is, 801 girls per 1,000 boys as per the 2011 census.

In this context of unregulated commercialisation of medical care, a clear pattern emerges. The working and living conditions of sugarcane cutting families contribute to incidence of minor gynaecological ailments, and these women land up for treatment mostly in private hospitals. These hospitals may offer some outpatient treatment, often unsatisfactory, followed by hysterectomy being offered as the "solution." Large sums of money (generally in the range of ₹ 30,000 to 60,000 (350–725 USD) required for this operation then must be raised, often through advances from the sugar factory *mukadams*.

Under-resourced and inadequate public health services contribute to people's dependence on private hospitals. In Beed district with population of over 2.5 million, among public health facilities, only the district hospital and women's hospital have full-fledged capability to conduct regular gynaecological operations like hysterectomies. However, a few dozen private hospitals in Beed district are known to carry out hysterectomies. More importantly, peripheral public health services such as primary health centres (PHCs) appear to have inadequate screening, basic treatment, and counselling facilities for women with gynaecological problems, which could potentially enable many of them to be treated at primary level without operations.

Given the weak public provisioning of both basic and specialised care, women's bodies have emerged as hunting ground for the unregulated, commercialised private medical sector, which aggressively promotes "lucrative" interventions, even when these are often not justified – sex selective abortions and hysterectomies being prime examples.

Pro-business neoliberal policies adopted by the state have also led to major deficits of governance concerning social protection of labour, as well as regulation of private health care. In 2014, the Maharashtra government announced the launch of a welfare board for sugarcane labourers, to provide financial security and improve working conditions of cane cutters. This move proposed that sugar factory owners be established as principal employer and cane cutting labourers as employees of the factories, whereby social security and welfare benefits would become available to cane cutters, while levy would be collected from the factory owners to support these measures.

However, opposition by sugar factories forced the labour department to abandon the welfare board, and instead after much delay they launched a scheme in 2018 for provision of some social entitlements to cane cutting workers. However, even this scheme launched in 2018 is yet to be implemented (Waghmode, 2018). Related policy deficits include failure of the public rural employment programme (NREGA), inability to address the persistent drought, and agrarian distress, worsening the situation for cane cutters (MAKAAM, 2020).

Similarly, Maharashtra government has miserably failed to legally regulate the private medical sector. Although an official draft of Maharashtra Clinical Establishments Act (CEA) was ready in mid-2014, despite completing five years into its tenure, this act is yet to be adopted by the government of Maharashtra. Deepak Sawant, then Health Minister of Maharashtra promised on the floor of state assembly that CEA would be enacted; however, the powerful private hospitals lobby apparently vetoed against this proposal. This appears to be another empty promise which was never meant to be implemented.

Hence, performance of thousands of unnecessary hysterectomies in Beed in last few years may be regarded as a tragedy which could have been foretold. Persistent failure of the state to fulfil its long-standing promises seems linked with its capitulation to powerful private interests. Formation of the Social security board for sugarcane cutting workers (promised in 2013) was opposed by sugar factories, while implementation of CEA to regulate private healthcare providers (drafted in 2014) seems to have been blocked by private hospitals. Betrayal of both these promises spanning five years of "good governance" have sacrificed public interest in favour of profiteering private operators, promoting the convergence of oppressions which underlie this tragedy.

Box: Political economy of sugarcane in Maharashtra

Sugarcane is politically a very important crop for the state of Maharashtra. Cultivation and processing of sugarcane has assumed a crucial position in the political economy of Maharashtra, with sugarcane growers and sugar factory owners forming strong alliances with political parties to protect their interests, at the cost of the smaller farmers as well as sugarcane harvesters. The area under sugarcane in the state is less than ten percent of entire cropped area, however it utilises more than 60 percent of the water, thereby displacing many other food crops that would have benefitted from the available irrigation sources. The predominance of this water-demanding crop, with water requirement of 1700–2500 mm through the perennial growth cycle, in a drought prone state like Maharashtra is remarkable.

Farmers preference for this crop is located in the socio-political history of Maharashtra. In the late nineteenth century, the British rulers started building water infrastructure projects in response to the famines and droughts which occurred rather frequently in the rain shadow areas of the state, aimed at providing food and employment to the local populations. However, later the Irrigation department started encouraging crops that consume more water and which would also be commercially remunerative. Sugarcane thus emerged as the crop of choice for large sections of farmers of Maharashtra from low rainfall regions of the state. The area under sugarcane has increased seven times since the early 1960s to the present, now covering many districts, while the area under cereals and other food crops has gone down in absolute terms.

The political class of the state rallied around this crop and has established co-operative sugar factories in various sugarcane growing areas of the state. At present there are 210 sugar factories in Maharashtra, of which close to 50 percent are private factories. Recently the government of India has introduced a green fuel policy, and to achieve related targets the requirement of ethanol is estimated at 13 billion litres annually, of which 6.5 billion litres should come from sugarcane. This would mean that the area under sugarcane is likely to increase further, aggravating the water situation even more (Economic Times, May 2, 2023).

Harvesting of sugarcane is the foundation of the sugar production process. Sugar yields are highest when the harvested cane reaches the sugar factories early for crushing, hence any delays in transportation result in reduced production of sugar. Therefore, the sugar factories ensure the presence of migrant harvesting workers, through advancing them money before the start of the sugarcane crushing season through a form of semi-bondage. Petty labour contractors who have signed agreements with the sugar factories form 'labour gangs,' working through intermediaries in the home districts. These labour gangs comprise primarily of couples, with each husband-and-wife pair of harvesters being referred to as a *Koyta* (term in Marathi language for sickle or machete that is used to harvest sugarcane).

The migrant labour that is engaged in harvesting is mainly drawn from the water stressed regions of the state, which offer hardly any local employment or viable farming options for most part of the year. Surplus labour is available in these districts because often lands have been expropriated, or lands in these regions are not productive due to lack of water and other resources. It is also reported by the workers that local employment opportunities expected from public works programmes are deliberately curtailed in these regions, so that migrant labour becomes available for sugarcane cutting in the sugarcane growing districts. During six months of the year starting from September to March—April, hundreds of thousands of migrant workers move from drought prone water stressed areas to sugarcane growing districts of Maharashtra, and even to other states. Due to the profit-driven operations of sugar factories these labourers are expected to toil at back-breaking work for more than 12 hours a day, being summoned at any time at the beck and call of factory transporters. Women workers being the worst affected in this situation due to unregulated hours of work, lack of basic amenities like water and sanitation, and the disproportionate burden of domestic responsibilities.

Knee-jerk responses by the state

After the Beed hysterectomy issue erupted in the media (for instance, Jadhav, 2019) in April 2019, the initial state response has been to ask for pre-authorisation of hysterectomies done by private hospitals in Beed district. Following notices issued by the National Commission for Women and Maharashtra State Commission for Women, an order issued by the district collector[2] stated that all private hospitals wishing to perform a hysterectomy, must approach a specified health official with related documents, to be scrutinised before providing official permission to operate.

Hysterectomies are now supposed to be done on only a defined day of the week, with information about all such operations to be notified to mentioned health officials. The order letter also mentions that private hospitals will be inspected, hysterectomy operations are to be reviewed, and also curiously proposes a "*guptpane dekhrekh*" or "secret monitoring" over these hospitals. It is said that the hospitals that do not follow these instructions will risk cancellation of their registration.

While the intention underlying this order may be admirable, both its legality and practicality are questionable. Lacking a Clinical Establishments Act in the state – which would have included standard treatment protocols and legal provisions to regulate the content of care provided by private hospitals – today the state has no effective legal instrument with which to regulate or monitor private medical care. Beyond a few generic "dos," and "don'ts," no standard protocols defining indications and non-indications related to hysterectomy are available with the government until now.

In such a situation, what would be the concrete basis for district level officials to either authorise or deny hysterectomy in any particular case? Further, since unnecessary hysterectomies are clearly not confined to Beed district, this order fails to address the emerging trend of women in Beed being referred to private hospitals in neighbouring districts for the operation, while leaving unaddressed large number of avoidable and unneeded hysterectomies being performed in many other districts across Maharashtra.

Paving a tortuous path to justice

A broad-based coalition of women's organisations and health activists including Maharashtra Mahila Arogya Hakk Parishad, Mahila Kisan Adhikaar Manch, Jan Arogya Abhiyan, Ekal Mahila Sanghatana, and Bharatiya Mahila Federation have challenged the Maharashtra government to effectively tackle the diverse root causes underlying this complex issue. Following a well-attended meeting of various organisations and affected women organised in Beed in May 2019, and these networks convened a state level consultation in June 2019 at Mumbai (TNN, 2019). Three women legislators, health officials and experts, women's groups and health activists participated. Several women sugarcane cutting workers who themselves have undergone hysterectomy spoke there, highlighting the pattern of child marriages and early pregnancies, arduous working and living conditions encountered during cane cutting work, intense pressure to not skip work even during menstruation, lack of facilities to enable menstrual hygiene, frequent gynaecological complaints emerging in such a scenario, the subsequent journey to private hospitals, usually culminating in removal of their uterus at considerable expense, and the aftermath of chronic health problems as well as frequent indebtedness which plague them today.

This event generated considerable media publicity, also prompting discussion in the state legislative assembly, which has led the government to form a seven-member committee. This committee also has the mandate to enquire whether hysterectomies performed in Beed are medically justified, while providing recommendations regarding measures to prevent unnecessary hysterectomies in future. The coalition of civil society networks involved in this issue has submitted a range of demands and suggestions to this committee, with the expectation that the state may finally be prodded into some action.

Emerging demands and questions

A key demand of the coalition is for Maharashtra government to enact the long overdue CEA, including standard treatment protocols to ensure rationality of procedures in private hospitals. Further, public health services at various levels must be majorly strengthened to ensure that screening and basic treatment for gynaecological ailments is available at PHC level, while more advanced forms of non-surgical treatment, as well as operative management, when necessary, should be made available in guaranteed manner at all secondary and tertiary facilities within the public health system.

It is also demanded by the coalition that all hysterectomies performed by private hospitals in Beed in the last three years must be subjected to sample clinical audit. This must be based on the development of standard protocols for gynaecological conditions, while clearly defining indications for hysterectomy. The entire system of public and private health care in Beed must undergo regular community-based monitoring, involving local women's organisations, health sector non-governmental organisations, and elected representatives, to ensure that declared measures and entitlements are actually delivered. This should be accompanied by widespread awareness campaigns regarding women's health issues, involving Accredited Social Health Activists (ASHAs) and Anganwadi workers.

A welfare board for sugarcane cutters is strongly demanded, since this would bring sugar factory owners under purview of accountability. Registering cane cutters for provision of social security benefits is also being articulated by unions and women farmers organisations. Improving living conditions at work sites and ensuring provision of health care, water, sanitation, education, subsidised rations, and safety at work sites are equally important. Further demands concerning women include access to reproductive healthcare, mechanisms for addressing sexual harassment, and payment of wages into women's individual accounts. The larger issue of forced migration can only be addressed through stringent measures to implement MNREGA, while ensuring convergence of the rural development, agriculture, and water resources departments to help build peoples livelihood assets, creating pathways for sustainable prosperity to people within their own villages.

Unnecessary hysterectomies are not unique only to the Beed district; similar "womb scams" have been documented in other districts of Maharashtra (Sardeshpande, 2014), and other states like Karnataka (Xavier et al., 2017) and Andhra Pradesh (Kameswari & Vinjamuri, 2013). These ongoing epidemics which deprive thousands of women of their wombs uncover deep pathologies of our social and medical systems, raising basic questions regarding the distorted model of market-driven "development" being aggressively pursued today. It is not just uteruses of women which have become "disposable" entities subjected to compulsive profiteering; the labour of migrant workers and the bodies of women are being ruthlessly exploited by unregulated, profiteering capital operating in the spheres of both production and social consumption. What initially appears as an issue linked with biological reproduction turns out to be a deeper crisis of social reproduction.

Epilogue

Subsequent to writing this commentary (originally published in *Economic and Political Weekly* (EPW) in July 2019), the Government of Maharashtra brought out a circular identifying the legal rights of sugarcane cutters as workers with right to minimum wages, eight hour working day, and payment for overtime, maternity benefits, weekly off, etc., while identifying the sugar factories as principal employers. The circular also recognises the health and reproductive rights of the workers, especially concerning women, and mandates provision of basic amenities at the work site like housing, water supply, and sanitation among others. Menstrual hygiene is closely linked to provision of these basic amenities.

Due to the demands raised by various unions and social organisations working with sugarcane cutters, a welfare board for the sugarcane cutters and transporters was set up in September 2019 under the Maharashtra labour department. This was subsequently handed over to the Social Justice Department of the Government of Maharashtra in February 2020. A further policy to recognise and register sugarcane workers was introduced in September 2021. Yet despite the introduction of these policies, there is very little implementation on the ground due to disruptions during the COVID epidemic (2020–2022) and political instability in the state.

Nevertheless, the women farmers coalition MAKAAM has continued activities with women sugarcane cutting workers at the local level as well as at the state level. This includes continued engagement with the state and following up implementation of various recommendations of the Dr. Neelam Gorhe Committee, especially with reference to curbing unwarranted hysterectomies. MAKAAM has conducted an extensive study across the eight labour source districts from where sugarcane cutters migrate out for work. This study highlighted the condition of the workers, especially the health issues of women workers.

Findings of the study were discussed with the women workers during the pandemic years 2020–2022. These consultations were important from the point of view of finalising the demands of the women cane cutters for recognition, registration, health rights, and right to basic amenities at the workplace.

Women sugarcane cutters have continued to organise themselves and are consistently engaging with the local and state governments through rallies, dialogues, and campaigns in over 80 villages across two labour source districts where MAKAAM has focussed its work. Women workers from several of these villages are now being recognised as workers and are being registered, they have been able to ensure the organisation of health camps where basic health parameters are checked. There is raised awareness about unwarranted hysterectomies, and women are coming forward to say no to these unnecessary operations at young age. Moreover, the women cane cutters organisation is also pressing its demand for provision of meaningful employment in their home districts so that they do not have to migrate. Demands are being made for revisiting the policies around water, agriculture and energy in Maharashtra which would focus on food and nutritional security, and creation of local employment which is ecologically sound. The regressive social structures and pernicious policy frameworks underlying the situation of woman sugarcane harvesting workers are now being challenged. A vibrant organisation of women cane cutters in the state is changing the terms of work and is changing women's lives!

Note: This chapter is based on a commentary which was published in *Economic and Political Weekly* (EPW) in July 2019 (www.epw.in/journal/2019/29/commentary/harvest-uteruses.html).

Acknowledgments

The authors are grateful to numerous women cane cutters and social activists from Beed, especially those associated to Jan Arogya Abhiyan or People's Health Movement – Maharashtra MAKAAM, a network of health activists which works on health rights and health policy issues and the Society for Promoting Participative Ecosystem Management SOPPECOM, an organization based in Pune working on issues related to water, agriculture, and people. They have provided them with deep insights and a grounded understanding of the issue. They also acknowledge valuable discussions with colleagues in Jan Arogya Abhiyan which contributed to analysis of the issue.

Notes

1 Mahila Kisan Adhikar Manch or MAKAAM (www.makaam.in) is a national network of women farmers. Its main objective is to recognize women as farmers and amplify their voice in policy making spaces.

2 Letter by District Collector, Beed to Medical superintendents of Rural/Sub-district/Women's hospitals and all private hospitals/nursing homes in Beed district, dated 16 April 2019.

References

Economic Times. (2023, May 2). India will achieve 20% ethanol-blended petrol target by 2025. *Economic Times*. https://economictimes.indiatimes.com/industry/energy/oil-gas/india-will-achieve-20-ethanol-blended-petrol-target-by-2025-maize-to-play-important-role-says-piyush-goyal/articleshow/99941800.cms?utm_source=contentofinterest&utm_medium=text&utm_campaign=cppst

Gadre, A., & Shukla, A. (2016). *Dissenting diagnosis*. Random House India.
IIPS. (2022). *National Family Health Survey-5, 2019–21*. International Institute for Population Sciences, Mumbai.
Jadhav, R. (2019, April 11). Why many women in Maharashtra's beed district have no wombs. *Hindu BusinessLine*. www.thehindubusinessline.com/economy/agri-usiness/why-half-the-women-in-maharashtras-beed-district-have-no-wombs/article26773974.ece
Kameswari, S. V., & Vinjamuri, P. (2013). *Medical ethics: A case study of hysterectomy in Andhra Pradesh*. Life-Health Reinforcement Group, Centre for World Solidarity and Knowledge in Civil Society.
MAKAAM Mahila KIsan Adhikar Manch. (2020). *Women agricultural labourers in India: MAKAAM position paper*. https://makaam.in/women-agricultural-labourers-factsheet/
Navachetana Sarvangin Vikas Kendra. (2018). *Report on health conditions and violence experienced by women in Kej and Dharur blocks of beed district in marathwada region of Maharashtra*. Presented to Maharashtra State Women's Commission, Mumbai.
PTI Beed. (2019, June 19). Maha Government panel to probe cases of hysterectomy in Beed. *Hindu*. www.thehindu.com/news/national/maha-government-panel-to-probe-cases-of-hysterectomy-in-beed/article28070453.ece
Sardeshpande, N. (2014). Why do young women accept hysterectomy? Findings from a study in Maharashtra, India. *International Journal of Innovation and Applied Studies*, *8*(2), 579–585.
Shiralkar, K., Kulkarni, M., Ghotale, V., & Gholwe, S. (2019). Migrant labour in Maharashtra's sugar industry. *Economic & Political Weekly*, *54*(20), 36–43.
TNN. (2019, June 13). Beed sees 14 times more hysterectomies, NGOs seek action against private hospitals. *Times of India*. https://timesofindia.indiatimes.com/city/mumbai/beed-sees-14-times-more-hysterectomies-ngos-seek-action-against-pvt-hospitals/articleshow/69763004.cms
Waghmode, V. (2018, June 11). U-turn: No welfare board, but a scheme for sugarcane farmers for now, says Maharashtra labour department. *Indian Express*. https://indianexpress.com/article/cities/mumbai/u-turn-no-welfare-board-but-a-scheme-for-sugarcane-farmers-for-now-says-maharashtra-labour-department-5212237/
Xavier, T., Vasan, A., & Vijayakumar, S. (2017). Instilling fear makes good business sense: Unwarranted hysterectomies in Karnataka. *Indian Journal of Medical Ethics*, *2*(1), 49–55.

19

CAREFUL SANITATION FOR SHARED WATER FUTURES

Kelly Dombroski

Introduction

Hygiene practices, and the sanitation infrastructure built around them, develop through particular interactions in particular places. When I first traveled to a mountainous minority ethnic area of China to do research while pursuing a master's degree in 2004, I had my first experience of mountain latrines. These structures were outhouses built over the side of mountains, with a plank or two removed so one could squat over the resulting hole, feel the breeze on your underside, and watch your stream of waste cascade down the mountain (see Figure 19.1). Despite being frightened of outhouses as a child, I was by that point comfortable with them generally, having done a lot of vineyard work back in my home country of Aotearoa New Zealand, not to mention "tramping" (hiking) where simple "long-drops" (pit latrines) were a step up from the "bush wee" (open air eliminations, usually behind a bush). But the open-air ultra-long-drop of the Sichuan Tibetan outhouse was new, and at first, scary and uncomfortable. Especially the time I peered through the slats of the outhouse and saw a pig far down the mountainside, rummaging about in the waste. But I could immediately see that it worked without too many problems in this remote area, at least until large quantities of disposable sanitary items such as menstrual pads became a problem as tourism increased. I learned then that hygiene practices and sanitation infrastructures are diverse, assembled differently in place and time, *and*, like any point of difference, we may feel uncomfortable when we interact with hygiene assemblages different from our own. But this does not mean that they are maladaptive. I learned that our "hard infrastructures" of sanitation – the visible material structures that help us manage our waste – are intimately entangled with "soft infrastructures" of sanitation – the embodied practices and social structures that govern how we engage with our waste.

The remainder of this chapter will unfold as follows: first I will detail how our hygiene practices are deeply embedded and embodied. I will delve into how "elite" minority world[1] practices of hygiene, and the sanitation infrastructures that support them, have come to exist all over the world. I will then move on to examine the nexus of water and sanitation infrastructure in Australia, and how the extremely high-water requirements of that infrastructure require deep changes in times of drinking water insecurity. Yet the embodied

DOI: 10.4324/9781003100379-23

sense of hygiene minority-world, residents are accustomed to hold back these changes. I will then discuss my research with a group of women who have attempted to re-orient and rehabituate themselves, their babies, and their families to different forms of hygiene that are less water intensive, and the soft infrastructure of a knowledge commons they created and maintained. I finish by asking: how do we create careful sanitation infrastructures that care for our shared water futures? How do we move away from elitist forms of sanitation and reconfigure ourselves and our relationships to engage more ethically with our own waste?

Elite sanitation: take it away

During that same fieldwork trip in rural China, I lived in a mountain village and traveled to tourist sites each day with the women and young people who were all ethnic Tibetan. Once, one of the young men overheard a tourist calling Tibetan people dirty. Unfortunately, this was not an isolated incident. And voicing the racist play on words between *zāng* (dirty) and *Zàng* (Tibetan) was too tempting for a group of Han Chinese tourists. This young man, Yubo, had heard it one too many times. He erupted in Mandarin, accusing Han Chinese of being dirty, since the toilets in their apartments are right next to the kitchen. Indeed, a new toilet and bathing block was under construction in the village at that time, and it was certainly not close to anyone's kitchen: it was a shared bathroom facility for the whole village. But Yubo's outburst revealed how different hygiene practices are deeply intertwined with different sanitation infrastructures.

The Han Chinese tourists were not actually referring to the long drops described in the previous paragraphs and shown in Figure 19.1, but were genuinely expressing racist

Figure 19.1 Mountain outhouse in Tibet. This one is extremely exposed, having no roof or doors.
Source: Becky Carruthers.

sentiments. I knew this because the tourists did not use the long drops, but a high-tech porta-loo, where a long plastic seat cover automatically scrolled when "flushed", effectively wrapping the excreta in plastic, and protecting the tourist from smells of other people's shit (see Figure 19.2). At the end of the day, workers trucked these long scrolls of plastic out of the national park, to be processed ineffectually, and dumped in landfill somewhere else in the county where it would not ruin the pristine environment of the protected area (Gaulke et al., 2009). The outhouses and long drops could not have managed the volume of their waste, and high-paying tourists expected "modern" sanitation infrastructure even in the most remote areas of the mountain. Their experience was not a backcountry tramping experience, where long drops and bush wees were part of the charm. For these tourists, sanitation infrastructure was a mark of modernity. And culturally, the wealthy tourists were not used to having to see and think about their own excrement. They were of a class where their excreta were taken away. Gay Hawkins writes of this global class as follows:

> Water flows in, shit flows out, where from and where to we hardly care. The thing is that the flows are maintained, that our bathroom *works* to protect us from encountering our waste, so that certain ethical and aesthetic sensibilities that are fundamental to the making of the purified private self will not be threatened.
>
> *(Hawkins, 2006, p. 61)*

She notes that those of us living with flush based, public sanitation infrastructure are distanced from our waste in the name of public health and infrastructure efficiency, but that this distancing affects our subjectivity, our sense of ourselves, at a deep embodied level.

Like anywhere in the world, there has always been a class in China who did not have to manage their own excrement. Historically in urban China, it was not flush toilets or

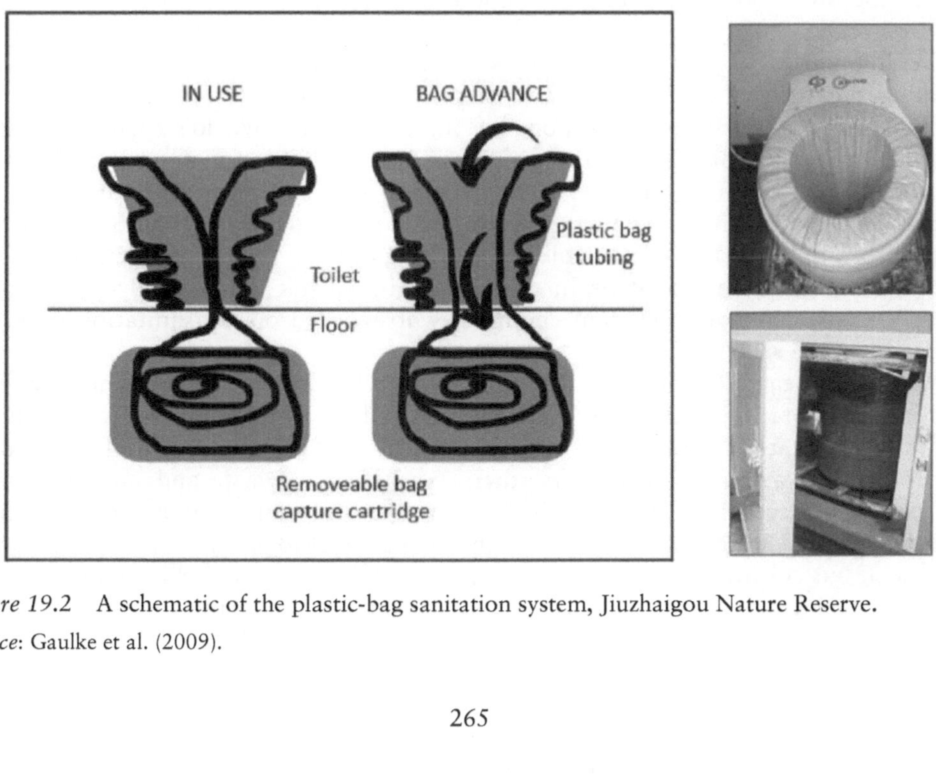

Figure 19.2 A schematic of the plastic-bag sanitation system, Jiuzhaigou Nature Reserve.

Source: Gaulke et al. (2009).

automated plastic rolls that took excrement away, but "honey pot collectors" (Yu, 2010), who collected urban night soil to process into fertilizer on farms outside the city. Urban elites used this system for thousands of years, and constituted what is now known as a "resource-oriented sanitation system" (Hashemi, 2020). The infrastructure for such sanitation systems is quite different from systems that are based on dumping or flushing excrement away. But as environmental experts have been saying for generations "there is no away",[2] and there never has been. All waste must be processed in some way, even if it is the great Pacific garbage island, the burning of "recycled" plastics revealed in Malaysia, the careful reuse of food scraps and excrement for composting, or the pig and microbes at the bottom of the mountain outhouse – it all goes *somewhere*. As Brian Thill writes "The soil itself is part of a new geology, as the beaches have been remade into plastiglomerate, their sands mingled with the pulverized microplastics of our petroleum age" (Thill, 2015, p. 3). Yet somehow, many of us in the minority world – that smaller percentage of the world who use up the majority of the resources – continue to assume that our waste habits and infrastructures are superior to those of places where waste is not an "away", but an intimate part of the material cycle of everyday life.

Reproducing norms through infrastructures: the case of the SDGs

The Sustainable Development Goals are a worthy cause, representing impressive global negotiating around justice, environment, equality, health. SDG 6 sets targets for clean water and sanitation. Yet like many global development agendas, the negotiating and decision-making was driven by groups who are undoubtedly elites. A variety of NGO and other groups were part of the negotiating process, but in the end, the decisions were mainly documented and researched by people who are part of the minority world. The Chinese tourists I mentioned previously are minority world – able to take a leisure trip to an expensive resort location. The in-country consultant or the educated NGO director are in the minority world. I am in the minority world – a university researcher with a PhD living in an OECD nation. It is likely that you, too, are in the minority world – reading an expensive academic book in English. As Oxfam has noted, the richest 10% of the world's population are responsible for 52% of the world's carbon emissions, and the richest 1% of the world's population is responsible for more than double of the world's poorest 50% (Harvey, 2020). The top 10% are those who earn more than $35,000 USD per annum. The top 1% are those who earn more than $100,000. It is certain that the UN staff responsible for working out the indicators for SDG 6 Clean Water and Sanitation would be in the richest 10%, and it is entirely possible that the UN statisticians or consultant engineers who worked on the sanitation indicators would be in the richest 1%.

What this means, I suggest, is that the people who decided on the sanitation goals and indicators for the SDGs are elites, who are likely to have an elite experience of waste and sanitation infrastructure. Let me return to Hawkins' notion of the relationship between waste and subjectivity: their experience of waste and sanitation infrastructure influences their understandings of hygiene and sanitation on a pre-conscious, embodied level. As Hawkins notes "the infrastructural logic of sanitation is not just technical, but cultural" (Hawkins, 2006, p. 46). What effect has this had on the SDG indicators, and what effect does this have on water-related sanitation infrastructure more generally?

For the SDG indicators, the most obvious elite bias is the statistics on "lack of access to sanitation". While the language has changed since I first started writing about this (Dombroski, 2015), the calculations remain the same. According to the SDG e-handbook,

> *Improved sanitation* facilities refer to sanitary facilities designed to hygienically separate excreta from human contact. They include flush or pour flush toilets to sewer systems, septic tanks or pit latrines, ventilated improved pit latrines, pit latrines with a slab, and composting toilets.[3]

The SDGs start off inclusive and are a step up from the Millenium Development Goals, which excluded pit latrines. The mountainside outhouses would likely be excluded, however, because they are not included in the list of possible improved sanitation infrastructures. However, many of the other toilets I encountered in my research would still not count in the statistics because of the way the indicators are calculated, where the statistics include:

> [T]he number of people (or households) using improved sanitation facilities *which are not shared* and where excreta are transported off-site, treated and disposed of[4];

and:

> [T]he number of people (or households) using improved sanitation facilities *which are not shared* and where excreta are treated and disposed of in situ.[5]

The calculations behind the indicators used for measuring how many people have access to safely managed sanitation *explicitly exclude people and households using shared facilities*. Undoubtedly, someone in the statistics department or a public health advocate will explain to me how bad shared facilities are, and also how bad mountain outhouses are since the excreta is out in the open and flies could land on it and who knows where that pig will go next? But when it comes down to it, someone – likely an elite person – made this decision about what counts, and it is now being used to collect *global* statistics on sanitation infrastructure that excludes community-based shared infrastructures. It even excludes that which is working well – such as that described by S.M. Waliuzzaman in his work on commoning waste infrastructures in a slum community in Dhaka, Bangladesh (Waliuzzaman & Alam, 2022; Dombroski et al., 2023). If decision-making (and funding) is with reference to SDGs, this actively disincentivizes investment in community-based shared infrastructure – such as the new shared bathing facilities in the Tibetan village, or the community-run shared toilet facilities in Dhaka. It also disincentivizes resource-oriented infrastructures since it is explicitly framed around separating excreta from human contact. The shift from historical resource-oriented systems to water-based systems has proven problematic in urban China, since it resulted in an *increase* in open defecation (Iossifova, 2020). As collection stations become unviable and disappear, migrant workers and others living without private toilets are left without places to drop off night soil. Open defecation thus includes the problematic "wrap and toss" system of plastic-bagging feces and tossing out wherever possible, used by people without access to sanitation infrastructure in some parts of the world (Jewitt, 2011).

There are diverse hygiene practices and diverse sanitation infrastructures to support them. Many of these are well-adapted to environments and cultures; some are not. While having global targets for clean water and access to sanitation are important, it seems to me that the sanitation infrastructures that have emerged from the hygiene practices of one group of people – elites – are implemented across the globe with little critical consideration of the appropriateness of this mode of hygiene. As Sarah Jewitt argues in her article "Geographies of shit", the minority world preference for expensive and wasteful flush toilet, water-based sanitation systems is problematic, especially when it is exported all over the world to environments where it simply does not make sense (2011). It is not just the infrastructure put in place, but the embodied subjectivities intertwined with it. This can actually be damaging, from both an environmental perspective and a human health perspective. Let us look at the kinds of hygiene practices and sanitation infrastructure implicated here.

Shifting waste assemblages: there is no "away"

During one period in my PhD studies, I lived in Sydney, Australia. Sydney has water-based flush-toilet systems that use drinking-quality water to flush and process excrement into a publicly managed sanitation system. Sydney Water reports that the average person in Sydney uses 200 liters of water *per day*, with about 12 liters per toilet flush and 65 to 100 liters per load of washing, not to mention watering gardens and filling up swimming pools.[6] This is one of the highest per capita rates of consumption in the world, despite being situated on one of the world's driest continents. The entire system requires that a certain level of water is flushed through it in order to work. But all kinds of things get flushed down that system to the mythical "away", including baby wipes, which are contaminated "disposable" materials that minority world parents use to clean up babies' bottoms. These wipes have coagulated into giant fatbergs, clogging Sydney's sewerage system several times. There is no "away", yet this elite sanitation system has fostered a mindset and embodied practice where waste must be unseen, not smelled, and certainly not discussed. Indeed, until people have children, they rarely have to deal with excrement (even their own), and the temptation to flush away contaminated materials is strong. If flushing away is not possible, throwing away with minimal engagement is next.

Most of us would quickly agree that the "wrap and toss" sanitation, involving plastic-bagging of excrement and tossing it "away", is problematic for public health and environmental reasons, but using disposable nappies is comparable to a wrap and toss approach. While manufacturers recommend that caregivers scrape and flush feces before disposing of the nappy, the practice is not widespread. Mostly, caregivers wrap the feces and contaminated disposable wipes in the nappy, sealing it tight (sometimes with a scented plastic bag around it), then toss it out in the household rubbish or nearest bin at the café or mall. For those raised on elite sanitation infrastructures, the incentive to throw such disturbing waste "away" is strong, but public infrastructure does not support this impulse. In Sydney, I witnessed parents out and about on the evening before rubbish collection trying to dump extra nappies in other people's less full curbside bins – a child wearing disposable nappies could easily fill a bin on their own each week. In other places, there are other challenges: in a river in Indonesia, one researcher fished out three hundred kilograms of disposable nappies in two hours, in an area where people felt uncomfortable about burning or burying items associated with babies, for spiritual reasons (Victory, 2019). Vanuatu has considered banning disposable nappies due to a lack of landfill space (Dombroski, 2019). Disposable nappies are resource- and water-intensive to produce and are predicted to take up to 500 years to break down in landfill conditions (Meseldzija et al., 2013). These are not environmentally or politically neutral items, yet their convenience is hard to contest.

Between 85 and 96% of Australian babies use disposable nappies (O'Brien et al., 2009; Thompson Brewster et al., 2022), and upwards of 95% of those nappies go to landfill. The average baby in Australia uses 4600–4800 nappies (Dey et al., 2016), producing about a ton of plastic waste each. Rethinking disposable nappies is an urgent matter. Sydney local governments have responded by getting involved in promoting cloth nappy libraries, where parents can borrow the expensive modern cloth nappies and try before they buy.[7] For those using cloth nappies, a much more intimate encounter with infant waste is necessary. The sanitary recommendations can be daunting: users need to pay attention to the disposal of feces, the washing of nappies, and the disposal of contaminated wash water. Best practice for modern cloth nappies (made of PUL, bamboo, and/or microfibers) require scraping out solid feces, dry bucketing, a prewash cycle, then a full wash at a higher temperature (60 degrees Celsius).[8] Traditional cotton cloth nappies in the minority world are usually said to require soaking in detergent, rinsing, and a hot wash. Not only then is increased contact with excrement required, but significant amounts of water and labor. In both the minority world and majority world, moving away from disposable nappies is not an easy choice, especially when water and time are scarce.

But change is required, nevertheless. While some circular economy advocates focus on the design and distribution of more environmentally friendly products – such as compostable disposable nappies, or reusable nappies with disposable inserts – they also acknowledge that modern designs are meaningless unless the waste infrastructure is able to handle them. This includes the "soft" infrastructures of care into which users incorporate these items. There are knowledges that come alongside practices, and practices have material implications just as new products have practical implications. Compostable nappies are useless unless there is some way and somebody to compost them. Cloth nappies are likewise enrolled in a wider hygiene assemblage that includes water, gendered labor, and the hinterlands of production (cotton plantations, bamboo harvesting, PUL production and more). Waste infrastructure thus involves shifts in material infrastructures *and* shifts in waste subjectivities and labor. Over the last decade, I have been interested in studying such shifts in subjectivity – and assemblages gathering around such shifts – with regards to the infant hygiene practice known as "elimination communication" (Dombroski, 2016). It reduces or eliminates the need for nappies and teaches us something interesting about how we might shift waste subjectivities and the assumptions that minority-world planners exported around the globe.

The remainder of this chapter will concentrate on a group of caregivers in Australia who have spent a considerable amount of time seeking to rehabituate themselves and their families into different forms of sanitation and excreta management. While many of these caregivers are not primarily interested in overcoming the cultural blind spots of Western sanitation norms, they *are* effectively making steps towards doing so by adopting an infant hygiene technique that is widespread in Asia and many other places in the majority world. This infant hygiene technique is called "elimination communication", "baby pottying", or "assisted infant toileting" (Wang et al., 2019; Sibonney, 2019; Dombroski, 2018; Sun & Rugolotto, 2004). In China, where I have also researched the practice, it is referred to simply as *baniao* or "holding out to urinate" (Dombroski, 2015, 2018).

Elimination communication: (re)habituating hygiene

Elimination communication is an infant hygiene and toileting technique whereby caregivers learn the idiosyncratic and common signs and signals for babies impending eliminations – both urine and feces – and hold them out in an appropriate spot to complete them. I first came across the practice without knowing what I was seeing. On my first trip to China as a young New Zealand Pākehā woman in my twenties, I noticed the bare babies' bottoms peeking out of split-crotch pants, out on the street. I did not know anyone with a baby at that point and was far less interested than I later became when I found I was pregnant with my first child in 2005, a few days before receiving notice that I had been awarded a PhD scholarship in Canberra, Australia. Later, my husband and I came to hear about an equivalent practice of nappy-free infant care from a friend who read about it in a newspaper article. By the time my daughter was born in Australia in 2006, we had a clear sense that this was something we wanted to try. Like many practitioners of elimination communication, we were motivated by environmental reasons to avoid the problems of disposable nappies and reduce the load of cloth nappies we would have to deal with in the height of the drought in Canberra. But unlike other practitioners, I already had funding and a field trip to Western China on the horizon – although I freely admit, infant toileting was far from my actual PhD research proposal topic. What I quickly came to learn, however, is that hygiene and sanitation systems are culturally and geographically situated. Let me set the scene for you.

Summer 2007, Xining, Qinghai, China. *I hoist my baby daughter onto my back, adjusting the straps of my mei-tai, jiggling, and swaying in the special dance that calms her and points her in the direction of sleep. I smile at a grandmother on a small stool, holding her grandchild – the child seems to melt into her arms, fast asleep. She stares at me: she can't believe what she has just witnessed. I had just "held out" my baby over a drain, her split-crotch pants open as I manipulated her gently into a squat position. A soft whistle, a yellow stream, a little wiggle to dry out the bottom. She's not staring at me because of that, however – that's pretty ordinary around here, where mothers and grandmothers pick up subtle cues from babies and help them to direct their "eliminations" to the appropriate spot. She's staring because it's the first time she's seen a foreign woman and a foreign baby do what she "knows" is right and normal. She catches my eye and nods in approval as she settles herself and her sleeping charge more comfortably. As I continue down the street, I begin planning a post to the OzNappyfree web forum, who would be just as excited as I am by the normalcy of toileting babies in China.*

Winter 2011. Bankstown, Sydney, Australia. *My toddler is toddling around barefoot on the concrete area outside the school hall, where inside her five-year-old sister has just performed in the school production. She toddles, wobbles, then plonks down hard on the concrete, fortunately protected by her cloth nappy, waterproof cover, and stretchy leggings. Refusing my outstretched hand that offers help in regaining her feet, she crawls across the concrete and pulls herself up on the brick wall to begin again. Midway through the next cycle she stops, looks at me, and says "unh".*

"Do you need to go potty?" I ask.

She looks around for a potty, which I take as a yes. Unfortunately, the toilet is inside the crowded school hall, and I would have to push through other parents watching the performance to get there. I make a quick decision – I pick her up and walk around behind the building. Looking around to see if anyone can see me, I quickly pull down her leggings and remove her nappy, holding her out on the grass. She immediately urinates, and with some difficulty I manage to put her nappy back on with her balanced on my knee. The tension goes out of my shoulders as I complete the operation without having to explain myself to anyone.

As you might gather from these ethnographic snippets, the differences between practicing elimination communication in Xining and Sydney were stark. They illustrate the cultural expectations around hygiene: in Xining, people place importance on babies' bottoms being kept clean and dry, while in Sydney, they place importance on the public spaces being kept clean and dry (Dombroski, in press). In Xining, a city of around a million people at the time of fieldwork, people managed sanitation in a variety of challenging situations, including living in toilet-less shops or storage rooms in shops with their babies, living in single rooms within shared courtyard housing, or living in older apartment blocks with shared bathrooms and washrooms. Sharing sanitation infrastructure was the norm, and holding out babies to urinate in public was completely acceptable – who could find a toilet quick enough for a baby to go? Caretakers held babies out over basins or newspaper (for feces), up against trees, over gardens, or over concrete footpaths or courtyards (for urine). Older toddlers could squat and relieve themselves in those same places, while family members manipulated them into an early morning defecation at home by feeding them large quantities of warm milk. People managed. Disposable nappies were only used at night, and cloths were tucked into the split crotch pants of smaller babies who could not signal clearly. There was no shame in a small accident, no one saw this as trouble, as one mother notes in an interview with me in Xining in 2009:

> If she got her clothes wet, then you just changed them. After all, she was just a child. It was natural for her to wet her clothes. Just wash the clothes and change her into some new ones. Wash her wet trousers and dry them in the sun. It was not troublesome, no trouble at all. This is just what you have to do. No trouble.

In Sydney, practicing elimination communication *was* seen as trouble – or at least, not as convenient as nappies. The need to remove layers of clothing and find a private space to hold a baby out added complications to the practice that were not present in Xining. But practitioners were keen to experiment with a practice that could work in Australian cultural and spatial conditions. As one mother summarized her journey in 2009, on the Australian nappy free web forum:

> I have tried to let [my daughter] guide me, but also needed to give her the environment in which she could learn and move forward, and [to] get over my own fear/impatience of dealing with public accidents/dealing with potties/finding toilets/asking shop owners to use their toilets etc. . . . we have toilet trained slowly, at the end of a great EC [elimination communication] journey that was filled with many many books, potty songs and a lot of fun.

For caregivers in both places, learning to be affected by the bodily and preverbal signals of babies' impending communications was key to the practice (Dombroski, 2018). But in one environment, how to do this was culturally embedded in practices of post-partum confinement and norms in public spaces with minimal private sanitation infrastructure, while in the other environment, caregivers had to experiment with and make up different ways of being with their baby and in public and private spaces where sanitation and hygiene were largely private matters. During the time of my research, the Australian caregivers contributed data and reflection in the web forum known as OzNappyfree (Dombroski, 2016). This knowledge commons was cultivated and cared for by a range of people and has shifted since that time into diverse social media platforms including TikTok, Instagram, Facebook, and YouTube. The practice of elimination communication and infant toileting is nothing new, in fact it is possible that up to 50% of the world's population have had experience with it. But what has interested me in this whole area of research is the fact that people in the minority world are *intentionally* rehabituating themselves to different relationships with sanitation, with great levels of detail as to how they did it.

Lay research collectives: experimenting and documenting

Given the challenges facing us as species, it is important to think about how change can happen at an embodied and habituated level. As Elizabeth Shove has traced in her work on cleanliness, comfort, and convenience, changes in cleanliness technology can lead to changes in social practice as well as norms and expectations of cleanliness (Shove, 2003). But the reverse could also be true: changes in social practices and norms might lead to different kinds of embodied relationships with practices of hygiene and sanitation and the infrastructures that support them. Infrastructures are cultural and social, "reproducing social difference as they are incorporated into everyday routine" (Power & Mee, 2020, p. 487). But as Abby Lopes and colleagues point out, "infrastructures both shape and are shaped by conventions of practice" (Lopes et al., 2018, p. 50, see also; Leigh Star, 2010), meaning that as conventions of practice change, infrastructures might too, albeit in a slow and lumbering way that might seem endless. This means that the work of rehabituating oneself and one's household into new ways of doing things is both important and difficult. Working against the tide of current practice and infrastructure requires a level of motivation and support that new parents might lack, yet the caregivers I researched were doing just this. In the time since I interviewed them, things *have* changed. In Australia and New Zealand (where I now live), it is now relatively easy to find information about elimination communication and reusable modern cloth diapers. Other alternative hygiene practices and products are becoming mainstream, including shampoo bars, menstrual cups and reusable period underwear, swimwear, and sanitary pads. How did we get to this point?

One contributing factor was the work of embodied experimentation and adaptation that elimination communication practitioners undertook in the early 2000s via online web forums. While these mothers experimented in their homes with some of the most intimate and normally private bodily habits, they also documented and shared these via the OzNappyfree web forums I researched. For example, someone might share exactly the way in which they got their child out of a car seat and took them potty in the back of a station wagon, down to what they kept in the back of the car, and how they negotiated pulling over with a pre-verbal child who did not understand the road rules on no-stopping motorways. Another might share how they balanced an infant on their lap while breastfeeding them over a potty-basin ready for a mid-feed "poo explosion". Participants collectivized these individual experiments by documenting and reporting on the forum intentionally. One mother wrote that the essence of "what we are all aiming for" is to "help people who have an interest in EC", in particular, "regaining and preserving, and sharing this lost knowledge" (web forum post, 2009). The knowledge commons that were developed via the web forum allowed people to search and read old posts and follow threads on matters of interest, including what other people's experiences were with infants of a particular age, or how they worked out certain practical problems like infant toileting during international flights. Others argued that this group was the "beginning of a move back towards early [toileting] as the norm" (web forum post, 2008), highlighting the transformative role possible when a small group of committed people try to do things differently. As such, group members saw themselves as the keepers and revivers of the knowledge commons around EC, much in the same way that Lopes and colleagues found people in the hot parts of Western Sydney willing to care for the knowledge commons of how to stay cool without air conditioning (Lopes et al., 2018).

I have three key points to make around this. Firstly, the knowledge produced by the forum was more than just a reclaiming of old knowledges. It was a practice of lay experimentation and documentation on issues of importance to the group. As such, it has much in common with the hybrid research collectives Michel Callon and Vololona Rabeharisoa have described in their work with patients of muscular dystrophy in France (Callon & Rabeharisoa, 2003), a point I have elsewhere (Dombroski, 2016; Dombroski, in press; Dombroski et al., 2018). These hybrid collectives include both lay and professional researchers and work to produce new knowledge. Callon and Rabeharisoa propose that:

> It might be fruitful to consider concerned groups as (potentially) genuine researchers, capable of working cooperatively with professional scientists. In so doing, they invent a new form of research, which we propose to call research "in the wild".
>
> *(Callon & Rabeharisoa, 2003, p. 26)*

Such a hybrid collective of lay and professionals produce new knowledges through experimentation and documentation, much in the same way it is produced in laboratory as described by Bruno Latour and Steve Woolgar in *Laboratory Life* (1986): through inscriptions, disputes, and resolutions.

The second point is that this knowledge is a form of commons, and like any commons, users need to maintain and care for it if it is to survive. As Gibson-Graham et al. (2013) note, a commons is a resource that has shared and wide access, use negotiated by a community, benefits distributed to a community and perhaps beyond, is cared for by community members, and is the responsibility of a community. The ownership of the commons, in Gibson-Graham et al.'s formulation, does not have to be collectively owned if these other aspects of the commons are in place. In the case of OzNappyfree knowledges of elimination communication in a minority world context, we can see that this was the case – Yahoo!, a large corporation, owns the forum, but all the other elements of a commons were in place. Access is shared and wide, and people were given access on application and introduction; use of the knowledge was negotiated by the community of users (information could be repeated, or users could search the archives); benefits were distributed to the community of users but also beyond as people shared their knowledge more widely and environmental benefits accrued to those who were not practicing EC; the forum was cared for by community members including forum moderators and others who cared for collective culture of the members through encouragement and tips; responsibility was likewise distributed between moderators and community members.

Thirdly, while this particular web forum knowledge commons is now no longer in use, the temporary nature of such a knowledge commons does not take away from its transformative value. Because communities necessarily care for their commons, the interpersonal aspects of the commons are also deeply important. As Neera Singh has argued in her work on caring for commons, the act of caring for commons has intersubjective effects. Indeed, she notes that commons are "a nurturing ground for subjectivity", where "commons and commoners are co-constituted through intersubjective communication and affective relations" (Singh, 2017, p. 755). I would add that such commoner subjectivities, while formed by the act of caring for commons, do not die as temporary or transitional commons morph and change communities, but might go on to build and care for other commons (Dombroski et al., 2019). With elimination communication, those commons have transitioned to different digital platforms as a new generation of parents learn the ropes of both parenting and elimination communication in the age of smartphones. While none of this implies infrastructural changes in terms of water infrastructure, hygiene, or sanitation in Australia, it *does* indicate a wider acknowledgement of the role of practices in sanitation – that is, it is not just a matter of water infrastructure. While OzNappyfree had 400–500 members from New Zealand in Australia in the age of internet forums, in 2022, the Facebook elimination communication group in Aotearoa New Zealand alone has 796 members. In the United States, there are at least two groups with 8000 plus members in each, while one in the UK has 8,600 and a global group has 20,689 members.

While men also practice EC at home with their children, the vast majority of forum and social media group members are women. What does this mean for water futures going forward? Like other forms of gendered care work, these practices have been ignored in more formal public forums. Women continue to take the lead in caring for this particular knowledge commons, and other parenting and "domestic" knowledge commons. The more visible practices like EC become in spaces like public forums, the more we can reduce the expectation that sanitation and water use practices everywhere will look the same – a point that I will elaborate further in the final section.

Careful sanitation for shared water futures

Widespread changes are necessary in the way we in the minority world think about water use, sanitation infrastructure, our waste subjectivities, and our "geographies of shit" (Jewitt, 2011, p. 608). In our specific places and societies, we must be aware of the kinds of hard and soft infrastructures that help us deal with our waste here and now, not just send it off to some mythical "away", where it is somebody else's problem. As I write, New Zealand's largest city has been subject to days of flooding, where heavy rain has overwhelmed the waste and stormwater systems and sewage contaminated "black water" sits across the impermeable concrete surfaces of the city. This flooding laid violently bare the myth of "away" and waste infrastructure has suddenly become everyone's concern. Hard infrastructures are important, and citizens must be involved in thinking and acting politically around them. But the work of elimination communication practitioners is also important: they have (re)habituated themselves and their children into different relationships with human waste, their children's bodies, and documented and commoned the process. They have something to teach us about minority world transformation. For too long, we have prioritized hard infrastructures, such as water-based sanitation systems, and ignored practice-based and resource-based systems. What if it is those of us with water-based infrastructures that have it wrong? What if there are simpler, everyday practices of hygiene and sanitation that use social infrastructures and practices to keep hygiene? What if the knowledge commons of practices such as elimination communication form part of the infrastructures of care that can help us rehabituate ourselves to less resource-intensive hygiene and sanitation? Neera Singh cites Felix Guattari with a related question:

> How do we change mentalities, how do we reinvent social practices that would give back to humanity – if it ever had it – a sense of responsibility, not only for its own survival, but equally for the future of all life on the planet?
>
> *(Guattari, 1995, p. 119, in Singh, 2017, p. 754)*

One way to do so, in the context of responsibility over water, would be paying closer attention to the everyday practices of care that affect water usage and our relationships with water, both dirty and clean. We need careful infrastructures that are built on everyday practices of hygiene that care for the planet and each other. Engineers, development practitioners, statisticians, and planners must take better account of ethnographic studies of everyday hygiene, sanitation, and care work in a variety of diverse places, minority and majority world (Chitondo & Dombroski, 2019). Those currently doing hygiene and sanitation care work differently (particularly women) should have more of a say in what safe and adequate future sanitation systems look like, with an openness to radical change. Finally, those who design such systems should rehabituate themselves to different relationships with waste, which might mean long-term visits in places with mountain outhouses or shared community toilet facilities. A shift in the gendered nature of these practices toward men performing some infant toileting care-work when it arises might mean better paid parental leave for all parents, indicating some of the ripple effects of these soft infrastructures.

Participating in caregiving, being deeply responsible for another life and the social networks of knowledge commoning that go along with it, is a deeply affective and transformative experience (Dombroski, 2018). When caregivers experiment in hygiene and sanitation within hybrid collectives, it might lead us to diverse ways of learning and doing sanitation infrastructure. In my work, I have centered the experiences of women who lead in commoning caregiving and creating hygiene knowledges without waiting for gender equality, while recognizing the importance of these practices for all genders, including men (Dombroski, 2020). Change is needed both in the minority world, where high water use and the "away" mentality are deeply concerning, and in the majority world, where access to clean water and appropriate sanitation infrastructure are deeply unequal.

Notes

1 The "minority world" is a shorthand term for those people, wherever in the world they are, that fall into the top percentage of indicators for income and other indicators of wealth and wellbeing.

2 For an excellent educational resource on the topic "There is no away" see www.nationalgeographic.org/activity/theres-no-such-place-away/.

3 See https://unstats.un.org/wiki/display/SDGeHandbook/Indicator+6.2.1.

4 Ibid.

5 Ibid.

6 See www.sydneywater.com.au/education/drinking-water/water-use-conservation.html.

7 See www.australiannappyassociation.org.au/cloth-nappy-services-from-councils-across-australia/.
8 See the Australian Nappy Association, www.australiannappyassociation.org.au/.

References

Callon, M., & Rabeharisoa, V. (2003). Research "in the wild" and the shaping of new social identities. *Technology in Society*, *25*(2), 193–204. https://doi.org/10.1016/S0160-791X(03)00021-6

Chitondo, M., & Dombroski, K. (2019). Returning water data to communities in Ndola, Zambia: A case study in decolonising environmental science. *Case Studies in the Environment*, *3*(1), 1–8. https://doi.org/10.1525/cse.2018.001552

Dey, S., Purdon, M., Kirsch, T., Helbich, H., Kerr, K., Li, L., & Zhou, S. (2016). Exposure Factor considerations for safety evaluation of modern disposable diapers. *Regulatory Toxicology and Pharmacology*, *81*, 183–193. https://doi.org/10.1016/j.yrtph.2016.08.017

Dombroski, K. (2015). Multiplying possibilities: A postdevelopment approach to hygiene and sanitation in Northwest China. *Asia Pacific Viewpoint*, *56*(3), 321–334. https://doi.org/10.1111/apv.12078

Dombroski, K. (2016). Hybrid activist collectives: Reframing mothers' environmental and caring labour. *International Journal of Sociology and Social Policy*, *36*(9), 629–646. http://doi.org/10.1108/IJSSP-12-2015-0150

Dombroski, K. (2018). Learning to be affected: Maternal connection, intuition and "elimination communication". *Emotion, Space and Society*, *26*, 72–79. https://doi.org/10.1016/j.emospa.2017.09.004

Dombroski, K. (2019, July 4). What other countries can teach us about ditching disposable nappies? *The Conversation*. https://theconversation.com/what-other-countries-can-teach-us-about-ditching-disposable-nappies-114604

Dombroski, K. (2020). Caring labour: Redistributing care work. In J. K. Gibson-Graham & K. Dombroski (Eds.), *The handbook of diverse economies* (pp. 154–160). Edward Elgar Publishing.

Dombroski, K. (In press). *Caring for life: A postdevelopment politics of infant hygiene*. University of Minnesota Press.

Dombroski, K., Diprose, G., & Boles, I. (2019). Can the commons be temporary? The role of transitional commoning in post-quake Christchurch. *Local Environment*, *24*(4), 313–328. https://doi.org/10.1080/13549839.2019.1567480

Dombroski, K., Healy, S., & McKinnon, K. (2018). Care-full community economies. In C. Bauhardt & W. Harcourt (Eds.), *Feminist political ecology and the economics of care* (pp. 99–115). Routledge.

Dombroski, K., Waliuzzaman, S. M., Conradson, D., Diprose, G., & Healy, S. (2023). Commoning for urban wellbeing in Majority and Minority Worlds. In A. Kamp (Ed.), *Wellbeing: Global policies and perspectives* (pp. 235–252). Peter Lang.

Gaulke, L. S., Weiyang, X., Scanlon, A., Henck, A., & Hinckley, T. (2009). Evaluation criteria for implementation of a sustainable sanitation and wastewater treatment system at Jiuzhaigou National Park, Sichuan Province, China. *Environmental Management*, *45*, 93–104. http://doi.org/10.1007/s00267-009-9398-1

Gibson-Graham, J. K., Cameron, J., & Healy, S. (2013). *Take back the economy: An ethical guide for transforming our communities*. University of Minnesota Press.

Guattari, F. (1995). *Chaosmosis: An ethico-aesthetic paradigm* (P. Bains & J. Pefanis, Trans.). Indiana University Press.

Harvey, F. (2020, 21 September). World's richest 1% cause double CO2 emissions of poorest 50%, says Oxfam. *The Guardian*. www.theguardian.com/environment/2020/sep/21/worlds-richest-1-cause-double-co2-emissions-of-poorest-50-says-oxfam

Hashemi, S. (2020). Sanitation sustainability index: A pilot approach to develop a community-based indicator for evaluating sustainability of sanitation systems. *Sustainability*, *12*(17), 6937. http://doi.org/10.3390/su12176937

Hawkins, G. (2006). *The ethics of waste: How we relate to rubbish*. Rowman & Littlefield.

Iossifova, D. (2020). Urban (sanitation) transformation in China: A toilet revolution and its socio-eco-technical entanglements. In M. Keith & A. de Sousa Santos (Eds.), *Urban transformations and public health in the emergent city* (pp. 102–122). Manchester University Press.

Jewitt, S. (2011). Geographies of shit: Spatial and temporal variations in attitudes towards human waste. *Progress in Human Geography*, *35*(5), 608–626. https://doi.org/10.1177/0309132510394704
Latour, B., & Woolgar, S. (1986). *Laboratory life: The construction of scientific facts*. Princeton University Press.
Leigh Star, S. (2010). This is not a boundary object: Reflections on the origin of a concept. *Science, Technology, & Human Values*, *35*(5), 601–617. 10.1177/0162243910377624
Lopes, A. M., Healy, S., Power, E., Crabtree, L., & Gibson, K. (2018). Infrastructures of care: Opening up "home" as commons in a hot city. *Human Ecology Review*, *24*(2), 41–59. http://doi.org/10.22459/HER.24.02.2018.03
Meseldzija, J., Poznanovic, D., & Frank, R. (2013). Assessment of the differing environmental impacts between reusable and disposable diapers. *Dufferin Research*, 1–11.
O'Brien, K., Olive, R., Hsu, Y. C., Morris, L., Bell, R., & Kendall, N. (2009). *Life cycle assessment: Reusable and disposable nappies in Australia*. University of Queensland School of Engineering.
Power, E. R., & Mee, K. J. (2020). Housing: An infrastructure of care. *Housing Studies*, *35*, 484--505. https://doi.org/10.1080/02673037.2019.1612038
Shove, E. (2003). *Comfort, cleanliness and convenience: The social organization of normality*. Berg.
Sibonney, C. (2019, May 24). Elimination communication: Can infants actually be potty trained? *Today's Parent*. www.todaysparent.com/baby/baby-development/elimination-communication-can-infants-actually-be-potty-trained/
Singh, N. (2017). Becoming a commoner: The commons as sites for affective socio-nature encounters and co-becomings. *Ephemera: Theory & Politics in Organization*, *17*(4), 751–776.
Sun, M., & Rugolotto, S. (2004). Assisted infant toilet training in a Western family setting. *Journal of Developmental & Behavioral Pediatrics*, *25*(2), 99–101.
Thill, B. (2015). *Waste*. Bloomsbury.
Thompson Brewster, E., Rounsefell, B., Lin, F., Clarke, W., & O'Brien, K. R. (2022). Adult incontinence products are a larger and faster growing waste issue than disposable infant nappies (diapers) in Australia. *Waste Management*, *152*, 30–37. https://doi.org/10.1016/j.wasman.2022.07.038
Victory, E. G. (2019, December 10). Indonesian rivers are flooded with diapers because of popular myth. *Vice*. www.vice.com/en/article/7x55ky/indonesian-rivers-flooded-diapers-myth
Waliuzzaman, S. M., & Alam, A. (2022). Commoning the city for survival in urban informal settlements. *Asia Pacific Viewpoint*, *63*(1), 97–112. https://doi.org/https://doi.org/10.1111/apv.12332
Wang, X. Z., Wen, Y. B., Shang, X. P., Wang, Y. H., Li, Y. W., Li, T. F., Li, S. L., Yang, J., Liu, Y. J., Lou, X. P., Zhou, W., Li, X., Zhang, J. J., Song, C. P., Jorgensen, C. S., Rittig, S., Bauer, S., Mosiello, G., Wang, Q. W., & Wen, J. G. (2019). The influence of delay elimination communication on the prevalence of primary nocturnal enuresis – a survey from Mainland China. *Neurourology and Urodynamics*, *38*(5), 1423–1429. https://doi.org/10.1002/nau.24002
Yu, X. (2010). The treatment of night soil and waste in Modern China. In A. K. C. Leung & C. Furth (Eds.), *Health and hygiene in Chinese East Asia: Policies and publics in the long twentieth century* (pp. 51–72). Duke University Press.

PART 4

Precarious livelihoods

20

WATER REUSE IRRIGATION, GENDER, AND POVERTY INEQUALITIES IN KAFR EL SHEIKH, EGYPT

Deepa Joshi, Amina Dessouki, and Alexandra Schindler

Introduction

> Extracting, containing, channeling, processing, leaking, and distributing different kinds of water, water infrastructure produces spaces, scales, meanings, and differences among communities.[1]

In this chapter, we explore the complexity of gendered challenges in water-reuse irrigation for women farmers using polluted drainage water for agriculture in Kafr El Sheikh, a rural agricultural area located at the tail end of the North Delta canal irrigation network in Egypt. Kafr El Sheikh has experienced a water crisis for decades and related challenges are not uncommon in Egypt's "hyper-arid" climate (Mahmoud et al., 2021, p. 191), where only 4% of the country's total land area is arable (FAO, 2016). Some argue that women's increased participation in irrigated agriculture is potentially positive as it provides new opportunities for women and challenges gender norms. However, this paper, based on ethnographic fieldwork with women farmers in Kafr El Sheikh, questions that assumption and demonstrates how conditions shaped by the colonial legacy of agriculture, irrigation modernization and efficiency, and patriarchy often prevent these women from working under equal conditions.

The lack of clean water and lack of access to agricultural cooperatives and water user committees are key reasons for the physical and psychological struggles that women farmers in Kafr El Sheikh suffer while trying to make a living under these challenging circumstances. The drainage water farmers have access to is polluted and its quality is well below the government's regulatory standards (see Table 20.1) (Barnes, 2014b, p. 186). While better quality irrigation water is central to the needs of many women farmers, it is essential to look at the difficult social, economic, and ecological conditions under which these women increasingly engage in agricultural work. These women's struggles to access irrigation water cannot be separated from the pressure of their everyday lives, which includes unpaid domestic work, health concerns, food insecurity, and experiences of gendered inequality.

The focus on irrigation efficiency, including the reuse of polluted drainage water for irrigation in Egypt stems from a colonial history of cotton cultivation in a dry, arid landscape. The oil boom in the Gulf regions in the 1970s, followed by other political events and

DOI: 10.4324/9781003100379-25

Table 20.1 Water quality in the Nile River and drains through Upper Egypt (between Aswan and Cairo).

	Salinity (ppm)	*Biological oxygen demand (mg/l)*	*Chemical oxygen demand (mg/l)*	*Fecal coliform bacteria count (MPN/100 ml)*
Nile River				
Range	169–240	1–5	5–24	50–3000
Average	203	2	13	608
Drains				
Range	200–1430	1–43	2–144	250–35,000
Average	565	7	33	8023
Standard set in Law No. 48 (1982) on the Protection of the Nile	500	6	10	1000

This data (for the year 2001) comes from the National Water Research Center, cited in Brown et al. (2003: 11 & 15). The report does not include data on water quality in the drains of the delta (Lower Egypt). So, in order to allow comparison between the river and the drains, I only include here data for Upper Egypt, between Aswan and Cairo. Since water quality deteriorates markedly through the delta, was that data to be available, the figures would illustrate an even more pronounced deviation of water quality from the standards.

Source: Barnes (2014b).

socio-economic challenges since then have resulted in a slow but steady male outmigration to larger Egyptian cities and neighboring countries for work, resulting in women's increased roles in agriculture. However, these changes are not complemented by an increase in women's visibility, voice, and participation in water user associations (WUAs) or local agricultural cooperatives (LACs) – institutions that are still controlled by men – a contradiction that the women farmers we spoke to in Kafr El-Sheikh emphasize. Women's increasing participation in agriculture happens in parallel with the growing shortages in the water supply due to climate change (Najjar, 2015). Women are thus disproportionately impacted by water scarcity because they are the ones left behind to manage both domestic and productive responsibilities that include agriculture and irrigation, despite their limited access to institutions and services.

In recent years, several initiatives have supported and encouraged women's engagement in irrigated agriculture. This includes enabling women's ownership of land (specifically in New Land, which refer to desert lands that were reclaimed by the Mubarak Resettlement Scheme in the 1990s), encouraging their engagement in irrigation management at the community level, and improved women's access to modern "irrigation technologies such as drip, sprinkler and *tatweer*"[2] (Najjar et al., 2019, p. 291). Because agriculture and irrigation, particularly in the Middle East and North Africa (MENA) regions are strongly patriarchal, these developments are seen as providing new opportunities for women and challenging gender norms. Current irrigation strategies in Egypt call for more "gender integration"[3] in irrigation interventions and see these as "opportunities for loosening of social norms" and for women's empowerment (Ibid, p. 309).

However, Lila Abu-Lughod (2008, 2011, 2013), writing about women in the MENA region, questions these narratives and measures of empowerment, centered on women's rights to land in relation to their engagement in productive work. Abu-Lughod argues that these approaches fail to adequately capture the cultural, political, and economic complexities that frame the changes in the lives of women. Feminist researchers Ahlers and Zwarteveen (2009, p. 409) have also cautioned against simply seeing mainstreaming women in water management as "offering possibilities for confronting" deeply entrenched gender inequalities. In many rural areas of Egypt, where men are forced to migrate for more profitable work, the question to ask is: what do women *gain* from engaging in agriculture when the risks are high and the gains are low? To what extent are poor, marginalized women able to navigate increased work in irrigated agriculture in highly gendered social and cultural contexts? And to what extent does the emphasis and investment in irrigation happen at the expense of other essential needs, such as education, health, and gender equality? (FAO, 2016).

Our findings show that women's increasing engagement in agriculture and irrigation and a contradictory inability to challenge the realities that shape their lives is only fully understood by mapping the "wider political-economic structures and historical dynamics" of irrigation that are shaped by colonialism, capitalism, and patriarchy (Ahlers & Zwarteveen, 2009, p. 409). By focusing on embodied, individual experiences of women's engagement in polluted drainage water irrigation and by discussing how the colonial legacy of agriculture, irrigation modernization, and efficiency co-exist with environmental and economic crises, this chapter addresses a contradiction in existing literature on women's involvement in irrigated agriculture in rural Egypt (Barnes, 2014b; Molle & Rap, 2013 Najjar et al., 2019; Sukkary-Stolba, 1985; Zwarteveen, 2008, 2011).

Methodology

This chapter is based on ethnographic research with mostly women farmers and irrigators using "drainage water" for agriculture in Kafr El Sheikh, a rural area located 134 km north of Cairo in the north Nile Delta area.[4] While the research project was initially focused on the impact of COVID-19 on the gendered dimensions of food systems, we soon became aware that the issues that female irrigators raised were indicative of much more complex and deeply gendered challenges in polluted drainage water irrigation in Egypt. This research takes an intersectional approach to gender and considers the complexity of individual experience and socio-historical factors that affect the women farmers' lives.

We conducted preliminary discussions and focus groups with an equal number of men and women (170 total) living in Kafr El Sheikh and irrigating agricultural plots at the head, middle, and tail regions of the Kitchener drain or Drain 7 (see Figure 20.1). From this group of respondents, we selected ten farmers (nine women and one man) with whom to conduct one-on-one semi-structured interviews to understand the everyday experiences of farming and irrigation in more depth.[5] With these ten respondents, our focus was to get a more comprehensive picture of their lives at home and in the fields, how they engaged with (or not) agricultural committees and water user associations, and their abilities to negotiate for improved water supply and more equitable agricultural practices.[6]

Feminist political ecology – contextualizing and understanding gender-power relations

General development policy recommendations in Egypt emphasize the importance of engaging women in water reuse governance and management. These policies include the Egypt Sustainable Agricultural Development Strategy towards 2030 (SADS), which was published in 2009 by the Ministry of Agriculture and Land Reclamation (MALR) and the National Strategy for the Empowerment of Egyptian Women 2030, which was published by the National Council for Women (MALR, 2009; NCW, 2017).

ReWater MENA[7] is the project under which the research for this chapter was carried out. The project aims to help increase the safe reuse of water in the Middle East and North Africa (MENA), to address obstacles to reuse in the region, and to promote safe reuse practices that improve food safety, health, and livelihoods. ReWater research identified that "in Egypt, women are increasingly engaging in irrigated agriculture, but are often unable to engage in water management strategies at local and policy levels" (Mateo-Sagasta, 2022, p. 2). This lack of engagement at the policy level is because women face different limitations.[8]

This chapter follows in the footsteps of feminist scholars who question initiatives that focus on simply "engaging women." This includes the tendency to interpret "gender" as "women," assuming simplistic female-male binaries, or planning to remedy these inequalities through technocratic projects gathering sex-disaggregated statistics, or by targeting women as uniquely able to engage in and benefit from productive labor markets (Cornwall

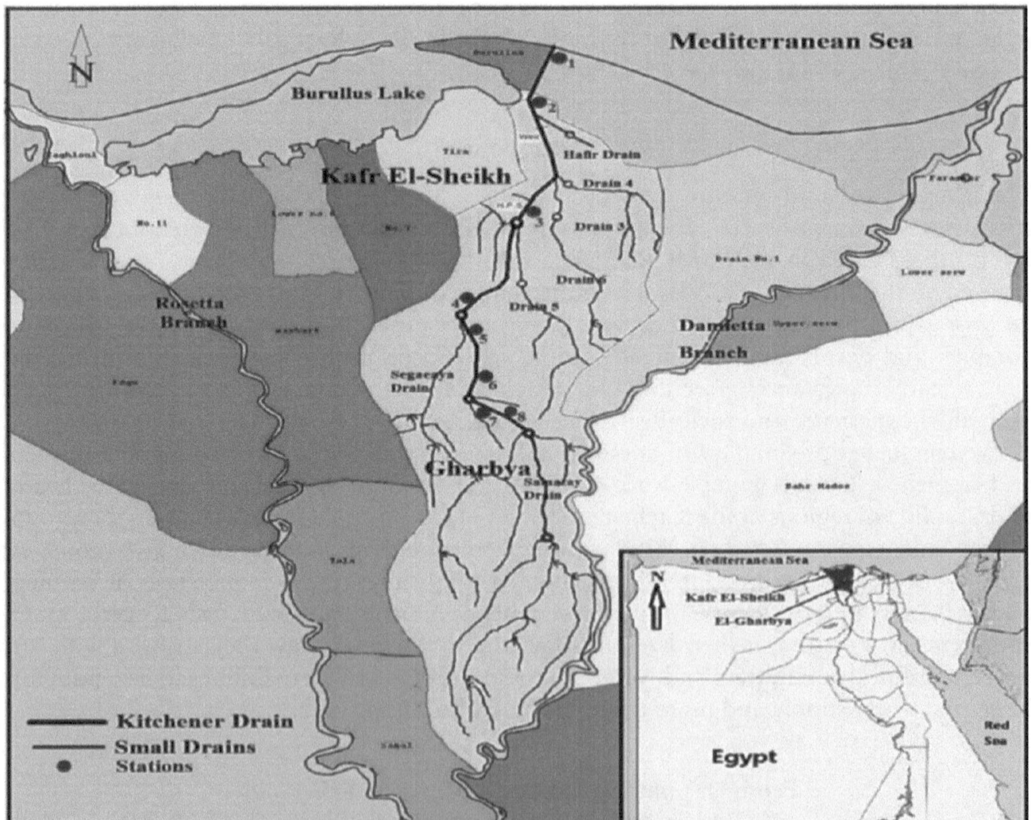

Figure 20.1 Map of Nile Delta showing sampling sites along Kitchener drain.

Source: El-Amier et al. (2018, p. 252).

& Rivas, 2015). In practice, these ways of "doing gender" rarely account for the complexities of deep-rooted intersectional inequalities. In addition, these initiatives rarely take into consideration the enormous and unaccounted-for burden that domestic work plays in many working women's lives (Reddy et al., 2021). Productive work adds to women's domestic and care burdens which are often solely women's responsibility, and are unaccounted for economically (Chant, 2008; Cornwall & Rivas, 2015; Leach et al., 2016). Standalone technical interventions do not easily enable the transformation of unequal gender relations that severely constrain women's opportunities, from the household up to the formal and informal spaces of governance of agriculture and irrigation (Kabeer, 1994).

A feminist political ecology framework provides a gender lens to this analysis by giving voice and visibility to the everyday experiences of marginalized women – in this case, highlighting the most marginalized farmers irrigating with drainage (reuse) water and connecting these narratives to the politics of irrigation efficiency in Egypt. It allows us to explore how gender, in relation to class, health, land ownership, and other relevant axes of power, shapes access to and control over land, water, and related institutions, and how these axes of power are situated in intertwined histories of colonialism, patriarchy, and capitalism. Viewed with this lens, the objective to integrate women in reuse irrigation without adequately understanding the complex social differences that play into the politics of water decision-making, management, and governance is considered reductionist. This gendered lens partially explains why extensive efforts, investments, and technological interventions have not yet solved the problem of a deepening water crisis in Egypt and how these challenges have increasingly been offloaded on marginalized women.

In feminist thinking, the recognition of women as active stakeholders needs to be seen through the possibility of women's agency (Arora-Jonsson, 2014; Bell, 2016). This requires understanding which women can and will find a place and voice in water governance, how deeply entrenched gendered norms shaped by patriarchy can be tackled at scale, and above all how externally predetermined agendas such as improving water efficiency have an impact on diverse groups of women. Without unpacking these complexities, it is deeply problematic to assume that (all) women can be easily engaged as active and functional water managers (De La Torre-Castro, 2019; Rao et al., 2019; Ravera et al., 2016; Rist, 2007). To ensure this, marginal voices need to be heard, and the everyday experiences of marginalized groups made visible and prominent to those writing policies and designing interventions. This requires paying attention to the nuanced and embodied gender-power dynamics present in everyday lived experiences (Elmhirst, 2015; Bell, 2016; Sundberg, 2017).

A brief history of water reuse irrigation in Egypt

The traditional method of irrigation in the Nile flood plains was by flooding and submersion. Starting in 1820, there was a historic shift in irrigation in the Nile floodplains. Extensive networks of irrigation canals established perennial irrigation to support extensive cotton, and later sugarcane production (Molle & Rap, 2013). Investments in irrigation development benefited colonial and local elites, who were given rights to reclaim and irrigate land in the Nile Delta region.

Later, with the construction and further development of the Aswan Dam, "85% of the cultivable agricultural land had perennial irrigation" (Molle & Rap, 2013). This made irrigation the largest consumer of water in a predominantly dry and desert environment and put the irrigation demand on total water at 86% (FAO, 2016). The extensive perennial irrigation necessitated a drainage system (El Quosy, 2017). Until the 1920s–30s, "freshwater was the only source of irrigation" (FAO, 2016, p. 11) and a network of drains were built essentially to "de-water" the irrigation command areas (Molle & Rap, 2013, p. 8). However, a dry period in 1928 resulted in the construction of the first pumping station and the reuse of drainage water for irrigation (FAO, 2016, p. 11). Water flowing out through the drainage systems was seen as a "major cause of irrigation efficiency" resulting eventually in an extensive network of drains and pumping stations (Mahmoud et al., 2021, p. 119) (see Figure 20.2). In the face of increasing demand for water and limited water supply, the emphasis on improving the reuse of drainage water has continued and is pivotal to the national priority of "irrigation efficiency."

Irrigation intensification in Egypt – initially for cotton and now for other niche crops including wheat, rice, sugarcane, and sugar beet – is said to be extremely efficient (around 85%). This is achieved by a parallel system of canal (fresh) and drainage (reuse) networks which allows for "water reuse" three to four times through an extensive network of drainage canals. However, this efficiency happens at a significant cost to the ecology and to those irrigating at the tail ends of the system, as we will see with the women farmers living at the tail end of Drain 7. The improvement and extension of irrigation canals and drainage continues, but poor management of these systems causes unreliability of water supply, and pollution from domestic and industrial polluted drainage water, making farming a challenging occupation (Molle et al., 2015). This is especially true in "downstream" areas where the reliance on drainage water is high. This causes health risks to both irrigators as well as domestic livestock.

These challenges have resulted in the increasing practice of smallholders owning land renting out their small plots of land to landless tenant farmers. Officially, on paper, tenant farmers are mostly men. But increasingly, it is women who actually farm these small plots of land (FAO, 2016). In Kafr El Sheikh, it is reported that of those who are primarily responsible for agriculture and irrigation, 78% are women. However, a very low percentage of these women actually own the lands or are able to participate in irrigation management (Najjar et al., 2019). Egyptian women lack visibility in the agricultural sector and have almost no influence in formal decision-making structures. As El Khorazaty argues, "despite the reported success of WUAs in improving on-farm water-user efficiency, and distribution of irrigation water while reducing the cost of irrigation, the irrigation needs of women were often subordinated to male interests" (El Khorazaty, 2021, p. 41).

Barnes (2014a) identifies that irrigation is a deeply political and embodied experience in Egypt. Understanding these issues requires analyzing "those whose lives are directly or indirectly tied to agricultural production . . . and who [amongst them] is able to access [what types of] water" (p. xii). Irrigation in Egypt is as much about the people who access the water and farm the land as it is about flows of water and water scarcity. Both issues have major implications for inequality in the access to, use of, and returns from water reuse irrigation. Barnes observes:

> [I]n 'international donors' efforts to forge a participatory management regime through the establishment of water user associations along the irrigation [and drainage] canals – the tensions, contradictions, risks, challenges, and emotional anxiety of those who struggle to irrigate with inadequate, poor-quality water remain largely invisible.
>
> *(p. xiii)*

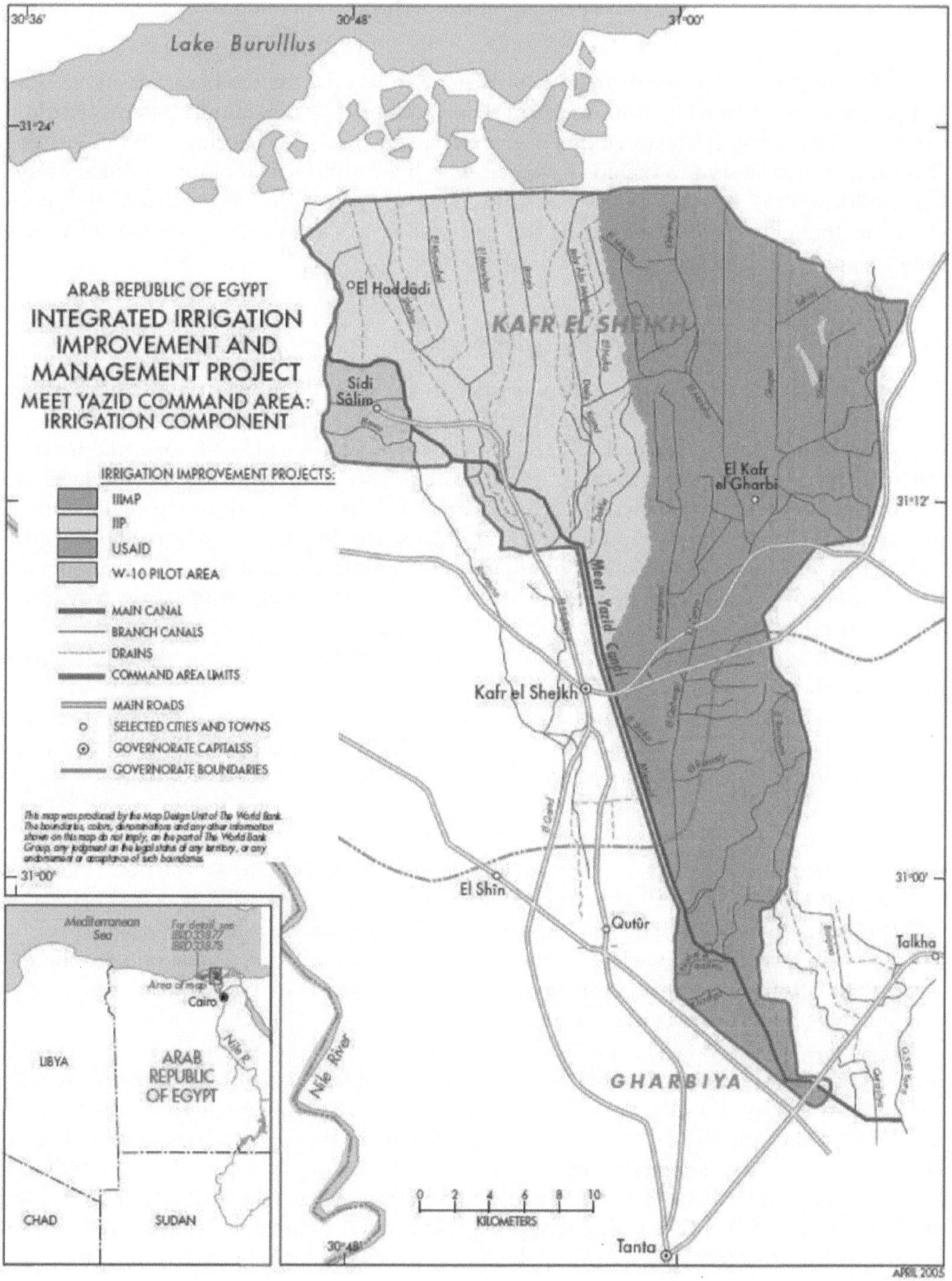

Figure 20.2 Kafr El Sheikh irrigation.
Source: Molle et al. (2015, p. 10).

Gender-power inequalities in land ownership, governance, and domestic labor

In the following section, we hear how, in their own words, the women farmers navigate complex and interrelated circumstances in their attempts to 1) sustain irrigated agriculture livelihoods, 2) access clean water, and 3) negotiate social relations at home and in farming. Our sample of case studies included women who lived and farmed alone as single headed households, others as wives. A few had land ownership through their husbands or inheritance, but the majority farming at the tail end of Drain 7 are landless tenants engaged in sharecropping agreements.

Land: ownership, sharecropping, and cultivation

The first issue that these farmers face is how to make a living by irrigated agriculture. A farmer's sense of security and well-being is usually tied to land ownership. However, given the specificities of Kafr El Sheikh's historical, ecological, and social context, to inherit a piece of land here is both an advantage and a burden because the land can function as one's material means of survival as well as one's impoverishment. On the one hand, it is their only source of livelihood because they have few job opportunities outside of farming. On the other hand, the environmental and economic constraints that accompany land ownership result in high operational costs and barely any gains. Women farmers in Kafr El Sheikh deal with even more difficult circumstances as it is much harder and rarer for them to own the land and they are also unable to participate in irrigation management.

Mona,[9] who is 28 years old and got married at 19, works on her husband's three-acre agricultural plot, which he rents on a one-quarter share-crop agreement. They have access only to drainage water as the source of irrigation. As she explains,

> The money from the crops goes mostly back into the land but the money from the milk is used to cover house expenses. This month we had to pay for just one month 320 EGP (20.47 USD)[10] for the water supply at home. I swear yes it was shocking! 320 pounds! We didn't have any money from the crop so we asked the water collector to wait a week until we had money. We don't even use water but sometimes there are fluctuations in prices and things get more expensive, so we have to pay more even though we don't use that much water.

As Mona explains, farming does not create an income, and instead often puts them into debt.

Fatima is a 30-year-old farmer who is married with two young children. Her husband was 75 and she was 17 when they got married. Her husband, who rents one feddan of land on a one-quarter sharecropping arrangement, is too old to work. Fatima handles all the agriculture and irrigation work in addition to her domestic responsibilities. As the only working member in her family, her workload is excessive. Fatima occasionally sells a small portion of the crops from her share and the family consumes the rest, even though they wish they could do otherwise. Her source of daily income is the cow owned by her landowner. She gets to keep the milk and he takes any newborn calves. Fatima cares for her 88-year-old sick husband and her son, who also had health issues. She herself has bone marrow cancer but cannot afford additional medical bills.

Fatima says that the produce from farming:

> Barely covers its own costs, it barely covers the food we eat. Nothing extra at all, I cannot even afford to take transportation to the Cancer institute to present my papers so I can continue getting treatment. . . . My husband is incapacitated and can't even work, as you can see he just sits there. He gets his pension of 1000 EGP (63.98 USD) but it doesn't cover much. Electricity, water supply, gas, we don't have – we do not make any money to pay for these.

In Kafr El-Sheikh, small plots of land of one to three feddans (or acres), are rented from landowners on a sharecropping basis. The tenants pay their rent in crops and the landowner gets three quarters of their yield while they get one quarter. This agreement requires the tenant farmers to provide labor, whether their own or supported by neighbors, although tenant farmers are rarely able to hire paid labor. The landlord alone has access to the WACs and LACs. Except in the case of single women-headed households, sharecropping agreements are made between men (landlords and tenants). Patrilineal inheritance limits women's control over contested resources such as arable land, water, and their own mobility.

However, everyday engagements with the landlords are often between women who work the fields, as in the cases of Fatima and Aliah Salama, both women who married as teenagers and whose husbands are unwell and unable to farm. As Fatima explains:

> The landowner tells me what to plant and when. The landowner is responsible for chemical fertilizers, what to plant, and irrigation. He tells me when to irrigate. He gets me the fertilizer and tells me how much to use. I do all the work, but he decides on everything.

Both Fatima and Aliah, although they do all the work in their fields, do not have control over what they grow and how they grow it.

Even when women do own land and hold titles, they still do not have easy access to local governance systems and decision making around the irrigation and cropping issues because of gendered norms about participation in these systems. When Aliah was asked if the land she worked on was her property and under her name she replied: "Yes, I bought it with a contract to my name, but in the Agrarian Reform Authority it is still under the previous owner's name. I bought it from him, but he gets me the agricultural inputs for me from the cooperative." As in Aliah's case and others, owning land does not always permit one to choose what crops to grow and when. On land distributed by the government after 1952, our observation was that despite policy to the contrary, local agricultural collectives still dictate what crops will be grown.

Water: inequality, toxicity, and endurance

The second issue that the farmers face is lack of access to clean water. Along the canal (freshwater) and drainage (reuse water) networks, farmers manipulate the flow of water through informal interventions to improve their access to water using diesel pumps. This leads to physical challenges and social conflicts for those at the tail end of canals, such as Drain 7. Opportunities for improved incomes, food security, and safety for the most marginalized tenant farmers, particularly these women who cultivate the least productive of agricultural plots located at the tail end of the canals, are slim.

Along Drain 7, the poor quality of water for irrigation was identified as a challenge by all women: they recognize its impact on crop productivity, incomes, their health, and the health of their animals. As Ashar explained, "We have to bear it, sometimes the water is blue, green or white. It smells bad. The water is bad all year round. For those with only drainage water access, the yields can be almost ¾ less than the normal productivity." For this reason, except for the most marginalized of households, the practice is to sell what is cultivated at low prices and pay higher prices for similar crops grown elsewhere with clean water.

Aliah swears that she will not consume what she grows, but she is unable to afford the cost transactions.

> The regulated market price for wheat was 720 EGP (46.15 USD) per bag, but guess how much I sold it for, 580 EGP (37.17 USD), because it is from the bad water. I bought wheat which I know is grown with clean water for 830 EGP (53.2 USD). So, I am at a loss. That's why I put all my effort into getting clean water.

She added, "These days I don't make any profit, I actually pay from my pocket. The bag of chemicals costs 500 EGP (32 USD), pesticides are very expensive, prices are burning up. Any land nowadays takes and doesn't give back." Ironically, for all the official investments in irrigation and the work and labor of these women, there is little income from agriculture for the poorest sharecroppers.

Health concerns were raised repeatedly by the women because living at the tail end of the drain means they get water that is polluted for most of the year. As Mona explains:

> We haven't eaten our own crops for the past 20 years. The amount of liver failure and kidney failure and cancer has increased because of what we eat. We live off drainage water. We try to sell our crops in the cities and buy cleaner food but it's expensive. I hope that my land gets clean water.

Many of the women recounted the health issues they and their families suffered. Their bodies, already overworked and exhausted from farming and domestic care workloads, are further taxed by their bodies' responses to the polluted water.

Tenant as well as land owning cultivators have few avenues to make their concerns heard and addressed. Women especially do not have any spaces in which they can participate: they are not represented, at least not effectively, in the WUAs, which determine distribution and maintenance, protection, and upgrading of physical equipment. They are entirely reliant on others organizing and arranging the distribution and maintenance of water. Significantly, in focus group discussions, when we asked the women if they would be interested in becoming members of the agricultural cooperatives and user associations, they unanimously answered in the negative. They are well aware of the gendered politics of these institutions and recognize their limitations. And more practically, none of them feel they have the time because of their domestic care workload.

The significant challenges of farming with low quality and polluted water came through clearly in the interviews, but so did the determination of the women to improve their lives for themselves and their families. These women continue to work hard, especially to ensure that their children can get an education and escape a lifetime of farm work in this polluted environment.

Gendered productive and domestic work

The third issue is the societal and cultural expectation that women farmers perform domestic care work alongside farming. While women's shifting roles in agriculture and irrigation in Egypt can be seen as transcending the sociocultural and gender boundaries imposed on them culturally, our research provides evidence of the conditions experienced by the most marginalized women farmers: lack of proper irrigation infrastructure and basic social services, economic hardships, and poor prospects or opportunities for other income. Women deal with the challenge of organizing and managing uncompensated domestic care work as well as experience of increasing demands on their time and labor for productive work. Domestic work is affective and elemental to well-being and livability (Gutiérrez-Rodríguez, 2010, p. 6), and yet, domestic work is perceived as banal and linked to normalized maternal instincts, even though it plays an integral role in the production process. To date, very few analyses of gender, agriculture and irrigation in Egypt speak about domestic work (El Khorazaty, 2021). The focus is usually on women's increasing engagements in productive work.

And yet, when women spoke of their experiences, most of them expressed that they spend an average of six hours per day on household work: taking care of children, extended family and animals, preparing meals for daily laborers working on their land, attending social obligations, producing/marketing dairy products, washing clothes and dishes, cleaning, and doing farm-work. Nahar, a woman in her mid-40s who works as a farmer alongside her husband, also needs to cook and feed hired labor contracted at specific times of production:

> Women here are heavily pressured. You could have 40 people working to harvest cotton and beetroot, you have to feed all of them, sometimes three times per day. The men go out to the farm early, you would go give them their breakfast there.

Such long hours of heavy work come at a high cost to women, and yet there is little attention to women's need for sleep or relaxation, and their well-being is absent from the equation.

These women, in a literal sense, embody work. As Mona says, "You have to be at the farm by 6 am to harvest. You start the day at dawn. You milk the cows, feed your children, feed the birds, make breakfast." Mona returns home at midday to make lunch, to check on the livestock, and then goes back to the farm.

> I work on the farm until sundown. After sundown, I go back home to make dinner. Then I tend to the animals till dark, and when it's dark I finally eat but I have to clean the dishes then sleep. It's hectic, I have to run around all day to finish everything.

As Mona explains, her work doesn't stop when she returns home from the farm, as she is also responsible for all the food and cleaning in her family home.

In this socio-cultural context, it is taken for granted that it is a woman's role to do the work of social reproduction and domestic work is kept outside the framework of recognition. Hiring external help entails setting a monetary value for the work of social reproduction, which goes against the cultural norm that recognizes this labor as "non-productive." As one of the male farmers expressed in one of our focus group discussions: "I would rather marry a second wife than hire external help for my household." This response epitomizes the patriarchal hierarchies and gendered norms embedded within families.

Given such binding responsibilities at home, most of the women we met do not wish to work in the fields, and especially in such difficult circumstances. Yet, our conversations with Mona, who finished her education but was forced to do farm work despite her parents' offer to cover costs of extra laborers, show how women are well conditioned to be compliant wives. As she explained, "I never wanted to be a farmer but my mother-in-law forced me to start working 14 days after I got married. I started working on the farm while I was pregnant, a few months after getting married." Mona's late mother had offered to pay for external labor, especially during her pregnancy, but Mona said she had to support her husband:

> I am now living in his family home, and I have to abide by their rules. I take care of my mother-in-law and I help out my husband on the farm. I pray every day, and hope to always be kind even to the people who do not do right by me. Even though I wish I could do something less exhausting, I feel the people in my community respect me for supporting my husband and working on the farm. Being a farmer is not looked down upon, even though I feel that this is not the life I deserve.

These women perform the identities that are associated with their gender role because it is what is culturally acceptable within their given community. Mona chooses to do farm work, despite having a secondary diploma, because she believes that farming is necessary to attain social membership. Mona describes her experience within a social structure that is (re)produced by "respecting the collective rhythm" of the group (Bourdieu, 1977, p. 162). Women are also present on the farm with their husbands to perform their social belonging. In so doing, they form networks of mutual cooperation with their neighbors which is the basic social fabric on which this community is built. Farming not only makes Mona a proper wife, it also makes her a proper neighbor as she fulfills her social obligation towards her community. Although she feels that this is not the life that she deserves for someone with a university degree, bearing the symbolic and physical violence that farming (with polluted drainage water) inscribes on her body and soul seems more practical than living as a social outcast.

Conversely, Dina, a woman farmer whose husband is a teacher, experiences a different reality. Dina dreams of finding a job and utilizing her university degree, but she could not find one. Unlike Mona, her husband does not stand in her way if she can find a job.

> I wanted to be employed but I couldn't find a job. I didn't used to work but now I have to, it is very hard, but I have to bear it for the kids. The house has needs, and we need more income, so I decided to do this. I told my husband why don't I farm a piece of land to help out with the house and kids? The owner had a farmer before us who quit and he randomly told my husband around the same time that he was looking for someone else to take his place, praise god for this coincidence. And so we did. I wish I had any other option. I would rather work in anything else. If there is any other good job I would do it, but I cannot find any. I work to help the household.

Mona and Dina are neighbors, they live two houses away from each other. Yet their realities, familial relationships, sentiments about the present, and prospects for the future are different. These two women's similar yet different experiences highlight the diversity of human experience, needs, and circumstances, even when performing similar roles (or subjectivities) as women farmers.

The diversity of experiences of these women exemplifies what Lila Abu-Lughod (2013) explains as the failure in development design to adequately reflect the deeply embodied and diverse realities of women's lives. Even when women live in proximity and share similar realities, the ability to navigate, adapt to and/or resist the structures that shape their life are different, as are the choices they make, which are influenced by a multitude of factors. For this reason, it is significant to observe that top-down interventions designed to empower women by integrating them into irrigation governance often misrepresent the complexities of women's lives and work and the choices they make around these issues. An in-depth understanding of the gendered dimensions of agriculture and irrigation requires not generalization, but a more nuanced analysis of the dailiness and particularities of people's struggles, which are often in a state of incoherent flux.

Conclusion

While agricultural drainage water is offered as a sustainable solution for irrigation efficiency, it can also (as we have seen with the farmers living on Drain 7 in Kafr El Sheikh) deeply impact farmers' well-being. The focus and emphasis made on integrating women into water governance happens without critically analyzing the wider context of why marginalized women increasingly engaged in drainage irrigation are struggling to navigate the predominantly masculine settings of irrigation and agriculture management, governance, and markets. There is very little attention to how and in what ways these women do or do not benefit from growing crops they mainly do not consume; from laboring to grow crops that have low yields and command low prices, or from being increasingly engaged in productive work in contexts where they have little to no access to decision making about the land, irrigated water, and overall land use.

How do women perceive this work which holds such high risks and offers so few gains? In our research, we noted that there was significant ambiguity in how women reacted to irrigating with drainage water. Some women resented this work, others saw it as their social obligation, and some even felt a sense of pride in the contribution they were able to make to their families. We noticed that the ambiguity was also because the experience of irrigation and agriculture work is not homogenous among all women; rather it is shaped by intersectional inequalities. Marital status and relationships as well as the health of their husbands and themselves, the support of children, other family members, and neighbors are significant, as is ownership of land and the position of their farms in the irrigation network. Regardless of individual experiences, it was clear that women struggle to manage domestic and productive work, extremely heavy workloads, and poor health, which they perceive to be caused by poor water quality.

The popular tendency is to view any productive work for poor women as indicative of their empowerment (Scott, 2016). However, women's productive work, as we saw along the tail ends of Drain 7, usually does not "confront [but rather reiterates] the structures of inequality and patriarchy, which are the grim conditions under which [poor] women exist" (Chakravarti, 2008, p. 10). In the recent past, Egypt has experienced "new and deep structural threats, which have converged into cycles of poverty, inequality and vulnerability" resulting in significant economic and food security hardships and "the inability to meet basic household needs" (Khouri, 2019). It is precisely these issues that push the most marginalized women from landless households to take on any work, including water-reuse irrigated agriculture in drainage locations.

Our findings support the implementation of national initiatives such as Heya Karima,[11] or the Decent Life Initiative, which aims to assist Egyptians living in poverty in rural areas with housing, medical and educational services, economic empowerment, and social and environmental interventions. The initiative's priority is to focus on Egypt's poorest communities, such as these women farmers in Kafr El-Sheikh, who do not have access to the means to improve their living situation. We argue here that alongside improved irrigation technology, interventions are urgently needed to improve the well-being, dignity, and rights of marginalized irrigators at the tail ends of the water-reuse systems.

Viewed from a feminist political ecology lens, this situation calls for a more careful consideration of current recommendations "to engage women" in water reuse irrigation. A bottom-up approach would allow investments, innovations, and interventions along the water-reuse irrigation value chain, with a more deliberate focus on the everyday experiences of marginalized smallholders and tenant farmers – including women. More practically, this could include the design and implementation of regulatory norms and technical interventions that ensure a more equitable distribution of freshwater that would enable less punitive sharecropping arrangements, including providing the tenant farmers with some autonomy in crop choices and more direct access for women to agriculture subsidies, credits, and extension. Integrating women in water user associations and agriculture committees will only work to women's benefit if there are structural changes in these institutions so that they are inherently more inclusive and driven by grounded realities concerning the control over crops, cropping cycles, and the recognition of who is the "real farmer."

Finally, it is quite clear that gender gains or empowerment will require looking beyond irrigation to explore more proportionate investments in social protection and social welfare for the most marginalized households, services that have been sidelined because of the massive investments in making irrigation more efficient. At the heart of these challenges is the political economy of irrigation and agriculture in Egypt, which is rooted in the production of particular crops and certain industries, as well as the wider Nile Basin politics. Until these are understood to be part of the wider context that shapes women's gendered dimensions of food and water systems, focusing on single issues like women's engagement in water management locally will do little to change the landscape for these women farmers.

Acknowledgement

We kindly acknowledge Dr Tina Wallace for expert review and editorial support.

Notes

1 Acevedo-Guerrero (2018, p. 1).

2 Tatweer is an irrigation system that enables "the distribution of water to agricultural land via a network of lined canals, thereby reducing water loss and ensuring more equitable distribution of water to farmers" (Najjar et al., 2019, p. 293).

3 See ReWater MENA project: https://rewater-mena.iwmi.org/.

4 The Kitchener drain (or Drain 7) in Kafr El Sheikh, where this research was conducted, is one of the most severely polluted drainage canals in Egypt and also the main source of irrigation water for smallholder and renter farmers. The area is a resource poor area where livelihood opportunities are severely limited and irrigated wastewater farming is essential to family incomes and food security. Policy regulations that clearly prohibit the use of untreated water are difficult to enforce here, as it is the main source of irrigation for approximately 460,000 feddans of agricultural land. Around 50 industrial facilities pour their waste into this drain, which also collects "domestic wastewater (poorly treated and/or untreated); uncontrolled municipal solid waste disposal into and along the banks of the drain; industrial wastewater discharges; and agricultural runoff (MWRI, 2018, p. 4).

5 The ten individuals selected were from different age groups and experienced different levels of poverty and vulnerability – both in the household and in relation to irrigation. They were selected through community leaders' guidance and based on their answers to a questionnaire during focus groups to represent a diversity of backgrounds and willingness to participate in multiple semi-structured interviews in their homes. These case studies were based on nine field visits to the research location, both in the summer and winter, by a team of IWMI and local researchers from Kafr El Sheikh University. The focus group discussions, questionnaires, and interviews were conducted in Arabic and the research was always done with prior organization and information.

6 COVID-19 impacted the drainage irrigation agri-markets and value chains in Egypt through imposed quarantines, partial lockdowns, movement restrictions and social distancing (Selim & Eltarabily, 2022). COVID-19 also impacted the nature and quality of the primary research, as all research was conducted during the pandemic. We had to limit the visits to a bare minimum, and all meetings were conducted outdoors, following official protocols. Our initial plan was to establish contact and familiarity with the individual case study respondents in person, and to then continue the discussions via mobile phones. However, this did not work in practice. Not everyone had a mobile phone, and if they did, the quality of conversation was poor. It became clear very quickly from the 15 focus group discussions conducted, as well as individual interviews with nine women and one man, that while there are many technical issues around the flows of the water, the daily lives of women farmers are shaped by far more than just the technicalities of irrigation.

7 ReWater MENA is a project run by the International Water Management Institute (IWMI) and funded by SIDA (www.sida.se/en).

8 One of the biggest disadvantages for female farmers is limited access to land rights. Land ownership for women in Kafr El Sheikh is around 2–6%, which is very low compared to other areas known as New Lands, where land ownership for women is around 20%. The World Food Program (WFP) provided food aid to the farmers in the New Lands on the condition that one-fifth of the land titles were given to women and that women were given leadership positions in water user associations (WUAs) and local agricultural cooperatives (LACs). The Old Lands, on the other hand, have been settled on and farmed for thousands of years and prevailing gender norms prevent an increase in land ownership and irrigation management for women. Old standards determine women's standing and role in these communities. Socio-cultural traditions, which are often supported by Shari'a law, continue to have a detrimental effect on women's empowerment, especially in terms of attaining inherited property ownership. Shari'a law dictates that men should inherit double the amount of land as women, so usually men inherit two thirds of the land, while women get one third. This is one of the reasons that Kafr El Sheikh has a low percentage of women who own land as well as a low rate of participation in WUAs and LACs. (Najjar et al., 2019, p. 293).

9 All names have been changed.

10 All conversions to U.S. dollars reflect the exchange rate at the time, not the current exchange rate.

11 Decent Life Initiative: www.hayakarima.com/index.html.

References

Abu-Lughod, L. (2008). *Writing women's worlds*. University of California Press.

Abu-Lughod, L. (2013). *Do Muslim women need saving*? Harvard University Press.

Abu-Lughod, L., & El-Mahdi, R. (2011). Beyond the women question in the Egyptian revolution. *Feminist Studies*, *37*(3), 683–691.

Acevedo-Guerrero, T. (2018). Water infrastructure: A terrain for studying nonhuman agency, power relations, and socioeconomic change. *WIREs Water*, *5*(5).

Ahlers, R., & Zwarteveen, M. (2009). The water question in feminism: Water control and gender inequities in a neo-liberal era. *Gender, Place & Culture*, *16*(4), 409–426. https://doi.org/10.1080/09663690903003926

Arora-Jonsson, S. (2014). Forty years of gender research and environmental policy: Where do we stand? *Women's Studies International Forum*, *47*(B), 295–308. https://doi.org/10.1016/j.wsif.2014.02.009

Barnes, J. (2014a). *Cultivating the nile*. Duke University Press.

Barnes, J. (2014b). Mixing waters: The reuse of agricultural drainage water in Egypt. *Geoforum*, *57*, 181–191. https://doi.org/10.1016/j.geoforum.2012.11.019

Bell, D. (2016). *Reordering the world*. Princeton University Press.

Bourdieu, P. (1977). *Outline of a theory of practice*. Cambridge University Press.

Chakravarti, U. (2008). Beyond the mantra of empowerment: Time to return to poverty, violence and struggle. *Institute of Development Studies Bulletin*, *39*(6), 10–17. https://doi.org/10.1111/j.1759-5436.2008.tb00506.x

Chant, S. (2008). The "feminisation of poverty" and the "feminisation" of anti-poverty programmes: Room for revision? *The Journal of Development Studies*, *44*(2), 165–197. https://doi.org/10.1080/00220380701789810

Cornwall, A., & Rivas, A. M. (2015). From "gender equality" and "women's empowerment" to global justice: Reclaiming a transformative agenda for gender and development. *Third World Quarterly*, *36*(2), 396–415. https://doi.org/10.1080/01436597.2015.1013341

De la Torre-Castro, M. (2019). Inclusive management through gender consideration in small-scale fisheries: The why and the how. *Frontiers in Marine Science*, *6*(156). https://doi.org/10.3389/fmars.2019.00156

El-Amier, Y., Zahran, M., Gebreil, A., & El-Salam, E. (2018). Anthropogenic activities and their impact on the environmental status of Kitchener drain, Nile Delta, Egypt. *Journal of Environmental Sciences*, *46*(3–4), 251–262.

El Khorazaty, N. (2021). *Egyptian women's agriculture contribution; Assessment of the gender gap for sustainable development* [Master's Thesis, American University in Cairo].

Elmhirst, R. (2015). Feminist political ecology. In A. Coles, L. Gray & J. Momsen (Eds.), *The Routledge handbook of gender and development* (pp. 82–90). Routledge.

El Quosy, D. E. D. A. H. (2017). The evolution of drainage water in egypt. In A. Negm (Ed.), *Unconventional water resources and agriculture in egypt* (pp. 3–16). Springer.

FAO. (2016). AQUASTAT country profile – Egypt. *Food and Agriculture Organization of the United Nations*. Rome, Italy. www.fao.org/3/i9729en/I9729EN.pdf

Gutiérrez-Rodríguez, E. (2010). *Migration, domestic work and affect: A decolonial approach on value and the feminization of labor* (1st ed.). Routledge.

Kabeer, N. (1994). *Reversed realities: Gender hierarchies in development thought*. Verso.

Khouri, R. (2019, March). Poverty, inequality and the structural threat to the Arab region. *Project on Middle East Political Science, Harvard Kennedy School Belfer Center for Science and International Affairs*. www.belfercenter.org/publication/poverty-inequality-and-structural-threat-arab-region

Leach, M., Mehta, L., & Prabhakaran, P. (2016, July, No. 13). Gender equality and sustainable development: A pathways approach. For the world survey on the role of women in development 2014. *UN Women*. www.unwomen.org/sites/default/files/Headquarters/Attachments/Sections/Library/Publications/2016/DPS-Gender-equality-sustainable-development.pdf

Mahmoud, E. M., Nour El Din, M. M., El Saadi, A. M. K., & Riad, P. (2021). The effect of irrigation and drainage management on crop yield in the Egyptian Delta: Case of El-Baradi area. *Ain Shams Engineering Journal*, *12* (1), 119–134. https://doi.org/10.1016/j.asej.2020.08.009

Mateo-Sagasta, J. (2022). Gender mainstreaming in water reuse: Guidelines for planners, investors, project designers and operators. *International Water Management Institute*. https://mel.cgiar.org/reporting/download/hash/ec2095d7d8148dec2e30f7cc87295452

Ministry of Agriculture and Land Reclamation (MALR). (2009). *The sustainable agriculture development strategy towards 2030*. Ministry of Agriculture and Land Reclamation (MALR) of Arab Republic of Egypt, Cairo. https://faolex.fao.org/docs/pdf/egy141040E.pdf

Ministry of Water Resources and Irrigation (MWRI). (2018). *Kitchener drain rehabilitation project*. www.ebrd.com/documents/comms-and-bis/esia-49417-sep.pdf

Molle, F., & Rap, E. (2013). *Brief retrospective on water user organizations in Egypt*. Working Paper. International Water Management Institute.

Molle, F., Rap, E., El-agha, D., Ismail, A., Abou El Hassan, W., & Freeg, M. (2015). Irrigation improvement projects in the Nile Delta: Promises, challenges, surprises. *Agricultural Water Management*, *216*, 425–435. https://doi.org/10.1016/j.agwat.2019.02.013

Najjar, D. (2015). Women's contributions to climate change adaptation in Egypt's Mubarak resettlement scheme through cactus cultivation and adjusted irrigation. In S. Buechler & A. M. Hanson (Eds.), *A political ecology of women, water and global environmental change* (pp. 141–161). Taylor & Francis (Routledge).

Najjar, D., Baruah, B., & El Garhi, A. (2019). Women, irrigation and social norms in Egypt: "The more things change, the more they stay the same?" *Water Policy*, *21*(2), 291–309. https://doi.org/10.2166/wp.2019.154

National Council for Women (NCW). (2017, March). *National strategy for the empowerment of Egyptian women 2030: Vision and pillars* (1st ed.). https://ncw.gov.eg/Images/PdfRelease/final-version-national-strateg-10202312133058257.pdf

Rao, N., Lawson, E. T., Raditloaneng, W. N., Solomon, D., & Angula, M. N. (2019). Gendered vulnerabilities to climate change: Insights from the semi-arid regions of Africa and Asia. *Climate and Development*, *11*(1), 14–26. https://doi.org/10.1080/17565529.2017.1372266

Ravera, F., Iniesta-Arandia, I., Martín-López, B., & Bose, P. (2016). Gender perspectives in resilience, vulnerability and adaptation to global environmental change. *Ambio*, *45*(Suppl 3), 235–247.

Reddy, A. A., Mittal, S., Singha Roy, N., & Kanjilal-Bhaduri, S. (2021). Time allocation between paid and unpaid work among men and women: An empirical study of Indian villages. *Sustainability*, *13*(5), 2671. http://doi.org/10.3390/su13052671

Rist, G. (2007). Development as a buzzword. *Development in Practice*, *17*(4/5), 485–491. https://doi.org/10.1080/09614520701469328

Scott, L. (2016). Advisory note on measures: Women's economic empowerment. *Said Business School*. Oxford University Press. https://seepnetwork.org/files/galleries/1583_Advisory_Note_on_Measures_Womens_Economic_Empowerment.pdf

Selim, T., & Eltarabily, M. G. (2022). Impact of COVID-19 lockdown on small-scale farming in Northeastern Nile Delta of Egypt and learned lessons for water conservation potentials. *Ain Shams Engineering Journal*, *13*(4), 101649. https://doi.org/10.1016/j.asej.2021.11.018

Sukkary-Stolba, S. (1985). Changing roles of women in Egypt's newly reclaimed lands. *Anthropological Quarterly*, *58*(4), 182–189. https://doi.org/10.1016/j.asej.2021.11.018

Sundberg, J. (2017). Feminist political ecology. In D. Richardson, N. Castree, M. F. Goodchild, A. Kobayashi, W. Liu & R. A. Marston (Eds.), *International encyclopedia of geography: People, the earth, environment and technology* (pp. 1–12). Wiley-Blackwell.

Zwarteveen, M. (2008). Men, masculinities and water powers in irrigation. *Water Alternatives*, *1*(1), 111–130.

Zwarteveen, M. (2011). Questioning masculinities in water. *Economic and Political Weekly*, *XLVI*(18), 40–48.

21

ALTERING WATER FLOWS IN THE DRAA VALLEY, MOROCCO

A feminist analysis

Lisa Bossenbroek and Hind Ftouhi

Introduction

While visiting Mezguita, the upper oasis in the Draa Valley, Morocco, in mid-December 2023, we[1] were struck by the dryness of the Draa riverbed. Since our first visit in early 2019, we had never seen tiny plants growing in the riverbed. The river water is regulated by the Eddahbi Mansour Dam. The farmers told us that the last release had been in July. They hadn't received any water for almost six months, and it also had not rained for months.

Water resources are essential for the oases. These are artificial, man-made environments that provide a stripe of greenery in often arid surroundings. They are further characterized by a stratified production system that combines date palms, fruit trees, and annual crops. These three strata are essential for local populations, who often depend on these crop variations to fulfill their livelihood needs.

The cultivated area expands and shrinks according to the availability of water resources and their management. As a result, historically, oases are not fixed spaces (Chaléard et al., 2017). Besides water availability, the hard labor of women and men is essential for maintaining irrigation systems and farming fields. The survival of oases thus depends on a delicate equilibrium between the availability of water resources and labor. Irrigation systems are laborious infrastructures. Canals need to be cleaned and maintained, and the crops and fields almost need daily care for their productivity. Such labor activities are organized in a gendered way. For example, women and men may be responsible for different crops and fields and have various activities (Cloud, 1985, 1988 cited in Feldstein et al., 1989). These differences are in flux, rooted in social organization, and supported by socio-cultural norms and values (Feldstein et al., 1989).

In the current context of altering water flows and increasing periods of droughts, the future of these age-old societies has become a matter of concern for researchers and policymakers alike. Yet, local realities tend to be overlooked by a public and political national discourse in which changes in water availability are usually naturalized and caused by less precipitation, and longer periods of droughts. In this discourse political dimensions of water re-allocations and policy decisions related to market dynamics are neglected. Therefore, the first aim of this chapter is to look beyond the natural façade of water scarcity and illustrate

DOI: 10.4324/9781003100379-26

how the current environmental situation in the Draa Valley in Southeast Morocco is not only a result of climate change but also results from political choices and market forces.

In addition, in this discourse around water scarcity, the local experiences of rural populations and farmers tend to be neglected. Little is known about how altering water flows affect their livelihoods, what it means for their daily lives and survival. In an attempt to partly fill this gap, the second aim of the chapter is to focus on the lived experiences of the local population. More specifically, we chose to focus on women's experiences in the oases as they form an important social category who plays an essential role in maintenance and the survival of the oases. This is due to their specific labor activities but also because of demographic changes. Historically, oases have been strongly marked by male migration (de Haas, 2003; Rössler et al., 2010; Aït Hamza & El faskaoui, 2010). This male migration, as illustrated in the study of Ait Hamza and El faskaoui's (2010) focusing on the Draa region, is strongly linked to environmental degradation. De-deagrarization[2] of men's activities seems to have increased the agricultural burden of women (de Haas, 2003, p. 104). As a result of these transformations, women continue to play an essential role in the continuity of oasis agriculture (de Haas, 2003).

The second reason why we focus on women's experiences is because research on oasis women is scarce (de Haas & van Rooij, 2010; Battesti, 2005). Since their labor is unpaid and is often confined to the private space, it is generally unaccounted for in research and not recorded in official statistics.

For our analysis we have been inspired by the existing literature on gender and environmental change. Within this body of work, gender relations, identities, and environmental change are conceptualized in different ways. For instance, as pointed out by Kulkarni and Rao (2008) in their research focusing on the gender impacts of droughts in Southeast Asia, the existing literature on this topic portrays "women as victims walking long distances in search of fuel, fodder and water" (Kulkarni & Rao, 2008, p. 10). The authors further argue that although some recent literature on the topic moves beyond the image of women as victims, and frame rural women as having the potential to regenerate the mother earth, most analysis on women and the environment remain somehow static: women's roles of nurturing and caring for the subsistence of their families are foregrounded (Ibidem). Other inspiring feminist scholars instead have illustrated how altering gendered relations and changing environments are closely interlinked (see for example, Elmhirst, 2011; Harris, 2006; Nightingale, 2006). In this emergent body of work, scholars moved away from a static understanding of gender and instead considered it as being performative: it is something one does and a process in which identities and relations are continuously contested, redefined and re-interpreted. Such a definition of gender as performative allows the recognition of gender identities as fragmented, provisional, and wrought through the interplay of culture, class, the environment, and other fields of power and regulatory frameworks (see also Resurreccion & Elmhirst, 2008).

Inspired by these theoretical considerations, we first provide in what follows an overview of the methods we mobilized for this study followed by a brief description of the research area. We then look beyond the natural façade of water scarcity and illustrate how the current situation of water scarcity in the study area is not only a result of climate change but also a result of political choices and market forces. We then move to the lived experiences of water scarcity and illustrate how water is embedded in women's daily chores and interactions, and contributes in shaping their daily lives and their identities. Afterwards, we analyze how these activities change in the context of increasing water scarcity, and what this means for women. In the last part, we illustrate how rural feminine identities are renegotiated in relation to particular ecological conditions and broader process of rural dynamics and what this means for the future becoming of the oases.

Methods

Our analysis draws on recurrent fieldwork visits conducted from 2019 to 2023 in the Draa Valley, in the southeast of Morocco. We conducted about 60 interviews with women and about 15 with men. We also talked to different institutional actors[3] to better understand the altering water flows in the valley. To complement the interviews, we organized several focus group discussions with women. Some women were interviewed recurrent times to follow their pathways and their newly taken-up activities. Interviews focused on daily activities, experiences of altering water flows, how this affects them, and newly emerging activities. The interviewed women were selected randomly. Yet, we paid attention to age, marital-, and life-cycle status in an attempt to consider the rich heterogeneity of oasian women and their experiences. Nevertheless, our analysis does not represent women's experiences across the oases in the Draa Valley. Instead, we intend to unpack various social realities and gendered performances of rural femininity.

In addition, based on the information we gathered a graphical artist made different sketches on the spot (see Figure 21.1). These were discussed with several women. It was

Figure 21.1 A sketch illustrating the organization of local livelihoods with a regular availability of water.

Source: Drawing made by Caroline Sarah Straub.

interesting to observe how the sketches served as mean of communication between the interviewed women and us (the two authors and the graphical artist). The women were interested in observing how we visualized their lives and their context, while in the meantime they added some key information that we had overseen.

Research area

The Draa Valley (see Figure 21.2) is characterized by a 200 km belt of six oases along the Draa River. The Draa River is managed through the Eddahbi Mansour Dam, which was built in 1972 and is located near the city of Ouarzazate. Initially, seven dam releases were planned yearly, but these have been reduced to four (Silva-Novoa et al., 2022) due to a combination of increasing droughts, lack of rainfall, and growing water demands of other water sectors. Each oasis has an underlying alluvial aquifer, whose recharge depends primarily on the water released from the dam (Klose et al., 2010). In addition, in 2023 the Agdz Dam near to the upper oasis Mezguita was completed, further altering the water flow in the Draa River and the recharge of the aquifers (Ministère de la Jeunesse, de la Culture et de la Communication, 2023).

Irrigated agriculture associated with livestock is essential to the livelihood of the people living in the area. Most households produce for their own consumption. Farming is characterized by a stratified production system and by collective management of irrigation water. Due to the valley's arid climate, farmers mostly rely on dam releases to irrigate their land. For the last two decades, however, they have increasingly fulfilled their needs for irrigation water by drawing groundwater from shallow individual wells.

Altering water flows: looking beyond the naturalization of water scarcity

Altering water flows are manifested in different ways. They can be characterized by changes in water quality. In the upper part of the Draa Valley groundwater sources are marked by salinity, primarily caused by geological factors (Warner et al., 2013). In the middle and lower part of the valley salinity increases due to lower rainfall and the increasingly arid climate (Beck et al., 2018; Williams, 1999). The diminishing water quality negatively impacts drinking water resources and agricultural yields (Kaczmarek et al., 2023).

In addition, altering water flows also manifest themselves through a decline in water availability. Yet, during our interviews with various institutional actors, the declining water availability was often only framed as a natural phenomenon caused by declining precipitation rates and more extended periods of little rainfall and droughts. Yet, this façade of the "natural" decline of water resources and the naturalization of water scarcity has been strongly criticized by different scholars (Mehta, 2007; Mehta et al., 2019). They argue that the naturalization of water scarcity results in the neglect of its anthropogenic dimensions. In line with these scholars, we don't deny the biophysical elements of water scarcity, yet, we are critical towards presenting water scarcity as only a natural phenomenon. Instead, we argue that the current state of water scarcity is also a result of water allocation politics, policy choices, and market forces.

Hence, in the Draa Valley, the "natural" decline of water resources overshadows water (re-)allocation processes, the growth of particular water-consuming sectors in the area and market forces. For example, Ouarzazate has become a famous filmmaking location

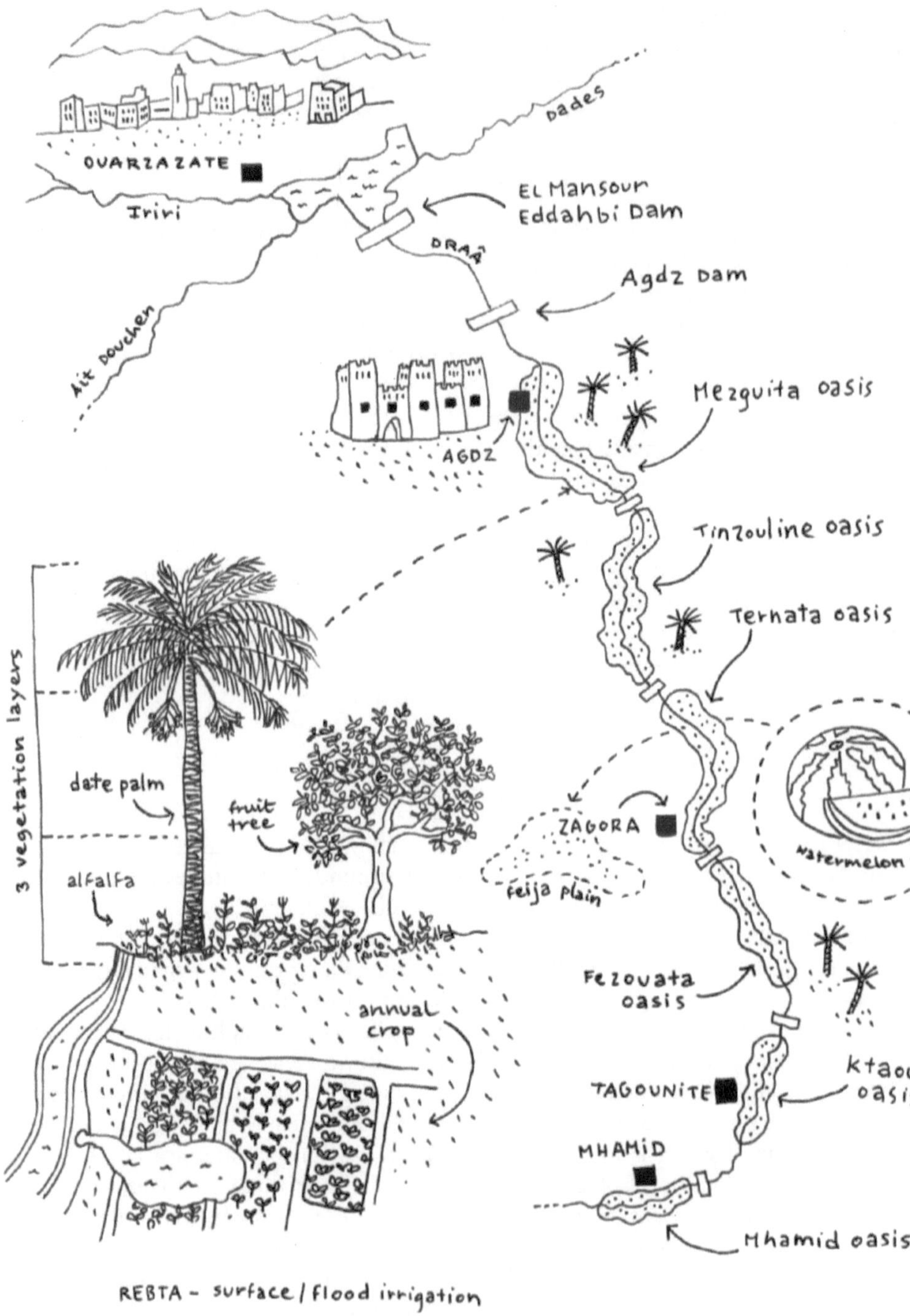

Figure 21.2 Illustration of the Draa Valley.

Source: Drawing made by Caroline Sarah Straub, co-produced by Lisa Bossenbroek and Hind Ftouhi.

and is characterized by processes of urbanization: from 32.1% in 2004 to 38.2% in 2014 (HCP, 2019). The development of the tourist sector and the urban population puts growing pressure on water resources (Chelleri et al., 2014; Karmaoui et al., 2019). In addition, in 2019, a solar power plant was completed next to the Eddahbi Manour Dam with a direct access to the stored water. Water is required for cooling one section of the solar panels in the station as well as for maintenance purposes (cleaning). Although the water requirements of the solar power plant are marginal compared to other sectors water demands, with the capacity of the reservoir declining and the negative impacts of climate change on the water supply, the power plant's water demand could become critical (Terrapon-Pfaff et al., 2020).

In addition, market forces and political choices also contributed to the watermelon expansion that began from 2000 onwards on rangeland outside the ancient oases in the valley. The growth of this crop fully relies on groundwater access, contributing to decreasing groundwater levels and increasing the competition among groundwater users in the area (farmers and inhabitants of the oasis, city dwellers, watermelon producers). Watermelons are a profitable crop that sell at high prices at a time of year when there is little competition; the climatic characteristics of the area – a desert zone – mean that watermelons in this area mature before those of other regions and are thus the first to reach national and international markets. The production of this crop is to some extent encouraged by a number of state programs and subsidies under the Green Morocco Plan (PMV) and the Generation Green strategy (GG) (Bossenbroek et al., 2023).

These different dynamics have contributed to altering water flows in the Draa Valley, resulting in increasing droughts and water scarcity affecting, as we describe in what follows, daily routines and local livelihoods in drastic ways.

Women and work in the oasis: a mixture of hardship, caring for their family, maintenance of social relations, and self-fulfillment

In this section, we describe women's activities in the oasis. The different testimonies allow us to draw up a portrait of oasis women who have always played an essential role in agriculture, the home, and the oasis. Some women characterize their activities as "*tamara*", drudgery, and hardship. On the other hand, through these different activities, they also maintain social relations, fulfill their identities as rural women, and find spaces of relative freedom. Water is essential for fulfilling their domestic and farming activities. Some women narrate about their activities in the past, especially women living in the downstream oases. These are more affected by water scarcity than women in the upstream oases.

Women work from dusk till dawn, and usually, every day. They wake up early and prepare soup or bread for breakfast. Then they leave to the "*jnans*", the gardens, often presented as an extension of the domestic space. The fact that they refer to gardens rather than a piece of land or a plot also illustrates the proximity of the house and the farming space. *Jnans* are small land parcels located approximately within walking distance of 20 minutes from the residential area. Women are responsible for harvesting fruit, cutting wheat, barley, and alfalfa for the livestock, weeding, and gathering wood and twigs to heat the oven to bake bread.[4] When it's time to harvest dates, they sort them in the *jnan* or at home. Once they have finished their work in the *jnan*, they return home and serve breakfast to the family. They also look after the family's livestock: one or two cows, a few sheep and/or goats, chickens, and a donkey.

The rest of the day is spent preparing lunch, dusting the house, washing clothes (often by hand), washing dishes, and looking after the children, sometimes combined with a second trip to the *jnan* in the afternoon to cut alfalfa to feed the animals. Women take up embroidery work and weave carpets if time is left over. For entertainment, they watch TV, use social media on their phones, or visit friends before preparing dinner. In families with several adult women, women take turns realizing the different household and farming activities.

Men are responsible for plowing, irrigating, sowing, maintaining the irrigation infrastructure in the oasis, pruning fruit trees, and harvesting dates. Men are also responsible for going to the souk (weekly market) and doing minor repair work in the house if needed. If the male family members have migrated, agricultural activities are reshuffled along gender lines. Several women we interviewed whose fathers or brothers have migrated have, for example, taken up irrigation activities or taken over the responsibility of going to the weekly souk.

Women's activities are essential for sustaining their families. Despite the arduous nature of some tasks, women are far from regarding them as mere drudgery, as illustrated by Najima, 32, married and mother of five daughters (aged under 12). When water was still abundant, she recalls how she used to work on her father-in-law's land in the Tarnata Oasis (see Figure 21.2). She loved working on the land. It enabled her to meet other women and change her daily routine: "We used to go to the jnans with members of our families. On the way, we'd meet other women who'd join us . . . we had a good time. It gave us a change of scenery".[5]

Similarly, Ilham, aged 40 and married, lives in the Ksar of Tissergate. She recounts: "We wake up early, prepare soup and bread. Then we go to the jnans and mow the alfalfa. We meet other girls and women when we're mowing . . . we have a laugh amongst ourselves". She adds: "If my neighbor needs a little alfalfa to feed her animals, I ask her to come with me, and I give her some".

In times of water availability, farming activities are essential for maintaining the orchards and the oasis. In addition, agricultural and domestic activities also contribute to weaving and welding social relationships of mutual aid, solidarity, and exchange among women. Ilham explains: "Before, when water was released from the dam, the water flowed into the seguias,[6] where the women went to wash carpets and blankets. We'd bring along dates, buttermilk, bread. . . . We would get together and have fun". She continues: "Even if we didn't want to do laundry, we used to go to the river or seguias just to meet other women and chat." Latifa, 39, divorced and mother of a young daughter, also living in the Ksar of Tissergate adds:

> When there's enough water in the oued (river), a seguia next to our house fills up with water, and all the women of the village get together to wash carpets and blankets. . . . We get together and spend the whole day in the seguia. We laugh and have fun.

Khadija adds another element to what these daily activities mean for women in the oasis and in particular in the jnans: "These activities are good for our health, allowing us to move and do sport; otherwise we'll be sick. They also reduce depression (*takhmom*)".

When talking about their daily activities, in times when water is available the terms "joy", "wellbeing", "gossip", and "keeping each other company" are recurrent in the testimonies of women in the different oases and indicate that the time spent with other women around the *seguias* is also a time of reunion, exchange, and conviviality, which alleviates their daily chores. In addition, their activities also give meaning to their daily lives and allow them to fulfill themselves as rural women, or as housewives like many identified themselves, who are capable "*bdraô*", "*someone who fights to feed her family*". Zineb, aged 48, further explains how:

> Here women work. I cannot stay in bed until 10 a.m. and only start to prepare breakfast then. At dawn, the soup should already be on fire and the bread ready . . . and if I'm done with the housework, I go to the gardens.

Zineb continues to talk about other younger women who leave the domestic chores and farming work to their mothers. According to Zineb "they can't be called women". In addition, remunerated employment or work outside the oasis is often poorly perceived as explained by elderly women: "At home in the Bled[7] it's frowned upon when women go out to work [outside the house]. They can only go to the jnans with other women to help with the crops, the harvest . . . ." As such, women's activities are framed by certain beliefs about what a rural woman should do and how she should behave. They are responsible for the good functioning of the household within the so-called private sphere. Their mobility is restricted and closely watched and controlled (see also Belarbi, 1995; El Harras, 1996). Women proudly take up domestic work, and actively engage in tasks close to the house such as animal husbandry activities, as these are or can be considered as an extension of their domestic chores. By engaging in these activities, they also re-affirm their rural womanhood and obtain social prestige and respect from the community, men and women alike (Bossenbroek, 2016).

Finally, the different activities carried out by women in the oasis and the *jnans* are rooted in a rural context marked by a gendered division of space where the public domain is usually attributed to men and the private domain is attributed to women. However, this division of space is not fixed and is instead negotiated daily. When water is available, women's different activities allow them to move around in the oasis with an aim. As such, they break down this gendered segregation of space. The various activities carried out in the oasis (going to the *jnans* and working in the *jnans*, washing carpets/clothes near the river or particular canals) give women access to the "outside world", where they conquer a space of relative freedom and meet up, exchange ideas, gossip, and have fun.

The meaning of the oasis that emerges from the different testimonies is a space of production, consumption, self-fulfillment, and leisure. Through their activities, which always involve water (to farm, to wash, to cook, etc.), women give life to this space and maintain it.

Changes in water availability: new livelihood challenges and altering labor activities

As described earlier in the chapter, due to altering water flows as a result of both biophysical circumstances, political choices and market forces, water has become increasingly scarce in the different oases. Local livelihoods are strongly affected. Consequently, women describe how their activities and social lives, as well as their sense of who they are, have drastically changed.

Water[8] is essential to irrigate the variety of crops that families grow. However, as Najima and her husband Salem explain, growing one's crops has become increasingly difficult due to the lack of water. Salem, aged 38, migrated when he was about 17. Today, he works as a crane driver. His family has one hectare of land dispersed among five or six plots. They used to grow dates, alfalfa, wheat, pepper, cabbage, okra, and squash. He explains how the dam's water releases have diminished, and its quality has worsened. They mostly rely on water from their two wells. The first well was dug when Salem was born in 1986. Back then, it was at a depth of ten to 12 meters. Today, the well has reached a depth of 30 meters. The second well, with a depth of 15 meters, was dug in 1995. Today, this well has a depth of 25 meters. Although they have deepened their wells, water remains little. Consequently, according to Najima, they hardly grow any crops and buy food products in the souk. Salem adds: "There are 10 persons who are left and farm [in his native village]. We used to be 80 or 90. We all used to farm".

Other testimonies from women confirmed that it has become more challenging to continue to produce food crops and that they increasingly rely on the *souk* (the local market) to satisfy their consumption needs. Consequently, food prices have doubled, sometimes even tripled over the years.

In addition to water scarcity affecting local livelihoods, the collected testimonies also illustrate how water scarcity is coupled with a change in women's daily routines and activities. There is hardly any work left in the gardens, and collective meeting points have disappeared as Halima testifies: "Today, due to the lack of water, the collective visits [among women] to the gardens, the river or the main canal stopped. Women stay at home and take care of their children. They don't have anywhere to go anymore". She continues:

> Everything has changed now . . . we can no longer go to the jnans because there's nothing left to do there. We stay home because we have nowhere to go, even if we want to go out. . . . There's nothing left in the jnans. There's no more alfalfa, dates, grass, only drought and dried-up jnans that are deserted.

Also, for Khadija and the women in her village, habits and routines have changed:

> We don't find any water for washing clothes. Nowadays, we only wash the most important things we want to wear. We don't do the big cleaning anymore.[9] . . . Women used to mow alfalfa and raised livestock. But most of them have now sold their herds. Before, we had something to keep us busy.

Inspired by a performative definition of gender and gendered identities (Butler, 2020, p. 519), "gender is in no way stable identity or locus of agency from which various acts proceed; rather, it is an identity tenuously constituted in time-an identity instituted through a stylized repetition of acts". As we lay out in the introduction in this understanding of gender, gender identities are not something fixed or static; therefore, they are never fully accomplished. They continuously need repetition and are tightly connected to other processes of change. Because of changing water availability, daily routines carried out by women, allowing them to perform and act as "*bdraô*" (capable) women, have changed. Their work in the oasis (in the *jnans* and washing clothes and carpets in the collective water spots) was important for themselves and to signal their competency to others in the community. Yet, today, for some of them, most work is confined to the house, and they cannot live up to this notion anymore: "Women stay at home. Some do embroidery work. Others are sleeping now".

Yet, whereas some women seem limited to perform their identities and face difficulties finding a new routine, others explore new livelihood options and occupational activities. As we argue in the next section, in doing so, they also renegotiate existing notions of rural femininity.

Changes in water availability, rural dynamics, and newly emerging feminine identities

Changes in water availability do not stand alone and are situated in a broader context of rural dynamics that characterize oases (see also Bossenbroek et al., 2023). From our interviews and observations, we noticed various elements of social change. For example, today, young women stay in school longer and are more educated than their parents. We were also told about the expansion of feminine mobility. Although until recently, the overwhelming majority of Moroccan labor migrants were men, the number of single and divorced women who migrate independently is increasing (de Haas, 2003, p. 107).

Moreover, all the women we talked to use some form of social media (YouTube, Instagram, WhatsApp, Facebook). Some follow YouTube channels "about recipes, decorations, social issues", explains Fatima, in her twenties, "or influencers like Sabah Benchouik", adds another young woman. This "peasant star of Instagram" (Style Trotter, 2021) is a young woman[10] from a village near Khémisset[11] who shares her daily life with "her followers". Her daily life used to be not very different from that of oasis women and was made up of various domestic and agricultural activities that resonate with the young girls who follow her. Over the years, the influencer has carved out a place for herself in the Moroccan social networking sphere and participates in TV series. Her activities have improved her living conditions, awakening dreams in rural girls to follow her example.

While due to water scarcity some rural feminine identities cannot be fulfilled anymore, these different societal transformations awaken new possibilities for being and becoming for rural women. We identified the following new activities, which we layout further down, through which women perform new rural femininities: coquetry stores, small-scale farming businesses, agricultural cooperatives, and cooperatives for handicrafts.

New coquetry boutiques and small-scale farming businesses

During our fieldwork, different new businesses were launched. Two initiatives were particularly inspiring and innovative. The first initiative is of Nawal, aged 38. She lives with her parents in the village of Mezguita. After quitting school at age 15, she followed a vocational training course in farming[12] in the city of Ouarzazate, about an hour drive from her native village. She chose this training because "I chose the domain I'm comfortable in. I don't want to be under someone's authority. . . . I want to set up my own project". Once she completed her training and the idea of her new project was clear in her mind, she found a financial arrangement and embarked on a rabbit breeding project with 400 rabbits. The idea was to sell them to hotels in the area. In parallel, she also opened a small boutique next to her parents' house. She sells handicrafts (little bags embroidered and traditional scarves she made herself), perfumes, and cleaning products in her shop. During the COVID-19 pandemic, all her rabbits died. Nevertheless, she did not give up. Last time we met her (December 2023), she started another project of breeding farm chicken. She bought some materials to renovate an ancient chicken coop located on a small plot owned by her family.

She explained that since the prices for eggs have increased, she has thought of selling them. Parallel to this new activity, she also takes care of a small parcel of land close to her sister's house, where she cultivates different crops, like carrots, onions, and date palms. Nawal proudly explained that her small plot allows her to experiment with different crops and that the produce ripens well before that of her father's *jnan*. To irrigate the vegetables, she arranged with her sister to use the tap water of her house.

Other young women have followed Nawal's example, like Islam aged 25, who also opened a small boutique. Both Nawal and Islam refer to their shops as "coquetry boutiques for village girls". Islam opened her store in May 2021 with the support of her husband: "At first we started selling soap and some cosmetics, then we switched to make-up". Today, she sells mainly make-up. Such boutiques enable village women to have at-home products for which they had to travel to the nearest urban center or depend on other family members to bring them back from the city.

Both women mobilized the help of male family members to (co-)finance the project or to supply products. Both boutiques are located in a small space linked to the house, overlooking the alleyway, preventing them from trespassing the public space considered "masculine". While they have a window to the outside world, the space remains attached to the house.

Feminine cooperatives in the domain of food processing and handicrafts

In addition to the previously-described individual projects, we observed how women created new feminine cooperatives over the last decade. This development model is widespread in Morocco and has been promoted since early 2000[13] by various institutional and private actors. The cooperatives we encountered were dedicated to food processing, valorizing local agricultural products, and selling handicrafts, like carpets and embroidery work. For example, in the oasis of Fezaouata a group of young women created a cooperative in 2013 to store dates and sell products derived from dates, such as syrup, jam, vinegar, and date powder to prepare coffee. Women of the cooperative also prepare and sell different types of couscous. They also have a cooling unit where farmers can store their dates for a fee. They buy the dates from local farmers and, as Samira puts it, "*including our fathers*". The cooperative currently consists of approximately 20 women of different ages and marital status. Another farming feminine cooperative in a small village near Zagora specializes in producing local bread. They sell 100 to 200 loaves a day. They get orders up to 600 loaves a day for weddings and funerals.

Renegotiating feminine identities and social relations

The previously-described activities allow women to perform new identities. They consist of professionalizing domestic and farming activities and extending their domestic roles in terms of responsibilities and location: from the family sphere, particularly assigned female spaces in the oases, to new spaces in the community and even in the public domain. They rely on their farm and domestic experiences, skills, labor, and knowledge to do so. Consequently, they also revalorize their experiences and work.

Most of the projects are small-scale. This can be explained by the fact that women face difficulties in accessing and mobilizing capital but also because of their existing domestic duties. Small-scale projects make it easier for women to manage their time and to continue to fulfill their other tasks. New feminine identities are carefully crafted next to existing domestic feminine identities, allowing them to continue performing as good housewives. Due to existing gender roles and responsibilities, women's activities start small and often remain small.

In a context where women face difficulties gaining access to employment opportunities, similar to other rural areas in different parts of the world (see also Bock, 2004; Little, 2016), these different activities can be regarded as new forms of feminine rural entrepreneurship. Although they are usually small-scale projects, they testify to their innovativeness, creativity, and courage in taking risks. Moreover, they create new spaces allowing the embracement and valorization of feminine knowledge, work that has been unpaid and has often been unaccounted for (see also Shortall, 2016) and perform new feminine identities.

The financial benefits were very much welcome when discussing the rewards of such activities. Although most women said they didn't earn much, their salaries helped make ends meet. Yet, more important seemed to be other forms of reward such as personal growth, new experiences, doing something of their own, and social goals. For example, Latifa, a member of the bread cooperative, testifies how her work in the cooperative changed her:

> It helped me a lot. You feel you've got an important place in society when you work and exchange with others. Before, we stayed at home and worked in our orchards. We've become another person. It [the work] gives you importance and a new personality.

The answers of other women we spoke to confirmed Latifa's testimony. In addition, some added that the new activities have also given them the possibility to be more mobile, as explained by Saida, a member of the date cooperative: "Before we stayed at home, we were shy, we couldn't talk to people, but today, the cooperative has given us the confidence to face people, to go and arrange administrative papers for example, we can go everywhere now." Finally, the cooperatives provide another (legitimate) space for women to gather outside the private sphere of their houses, where they can meet daily with other women.

They also allow, to some extent, to gain more respect and slightly modify existing social-gender relations in the family and the community. Nawal, for example, proudly explained during one of our interviews that although she is the youngest of the family, she holds the key to the room at home where all the goods are stored. In addition, she also manages the family budget: "I manage how much we spend and how much we save to protect ourselves against the vagaries of life". For her part, Samira, president of the date cooperative, explains how their parents' attitudes towards them changed as the project gained momentum, especially when the members of the cooperative started buying dates produced by their parents.

Yet, our interviews also illustrate how women carefully have to negotiate the sociocultural gender norms that codify their activities and their access to the public space, which is not always easy and comes with resistance. For example, when talking to Samira, she explains how:

> Our families gave us their trust. But we have some negative reactions from the inhabitants. Some say that women should go out and confront the people. Whereas others say that the place of women is in the house, next to her children.

Indeed, performing new feminine rural identities requires women to find "*a good equilibrium*" between their different roles. Married women especially face difficulties balancing their roles as mothers, spouses, daughters-in-law, and cooperative members and/or entrepreneurs.

Conclusion

Our entry point for this study was women's activities, identities, and experiences in relation to altering water flows. But as we tried to unpack how labor activities changed and how women took up new activities re-articulating their feminine rural identities, we noticed that altering water flows cannot be studied on their own and in isolation of other (social) dynamics. It is only by linking altering water flows to wider social transformations that the lived experiences of rural women and changing rural feminine identities can be understood.

Women have always played an essential role in oasis agriculture. In the past, when water availability was not an issue, and still nowadays in times of water availability, they walk daily to the fields, harvest fruits, cut alfalfa, wheat, and barley, and collect wood and twigs to heat the oven. Despite its hardship, these activities allow them to keep busy, fulfill themselves, and create and maintain social relationships. It also enabled them to find spots of relative freedom: meeting other women in the fields, exchange goods, laugh, gossip, and have fun. However, farming has become more difficult due to less water availability and diminishing water quality. For this reason, some women are increasingly confined to the private feminine domain in this newly emerging environmental context. In contrast, others take up new opportunities to make a living and give sense to their daily lives. Some opportunities are supported by the state, such as training and the development of cooperatives. Yet, many initiatives are initiated without any support. By taking up new activities, the interviewed women perform new rural feminine identities of entrepreneurship, creativity, innovativeness, and financial autonomy that they carefully craft around their existing "traditional" rural feminine identities of housewives, daughters (-in law), spouses, and mothers who are responsible for the domestic chores (including the *jnans* and livestock) and taking care of family and children.

Through these new activities, rural women create new spaces where they valorize feminine knowledge and their work, which has been unpaid and often unaccounted for. In addition to the relatively small financial benefits that are much welcomed, the rewards of such activities are related to personal growth, new experiences, and social goals. Nevertheless, having the possibility to engage in such activities is not the same for every rural woman. Married and older women especially face difficulties performing new rural feminine identities. For young, educated women, engaging in new activities is a logical next step after accomplishing their studies or training. The performance of new rural feminine identities emphasizes new social differences based on marital status, education, and age.

Finally, in a context where the becoming of the oases is under threat due to less water availability, the acceleration of processes of desertification, and the deagrarization of men's activities (de Haas, 2003, p. 104), our analysis offers a slightly alternative reading of the becoming of the oasis. Our study illustrates how women continue to give life to these magical places and maintain them. Through their activities, they contribute to the dynamization and diversification of these areas.

Notes

1 The authors of the article.
2 The term deagrarization is used by Hein de Haas to indicate the shift of men's activities in the oasis. Whereas men used to play an important role in farming, their activities gradually shifted to other non-farming domains often outside the oasis.
3 We talked to representatives of the Water Basin Agency Drâa-Oued-Noun, Regional Office of Agriculture of Ouarzazate, National Agency for the Development of the Oasis and Argan Zones, National Office of Water and Electricity.
4 Most women only bake bread in the oven. Generally, LPG gas bottles are used to cook.
5 The names of the interviewees were changed for confidentiality.
6 Earthen canals.
7 "Bled" refers to one's home village.
8 Women reported less problems related to drinking water. They experienced some cuts or worsening quality. In that case drinking water shortages are usually quickly taken up by local (drinking water) associations who then provide potable water that people can buy from water tanks.
9 "Grand menage", or the big cleaning, refers to cleaning the house completely, including the carpets. This is usually done one or two year a year and before Aids (important celebrations).
10 www.instagram.com/sabah_benchouikh/.
11 Located in northern Morocco.
12 The training was part of a program called AFAK, "for the qualification and socio-professional of young people not attending school" put in place by the Ministry of National Education, Vocational Training, Higher Education and Scientific Research (www.aecid.ma/wp-content/uploads/2019/02/a7dc25eaf316b5e11c1f41aaaefb6c17.pdf).
13 The number of cooperatives rose from 62 in 1957 to 2,000 in 1983, reaching 47,609 in 2021 (ODCO, 2021; ODCO, no date, cited in M'Barki and Schmitz, 2023). Their number especially increased after 2005 after the launch of INDH (L'Initiative Nationale pour le Développement Humain) promoting the creation of cooperatives. In addition, the Green Morocco Plan (2008–2019) has enabled the creation of 9,951 cooperatives to facilitate the pooling of production resources, cost optimization, access to new markets and the realization of investments difficult to implement individually (Ministry of Agriculture, 2020 cited it M'Barki and Schmitz, 2023).

References

Aït Hamza, M., & El faskaoui, B. (2010). Les oasis du Drâa au Maroc: Rupture des équilibres environnementaux et stratégies migratoires. *Hommes & Migrations, 1284*, 56–69. https://doi.org/10.4000/hommesmigrations.1241

Battesti, V. (2005) *Jardins au désert, Évolution des pratiques et savoirs oasiens. Jérid tunisien*. Éditions IRD.

Beck, H. E., Zimmermann, N. E., McVicar, T. R., Vergopolan, N., Berg, A., & Wood, E. F. (2018). Present and future Köppen-Geiger climate classification maps at 1-km resolution. *Scientific Data, 5*(1), 1–12.

Belarbi, A. (Eds.). (1995). *Femmes rurales*. Editions Le Fennec.

Bock, B. B. (2004). Fitting in and multi-tasking: Dutch farm women's strategies in rural entrepreneurship. *Sociologia Ruralis, 44*(3), 245–260.

Bossenbroek, L. (2016). *Behind the veil of agricultural modernization: Gendered dynamics of rural change in the Saiss, Morocco* [Doctoral dissertation, Wageningen University and Research].

Bossenbroek, L., Ftouhi, H., Kadiri, Z., & Kuper, M. (2023). Watermelons in the desert in Morocco: Struggles around a groundwater commons-in-the-making. *Water Alternatives, 16*(1), 87–107.

Butler, J. (2020). Performative acts and gender constitution: An essay in phenomenology and feminist theory. In *Feminist theory reader* (pp. 353–361). Routledge.

Chaléard, J., Planel, S., & Ruf, T. (2017). *General conclusion: The Oases challenged by glocalisation*. In E. Lavie and A. Marshall (Eds.), *Oases and globalisation*. Springer International Publishing.

Chelleri, L., Minucci, G., Ruiz, A., & Karmaoui, A. (2014). Responses to drought and desertification in the Moroccan Drâa Valley Region: Resilience at the expense of sustainability? *The International Journal of Climate Change: Impacts and Responses, 5*(2), 17.

Cloud, K. (1985). Women's productivity in agricultural systems: Considerations for project design. In K. Overholt & W. Hartford (Eds.), *Gender roles in development projects: A case book*. Kumarian Press.

Cloud, K. (1988). *A teaching module on women and agriculture: Household level analysis*. University of Illinois.

de Haas, H. (2003). *Migration and development in Southern Morocco: The disparate socio-economic impacts of out-migration on the Todha Oasis Valley*. Nijmegen University.

de Haas, H., & van Rooij, A. (2010). Migration as emancipation? The impact of internal and international migration on the position of women left behind in rural Morocco. *Oxford Development Studies*, *38*(1), 43–62. https://doi.org/10.1080/13600810903551603

El Harras, M. (1996). Féminité et Masculinité dans la société rurale marocaine; Le cas d'Anjra. In R. Bourqia, M. Charrad & N. Gallagher (Eds.), *Femmes, culture et société au Maghreb*. Afrique Orient.

Elmhirst, R. (2011). Introducing new feminist political ecologies. *Geoforum*, *42*(2), 129–132.

Feldstein, H., Poats, S., Cloud, K., & Norem, R. (1989). Conceptual framework for gender analysis in farming systems research and extension. In Feldstein H., Poats S. (Eds.), *Working together gender analysis in agriculture* (pp. 7–26). Kumarian Press.

Harris, L. M. (2006). Irrigation, gender, and social geographies of the changing waterscapes of southeastern Anatolia. *Environment and Planning D: Society and Space*, *24*(2), 187–213.

Haut Commissariat au Plan (HCP). (2019). *Monographie de la province de Ouarzazate*. https://www.hcp.ma/draa-tafilalet/Monographie-de-la-province-de-Ouarzazate-de-l-annee-2019_a270.html

Kaczmarek, N., Mahjoubi, I., Benlasri, M., Nothof, M., Schäfer, R. B., Frör, O., & Berger, E. (2023). Water quality, biological quality, and human well-being: Water salinity and scarcity in the Draa River basin, Morocco. *Ecological Indicators*, *148*, 110050.

Karmaoui, A., Minucci, G., Messouli, M., Khebiza, M. Y., Ifaadassan, I., & Babqiqi, A. (2019). Climate change impacts on water supply system of the Middle Draa Valley in South Morocco. In *Climate change, food security and natural resource management* (pp. 163–178). Springer.

Klose, A., Busche, V., Klose, S., Schulz, O., Diekkrüger, B., Reichert, B., & Winiger, M. (2010). Hydrological processes and soil degradation in Southern Morocco. In P. Speth, M. Christoph & B. Diekkrüger (Eds.), *Impacts of global change on the hydrological cycle in West and Northwest Africa* (pp. 198–253). Springer. https://doi.org/10.1007/978-3-642-12957-5_14

Kulkarni, S., & Rao, N. (2008). Gender and drought in South Asia: Dominant constructions and alternate propositions. *Droughts and Integrated Water Resource Management in South Asia: Issues, Alternatives and Futures*, *2*, 70.

Little, J. (2016). Gender and entrepreneurship. In *Routledge international handbook of rural studies* (pp. 357–366). Routledge.

M'Barki, R., & Schmitz, S. (2023). Les coopératives au Maroc, entre bricolage et développement rural: le cas des regroupements de fromage de chèvre de Chefchaouen. *Belgeo. Revue belge de géographie*, *1*.

Mehta, L. (2007). Whose scarcity? Whose property? The case of water in western India. *Land Use Policy*, *24*(4), 654–663.

Mehta, L., Huff, A., & Allouche, J. (2019). The new politics and geographies of scarcity. *Geoforum*, *101*, 222–230.

Ministère de l'agriculture. (2020). *Le Plan Maroc Vert, Bilan et Impacts 2008–2018*. Rabat.

Ministère de la Jeunesse, de la Culture et de la Communication. (2023, January 22). *Zagora: M. Baraka visite le barrage d'Agdez et lance deux projets hydraulique et routier*. www.maroc.ma/fr/actualites/zagora-m-baraka-visite-le-barrage-dagdez-et-lance-deux-projets-hydraulique-et-routier#

Nightingale, A. (2006). The nature of gender: Work, gender, and environment. *Environment and Planning D: Society and Space*, *24*(2), 165–185.

ODCO. (2021). *Coopératives par secteur en 2021*. Office du développement de la coopération. www.odco.gov.ma/fr/content/situation-au-31-décembre-2021

ODCO. (n.d.). *L'historique des Coopératives au Maroc*. Office du développement de la coopération. www.odco.gov.ma/fr/content/l'historique-des-coopératives-au-maroc

Resurreccion, B. P., & Elmhirst, R. (2008). *Gender and natural resource management: Livelihoods, mobility and interventions* (1st ed.). Routledge. https://doi.org/10.4324/9781849771436

Rössler, M., Kirscht, H., Rademacher, C., Platt, S., Kemmerling, B., & Linstädter, A. (2010). II-5.4 Migration and resource management in the Drâa Valley, Southern Morocco. *Impacts of global change on the hydrological cycle in West and Northwest Africa*. https://www.idos-research.de/en/others-publications/article/migration-and-resource-management-in-the-draa-valley-southern-morocco/

Shortall, S. (2016). Gender and identity formation. In *Routledge international handbook of rural studies* (pp. 400–411). Routledge.

Silva-Novoa, L. M., Bossenbroek, L., Schilling, J., & Berger, E. (2022). Governance and sustainability challenges in the water policy of Morocco 1995–2020: Insights from the Middle Draa Valley. *Water*, *14*(18), 2932.

Style Trotter. (2021). *Sabah Benchouikh: La Paysanne Star D'instagram*. https://shoelifer.com/lifestyle/reseaux/sabah-benchouikh-paysanne-star-instagram/

Terrapon-Pfaff, J. C., Ortiz, W., Viebahn, P., Kynast, E., & Flörke, M. (2020). Water demand scenarios for electricity generation at the global and regional levels. *Water*, *12*(9), 2482. https://doi.org/10.3390/w12092482

Warner, N., Lgourna, Z., Bouchaou, L., Boutaleb, S., Tagma, T., Hsaissoune, M., & Vengosh, A. (2013). Integration of geochemical and isotopic tracers for elucidating water sources and salinization of shallow aquifers in the sub-Saharan Drâa Basin, Morocco. *Applied Geochemistry*, *34*, 140–151.

Williams, W. D. (1999). Salinisation: A major threat to water resources in the arid and semi-arid regions of the world. *Lakes & Reservoirs: Research & Management*, *4*(3–4), 85–91.

22

WATER, WOMEN AND FISHING LIVELIHOODS IN SOUTH AND SOUTHEAST ASIA

Holly M. Hapke, Nikita Gopal, Kyoko Kusakabe and Gayathri Lokuge

Introduction

This chapter considers issues of gender, water resources, and water governance in the context of fishing livelihoods in South and Southeast Asia. Fishing and gleaning-based livelihoods are highly dependent on both fresh and marine water resources and ecologies, yet these resources and ecologies are under pressure from climate change and coastal and reservoir area developments such as urbanization, tourism, aquaculture and agriculture, dam and port development, and industrialization (Gownaris et al., 2017; Dugan et al., 2006; Adeyemo, 2003; Jensen, 2001). Warming ocean waters, pollution from industry and port activity, dredging, and over extraction or mismanagement of water from reservoirs, rivers, ponds, and lakes all adversely impact water ecologies and degrade water resources that support fishing and gleaning livelihoods. The livelihood implications of these impacts are necessarily gendered and as will be illustrated later, are further informed by other intersecting social identities – class, caste, religion, language/ethnicity, and age.

The chapter explores the relationship between gender, water resources, and fishing livelihoods and asks: in what ways are political, economic, and ecological developments impacting water resources and fishing livelihoods, and how are these gendered? How might different groups of men and women be impacted differently? What are the political, economic, cultural, and social implications of women's changing relationships to water resources and fishing livelihoods? We consider these questions through an exploration of three case studies. The first case study is from the Vembanad Lake region in Kerala State on the southwest coast of India, where urbanization and related economic development are polluting the lake and creating competition for access to water. We examine how these developments have affected women's access to water resources which support the small-scale fishing activities in which they have historically engaged. Second, we consider the impact of militarization and identity politics in Sri Lanka on women of different ethnicities and their relationships to diverse spaces in and around water linked to fishing and gleaning livelihoods. Third, we look at the case of inland Cambodia, where a changing economy and the emergence of new livelihood options are reworking women's relationship to the fish economy. Each of these cases demonstrates the close intersection between

DOI: 10.4324/9781003100379-27

gendered livelihoods, gendered access to water resources and the norms that govern these resources.

A growing body of literature illustrates how access to water is gendered and how differentiated access enjoyed by men and women is often mediated by gendered labor practices, socio-cultural expectations (e.g., related to notions of masculinity and femininity), as well as intersectional differences (e.g., race, religion, class, age, and so forth) (Fröhlich et al., 2018; Harris et al., 2017; Meinzen-Dick et al., 1997). Intersectionality is used here to refer to the interconnected nature of social categories and identities such as race, class, caste, religion, gender, sexuality, nationality, and age as they apply to an individual or group, that create overlapping and *interdependent* systems of discrimination or privilege. As an analytical framework, intersectionality describes how different systems of inequality "intersect" to create unique dynamics, effects, and experiences for different individuals and groups (see Runyan, 2018).

In South and Southeast Asia, fishing-based livelihoods are characterized by distinct gendered divisions of labor that situate men and women in water access rights regimes in different and often disparate ways. These differences become significant in the context of economic and ecological change, as men's and women's livelihoods are impacted in different ways. In the case of Vembanad Lake, urbanization, commercialization, and other development activities are rewriting traditional access rights and creating barriers to women's access to water resources, which they use for small-scale fishing operations. In Trincomalee, Sri Lanka, the failure of the state to recognize women's work in gleaning as part of the fisheries sector, alongside legacies of ethnic and religious strife and war, effectively denies them the right to access water bodies. In the context of military occupation, this lack of rights undermines women's ability to negotiate access to lagoons and earn a living from gleaning. Finally, Cambodia's case shows how the competing interests between agriculture and fishing, along with changing regimes of property ownership, commercialization, and water resource mismanagement degrade water resources and undermine fishing-based livelihoods in gendered ways.

Case study: women fishers in Vembanad Lake, Kerala, India[1]

Vembanad Lake is part of one of the largest wetland ecosystems in the country and includes the deltaic regions of rivers flowing into the lake and the adjoining wetlands. This wetland supports multiple activities, including agriculture and fisheries. The lake flows through three districts (Alappuzha, Kottayam, and Ernakulam) and is fed by several rivers (Manimala, Meenachil, Pamba, Achenkovil, and Muvattupuzha). The fisheries in the lake and in the wetlands are small in scale, with capture fishing carried out using small canoes and nets, many of which are made by the fishers themselves. Women especially use various Indigenous fishing methods that are uniquely ingenious in fabrication and operation. The alternate rice-fish culture system is a traditional practice in this wetland, and farmers in the region practice the only below sea-level rice cultivation in India (in the Kuttanad region, Alappuzha district).

This case considers women's access to water in two villages in the Vembanad wetland system situated in Ernakulam district. The first is village 'V', which occupies an island close to Kochi-Ernakulam city. The village has about 45 families all belonging to the Pulaya caste, traditionally agricultural laborers. This village, like several others in the region, had *paadams* or rice paddies on which *pokkali* rice, a tall rice variety suitable for water-logged areas, was cultivated organically and alternated with fish. Over time agriculture saw a decline and the *paadams* came to be used only for fish culture. Today in the village there is no rice cultivation. After several years of the land lying fallow, a few families got together and tried to re-start the *pokkali* paddy cultivation in 2018 (mainly the men decided). Different people owned the land, but *cultivation* was carried out collectively in an area called a *padashekharam* – a collection of *paadams*. However, in 2018 the region experienced significant flooding which inundated the area, and ever since there has been no cultivation of rice. Several streams of the Vembanad Lake also crisscross this village. Village women utilize fish resources in the *paadams* and streams mainly for household consumption and for additional income when there is surplus catch.

In earlier days, women of the village gleaned for fish, shrimp, and crabs in the inundated *paadams*, using their hands and legs to assemble the fish and then capture them live. They stored the fish in arecanut leaf spathes, and later began using more durable aluminum pots that were kept afloat in the water. After taking a share for household 'curry' (consumption) purposes, they stored the rest of the fish in a small, enclosed net that was tethered to a pole and kept under water until it was marketed. Older women of the village performed this type of fishing when they were younger, but there are very few women in the village doing this type of fishing now because of the physical effort that this activity entails. Instead, they use small dip nets, cast nets, gill nets, and hook-and-lines to capture fish from the streams and from *paadams*. Women in groups also pull woven tender coconut leaf fronds working from one end of the *paadams* to the other and push the fish with their feet and assemble them before capturing. The women also assist their husbands in fishing in the lake. The cast nets that the men use are heavy, and women find it difficult to deploy the net, but they assist by holding the canoes steady when the nets are cast by men.

The village is very close to the city, but families in the village can access it only by small canoes that all families own. The proximity to the city has also meant that this area has become prime real estate. Several acres of land in the village were procured by private realtors, though no construction has been started as yet. However, there is uncertainty with regard to potential legal issues that may arise, making any investment in fish culture precarious. Through close proximity to the city, the lake is more prone to pollutants, both from domestic and commercial establishments. This pollution has permeated the water bodies in the village and the waterways that are connected to the lake. The resulting decline in fish catches in the lake and *paadams*, impacts the incomes of village families and livelihoods of the women in particular, as the men still go fishing in the lake. Pollution of the waters in the streams has also resulted in health issues, mainly skin rashes, that make the women reluctant to enter the waters.

In another village, 'E', women glean and fish in *kettus*. *Kettus* are also fields where a rice-fish system of cultivation is followed. Water for the culture is usually from the extensive backwater system that reaches the *kettus* through a sluice gate mechanism. The rice season is usually from May–June to September–October and the shrimp/fish season follows from November to April. Women have access to these fields in between the two activities, that is, rice cultivation and fish/shrimp culture. However, they must give a share of what they catch to the owner of the *kettu*. When the owners begin to cultivate rice, *kettus* become out of bounds for women and they move to the streams or canals that feed the *kettus* for fishing. In earlier times, owners allowed women to glean even when the rice crop was standing, but this is a rare occurrence now. In the fish/shrimp season, between November and April, women manage to find some seasonal employment in these fields. Since the *kettus* are private property, women have little or no access to these potential fishing areas, even if the *kettus* are near their residences. They need the permission of the owners to fish in them. After the fish/shrimp is harvested, in about mid-April, there is the *kettu kalakkal* (*kalakkal* loosely translates to mixing up or churning). At this time, the *kettus* are opened up to whoever wants to glean in them for any un-harvested fish or fish that enters through the sluice gates. During *kettu kalakkal* the catch need not be shared with the owner.

Over time the area under paddy-cum-fish/prawn culture has declined, leading to access restrictions and shrinkage of water bodies. This has forced women to either find alternate *kettus* or move on to feeder canals to glean and fish. The *kettus* are increasingly being used for monoculture of prawns and this has impeded women's hitherto open access to these resources, thus impacting their livelihood. A *kettu* owner interviewed who still practices the rice-fish/prawn system allows women to glean in his land, 'as was the practice in these areas' but also opined that 'this should not be continued as all produce in my property belongs to me rightfully'. In the years to come, women (and men) might lose this traditional access, as there is no legal provision supporting such practice and it is only a form of traditional access right that is still honored in the village. Already several owners are reluctant to open up the *kettus* to the women. The shift from alternate rice-fish/prawn to the more profitable year-round culture of prawn, a development that appears to encourage privatization at the expense of traditional rights and livelihoods, was one of the major causes of this shift.

Case study: women gleaners of Trincomalee on the east coast of Sri Lanka[2]

This case focuses predominantly on Muslim fisher women from low socioeconomic classes who glean for clams and mussels, engaging in 'marginal' livelihoods, working hard, smelling of the lagoon, mud, and fish, and earning a lower income in comparison with men. Gleaning is passed down over generations: participants stated that 'our mothers and grandmothers used to do this' and young girls helping their mothers and aunts clean and extract the meat from the shells once the catch is taken to the lagoon shores is common practice. Often, these women sell their own catch, or market it in the neighboring villages through mobile vendors.

Women gleaners marginality derives from where they live and work: most live close to the low-lying lagoon areas of the provincial capital city Trincomalee, on the east coast of Sri Lanka; areas that are flooded during the rainy season and smell of drying mud during the rest of the year. Their marginality also comes from their socio-economic status: gleaning for clams and mussels (*matti* in Tamil) is traditionally associated with lower-income Muslim women in Trincomalee, who represent the second largest ethno-religious minority group in Sri Lanka. There are a few Tamil women, from the largest minority group in the country, who engage in gleaning, but there were none from the majority ethnic group, Sinhalese. While there is a general sense of social stigma attached to women who engage in fishing related activities in Sri Lanka, as they are looked down upon by men and women of higher classes, especially Sinhalese (Stirrat, 1988; Lokuge & Hilhorst, 2017), this stigma is more pronounced towards the Muslim women discussed in this case, as shown by the extract below.

> People look down upon those who are doing the matti [clam and mussels] collection, as they collect the matti in the muddy area. Others say that there will be a mud smell coming from the matti collectors. As we are the poorest people; rich people don't respect us.
>
> *(Female gleaner, Muslim)*

Those who manage to achieve a higher income level, in some cases through migration to the Gulf region as domestic workers, stop gleaning in the lagoon and strive to support their relatives financially, so that their relatives can also stop gleaning.

The kind of fishing women gleaners perform reinforces their marginality. Fishing is a gendered activity in Sri Lanka, with men largely working in capture fisheries in the coastal and deep seas, while women work in fish processing and marketing to a limited extent. However, in the lagoons and shallow seas, gleaning for clams and mussels and catching small fish, prawns and crab using hands, legs, and cages, for both home consumption and sale is common practice for women who live close to the lagoon systems on the east coast.

Despite engaging in gleaning for generations, these women are not recognized by state authorities to be fishers and are not formally registered. The system which frames fisheries management and governance as a relationship between the government and a rights holder, who is usually male (Jentoft, 2000), renders these women invisible. Official statistical data collected by the Department of Fisheries and Aquatic Resources of Sri Lanka is not disaggregated by sex and these women are not part of any fisheries registries.

Additionally, none of the women who glean stated that they belong to a fisheries society. This lack of membership in fisher collectives results in a lack of representation in decision-making spaces at the community level. On top of their invisibility in government fisheries structures, they do not have identity cards (which are generally issued to male fishers and members of fisher societies) to prove their identity as fishers and therefore as users of coastal and lagoon spaces and resources. This invisibility and marginality, in turn, mediates their access to often contested and restricted water systems such as lagoons, forest areas, and shallow seas in Trincomalee.

The areas where these women engage in gleaning in Trincomalee were fiercely contested by the government of Sri Lanka military and the Liberation Tigers of Tamil Ealam (LTTE) for about three decades until the war ended in 2009 with the military defeat of the LTTE. This fighting resulted in several waves of mass scale displacement in the area, with the latest taking place in 2006. After 2009, there has been no overt war-related violence in the area, but the military maintains a heavy presence, concentrated around military bases and smaller outposts and especially around forested areas, lagoons, and along the coastal belt of Trincomalee, which are seen as important for strategic security of the country. As a result, women (and men) who access these spaces for their livelihoods have to engage in constant negotiations with the military. Given the history of violence attached to the military and memories of displacement as a result of the war, Muslim women reported feeling apprehensive about engaging with the military, which is largely made up of majority ethnic Sinhalese:

> Since last Friday we haven't gone to the forest for any purpose because the Navy has come to the forest. We don't know the reason why they are there. We are scared to go now. They are bad people, and women can't trust them.
>
> *(Female gleaner, Muslim)*

The lack of official recognition by the government of these women who engage in fishing, combined with their lack of membership in community-level fisher collectives further impedes their ability to negotiate access to these water spaces to engage in their livelihoods. For example, after a young man drowned in the adjacent seas, the Navy banned a group of about nine Muslim women who collected clams in the shallow seas off Trincomalee from using their *vallam* (non-mechanized craft) for gleaning. Women, who are not part of a fisheries society, had to negotiate with the military, whom they do not "completely trust", for access to these shallow sea areas. Furthermore, they had to speak in the Sinhalese language, which for many is not their mother tongue. In this case, the women resorted to conducting their negotiations through their male relatives, bowing down to the gendered and ethnic social hierarchies and power dynamics in order to continue their livelihood. In another similar incident, a group of Tamil gleaning women explained how they were arrested by the police for damaging the mangroves, which they felt could have been prevented if they had an identification document that stated they were engaging in fisheries-related activities.

Working in marginal geographical spaces, belonging to marginal social spaces, rendered invisible by both state and community structures and institutions, these women continue their struggles to access their livelihood spaces.

Case study: inland fisheries, Cambodia[3]

In inland Cambodia, fishing is an important secondary source of income. Often, farmers are not able to produce enough rice for the whole year, and fishing and the income it provides have been a crucial source of food security. However, access to fish has become contentious with the introduction of intensive agriculture to the area. This is the case for a village in Banteay Meanchey province near the Thai border, with around 150 households, of which 50 households were full-time fishers as of 2016. There is a 20 km-wide reservoir in this village, which used to be surrounded by forest, but with deforestation, it has disappeared. Until the 1990s, both women and men fished in the reservoir. Women fished mainly in the forest area using scoop nets and fishing rods, catching fish for home consumption, while men fished for larger fish destined for sale in the deeper water using gill nets and cast nets. Men sell fish at the market in the village; from there fish are often sent to other villages. Seventy percent of women process fish, mostly for home consumption but also for selling. Only six women in the village are professional fish processors who process fish for sales. These women also sell fish in Thailand to Cambodian workers working in Thailand. The village produces wet season as well as dry season rice. In years past, they planted rice three times a year, but more recently the commune advised them to plant only twice a year due to water shortages.

Around 2014–15, the reservoir began to dry up due to increased production of dry season rice. A Thai company bought about 100 ha of land from farmers in this village, dug a canal and rented some of the land back to farmers for dry season rice cultivation. Previously, reservoir water was used only as supplementary irrigation, but the increased intensity of dry season rice cultivation has put pressure on the reservoir. At the same time, the canal dug by the Thai company was not well constructed, and some paddy fields did not get enough water. Water management was poorly implemented: paddy fields were flooded when the water level was high and dried up when it was low. In addition to the Thai company's land, a Chinese company also bought up land to produce rice farther away from the reservoir. However, the company built a large canal around its land and transports water through the canal using a large water pump, ensuring consistent availability of water for dry season rice.

Rice farmers did not passively observe this depletion of water: villagers went to the government-run water management committee to voice their concerns. The water management committee did not include villagers but was composed only of commune/village authorities. The complaint became violent, and vehicles were set on fire. After this incident, the committee began to listen to people's complaints and release water when requested.

All these developments mean that there is less water in the reservoir and consequently, fewer fish. Reduced water levels are compounded by deforestation around the reservoir, decreasing spawning sites for fish, and an increased number of fisherpeople because of population growth. Moreover, some fishers have resorted to using illegal fishing methods, such as the use of electricity, to make up for the lower catch. By 2017, the decrease in fish catch was felt keenly by all the fishers. Before, fishers said that they could earn 30,000 riels (USD7.5) per day from fishing; now, villagers report earning only 10,000 riels (USD5) per day. Still, fishing remains an important income source, as many reported they can earn at least 5000 riels (USD1.25) per day.

When there is less water, less forest, and less fish, women are not able to fish at the edge of the reservoir as they used to. In general, fishers need to go to deeper areas of the reservoir to secure a catch, but many women stopped fishing because going far away from the shore to fish was too time consuming for them, as they needed to juggle their time with household work and other care work. Women also said that the middle of the reservoir is too windy, and they feel scared to go by boat.

Such changes in fish availability and the way women are involved in fishing created differentiation among women. Women whose husbands fish in the reservoir will wait at the shore with food for their husbands, who are out on the reservoir the whole night to fish. They will wait to receive fish and then go to sell the fish. Some poor women who do not have men in the house to fish continue to glean, collecting snails and crabs along the edge of the reservoir. Such gleaning activities are done only by women. The prices of such snails are low, and only a few lower income women without other options sell the snails, such as those women whose husbands are sick. Women who have other sources of income only collect such snails for home consumption. The villagers said that if they were not desperate, they would not be collecting snails for sale, demonstrating that many villagers feel that such gleaning is degrading.

Still, even households that have male members to go fishing in the reservoir face difficulties in making ends meet. Around 10% of women support the household income through tailoring and weaving, and both women and men perform construction work to supplement their household income. But women are paid only 20,000 riels per day, while men are paid 30,000 riels.

A major way to compensate for the loss of income from fishing is labor migration. There are around ten to 20 households that have members who migrated to Korea, and many have migrated to Thailand.

> Even if they release fish in the reservoir, if water level is low, fish do not increase. So, again, people have to go to Thailand to earn. Nowadays, when there are not many fish, people just go for labor migration. Migration started to increase around two years ago. In 2015, the water level was low, and since then, labor migration increased. Before, people did not want to go to Thailand to work. But since the income in the village has decreased, they started to go. If they go to Thailand, they will earn cash income. They will come back during the New Year, and then go again.
>
> *(Focus group discussion with men fishers)*

Often, couples migrate for labor together. Women expressed that they did not want to send their husbands alone for migration, since men are not able to save money. One woman complained that after her husband went to Thailand, he stopped remitting, so she is thinking of going to Thailand to join him.

With the decrease in water in the reservoir and the decrease in fish catch, both women and men struggle to do various things to make up for the loss. However, the possibilities are quite limited. Especially for women, there are less options compared to men. The woman vice head of the village said that the problem is that the fishers do not have enough information to cope with and adjust to the changes. Especially for women, such lack of access to resources and information is evident. Women do not attend trainings or field trips to visit other villages as much as men. As she said:

> Men will criticize women for going to meetings and go to various places to learn and participate and say those who travel around are bad women. Such criticism makes women feel discouraged or scared to go out of the house and make women not able to get information. Women do not go out to study. If their husband gets angry or becomes jealous, women quickly quit participating. Men will say that women going here and there are bad women. Women are afraid to be divorced for this, so they quit [participating in training]. Such lack of knowledge and information makes women vulnerable.

She herself had problems with her husband – he was not agreeable to her traveling. However, she persevered and now he is supportive of her seeking out training. He stated, 'It is a problem if women do not have any information and do not know anything. When a woman is widowed, then she does not know anything and does not know what to do'.

This case shows how the decrease in water in the reservoir combined with other factors lead to difficulties for fishers continuing their livelihoods as before. Women are pressured to explore other income sources to make ends meet but have fewer options to cope with change because of their lack of access to information and knowledge. Poorer women are more affected, since they lack other sources of income and depend on gleaning around the fringes of the reservoir for their everyday survival.

Conclusion

The foregoing case studies illustrate the different ways women's fishery-based livelihoods are tied to water resources. Collectively, they reveal the complex manner in which water resources and ecologies, resource rights, and livelihoods are intertwined in uniquely gendered ways. In each case study setting, gendered divisions of labor assign women specific tasks, which along with patriarchal social, cultural, and institutional norms, shape women's mobility, livelihood options, rights, and access, situating them in water ecologies in particular gendered ways. Therefore, when economic development or ecological change unfolds, women are impacted by such developments in particular ways. Urbanization, commercial real estate development, pollution, agricultural development, and military occupation, have reconfigured water resource access rights in ways that are gendered and further shaped by class, caste, language, ethnicity, religion, and age.

In the Vembanad Lake region of Kerala, pollution has degraded water resources, damaging the health and sustainability of fishery resources. Real estate development, privatization, and commercialization have further impeded women's traditional access to water resources that provide the foundation for their fishery-based livelihood activities. In Trincomalee, Sri Lanka, military occupation, fear of violence, minority social status, and lack of formal rights to lagoons and waterways for fishing and gleaning, have deprived women of access to water resources to pursue their traditional livelihood of gleaning for mussels and clams. In Banteay Meanchey province, Cambodia, agricultural development, namely intensified dry season rice cultivation relying on reservoir fed irrigation, combined with deforestation, inefficient infrastructure, and poor water management practices, have reduced reservoir water levels, adversely impacting fishing and gleaning activities by reducing fish stocks. Women, who fish and glean at the edge of the reservoir, feel these effects most severely.

In all three case studies, women engage in a form of fishing known as gleaning, which takes place in rivers, streams, and on the edges of larger bodies of water. Although some of this activity goes toward commercial exchange, most is primarily for household consumption. As a result, women's involvement in the fisheries sector is ignored, undervalued, and in fact rendered 'invisible'. Thus, women enjoy few, if any, formalized rights of access to water resources on which fishing and gleaning livelihoods depend, a situation further compounded by exclusion from fisher organizations and water resource governance institutions that could represent and ensure their rights.

A third theme that emerges from these case studies is the way gender is intersected by other identities such as caste, ethnicity, religion, class, and age. In Kerala and Sri Lanka, fishing is a caste- and/or ethno-religious-based occupation. This means women from different caste or ethno-religious groups are connected to water resources in particular ways. In all three countries, poor women and elderly women are the most reliant on fishing and gleaning activities as they have fewer alternative livelihood options. When water resources are degraded or impacted by economic change, the impacts of such change are felt most acutely by particular groups of women – in these case studies, by poor, elderly women, and women from particular caste or ethno-religious groups. These situations point to the strong relationship between gender, water resources, and fishing livelihoods, and the way relationships with and access to water can be said to be gendered.

Notes

1 This study funded by Indian Council of Agricultural Research (ICAR) was carried out by Nikita Gopal with the assistance of Jiswin Joseph (Field Assistant), Sruthi P and Rakesh M Raghavan (Technical Assistants), ICAR-CIFT.

2 Data was collected by Gayathri Lokuge for her PhD thesis, with the assistance of Tharshini Kugan and Sathiyavani Subramaniyam (Research Assistants) in 2013–2014.

3 This study took place between 2016 to 2019, under the research project Migration and Collectives/Networks as Pathways out of Poverty: Gendered Vulnerabilities and Capabilities amongst Poor Fishing Communities in Asia, led by Ragnhild Lund and funded by Norwegian Research Council. The Cambodian study was conducted by Kyoko Kusakabe together with researchers from Cambodian Institute for Research and Rural Development (CIRD).

References

Adeyemo, O. K. (2003). Consequences of pollution and degradation of Nigerian aquatic environment on fisheries resources. *The Environmentalist*, *23*, 297–306.

Dugan, P., Dey, M. M., & Sugunan, V. V. (2006). Fisheries and water productivity in tropical river basins: Enhancing food security and livelihoods by managing water for fish. *Agricultural Water Management*, *80*, 262–275. https://doi.org/10.1016/j.agwat.2005.07.017

Fröhlich, C., Gioli, G., Cremades, R., & Myrttin, H. (Eds.). (2018). *Water security across the gender divide*. Springer. https://doi.org/10.1007/978-3-319-64046-4

Gownaris, N. J., Pikitch, E. K., Aller, J. Y., Kaufman, L. S., Kolding, J., Lwiza, K. M. M., Obiero, K. O., Ojwang, W. O., Malala, J. O., & Rountos, K. J. (2017). Fisheries and water level fluctuations in the world's largest desert lake. *Ecohydrology*, *10*, e1769. https://doi.org/10.1002/eco.1769.

Harris, L., Kleiber, D., Goldin, J., Darkwah, A., & Morinville, C. (2017). Intersections of gender and water: Comparative approaches to everyday gendered negotiations of water access in underserved areas of Accra, Ghana and Cape Town, South Africa. *Journal of Gender Studies*, *26*(5), 561–582. https://doi.org/10.1080/09589236.2016.1150819

Jensen, J. G. (2001). Managing fish, floodplains and food security in the Lower Mekong Basin. *Water Science and Technology*, *43*(9), 157–164.

Jentoft, S. (2000). The community: A missing link in fisheries management. *Marine Policy*, *24*(2), 53–60. https://doi.org/10.1016/S0308-597X(99)00009-3

Lokuge, G., & Hilhorst, D. (2017). Outside the net: Intersectionality and inequality in the fisheries of Trincomalee, Sri Lanka. *Asian Journal of Women's Studies*, *23*(4), 473–496. https://doi.org/10.1080/12259276.2017.1386839

Meinzen-Dick, R., Brown, L. R., Feldstein, H. S., & Quisumbing, A. R. (1997). Gender, property rights, and natural resources. *World Development*, *25*(8), 1303–1315. https://doi.org/10.1016/S0305-750X(97)00027-2

Runyan, A. S. (2018). What is intersectionality and why is it important? *Academe*, *104*(6).

Stirrat, R. L. (1988). *On the beach: Fishermen, fishwives, and fishtraders in post-colonial Sri Lanka*. Hindustan Publishing Corporation.

23

WET'SUWET'EN WOMEN LEADING THE DEFENSE OF RIVERS AND WATER FROM ABUSES COMMITTED IN CONNECTION WITH MEGAPROJECTS. THE PERSISTENT LEGACIES OF THE PAST IN CANADA

Nancy R. Tapias Torrado

Introduction

"The ancestors would say, we are the land, and the land is us. Because everything that we are, every part of us, is coming from the land. It's coming from our relationship with the animals and with the plants and the water."[1] When Wet'suwet'en leader Sleydo' Molly Wickham (Sleydo') explained her experience and connection with the Yintah (Wet'suwet'en territory) with those words,[2] it was impossible not to think of Wedzin Kwa (river).[3] When she arrived to meet me in the Gidimt'en camp, she was visibly distressed.[4] Coastal GasLink (CGL) was starting to drill under Wedzin Kwa, a sacred river for the Wet'suwet'en people,[5] whose "ties to this river are so ancient and deep that [their] Ancestors saw themselves as being part of it" (Morin, 2016, p. 1). Sleydo' was deeply concerned because this river is fundamental for her people. It is a main source of salmon – a staple food for the Wet'suwet'en. It is one of the last sources of pristine waters in their territory, there were outstanding environmental concerns, and the Wet'suwet'en Hereditary Chiefs claim that they have never granted consent for it. Furthermore, the river's area impacted by the pipeline construction is where salmon spawn, precisely at that time of the year when drilling and blasting were starting (Simmons, 2022b). Also, that part of the river is connected to Tsel Kiy Kwa (Lamprey Creek), which is home to "blue-listed" lamprey eels (Gidimt'en, 2022). Thus, for Sleydo' this was a very worrying and painful moment. For CGL, in contrast, this was a necessary step to advance the pipeline that crosses over the north of the British Columbia Province (BC), going from the Dawson Creek area to the Liquefied Natural Gas (LNG) Canada export terminal in Kitimat.

DOI: 10.4324/9781003100379-28

The CGL project, however, is only one of several projects affecting the Yintah, which include logging and mining. Wet'suwet'en leader Chief Howihkat Freda Huson told me they had to choose at a critical moment. "We went for the more imminent [projects]. So, we decided to go for the pipelines. . . . Four projects that could wipe out the territory."[6] For over a decade, Wet'suwet'en women have been leading the defense of the Yintah from the construction of pipeline projects. They and the Wet'suwet'en Hereditary Chiefs, the highest-ranking authorities in their ancestral law and socio-political structure, have consistently argued that they have never granted their consent and have called on the authorities and corporate actors to stop the project in the Yintah. Yet, at the end of 2022, CGL completed 75 percent of the construction (CGL, 2022) and started to drill under Wedzin Kwa, claiming that they "have agreements with all 20 of the First Nations along the project route, and . . . receive strong support" (CGL, 2023b).

In the face of such a contrast, an explanation is necessary. More specifically, what explains that the Wet'suwet'en women-led mobilization has not had the desired impact on corporate behavior? This is a critical question for several reasons: the Wet'suwet'en Indigenous women-led mobilization is one of the country's most visible, strong, and supported social mobilizations; CGL claims to be committed to positive relationships with Indigenous peoples (CGL, 2023b); and Canada is regarded as a leader in legal pluralism and human rights (Macaulay, 2022), and "serves as a global model for stability, [and] sustainable prosperity" (World Bank, 2021). Also, Canada claims "to be committed to a renewed nation-to-nation relationship with Indigenous peoples based on recognition of rights, respect" and reconciliation (Crown-Indigenous Relations and Northern Affairs Canada, 2022).

I first heard in detail about this Indigenous women-led mobilization in an extensive conversation with Wet'suwet'en leader Marlene Hale and Wayuu leader Jakeline Romero Epiayuu (from Colombia), in 2019 (Tapias Torrado, 2022c).[7] In this conversation, I found several overwhelming similarities between the two social mobilizations; and, of course, some important differences.[8] For Marlene, it made sense and was important to use the "braided action" theoretical framework, which emerged from my work in Latin America (Tapias Torrado, 2020), to understand the similar reality faced by her people in Canada. I was immediately engaged, but I wondered if my learnings from Latin America could be useful and applicable in a country like Canada.

In this chapter, I share some of my main findings using the "braided action" theoretical framework to understand the impact (or not) of the Wet'suwet'en women-led mobilization on corporate behavior regarding the CGL pipeline project in British Columbia, Canada. I first introduce the "braided action" framework, then briefly present the Wet'suwet'en women-led mobilization. Next, I focus the discussion on the adverse conditions that affect "the braid of action"; more specifically, I illustrate how colonial legacies persist, affecting the Wet'suwet'en women-led mobilization and its impact. I end with some concluding remarks.

The "braided action" framework – a brief overview

The "braided action" theoretical framework offers an explanation of the "success" (or not) of Indigenous women-led mobilizations challenging the abuses committed in connection to megaprojects in Latin America in the face of extreme violence (Tapias Torrado, 2020, 2022b).[9] By "success" I mean a substantial change in corporate behavior that is favorable to the social movement (i.e., the withdrawal or cancellation of the corporation's participation in a megaproject in which human rights have been abused). By "non-success," I mean there is no such change, even if there are other changes (e.g., a temporal suspension or statements expressing commitment to Indigenous rights). Although I have identified a continuum of successes, I use this dichotomous measure since even very small changes in behavior indicate an important level of success, given the power disparities among the actors involved in this case, as I explain in the next sections.

This framework is a metaphor and a theory that draws on the knowledge and experience of Indigenous women leaders and their peoples, the sociology of social movements, and legal and intersectional studies (Tapias Torrado, 2020). It emerges from a multilevel qualitative comparative analysis of cases of "success" and "non-success" in Colombia, Ecuador, Honduras, Mexico, and Peru, selected from an original database of cases of Indigenous women-led mobilizations challenging abuses related to megaprojects in the Americas, which included from an early stage, the Wet'suwet'en experience.

The "braided action" framework adapts the "Archimedes lever theory," which was developed to study accountability efforts regarding economic actors' participation in past atrocities (Payne et al., 2022, 2020; Tapias Torrado, 2022a). The "braided action" studies contemporary cases of Indigenous women-led mobilizations that impact (or not) the practice of economic actors participating in megaprojects where human rights have been abused. Thus, this framework has the following essential parts: the weight to be lifted with the lever; the force applied to keep it down; the placement of the fulcrum; and the braid lashed to the lever which pulls it from the territory to lift the weight up.

In essence, I argue that in a minimally favorable human rights context (fulcrum), Indigenous women-led mobilizations claiming respect and protection for their ancestral territories and most fundamental rights (weight to be lifted) can harness a mobilization power in "a braid of action," achieving "success" over much more materially powerful forces (veto power or the right-against-rights) (Tapias Torrado, 2023). The "braid of action," an essential tool and unique "weapon of the weak" (Scott, 1985), lashes to the "lever" to bring it down and lift human rights up. The "braid" has four "strands." The first three are internal to the movement, and the fourth is a reaction to an external factor. They are: (1) transforming the territory into mobilization power; (2) Indigenous women's effective leadership; (3) human rights framing; and (4) reacting to a grave violation overtly involving corporations. These four "strands" interweaved in a tough "braid of action" are capable of "harnessing" mobilization power. However, if some of the "strands" are "weak" or "frayed," the "braid" cannot achieve a favorable change in corporate practice.

This framework is a dynamic model,[10] and it does not operate in a vacuum. Thus, it starts by acknowledging power relationships and conditions that affect the braid, which will be the focus of the section entitled "A legacy of silencing practices that persists." First, it is necessary to present the Wet'suwet'en women-led mobilization briefly.

Some key aspects of the Wet'suwet'en women-led mobilization

"We got brought up on the land, shown how to live off the land . . . [Very often,] we would go to the territory. And the more we went out, the more destruction we saw," says Wet'suwet'en leader Freda Huson, who then went to live full time in the Yintah, establishing the Unist'ot'en Camp.[11]

The camp was a fundamental step in the Indigenous women-led mobilization. It was one of the many actions that the Wet'suwet'en people have taken to defend the Yintah and their rights, including several legal actions (Daly, 2004). Such as those that led to the 1997 ground-breaking "Delgamuukw-Gisday'wa" decision, which recognized that "Aboriginal titles" had not been extinguished when British Columbia entered confederation in 1871 (1997). Thus, building on a long history defending the Yintah, on their ancestral law, and clans system, and on the certainty that this is unceded and non-surrendered territory, the Unist'ot'en Camp was established.

The camp was set by Wedzin Kwa, at the entrance of the Unist'ot'en territory. The Unist'ot'en group is part of the Dark House of the Gilseyhu Clan. The Wet'suwet'en is a matriarchal and matrilineal people organized in five clans: Gilseyhu (Big Frog), Laksilyu (Small Frog), Gitdumden/Gidimt'en (Wolf/Bear), Laksamshu/Lihkts'amisyu (Fireweed), and Tsayu (Beaver clan). The clans are organized in thirteen houses, each led by a chief (a man or a woman). The Hereditary Chiefs are the head chiefs of each clan.[12] The most important decisions are taken by all clans in the *balhats* or feast system, which is a structural part of the Wet'suwet'en culture and legal system. Each clan is accountable for its own part of the land, for protecting it and sustainably managing it, and for making sure there is always abundance in it, as it impacts all. It is a system built upon *waggus*, respect – a core principle for the Wet'suwet'en.

The Unist'ot'en Camp was established at kilometer (km) 66 of the Forest Service Road west of Houston from Highway 16, which is known as the Highway of Tears given the many women who have "disappeared" or found dead along it, most of them Indigenous women (NIMMIWG, 2019). In 2009, it started as a checkpoint. Then, a log cabin was built at the exact location of the CGL, Pacific Trails and Northern Gateway pipelines, proposed (at the time) by Trans-Canada, Chevron, and Enbridge, respectively (Unist'ot'en, 2017a).[13] The camp became a center where hundreds attended workshops and cultural and land based practices, such as water ceremonies. The camp became a center to connect with the Yintah and to revitalize culture and health. In 2015, the Healing Centre was established as its core initiative, with the goal of "healing the people, healing the land" (Unist'ot'en, 2017b).

Such healing and connection were strongly felt by Wet'suwet'en leader Jennifer (Jen) Wickham. She explains:

> Unist'ot'en was having a summer action camp, so I came up . . . and that's when I met Wedzin Kwa. I drank directly from the river. I swam and bathed. . . . And that's when I decided to move and help protect her in any way that I could.[14]

Jen is the communications coordinator of the Gidimt'en Camp, and Sleydo's sister. They both are part of the Cas Yikh (Grizzly) House, of the Gidimt'en Clan. In 2012, after meeting Wedzin Kwa, they moved with their family to the Yintah to protect it.

At that time, all the proposed pipelines were trying to find their way in, despite the Hereditary Chiefs explicit opposition (Office of the Wet'suwet'en, 2013). And, in 2014 the British Columbia government approved the Enbridge Northern Gateway and the CGL pipeline projects.

CGL aims to transport between 2.1 to five billion cubic feet of natural gas per day to the export terminal LNG Canada in Kitimat (CGL, 2023a), to then transport it in container ships to Asia and "customers around the world" (LNG Canada, 2023). CGL is owned by TC Energy (formerly TransCanada) (CGL, 2023a), and has almost double its costs, from $6.6 billion to $11.2 billion (Bennett, 2022). LNG Canada is a joint venture among Shell, PETRONAS, PetroChina, Mitsubishi, and KOGAS. It is a $40 billion project, the "largest private sector investment project in Canadian history" (Prime Minister of Canada, 2018). Although the provincial government approved an alternate route for the pipeline (CGL, 2023a), it was not the Wet'suwet'en's proposal (*Coastal GasLink Pipeline Ltd. v. Huson*, 2019, para. 59).[15] The CGL pipeline is a 670-kilometers long pipeline that crosses "about 625 streams, creeks, rivers and lakes, including vital fish habitat" (UN WG Business et al., 2023, p. 3).

After the provincial government's approval, the Hereditary Chiefs maintained that they have not granted consent for that project in the Yintah and insisted on their many concerns, including those related to negative impacts on water quality, fish habitat and wildlife abundance (Office of the Wet'suwet'en, 2013, 2018).[16] Yet, the Unist'ot'en Camp was notified of an injunction and a civil lawsuit, sought by CGL against them (*Coastal GasLink Pipeline Ltd. v. Huso*, 2019, paras. 1–2; Unist'ot'en, 2018). The Gidimt'en checkpoint was then set at km 44 (Gidimt'en Checkpoint, 2018). The injunction was then modified to include it. On 7 January 2019, the Royal Canadian Mounted Police (RCMP) enforced it at km 44, arresting Sleydo' and 13 other land defenders, and establishing a temporary exclusion zone, thus only the RCMP could enter the area, not even the media or legal observers. An investigation by The Guardian later revealed that the RCMP "were prepared to shoot Indigenous land defenders" and "to arrest children and grandparents" (Dhillon and Parrish, 2019).

Sleydo' and her family tried to return to the Yintah, but the RCMP did not let them pass although the injunction was only to prevent "blocking the roads when construction activities are underway" (Michael, 2019; Supreme Court of B.C., 2019, para. 208). CGL continued the construction, even when there were archaeological concerns (Huson et al., 2019; Unist'ot'en, 2019).

The camp was re-established in February 2019, but the temporary injunction became permanent (Supreme Court of B.C., 2019). The Hereditary Chiefs rejected that decision, as a "criminalization of their ancestral law" (Unist'ot'en, 2020a); and, on 4 January 2020 the Chiefs issued an eviction notice to CGL (Unist'ot'en, 2020c). But, as CGL resumed work (CGL, 2020a), a new camp was erected at km 39, and the RCMP extended the exclusion zone to include it (Gidimt'en Yintah Access, 2020; Walia, 2020). The Hereditary Chiefs formally requested to meet with BC Premier John Horgan and Canadian Prime Minister Justin Trudeau, but they never met them (Meissner, 2020). After many calls for respecting Indigenous rights, including those of the United Nations Committee on the Elimination of Racial Discrimination (UN CERD) and the BC Human Rights Commissioner (BC HRC) (BC HRC, 2020; UN CERD, 2019), and some expressions of support for the project (Pawliw, 2020; The Canadian Press, 2020a), a period of dialogue was agreed to de-escalate the situation (Office of the Wet'suwet'en, 2020).

Despite that intention, the situation deteriorated. Between 6 and 10 February 2020, RCMP officers broke into all the Wet'suwet'en Camps arresting Wet'suwet'en women leaders, land defenders, some journalists, and legal observers. Also, the RCMP extended the "exclusion zone" to km 3, denying access to the Yintah to the Hereditary Chiefs (Hosgood, 2020; Jones, 2020; Patterson, 2020; Unist'ot'en Camp, 2020).

On 11 February, CGL expressed "disappointment" about the need for enforcement and announced the "recommencement of construction . . . [and that] will redouble efforts to engage with the Hereditary Chiefs" (CGL, 2020b). The Hereditary Chiefs, however, were insisting on a nation-to-nation dialogue, but without success (Bellrichard, 2020).

At this point hundreds had already expressed solidarity with the Wet'suwet'en.[17] Yet, the February events detonated an unprecedented show of support. The response to the hashtag "Shutdown Canada" was massive (CBC, 2020), and several parts of the rail service were suspended across the country (BBC, 2020).

In a tense context, on 1 March 2020, the Wet'suwet'en, the Canadian Minister of Crown – Indigenous Relations and the BC the Minister of Indigenous Relations and Reconciliation reached a memorandum of understanding (MoU). It started by stating: "Canada and BC recognize that Wet'suwet'en rights and title are held by Wet'suwet'en houses under their system of governance . . . throughout the Yintah" (Crown-Indigenous Relations and Northern Affairs Canada, 2020). This was an incredibly important step, but the pandemic arrived and broke the momentum, also badly impacting the Wet'suwet'en (Forsythe et al., 2020). The MoU marked the beginning of a process that should have led to legally binding agreements, but it was not binding in itself. It did not include or directly address the CGL project. The pipeline work continued (Unist'ot'en, 2020b). The Hereditary Chiefs persisted in their call, and on 3 July 2020, they met in the Yintah with David Pfeiffer, CGL's president at the time, who apologized to them (Gidimt'en, 2021g). Two months later, Pfeiffer resigned.

In September 2021, the Coyote Camp was established on the drilling site on the banks of Wedzin Kwa (Gidimt'en, 2021b). Also, in October 2021, Chief Dsta'hyl Adam Gagnon, enforcement officer of the Lihkts'amisyu Clan, deactivated a CGL excavator used in their territory without their consent (Fireweed Solidarity, 2021), but he was later arrested by the RCMP, released the day after, and is now facing a criminal proceeding. Thus, the eviction order issued by the Hereditary Chiefs in January 2020 was read once again to CGL (Gidimt'en, 2021d). Furthermore, Indigenous Haudenosaunee came to support Coyote Camp (Gidimt'en, 2021c). Between 18 and 19 November 2021, about 50 armed RCMP officers arrested Sleydo', two Wet'suwet'en elders, three legal observers, one journalist, and more than a dozen land defenders (Gidimt'en, 2021e, 2021f). Many of them continue to face charges of civil contempt for breaching the injunction granted to CGL. They have applied "to have these criminal charges stayed" arguing widespread human rights violations (Gidimt'en, 2023a).

After that, the main focus of the mobilization has been on investors and financiers of CGL, specially the Royal Bank of Canada (RBC) (Gidimt'en, 2021f, Chiefs of Cas Yikh, 2021). Among other actions, in 2022 and 2023, the Hereditary Chiefs tried to directly raise their concerns before the RBC's annual meetings; but were unsuccessful. The Hereditary Chiefs claim they have not been treated with respect: for example, on the last occasion they were segregated to a secondary room, but RBC said they were treated fairly and respectfully (Woodside, 2023).

On 18 January 2023, after 37 warnings and 17 orders, BC's Environmental Assessment Office (EAO) issued a third monetary fine to CGL, as there were "continued deficiencies" in the "measures to protect sensitive wetlands and waterways from sediment caused by erosion, which can negatively impact water quality and fish habitat" (EAO BC, 2023). Still, CGL continues to advance the project, and "16 First Nations signed an equity option agreement which gives . . . [them] a share in the profits" (LNG Canada, 2023). Also, although there is now a systemic investigation of the RCMP "E" Division Community-Industry Response Group (C-IRG), which has operated in the Yintah in the last few years, its violence persists. RCMP are arresting Wet'suwet'en women leaders and land defenders at the time I am writing these lines (CRCC, 2023; Gidimt'en, 2023b). Yet, the Wet'suwet'en women-led mobilization continues, insisting on their call to stop the project in the Yintah. The Unist'ot'en and the Gidmt'en Camps are active and strong.

The strength of the Indigenous women-led mobilization has been incredible, as well as the "veto power" used to advance the pipeline project. Understanding the conditions that affect the "braid of action" is then a necessary step.

A legacy of silencing practices that persists

Looking at the Wet'suwet'en women-led mobilization through the "braided action" framework reveals a power disparity among the different actors involved. It also allows us to see the different parts of the model. The placement of the fulcrum is minimally favorable, as there have been important legal and institutional advancements for the protection of Indigenous rights in the last few years, such as Canada's full endorsement of the United Nations Declaration on the Rights of Indigenous Peoples (UNDRIP) in 2016. It also shows the strength of "veto power" and of the "braid," but also how the "braid" has been affected by factors external and internal to the movement, such as the social divisions created by colonial law. Thus, given the extent of findings and the importance of starting with the conditions affecting the "braid," here I only focus on one key aspect of a possible explanation to the posed question: the persistent legacies of the past.

Indigenous women-led mobilizations take action in the face of very imbalanced power relationships and adverse conditions that affect their mobilization processes and impacts. The "braid" formation, the actions that compose its four "strands," and its interaction with other parts of the model do not occur in a vacuum. The "braid" is affected by the conditions in which it operates. It is affected by silencing actions and silencing conditions, whose intersection I call silencing practices (Tapias Torrado, 2022c). Silencing conditions include oppressive historical legacies that persist through contemporary patterns of oppression. Indigenous Lenca leader Berta Cáceres (from Honduras) eloquently explained that their struggles defending Indigenous territories and rights from abuses connected to megaprojects were conditioned by "a capitalist, racist and patriarchal depredation" (Cáceres, 2015). Thus, silencing conditions undermine the actions and impact of Indigenous women leaders and their mobilizations. They also make them more vulnerable to silencing actions, which are those of the "veto players" aiming to halt the defense of Indigenous territories and rights. Hence, the experience of Indigenous women leaders and human rights defenders supporting them is shaped by who they are (intersecting identities), by what they do to defend Indigenous dignity, territory, and rights, and by the context and conditions in which they mobilize (Tapias Torrado, 2020).

Colonial oppressive legacies have seriously affected Indigenous peoples in Canada, and the Wet'suwet'en women-led mobilization is no exception. On the contrary, the mobilization process has made visible how those adverse conditions persist. While Freda Huson was brought up on the land and shown how to live off the land, not all the Wet'suwet'en had a similar experience. Sleydo', for example, said: "I never had that growing up. . . . I'm like fighting extra hard so that my kids have something from the time they were born until they become adults."[18] This is linked to the fact that her mother was a victim of the "60s scoop," a state policy that removed Indigenous children from their parents to put them into "white foster care."[19] Sleydo's grandparents died while her mother was away from them before Sleydo' was born. They both went to the Lejac Indian Residential School, which was part of the residential school system:

> [C]reated for the purpose of separating Aboriginal children from their families, in order to minimize and weaken family ties and cultural linkages, and to indoctrinate children into a new culture – the culture of the legally dominant Euro-Christian Canadian society, led by Canada's first prime minister.
>
> *(TRCC, 2015, p. v)*

Also, only after 1985 Bill C-31, which aimed to address gender inequality in the Indian Act based on the Canadian Charter of Rights and Freedoms, Sleydo's mother managed to recover her "Indian status." Under the Indian Act, a married woman would lose her "Indian status" – with all the rights attached to it – if marrying a "non-Indian" man or a "non-status Indian,"[20] or she would become her husband's status if marrying an "Indian man" from another "band," separating her from her "Nation." Thus, the words and experience of Sleydo' and her family reveal serious damages caused by a colonial system that used racist, patriarchal, and capitalist oppressions, which extend to their current situation and the Wet'suwet'en women-led mobilization.

The final report of the Truth and Reconciliation Commission of Canada provides a detailed account of a colonial system that regarded Indigenous people as inferior and savages. This Commission was established after the largest class-action lawsuit in Canada's history. Its final report powerfully starts by stating:

> For over a century, the central goals of Canada's Aboriginal policy were to eliminate Aboriginal governments; ignore Aboriginal rights; terminate the Treaties; and, through a process of assimilation, cause Aboriginal peoples to cease to exist as distinct legal, social, cultural, religious, and racial entities in Canada.
>
> *(TRCC, 2015, p. 1)*

Residential schools, which existed from the 1870s to 1997, were a central element of this policy. But they were only one element of that policy of assimilation and elimination of Indigenous peoples and rights. Other elements under the Indian Act included, for example, the prohibition of the *Balhats* (the 1884 "potlatch law"), barring Indigenous people from organizing in political groups and from hiring lawyers and legal counsel to defend their rights in the legal system (Section 141), and the establishment of "Indian reserves." Indigenous people were forced to move from their traditional territories into smaller, regulated, and limited areas called "reserves," which were also a way to free lands for the colonial project. The Crown holds the reserves' legal title (Government of Canada, 2023). They are governed by the "bands." The "bands" or "Indian Band Councils" refer to the imposed governing structures created by the 1869 Gradual Enfranchisement Act and then regulated by the 1873 Indian Act, which remains valid (1876; Hanson, 2009). The "bands" represent the "reserves" (not the ancestral territories) and are bodies of elected councilors funded by and articulated with the Canadian government. In this vein, establishing camps as a core tactic of the Wet'suwet'en women-led mobilization to defend the Yintah is of critical importance, as it is also the camps' endorsement by the Hereditary Chiefs in *balhats*. They are an explicit form to resist oppression and reassert Indigenous identity, culture, governance, and law in the Yintah, despite the system that has tried for so long to eliminate them. This way of mobilizing contributes to understanding why the Hereditary Chiefs have insisted so many times that no consent or meaningful consultation has taken place, arguing that the State authorities had "circumvented the hereditary system . . . [dealing] exclusively with bands," creating confusion and exacerbating divisions (Office of the Wet'suwet'en, 2018).

The prohibition of the *balhats* and the burning of their regalia and sacred objects did not only intend to eliminate Indigenous culture and spirituality; it also aimed to undermine Indigenous economic systems. As Morin states,

> The colonial government and churches targeted the *balhats* for many reasons. It was believed to work against capitalism. Western economic ideas were and still are based on the private ownership of land and the individual accumulation and concentration of wealth. The feast system openly distributed wealth to all members of the community and functioned separately from the Euro-Canadian economy, until our people started buying food and trade goods from trading posts and stores.
>
> *(2016, p. 254)*

There was a fundamental change that went from redistribution to private ownership, accumulation, and competition. The assimilation process "equated Euro-Canadian socio-economic standards and materialism with success, progress and civilization" (Morin, 2016, p. 218). Such an idea of progress resonates with the words of BC Premier Horgan who, after refusing to meet with the Hereditary Chiefs, visited the LNG Canada site in Kitimat and from there said:

> I'm not going to drop everything . . . when someone is saying they need to speak with me . . . I'm not being disrespectful, I'm just saying be realistic here . . . I think it is disingenuous to suggest that a handful of people can stop progress.
>
> *(Watson, 2020)*

With this same idea of advancing and protecting "progress," in 2017, the RCMP created the C-IRG (RCMP, 2022). More specifically, as the media later revealed, it was irregularly created "to protect pipelines, specifically the Trans Mountain expansion and Coastal GasLink" (Forester, 2022).

Within the imposed patriarchal system, matriarchal and matrilineal cultures such as the Wet'suwet'en were also obstacles to that type of "progress." Mélanie Morin explains in her book *The Ways of Our Ancestors* that "the missionaries and government soon realized that to destabilize our Ancestors' society they needed to change our matrilineal system, and reduce the roles and value of women" (Morin, 2016, pp. 260–261). Thus, the relationship and approach to women changed. Their power and respect were severely reduced, "women were no longer as free as before. Women were shamed and targeted for punishment and violence" (Morin, 2016, p. 256). Grave violence against Indigenous girls and women persists (for example the Highway of Tears), including the negative effects of the "Indian status" regulation. Sleydo', for example, has been questioned about her Indigenous identity. Even within her people, some have questioned her legitimacy as a leader because she did not grow up in Wet'suwet'en land. Colonial regulations created divisions and tensions that the development of megaprojects in the Yintah has exacerbated. After the 2021 arrests, a statement from the Gidimt'en Camp claimed:

> CGL [was] seeking a number of conditions of release. . . . CGL is also asking Sleydo' to provide documentation to 'prove' she is Wet'suwet'en, and is seeking conditions

that would bar her from returning to her home on Wet'suwet'en Yintah where [she and her family] live. CGL is also challenging Chief Woos's daughter Jocelyn Alec's status as a Wet'suwet'en person because she has Indian Act status with her mother's First Nation.

(Gidimt'en, 2021a)

The intensely violent system that forced assimilation on Indigenous peoples is not only a matter of the past. Colonial patterns of oppression persist, generating adverse conditions for the "braid of action." For Sleydo' and many other Wet'suwet'ens I interviewed, the many aspects of the colonial system that affected them in the past still shape their lives. Thus, the Wet'suwet'en women-led mobilization not only challenges the abuses connected to the construction of pipelines and other megaprojects in the Yintah: they are challenging the core of a dominant extractive model that comes from a colonial time.

Concluding remarks

Drawing on the "braided action" framework and the knowledge and experience of the Wet'suwet'en people, it becomes clear that to live and to act in the Yintah is of fundamental importance for the Indigenous Wet'suwet'en people and the strength of their women-led mobilization to defend their territory and rights. Wet'suwet'en leader Dr. Karla Tait eloquently explains:

> The water ceremony is about acknowledging water as a life giver, as a living relative, as a part of us. . . . So, it is about re-establishing that strong connection to water as a life giver and asking that water to take care for you . . . your whole self. Especially when it comes to our negative self-talk as Indigenous people and some of the ways that that ongoing oppression impacts the way that we think about ourselves and about one another. . . . [It is like] shedding some of the things that have been put on us as Indigenous people and leaning into the land to help lift those pieces so that we can feel strong and connected. . . . With contaminated water, we wouldn't be able to experience that. Drinking the clean, living water of Wedzin Kwa, living on the land and working in closer connection to our land . . . is decolonizing and centered on our Indigenous values and beliefs.[21]

When I first heard of "healing the people, healing the land" as the main goal of the Healing Centre, a central initiative of the Unist'ot'en Camp (Unist'ot'en, 2017b), I never imagined the wide scope of its meaning and impact. This and all the other camps that the Wet'suwet'en women-led mobilization has established in the Yintah, with the full support of the Hereditary Chiefs and their endorsement in *balhats*, are not only a mobilization tactic or a form of resistance. They are a crucial way to reaffirm, revitalize and strengthen the connection with the power in the Yintah, something the colonial system tried to eliminate with extreme violence for a long time.

Indigenous people have internalized colonial oppressions, and those same oppressive structures have shaped institutions and laws that persist. Undeniably, Canada has taken important steps to address the system that tried, by all means, to "kill the Indian in the child" and assimilate and eliminate Indigenous peoples in the country (TRCC, 2015). However, the situation faced by the Wet'suwet'en women-led mobilization and the state and corporate responses given to the Hereditary Chiefs are incoherent with those steps. They coincide with and go beyond many aspects of similar cases I have studied in Latin America. There, as in Canada, Indigenous peoples are among the most disadvantaged groups. Their leaders, especially Indigenous women leaders, who claim respect and protection for their dignity, territories, and rights, are some of the most criminalized and attacked.

In this chapter, I briefly discussed some of the conditions affecting Indigenous women leaders, their mobilization, and their impact. They already revealed key aspects of the significant and deep challenges that have prevented their "braid" to lift the "weight" up or "succeed" in their mobilization and demands. But they have also revealed the incredible strength of the Wet'suwet'en people and their women leaders, the centrality of water, and the connection to Wedzin Kwa and the Yintah as part of their mobilization power.

Acknowledgement

I am very grateful for the Postdoctoral Fellowship I received from the FSPD of the UQAM to do this work, and for the support of my supervisor Prof. Bernard Duhaime and the FSPD Vice-dean on Research Maya Jergen. I also want to express my gratitude to Prof. Luis C. Sotelo, his support was crucial during fieldwork. I am particularly grateful to every single person who participated in my postdoctoral research project. I am grateful to Wet'suwet'en leader Marlen Hale for her very important support throughout the project, and with Wet'suwet'en matriarch Charlotte Euverman for supporting me during fieldwork. Special thanks to Dr. Karla Tait, Brenda Michell, Freda Huson, Gayline Morris, and Sleydo.'

Notes

1 Interview with Sleydo', 3 October 2022, Yintah.

2 Wet'suwet'en people refer to the Yintah as their land or territory. Drawing on its meaning and my work in Latin America, I use the terms Yintah and territory.

3 Wedzin Kwa is the Wet'suwet'en name of this river, but it was officially named the Morice River, after colonial missionary priest A.G. Morice (Mulhall, 1986).

4 For the project's methodology, Indigenous women leaders' life histories, voices, and experiences are at the center. I draw on the methodology I applied in my doctoral investigation, which combines six qualitative methods: research-action (Fals Borda, 1978), public sociology (Burawoy, 2010), multi-sited ethnography (Marcus, 1995, 2011), life-history (Goodson, 2001; Marshall, 1998), process tracing (George and Bennett, 1997, 2005) and qualitative comparative analysis (Mill, 1843; Ragin, 2014). After a very thorough process, ethical clearance was granted to my research project (No. de certificate: 4899_e_2021) and then renewed (No. de certificate: 2022–3732), by le Comité Institutionnel d'éthique de la recherche avec des être humains de l'UQAM (CIEREH). Following the ethical protocol approved, the methods application in this case included: 20 in-depth interviews with Wet'suwet'en leaders, Hereditary Chiefs, matriarchs, elders, and supporters, ten days of fieldwork in the Yintah, and the perusal of relevant state and non-state actors publications. Despite many attempts to interview Wet'suwet'en people supporting CGL, only one person accepted to briefly talk with me over the phone but not to be interviewed or recorded. In May 2023, I sent formal letters to relevant state and corporate actors asking for more details on their view of the situation, and at the time of writing I was waiting for their response.

5 I have decided to spell the Wet'suwet'en name and words as the Indigenous women leaders do. I acknowledge the reviewed spelling in the "Distinctly Witsuwit'en Orthography," endorsed by the Hereditary Chiefs in 1993 (Morin, 2016, p. II).

6 Interview with Freda Huson, 7 October 2022, Yintah.

7 Prominent leader of the Wayúu Women's Force (FMW) Jakeline Romero Epiayuu visited Canada for academic and advocacy purposes. Jakeline was also interested in learning from Indigenous

peoples facing the challenges of mega-projects in Canada. I facilitated a face-to-face meeting between Jakeline and Marlene Hale, a Wet'suwet'en leader and chef. FMW has been struggling against numerous human rights violations committed in the Wayúu territory, in La Guajira department, Colombia, including those related to the coal mine El Cerrejón.

8 For example, both Indigenous peoples are matrilineal and are organized in clans, and Indigenous women leaders have, in both cases, faced extreme violence. However, it is impossible to ignore that while Canada did not even have a war of independence, Colombia has had the longest internal armed conflict in the continent.

9 By megaprojects I mean large-scale, extractive, development or investment projects that have major impacts and involve risks, substantial resources and time; and their magnitude is considered within context (Gellert and Lynch, 2003; IACHR, 2016).

10 To explain the dynamics of this model, I build upon the learnings of cases and analytical sociology (Hedström and Swedberg, 1998; Tapias Torrado, 2020).

11 Interview with Freda Huson, 7 October 2022, Yintah.

12 Becoming a Hereditary Chief is an extensive process that starts in childhood and follows the Wet'suwet'en's clan system, law, and culture (Morin, 2016, pp. 36–41). The Hereditary Chiefs system is different from the imposed governing system of the Indian Band Councils, whose councilors are elected by band members, and was created by the Indian Act, as I explained.

13 In January 2022, the Pacific Trail Pipeline Project was purchased from Chevron and Woodside Petroleum by Enbridge. It seems that the Pacific Trail's pipeline will be built in parallel to CGL's. The Enbridge Northern Gateway, for the export of diluted bitumen, appears to be suspended but not cancelled (Simmons, 2022a).

14 Interview with Jennifer Wickham, 8 October 2022, Yintah. The poem she wrote to the river "Engussi Wedzin Kwah" can be accessed here: https://unistoten.camp/engussi-wedzin-kwah/.

15 CGL says that it cannot be changed (CGL, 2023b).

16 This concerns resonate with the significant impacts of megaprojects on the human right to water, which implicate extensive use of land, modification of water resources and long-implementation periods (UN Special Rapporteur on the Human Rights to Water and Sanitation, 2019, p. 2).

17 Many arrested supporters also complained about the use of force by police during arrests (The Canadian Press, 2020b).

18 Interview with Sleydo', 3 October 2022, Yintah.

19 Interview with Sleydo', 3 October 2022, Yintah.

20 Until 1961, others could also lose their "Indian status" (e.g., those graduating from university) (Crey and Hanson, 2009).

21 Interview with Karla Tait, 7 October 2022, Yintah.

References

BBC. (2020, February 14). Trudeau calls for "dialogue" to solve crippling blockades. *BBC News*. www.bbc.com/news/world-us-canada-51511326

BC HRC. (2020, January 10). Statement from B.C.'s human rights commissioner on UN CERD's decision statement. *British Columbia' Office of the Human Rights Commissioner*. https://bchumanrights.ca/news/statement-from-bcs-human-rights-commissioner-on-un-cerds-decision-statement/

Bellrichard, C. (2020, February 11). Coastal GasLink returning to work in injunction area in Wet'suwet'en territory. *CBC News*. www.cbc.ca/news/indigenous/coastal-gaslink-back-to-work-wetsuweten-1.5460328?fbclid=IwAR0u1HmN1quDp93DKPSqnvFkc0HAGstZk9dDS88LQ_LqZNUilU-t186PI-Y

Bennett, N. (2022, September 29). LNG Canada 70% complete. *Business in Vancouver – BIV*. https://biv.com/article/2022/09/lng-canada-70-complete

Burawoy, M. (2010). Southern windmill: The life and work of Edward Webster. *Transformation: Critical Perspectives*, *72/73*, 1–25. http://doi.org/10.1353/trn.0.0062

Cáceres, B. (2015). *Berta Caceres acceptance speech, 2015 Goldman Prize ceremony* [Video]. www.youtube.com/watch?v=AR1kwx8b0ms

The Canadian Press. (2020a, January 21). Indigenous LNG supporters chide human rights advocates over pipeline comments. *CTV News*. https://vancouverisland.ctvnews.ca/indigenous-lng-supporters-chide-human-rights-advocates-over-pipeline-comments-1.4777569

The Canadian Press. (2020b, January 30). Indigenous young people file complaints after Victoria police arrests. *The Province*. https://theprovince.com/news/local-news/indigenous-young-people-file-complaints-after-victoria-police-arrests/wcm/032931b9-b605-427c-9449-d88e2fa10059
CBC. (2020). Support for the Wet'suwet'en sparks nationwide action. *CBC*. www.cbc.ca/player/play/1696483395513
CGL. (2020a, January 20). Coastal GasLink comments on Unist'ot'en actions. *Coastal GasLink*. www.coastalgaslink.com/whats-new/news-stories/2020/2019-01-05/
CGL. (2020b, February 11). Recommencement of construction near Houston, B.C. *Coastal GasLink*. www.coastalgaslink.com/whats-new/news-stories/2020/recommencement-of-construction-near-houston-bc/
CGL. (2022, November 29). Construction update. November 29, 2022. *Coastal GasLink*. www.coastalgaslink.com/siteassets/pdfs/construction/updates/cgl-construction-update-november-2022.pdf
CGL. (2023a). About Coastal GasLink. A pipeline to support the liquefied natural gas (LNG) industry. *Coastal GasLink*. www.coastalgaslink.com/about/
CGL. (2023b). *Coastal GasLink FAQs Indigenous engagement and consultation*. www.coastalgaslink.com/about/faqs/#indigenous
Chiefs of Cas Yikh. (2021, October 4). Letter to investors and financiers of CGL and LNG Canada. *Gidim'ten Checkpoint*. www.yintahaccess.com/news/2021/10/4/wetsuweten-resistance-to-cgl-pipeline-occupation-september-2021
Coastal GasLink Pipeline Ltd. v. Huson. 2019 BCSC 2264. www.coastalgaslink.com/siteassets/pdfs/whats-new/2019/2019-12-31-coastal-gaslink-comments-on-injunction-decision/judge-church-coastal-gaslink-pipeline-ltd.-v.-huson-december-2019.pdf
CRCC. (2023, September 3). CRCC launches systemic investigation of the RCMP "E" division community-industry response group (C-IRG). *CRCC*. www.crcc-ccetp.gc.ca/en/newsroom/crcc-launches-systemic-investigation-rcmp-e-division-community-industry-response-group-cirg
Crey, K., & Hanson, E. (2009). *Indian status*. Indigenous Foundations. First Nations and Indigenous Studies UBC. https://indigenousfoundations.arts.ubc.ca/indian_status/
Crown-Indigenous Relations and Northern Affairs Canada. (2020). *Memorandum of understanding*. Office of the Wet'suwet'en. www.wetsuweten.com/files/DRAFT_MOU_AGREEMENT.pdf
Crown-Indigenous Relations and Northern Affairs Canada. (2022). *Truth and reconciliation commission of Canada*. www.rcaanc-cirnac.gc.ca/eng/1450124405592/1529106060525#chp1
Daly, R. (2004). *Our box was full: An ethnography for the Delgamuukw plaintiffs*. UBC Press.
Delgamuukw v. British Columbia. 1997 3 SCR 1010. https://scc-csc.lexum.com/scc-csc/scc-csc/en/item/1569/index.do
Dhillon, J., & Parrish, W. (2019, December 20). Exclusive: Canada police prepared to shoot Indigenous activists, documents show. *The Guardian*. www.theguardian.com/world/2019/dec/20/canada-indigenous-land-defenders-police-documents
EAO BC. (2023, January 25). Third fine issued for Coastal GasLink project. *BC Gov News*. https://news.gov.bc.ca/releases/2023ENV0004-000088
Fals Borda, O. (1978). El problema de como investigar la realidad para transformarla. *Federación de Analisis de La Realidad Colombiana*, 1–38. https://www.redalyc.org/journal/122/12270216010/html/
Fireweed Solidarity. (2021, October 18). Chief Dsta'hyl deactivates coastal gaslink excavator. *YouTube*. www.youtube.com/watch?v=UPBDZcXv39Y
Forester, B. (2022, June 16). Behind the thin blue line: Meet a secretive arm of the RCMP in BC. *APTN News*. www.aptnnews.ca/ourstories/cirg/
Forsythe, M., Ganon, L., Wickham, M., & Huson, F. (2020, November 30). Letter. Re: Urgent concerns from Wet'suwet'en Ts'ako ze' and Skiy ze'. *Unist'ot'en Camp*. https://unistoten.camp/wp-content/uploads/2020/11/Re_-Urgent-Concerns-from-Wet'suwet'en-Ts'ako-ze'-and-Skiy-ze'.pdf
Gellert, P. K., & Lynch, B. D. (2003). Mega-projects as displacements. *International Social Science Journal*, *55*(175), 15–25. https://doi.org/10.1111/1468-2451.5501009_1
George, A. L., & Bennett, A. (1997, October 17–19). *Process tracing in case study research* [Conference Presentation]. MacArthur Foundation Workshop on Case Study Methods.
George, A. L., & Bennett, A. (2005). *Case studies and theory development in the social sciences*. Mass. https://doi.org/10.1017/S1537592707070491
Gidimt'en. (2020). *History and timeline*. Gidimt'en Yintah Access. www.yintahaccess.com/historyandtimeline

Gidimt'en. (2021a, November 22). *Arrested land defenders appear in Court Today*. Gidimt'en Yintah Access. www.yintahaccess.com/news/court date
Gidimt'en. (2021b, October 18). *Coyote Camp*. Gidimt'en Yintah Access. www.yintahaccess.com/news/coyote-camp-update
Gidimt'en. (2021c, November 16). Gidimten takes land back! [Video]. *YouTube*. www.youtube.com/watch?v=XOWdmzdbNEE&t=139s
Gidimt'en. (2021d, November 14). Pipeline workers served mandatory evacuation from Gidimt'en land [Video]. *YouTube*. www.youtube.com/watch?v=Bjxwo-ZcFU8&t=1s
Gidimt'en. (2021e, October 19). Sleydo' Molly Wickham delivers a message to the investors for the Coastal GasLink. *YouTube*. www.youtube.com/watch?v=Pm6yfEyWkp4&t=17s
Gidimt'en. (2021f, November 19). Violent RCMP raid on unarmed Gidimt'en checkpoint. *YouTube*. www.youtube.com/watch?v=k4eO1ecdeEA&t=5s
Gidimt'en. (2021g, October 7). Wet'suwet'en hereditary chiefs meet with coastal GasLink president. *YouTube*. www.youtube.com/watch?v=TLfvZs_Slzc&t=190s
Gidimt'en. (2022, December 23). *Tsel Kiy Kwa*. Gidimt'en Yintah Access. www.yintahaccess.com/news/tselkiykwa
Gidimt'en. (2023a, February 28). *Charter violations*. Gidimt'en Yintah Access. www.yintahaccess.com/news/2023/2/28/charter-violations
Gidimt'en. (2023b, April 4). *Connecting the dots*. Gidimt'en Yintah Access. www.yintahaccess.com/news/2023/4/4/connecting-the-dots-cgl-amp-rbcs-corporate-colonialismnbsp
Gidimt'en Checkpoint. (2018, December 17). *Hereditary Chiefs erect new checkpoint on Gidimt'en (Cas Yikh) Territory*. https://static1.squarespace.com/static/5c51ebf73e2d0957ca117eb5/t/61664eec879be543a232b02a/1634094829043/PR+DEC+17+2018.pdf
Goodson, I. (2001). The story of life history: Origins of the life history method in sociology. *Identity: An International Journal of Theory and Research*, *1*(2), 129–142. https://doi.org/10.1207/S1532706XID0102
Government of Canada. (2023). *Additions to reserve*. www.sac-isc.gc.ca/eng/1332267668918/1611930372477
Hanson, E. (2009). *The Indian Act*. Indigenous Foundations. First Nations and Indigenous Studies UBC; Indigenous Foundations UBC. https://indigenousfoundations.arts.ubc.ca/the_indian_act/
Hedström, P., & Swedberg, R. (1998). *Social mechanisms: An analytical approach to social theory*. Cambridge University Press.
Hosgood, A. F. (2020, February 7). Armed RCMP Launch Raid on Second Wet'suwet'en Camp Supported by Helicopters, Police Dogs. *The Tyee*. https://thetyee.ca/News/2020/02/07/Second-Raid-Wetsuweten/?fbclid=IwAR3L4LnYiv5BcmIfR_eFVksMmQO9AZcR0hGsg3KdyP0UesK3RR_rMdz_cl0
Huson, F., Armstrong, C., Deal, M., Blake, M., Martindale, M., Ferris, N., Lepofsky, D., Menzies, D., Mills, A., Reimer, R., D'Andrea, C., Metcalfe, J., Conolly, J., Jamieson, R., Johansen, P., Rowley, S., Supernant., K., Connaughton, S., Angelbeck, B., & Quintana Morales, E. (2019, February 16). *Open letter to archaeology branch: Stop the destruction of sites at Unist'ot'en*. https://docs.google.com/document/d/1HsWDHM74NikhTEwhsTsLqnTdx4ip4_WCNbIrZmpUUoY/edit?fbclid=IwAR32OuAYW-2i1fykBLIlAkZk95H6-9qTToHHe62K3Dvl5DOCNMKs8g5KfhU
IACHR. (2016). *Indigenous peoples, communities of African descent, extractive Industries*. Organization of American States. www.oas.org/en/iachr/reports/pdfs/ExtractiveIndustries2016.pdf
Indian Act (R.S.C., 1985, c. I-5). https://laws-lois.justice.gc.ca/PDF/I-5.pdf
Jones, A. M. (2020, February 10). Journalists say RCMP blocked efforts to cover police raids on Wet'suwet'en camps. *CTV News*. www.ctvnews.ca/canada/journalists-say-rcmp-blocked-efforts-to-cover-police-raids-on-wet-suwet-en-camps-1.4806267
LNG Canada. (2023). *Finish line coming into focus for LNG Canada, Coastal Gas Link*. www.lngcanada.ca/news/finish-line-coming-into-focus-for-lng-canada-coastal-gas-link/
Macaulay, M. M. (2022). Keynote Speech. Canada and the Inter-American Human Rights System. *Revue Québécoise de Droit International, Special Issue, June 2022 S'ouvrir aux Amériques pour mieux protéger les droits humains et s'engager dans la réconciliation*, 15–25. https://doi.org/https://doi.org/10.7202/1091076ar
Marcus, G. E. (1995). Ethnography in/of the world system: The emergence of multi-sited ethnography. *Annual Review of Anthropology*, *24*, 95–117. https://doi.org/10.1146/annurev.an.24.100195.000523

Marcus, G. E. (2011). Multi-sited ethnography: Notes and queries. In M.-A. Falzon (Ed.), *Multi-sited ethnography. Theory, praxis and locality in contemporary research*. Ashgate.
Marshall, G. (1998). "Life-history." A dictionary of sociology. *Encyclopedia.Com*. www.encyclopedia.com/doc/1O88-lifehistory.html
Meissner, D. (2020, January 14). Gas pipeline will be built despite protests, Horgan says – Victoria Times Colonist. *Times Colonist*. www.timescolonist.com/business/gas-pipeline-will-be-built-despite-protests-horgan-says-4678388
Michael, T. (2019). RCMP occupy Gitdumt'en territory. *YouTube*. www.youtube.com/watch?v=VN-h1zyICU8&t=382s
Mill, J. S. (1843). *A system of logic, ratiocinative and inductive: Being a connected view of the principles of evidence and the methods of scientific investigation* (Vols. 1–2). John W. Parker.
Morin, M. (2016). *Niwhts'ide'nï Hibi'it'ën* (2nd ed.). School District No. 54 and Witsuwit'en Language and Cultural Authority.
Mulhall, D. (1986). *Will to power: The missionary career of father morice*. University of British Columbia Press. https://doi.org/10.2307/1860116
NIMMIWG. (2019). *Reclaiming power and place: The final report of the national inquiry into missing and murdered Indigenous women and girls*. National Inquiry into Missing and Murdered Indigenous Women and Girls. www.mmiwg-ffada.ca/wp-content/uploads/2019/06/Final_Report_Vol_1a-1.pdf
Office of the Wet'suwet'en. (2013, December 17). *Wet'suwet'en chiefs reaffirm opposition to pipelines*. www.wetsuweten.com/files/Media_Release_-_December_17,_2013.pdf
Office of the Wet'suwet'en. (2018, October 1). *BC and Canada ignore Wet'suwet'en title holders to push pipeline agenda*. Unist'ot'en Camp. http://unistoten.camp/wetsuweten-hereditary-jurisdiction-and-opposition-to-pipelines/
Office of the Wet'suwet'en. (2020, January 30). *Statement by the Wet'suwet'en Chiefs*. www.wetsuweten.com/files/January_30,_2020_Statement_by_the_Wetsuweten_BC.pdf
Patterson, B. (2020, February 6). RCMP raid on Wet'suwet'en territory taking place now. *PBI*. https://pbicanada.org/2020/02/06/rcmp-raid-on-wetsuweten-territory-taking-place-now/?fbclid=IwAR3AdiqZt84MNkbzGlMZ3YCIjwwknIFWkRk2uWCOj9qATATqGdnTe2LjU-0
Pawliw, B. (2020, January 29). Skin Tyee Hereditary Chief shows support for Coastal Gas Link pipeline during rally. *My Bulkley Lakes Now*. www.mybulkleylakesnow.com/39434/skin-tyee-hereditary-chief-shows-support-for-coastal-gas-link-pipeline-during-rally/
Payne, L. A., Bernal-Bermúdez, L., & Pereira, G. (2022). In L. A. Payne, L. Bernal-Bermúdez & G. Pereira (Eds.), *Economic actors and the limits of transitional justice: Truth and justice for business complicity in human rights violations*. Oxford University Press. https://doi.org/10.5871/bacad/9780197267264.001.0001
Payne, L. A., Pereira, G., & Bernal-Bermúdez, L. (2020). *Transitional justice and corporate accountability from below: Deploying Archimedes' lever*. Cambridge University Press.
Prime Minister of Canada. (2018, October 2). *Government of Canada welcomes largest private sector investment project in Canadian history*. https://pm.gc.ca/en/news/news-releases/2018/10/02/government-canada-welcomes-largest-private-sector-investment-project
Ragin, C. (2014). *The comparative method: Moving beyond qualitative and quantitative strategies*. University of California Press.
RCMP. (2022). *RCMP in British Columbia – Community-Industry Response Group (C-IRG)*. Royal Canadian Mounted Police. https://bc-cb.rcmp-grc.gc.ca/ViewPage.action?siteNodeId=23&languageId=1&contentId=66492
Scott, J. C. (1985). *Weapons of the weak: Everyday forms of peasant resistance*. Yale University Press.
Simmons, M. (2022a, September 8). The northern B.C. pipeline you've never heard of -Enbridge's Westcoast Connector. *The Narwhal*. https://thenarwhal.ca/enbridge-westcoast-connector-bc-pipeline/
Simmons, M. (2022b, October 19). Coastal GasLink drills under Wet'suwet'en river as salmon spawn. *The Narwhal*. https://thenarwhal.ca/coastal-gaslink-spawning-salmon/
Supreme Court of B.C. (2019). *Coastal GasLink Pipeline Ltd. v Huson*, 2019 BCSC 2264 (CanLII). https://www.canlii.org/en/bc/bcsc/doc/2019/2019bcsc2264/2019bcsc2264.html
Tapias Torrado, N. (2020). *Indigenous women leading the defence of human rights from the abuses by mega-projects in Latin America, in the face of extreme violence*. ORA. https://ora.ox.ac.uk/objects/uuid:3a1393b3-1a8b-4341-bb6e-7a33897db5c7

Tapias Torrado, N. (2022a). Honduras: !Berta vive, la lucha sigue! Corporate accountability for attacks against human rights defenders. In L. Payne, G. Pereira & L. Bernal-Bermúdez (Eds.), *Economic actors and the limits of transitional justice: Truth and justice for past business complicity in human rights* (pp. 214–235). Oxford University Press.

Tapias Torrado, N. (2022b). Indigenous women leading the defense of human rights from abuses related to mega-projects: Impacting corporate behavior – overcoming silencing practices. *Revue Québécoise de Droit International*, 105–120. https://doi.org/10.7202/1091082ar

Tapias Torrado, N. (2022c). Overcoming silencing practices: Indigenous women defending human rights from abuses committed in connection to mega-projects: A case in Colombia. *Business and Human Rights Journal*, 7(1), 29–44. https://doi.org/10.1017/bhj.2021.54

Tapias Torrado, N. (2023). The violent rollback of indigenous and environmental rights: The emblematic case of lenca leader berta cáceres in Honduras. In L. A. Payne, J. Zulver & S. Escoffier (Eds.), *The right against rights in Latin America* (p. 320). Oxford University Press. https://global.oup.com/academic/product/the-right-against-rights-in-latin-america-9780197267394?prevNumResPerPage=20&prevSortField=1&view=Grid&facet_narrowbybinding_facet=Hardback&facet_narrowbyprice_facet=50to100&lang=es&cc=cz#

TRCC. (2015). *Honouring the truth, reconciling for the future. Summary of the final report of the truth and reconciliation commission of Canada*. Truth and Reconciliation Commission of Canada. https://publications.gc.ca/site/eng/9.800288/publication.html

UN CERD. (2019). *Decision 1 (100) of the UN committee on the elimination of racial discrimination*. www.ohchr.org/en/treaty-bodies/cerd/decisions-statements-and-letters

Unist'ot'en. (2017a). *Background of the Campaign*. Unist'ot'en Camp. https://unistoten.camp/no-pipelines/background-of-the-campaign/

Unist'ot'en. (2017b). Unist'ot'en Camp. *Unist'ot'en Healing Centre*. https://unistoten.camp/come-to-camp/healing/

Unist'ot'en. (2018). Unist'ot'en Camp. *Action Alert – Application for Injunction Served to Unist'ot'en Camp*. https://unistoten.camp/action-alert-application-for-injunction-served-to-unistoten-camp/

Unist'ot'en. (2019, August 27). Coastal GasLink Bulldozes Kweese Trail [Video]. *YouTube*. www.youtube.com/watch?v=3OFnANtjZSY&t=246s

Unist'ot'en. (2020a). *Wet'suwet'en hereditary chiefs reject the BC Supreme Court decision to criminalize Wet'suwet'en law*. http://unistoten.camp/wetsuweten-hereditary-chiefs-reject-the-bc-supreme-court-decision-to-criminalize-wetsuweten-law/

Unist'ot'en. (2020b). *CGL is still working amidst COVID-19 pandemic*. Unist'ot'en Camp. https://unistoten.camp/covid19/

Unist'ot'en. (2020c, January 5). *Coastal gaslink evicted from Unist'ot'en territory* [Video]. YouTube. www.youtube.com/watch?v=lnpztCgP_oU&t=151s

Unist'ot'en Camp. (2020, February 10). *Unist'ot'en matriarchs arrested: Compiled live stream footage*. Facebook. www.facebook.com/watch/?v=189902172377320

UN Special Rapporteur on the Human Rights to Water and Sanitation. (2019). *The impact of mega-projects on the human right to water. Report to the 74th session of the UN General Assembly. A/74/197*. www.ohchr.org/en/documents/thematic-reports/a74197-impact-mega-projects-human-rights-water-and-sanitation-report

UN WG Business, UN SR on Cultural Rights, UN SR Environment, UN SR Freedom of Opinion and Expression, UN SR Freedom of Peaceful Assembly and of Association, UN SR Human Rights Defenders, UN SR Indigenous Peoples, UN Independent Expert on Sexual Orientation and Gender Identity, & UN SR Violence Against Women and Girls. (2023). *AL OTH 123/2022 – Letter to Mr. Poirier, TC Energy President*. https://spcommreports.ohchr.org/TMResultsBase/DownLoadPublicCommunicationFile?gId=27697

Walia, H. (2020, February 6). *Open letter from BCCLA re: Police actions on Wet'suwet'en territories*. British Columbia Civil Liberties Association. https://bccla.org/wp-content/uploads/2020/02/2020-02-06-Wetsuweten-Solidarity-Statement-BCCLA.pdf

Watson, B. (2020, January 20). "I'm not being disrespectful, I'm just saying be realistic": Horgan defends not meeting Wet'suwet'en chiefs. *CBC News*. www.cbc.ca/news/canada/british-columbia/i-m-not-being-disrespectful-i-m-just-saying-be-realistic-horgan-defends-not-meeting-wet-suwet-en-chiefs-1.5433495

Woodside, J. (2023, April 6). *'How dare you dismiss this?' Indigenous, climate leaders shut out of RBC meeting room*. National Observer. www.nationalobserver.com/2023/04/06/news/indigenous-climate-leaders-rbc-shareholder-meeting
World Bank. (2021). *Canada overview*. World Bank. www.worldbank.org/en/country/canada/overview

24

DOMESTICITY, MASCULINITIES AND FEMININITIES

Complicating gender and dealing with water in Pemba, Mozambique

Sandra Manuel, Margarida Paulo, Tatiana Acevedo-Guerrero, Danícia Munguambe and Amanda Matabele

Introduction

Aedes mosquitoes breed in stored and stagnant water and need human density to feed; they are the primary vectors of dengue, chikungunya and Zika, diseases that have been increasingly affecting populations in the Global South (Beaty et al., 2016). Neighbourhoods where water service provision is intermittent are particularly vulnerable to mosquito-borne diseases as water is stored in/around households (Acevedo-Guerrero, 2022). Once water is stored, mosquitoes can access it and lay their eggs (Overgaard et al., 2017). After the report of an outbreak of dengue in Pemba, the capital of Cabo Delgado in northern Mozambique, the National Institute of Health developed epidemiological surveillance systems and started conducting awareness campaigns (Abílio et al., 2018). Local health authorities also carried out prevention programmes. These entailed domestic visits by health officers to instruct residents on how to reduce "breeding sources" (Oludele et al., 2017).

Literature has long shown how in many parts of the world, including in African countries, women are responsible for securing and fetching water for the household (O'Reilly, 2016; Sultana, 2011; Truelove, 2011; Zwarteveen et al., 2012). Therefore, interventions to prevent vector-borne diseases such as dengue have historically targeted women. These interventions promote daily water management practices such as covering containers/tanks where water is stored and periodically washing and brushing storage containers/tanks (Acevedo-Guerrero, 2022; García-Betancourt et al., 2015; Nading, 2014). This has also been the case in Pemba: the different water management campaigns developed to explain the links between water storage practices and possible mosquito breeding targeted women specifically. However, after conducting a few weeks of fieldwork in Pemba studying these campaigns, we (the authors of this chapter) started questioning this decision of targeting women exclusively. In fact, we realised that the linkages between gender and water were quite complex. Both women and men are associated with water management in different ways. Firstly, many men in Pemba perform everyday domestic work. That is, they work for daily wages in homes where they are in charge of chores such as cooking, cleaning and laundering, which all involve dealing with stored water. These men are male migrants who

DOI: 10.4324/9781003100379-29

belong to the *Makhuwa* ethnic group and come from the rural district of Chiúre (about 149 km from Pemba). They migrate to the city in search of stability and better economic opportunities. Secondly, many women are in charge of selling water at the neighbourhood level: they are responsible for taking care of storage tanks and water treatment tasks, selling the extra water that their household does not need and managing the earnings that come from the sale of water.

These insights prompted us to conduct an in-depth study to understand the *lieu* of gender in everyday Pemba. In this chapter, we present the results, documenting how, although water-related activities are clearly gendered, this genderness is deeply multidimensional. The chapter is structured as follows: after describing the methodology we used to conduct this research, we contextualise gender relations in northern Mozambique taking a historical perspective. Subsequent sections dig into the case of *Makhuwa* migrants working as domestic workers in Pemba and the role of community women as small-scale water vendors. A final section offers concluding remarks on the importance of carefully considering situated gender and water relations when developing diseases prevention campaigns and programmes.

Methodology

The research adopted qualitative methodologies to study the daily life of families in three different neighbourhoods of Pemba: *Alto-Gingone, Cariacó* and *Natite*. Fieldwork was conducted during the last trimester of 2017. Data were collected using direct observations, fourteen semi-structured interviews and six focus group discussions, involving a total of 39 participants.[1]

During data collection, several aspects of everyday life were documented, including predominant house types and building materials in the three neighbourhoods. Daily household chores were observed and analysed, especially those related to water: water collection, storage and management, use of tanks in the backyards and of fountains in the neighbourhoods, interactions between people selling and buying water.

After observing the neighbourhoods, some houses with characteristics of interest to the research (for example, houses with a tank in the yard where water was sold, and houses with no working water taps) were chosen. Participants were selected according to their availability to participate in the studies and targeting different categories of people: household members selling water to neighbours; people carrying water; people buying water, amongst others. During the interview, people felt free to share their experiences regarding the storage, sale and buying of water. The interviews lasted an average of forty minutes each.

The semi-structured interviews and group discussions helped to clarify issues that emerged during direct observations. Focus groups were held with men and women separately. Community leaders, who are knowledgeable about water problems, helped select participants. During the focus group discussions, questions were asked in Portuguese. The participants answered in either Portuguese, *e-Makhuwa* or *Xi-Maconde* languages: a local research assistant helped with the translation.

The researchers were equipped with mobile phones to record conversations and take photographs. Photographs allowed the team to document the shapes, locations and conditions of tanks, buckets and other recipients where water was stored in and around the home; taps, pipes as well as the different mechanisms to access water.

Content analysis of all collected material was performed. Similar and contradictory explanations were grouped. This data provided an overview of common understandings concerning water and neighbourhood everyday dynamics. At the same time, it also shed light on how not all interviewers agreed on which activities men and women perform in the public and domestic spheres. Secondly, data gathered through direct observations was relevant to provide an overview of water management practices, household dynamics and attitudes towards mosquitoes. Thirdly, all data provided by interviewers was analysed by identifying the most and least shared topics.

The study sites

Data collection was carried out in the neighbourhoods of Cariacó, Alto-Gingone and Natite, in the city of Pemba, Cabo Delgado Province, in the country's northern region. These are the city's most populous neighbourhoods. They house over 50% of the total urban population and have significant problems accessing basic sanitation and regular water supply.[2]

The municipality of Pemba is located on the east coast of Africa, south of Cabo Delgado Province. To the north, it is bordered by the Indian Ocean, to the south by the Macufi district, to the east by the Indian Ocean and to the west by the Pemba-Metuge district. The prevailing climate in this municipality is tropical humid, with an average annual temperature of 25.8°C, an average annual rainfall of 878.4 mm and air humidity of 78.8% (Macuácua, 2013). Historically, the city of Pemba emerged to facilitate access to inland resources and as a port for exporting inland products during the colonial domination of Mozambique from Portugal.

Generally, water in urban Mozambique and specifically in Pemba city, has been intermittent, especially as the country urbanises (Bayona Valderrama et al., 2020). Mozambican cities have been a refuge for displaced people during the 16 years of the civil war (1976–1992), as well as for people migrating from rural areas in search of economic opportunities. Such migration to cities led to a disorganised occupation of urban spaces, making it difficult for the state to build proper infrastructures and provide essential services such as water (Bayona Valderrama et al., 2020). This is definitely the case in the city of Pemba.

Contextualising gender in northern Mozambique

Grounded in substantial and opposing gender differences, western normative models of gender associate femininity with the domestic space, meaning unpaid reproductive housework and masculinity with the public space, meaning productive paid work conducted outside the house (Gorman-Murray, 2013). The conventional division of public and private spheres is at the core of the segregation between productive and reproductive labour. The dichotomic labour profiles were inspired by the Victorian ideal of the middle-class family in which the rational husband works in the public sphere to support the family, being respected and worshipped by his wife. The wife manages the domestic tasks and is in charge of all the emotional, nurturing and childcare tasks in the private sphere (Gorman-Murray, 2013; Rezeanu, 2015). Functionalist thinkers explained gender differences by anchoring them in biological differences between sexes. Thus, they stated that men and women have opposing but complementary psycho-socio-cultural characteristics, later conceptualised as the essence of masculinity or femininity (Rezeanu, 2015).

Feminist work has made visible the ways in which ideas around femininity and masculinity are socially constructed. Critical readings of gender dichotomies highlight that binary logics are anchored in the Western culture, which organises the social world through a specific perception of human bodies (Oyewùmí, 1998). In a comparative study of Nigeria's Western culture and Yoruba culture, Oyeronke Oyewùmí (1998, p. 5) argues that:

> The cultural logic of the West is actually a "biologic" encapsulated in an ideology of biological determinism – the idea that biology provides the rationale for the organisation of the social world . . . those with certain kinds of bodies are superior to others by virtue of having certain favoured body parts; men are superior to women and whites to blacks. This way of organising the social world is not inherent in nature; it is produced by a specific conception, a cognitive schema, a way of seeing the world.

The above highlights a specific way of seeing the world. However, gender itself, as a social category, is not transcultural:

> The idea that people have their own historically sedimented frames of reference, and come at events [the social world] with their own ways of thinking and feeling which mean that people of different cultures organise their world in distinct ways.
>
> *(Oyewùmí, 1998, p. 2)*

In this regard, theorizations around gender have failed to give testimony of the richness of societal constructions across time and place. These constructions go beyond gender dichotomies and include the experiences of hijras in India, which are neither man nor woman (Nanda, 1990); women marriages and female husbands in Igboland (Amadiume, 1987; Chacha, 2004); and the absence of gender in the Yoruba language amidst the relevance of seniority (Oyewùmí, 1997).

Through an analysis of cultural contexts that differ from the West, some authors have made calls for a more flexible concept of *gender*, in order to capture nuances across cultural contexts. In her analysis of northern Mozambique, home of the Makhua people, Arnfred (2001, p. 6) has highlighted the flexibility and situatedness of gender notions, while at the same time talking of female and male positions, and of positions of power as linked to specific capacities:

> Some positions linked to certain capacities in Makhuwa society are ungendered. The word for a counsellor (for male/female initiation rituals) is **olaka**, an un-gendered noun. And similarly, the word for healer/diviner, **kulukana**, is also un-gendered. Either men or women may occupy these positions. Other positions are gendered. Somebody who is an expert on food and sex occupies a female position. Generally, such a person is a biological woman but occasionally may be a man.
>
> *(Arnfred, 2001, p. 6)*

In light of these premises about the concept of gender, in the next section, we explore the ways in which gender unfolds in Pemba, often in ways that do not stick to strict dichotomic models.

Domestic masculinities: *Makhuwa* migrants working as domestic workers in Pemba

The *Makhuwa* are a Bantu ethnic group found in northern Mozambique and in the southern border provinces of Tanzania. Traditionally, the *Makhuwa* men perform paid domestic work, including cleaning, cooking, washing and caring for children as a duty. These employees are in constant contact with water stored within the home. This water is used for cooking, washing bathrooms, washing clothes and for cleaning tasks.

Makhuwa men tend to work for different employers every weekday. Usually, they do not have a formal contract but get daily payments from their employers. Such work is perceived as honourable. Chiúre migrants do not see themselves as "failed patriarchs". Instead, they think about themselves as successful in comparison to those who stayed back home and are not able to get an income to provide for themselves and for their families due to the unavailability of jobs and the decline of the agricultural sector in Chiúre:

> Before moving to the neighbourhood of Cariacó, I lived in Chiúre, where I worked in the field, farming. However, we were suffering as we were not producing enough, it was difficult to live. With the help of a friend who had moved to Pemba months before, I managed to get a job as a domestic worker in the house of a health technician. With what I earn, I can rent a room, feed myself and send some money to my family in Chiúre.
>
> *(35-year-old man, working in a household at Cariacó)*

The regional matrilineal system plays a significant role in constructing a notion of masculinity that positively embraces men as domestic workers. In matrilineal societies, men are literally "*women-defined men*" (Geffray, 2000; Arnfred, 2001). In such societies, ancestry and family names are passed down through the female line, along with house, land, and livestock. Men grow up and are socialised in groups defined by a female core. First, the man lives and is socialised in his mother's group and after marriage, a *Makhuwa* husband subjects himself to his wife's family and becomes a labour force in it (Watson-Franke, 1992).

In the specific matrilineal system of the *Makhuwa*, in northern Mozambique, marriage has historically been defined by matrilocal residence, that is, the man would move to his wife's family house (Osório, 2006). The young husband living in the house of the in-laws is thus supposed to be responsible for social reproduction.[3] The man demonstrates that he is fertile with the first child's birth. Afterwards, he can decide to have his own house and move out from his in-laws. Children born from this relationship belong to the woman's family. Traditionally, the father would have no authority over them. However, he would exercise some control over his sister's children. Concerning productivity, married *Makhuwa* men are expected to work in the in-laws' fields and perform all the household chores including cleaning around the huts, fetching water, grinding sorghum and all other tasks asked of him (Geffray, 2000; Arnfred, 2001; Medeiros, 2007, Humbane, 2018).

Thus, men in matrilineal societies are socialised to be acquainted with and regular performers of domestic chores, both in their mother's and wife's household. Culturally, the image of a man is linked to household chores and domestic spaces. The performance of house chores is socially constructed as masculine and therefore is perceived as "natural", and not as something that diminishes their masculinity.

"Hence, socialisation in (mainly rural) matrilineal *Makhuwa* families links domestic work to the idea of "crafting manhood". Additionally, for *Makhuwa* men from rural areas like Chiúre, migrating to work as domestic workers is perceived as something honourable, as it is linked to the ability to venture into new challenges, having the resilience to conquer hardship and being able to bear the difficult conditions of the city. Concomitantly, migration to urban areas allows them to provide for their families and to have access to commodities such as smartphones, which grants them respect in their communities back home (Quartararo & Falcinelli, 2013; Raimundo, 2008, Humbane, 2018).

However, in northern Mozambique, colonialism, capitalism, modern economy, Christianity and Islam have – throughout time – affected matrilineal systems in many ways, fostering patrilineal family arrangements (Bonate, 2006; Humbane, 2018). This is particularly true in urbanised areas. For instance, in Pemba, *Makhuwa* urban communities historically living along the city's coast have incorporated patriarchal arrangements into their everyday living, including married couples moving to the husband's land or far away from the wife's family members. They live *and* understand their matrilineality differently from, for instance, the *Makhuwa* who live in rural Chiúre.

Moreover, in their daily interactions, rural *Makhuwa* men from Chiúre who work as domestic workers and *Makhuwas* from Pemba frequently do not get along. They clearly have different understandings of gender roles and relations. For instance, urban *Makhuwas* see men's performance in domestic work negatively. Such negative opinion is related to the fact that urban *Makhuwa* men associate domestic work with the female domain; but they also associate it with age. They believe that older and married men should be employed in non-domestic professions. Thus, rural men from Chiúre who work as domestic workers are referred to by local residents as *boys* or *kids* and are hardly ever called by their first names. Local residents also use derogative, and non-verbal condemning facial gestures to express their indignation towards male domestic workers.

We argue that the negative connotation of men performing domestic work emerged in colonial Mozambique. Throughout the colonial occupation, the Portuguese administration forcibly recruited Mozambicans between the ages of 15 and 60 to engage in between 90 and 180 days of underpaid, forced labour in agriculture, locally designated as *chibalo*. To avoid forced labour, men migrated legally or illegally to South Africa to work in gold mines. These men would be exempt from *chibalo*, given the bilateral agreements between the South African Union and the Portuguese colonial government in Mozambique (Penvenne, 1995; Zamparoni, 1998). In addition, some men sought out the cities to volunteer as cooks and/or cleaners in the hospitality industry (hotels and restaurants) and as domestic workers in the residences of white Portuguese colonisers (Zamparoni, 1999, p. 147).

Thus, during colonial times domestic waged work performed by *Makhuwa* men became the norm in Northern Mozambique. This happened also because European white women often preferred male domestic employees as they perceived African black women as lazy and slow learners (Penvenne, 1995). They also feared their husbands would get sexually involved with black women and "the yellow peril", that is, the miscegenation between white men and black women (Penvenne, 1995). However, such male domestic workers were referred to as *blacks*, *boys*, assigned biblical names, or names of cleaning products (like "soap") (Meneses, 2010; Penvenne, 1995). This colonial legacy still characterises present-day Africa: male domestic employees are seen in a pejorative, derogatory way.

Generally, in Southern Africa, including southern and central Mozambique, the feminization of domestic labour is a post-independence phenomenon (Zamparoni, 1999; Pape,

1993; Miles, 1999; Le Roux, 1999; Cock, 1980). However, as has been noted, in northern Mozambique, domestic wage work continues to be exercised by men.

Femininity and small private water suppliers

Many homes in Pemba have indoor or outdoor water connections provided by the state water utility FIPAG.[4] Despite the existing connections, infrastructure is frail and poorly maintained and water shortages in Pemba happen due to intermittent supply. Sometimes, neighbourhoods experience consecutive days, weeks or even months without water:

> In the neighbourhood, there are many houses with a tap, but water does not always come out of it. Sometimes we can stay for one, two weeks or even a month without FIPAG providing us with water.
>
> *(36-year-old woman, resident in Alto-Gingone)*

> FIPAG has five units containing water to distribute to the city. They (FIPAG) weekly alternate the units that provide us with water: in week one, the water may come from unit A. In the following week, it may come from unit B, and so forth. However, in practice, things do not work well because on the days that water should come out from the unit that supplies us, it may not happen, and we get no water.
>
> *(42 years old woman, resident in Natite)*

A consequence of this irregular water supply is that the more economically stable households in the neighbourhoods of Cariacó, Alto-Gingone and Natite store large quantities of water, whenever it is available, in outdoor cement tanks for their own consumption but also to sell it to their neighbours.

Selling water is becoming widely shared in less advantaged urban neighbourhoods in Mozambican cities, including Maputo (Alda-Vidal et al., 2017) and Pemba (see Zuin et al., 2011). This is due to accelerated urban growth and the fragility of water supply systems (Kjellén & Mc Granahan, 2006). In Pemba, a licence to sell water is not required, nor is a supply contract between the supplier and its customers. The suppliers only need to have a tank with sufficient capacity to store water; the customers pay a price per bucket or gallon.

> The requirements to be a private supplier are to have a tank and a tap. It is difficult to have a water tank. You need money to buy many sacks of cement. You need to have enough or maybe a lot of money to build and prepare the tank.
>
> *(35-years -old man, resident in Cariacó)*

There are two types of cement tanks, namely square and cylindrical tanks. The square tanks usually have the capacity to store between five hundred to one thousand and five hundred litres of water. They are about a meter high, have a dimension of eight square meters and are built in backyards. Comparatively, the cylindrical tanks have a capacity of one hundred litres and are built in the backyards and inside the homes (in bathrooms and kitchens) to avoid water being stolen in the evening or when no one is in the house (see Figures 24.1 and 24.2).

Figure 24.1 Square cement tank in Natite, Pemba.
Source: The authors.

Figure 24.2 Cylindrical cement tank in Cariacó, Pemba.
Source: The authors.

The residents who do not have a water supply contract or have storage problems due to intermittent supply buy water from the neighbours who own the tanks. We learned from community leaders that years ago, water was offered for free in mosques (as part of religious tradition) but also by neighbours to the households in need. However, from 2013 onwards, as the demand for water increased, many started selling the water that was extra after covering their household's needs. That was the only way to pay for the increasingly expensive water bill, as two private water suppliers mentioned:

> Three years ago, we would just offer water to our neighbours; we would not sell it. However, this year my parents say that we cannot offer it. The water bill is expensive, so they say we must sell it to pay for it. The rest of the earnings are used by my father to drink beer (sic).
>
> *(19-year-old woman, resident in Cariacó)*

> We started the water business in 2013 when there was a water crisis here in the neighbourhood. We would go for weeks or months without water coming out of the tap. Before selling it, we gave water to our neighbours for free, but with that crisis, the number of people needing water increased, so we decided to start charging because the water bill increased.
>
> *(40-year-old man, resident in Natite)*

In these economically stable households, men build the tanks, but it is their wives, sisters or daughters who are responsible for managing them. Men accumulate money from their salary to buy the material needed for the construction of the tank and then pay someone to build it. In most cases, they work all day outside the house, sometimes even outside the district or province.[5] Thus, the women are responsible for managing the water business by serving the customers, looking after and refilling the tanks, receiving and managing the money earned.

> We have been selling water since the beginning of 2016, when my husband went to work in Nampula, and I decided to build a tank for us to get some money for the immediate needs of the house. As I don't work [outside the house], I stay at home to monitor the tank because we depend on selling water to earn some money to pay for our own water and energy.
>
> *(46 years old woman, private supplier, resident in Natite)*

In Natite, Alto-Gingone and Cariacó, these private suppliers charge different prices according to the period of the year and the quantity of water available in the tanks. Generally, the prices range from three to ten meticais for a 20-litre bucket or gallon (about 1USD). The scarcer the water, the higher the price charged, as one private supplier commented:

> We charge 5 meticais for a 20-litre bucket, but when there is a shortage of water or when the tank is low, we sell for 10 meticais because sometimes we also buy water from other neighbours' tanks.
>
> *(34-years-old woman, Private supplier, of about, Alto-Gingone)*

The statement given by this supplier illustrates practices in periods of water scarcity and the period when the business becomes more profitable due to increased demand. When there is a period of water scarcity in the neighbourhood and the city, water is sold at a higher price than in times of abundance. Interestingly enough, when there is water scarcity, suppliers buy water from each other, shifting from being suppliers to being customers.

The purpose of the profit collected through the sale of water depends on each supplier. However, for most suppliers, the money is intended for their livelihood, to cover the needs of households such as food and transportation costs as well as to pay the water bill at the end of the month. In the case of households where one or more members have a stable salary, the money is used to supplement the household's income, as two private supplies pointed out:

> I use the water money to pay for my own water bill, for the children's school fees, and to buy food when my salary is not enough.
>
> *(36-year-old woman supplier, resident in Alto Gingone)*

> My salary always runs out before the end of the month, so then we buy food with the money we get from selling water, until I get the next salary.
>
> *(27-year-old man supplier, resident in Natite)*

The statement emphasises that selling water is a way to complement the household's income (Alda-Vidal et al., 2017). Therefore, it is an activity often accompanied by another, such as formal employment. In addition to complementing household income, the sale of water places households in a position of prestige within their community. In fact, these households facilitate access to water to other families in the community.

Communities in Natite, Alto-Gingone and Cariacó believe that men should be responsible for providing for their household and women should take care of the house and of domestic activities. However, as mentioned, it is women's responsibility to manage the water tanks and to sell the extra water. After all, since most of the time women are the ones who stay closer to the domestic space, they are in contact with the neighbours who come to look for water. These women make sure that the tanks are filled and that everyone who draws water from them pays for it. Concerning this point, a man stated:

> This matter of tanks is with the lady at home. I go to work and stay away most of the day, so she is the one who controls the tank at home.
>
> *(45-years -old man, resident in Cariacó)*

This statement reinforces the idea that there is a separation between spaces: the man in the public space works and earns a salary to provide for the family and the woman in the domestic space does not work but cares for the home. The involvement of women in selling and supplying water is seen as an extension of their domestic activities, but it also somehow challenges the idea that women do not contribute to the household's income and do not manage finances. In fact, the women usually manage the household income and the money earned from selling water. They are the ones who know how much money can be spent on different needs such as food, children's education, water and energy bills, and to pay domestic workers.

At the same time, it is important to highlight that, in Pemba, the belief that it is women who take care of domestic chores and perform unpaid domestic work coexists with the reality of migrant *Makhuwa* men working in domestic service. In fact, women in the studied neighbourhoods find time to sell water due to the shift in their household care duties, on behalf of these domestic workers.

Concluding remarks

Starting from an understanding of gender as a social construct, in this chapter, we have shown how gender dynamics in Pemba are nuanced. Understanding of gender needs to be context and history-specific rather than being based on binary thinking that associates femininity with domestic, unpaid work and masculinity with non-domestic paid work. In Pemba, there are *Makhuwa* migrant men who work as domestic employees in different households in the neighbourhoods, and are in charge of tasks such as cleaning, cooking and taking care of water in and around homes – which are nowadays most often seen as female activities. At the same time, although men in Pemba are historically responsible for earning a salary and providing for their households, many women are in charge of managing and selling the extra water stored in the household's tanks. This includes being aware of the quantity and quality of the stored water and managing the earnings that come from this business.

This analysis highlights the practical problems of universalizing culturally specific gender concepts into a context where traditional matrilineal cultures mingle with patriarchal ones that spread with colonialism. Our analysis hopes to contribute to gender studies to question rigid gender conceptualizations, especially those anchored in Western culture and often uncritically universalised (Oyewùmí, 1997, 1998). By providing ethnographic accounts of everyday dynamics in Pemba, we show multiple ways in which gender unfolds in relation to everyday dealings with water, in relation to how people organise their lives and understand the world (see Carney, 1993, Sultana, 2011, Truelove, 2011).

In this regard, this study attempts to engage with policymakers and implementers of specific water-related campaigns. Post-colonial governmentalities have been trapped in the conundrum of operating within colonial and western frames rather than acknowledging and embracing local dynamics and their diversity, particularly in relation to gender (Houweling, 2016). Our results demonstrate the challenges of using universal concepts that are contextually unrooted. We call for increased liaison between gender and water research and the design and implementation of policies and programmes, such as disease prevention programmes. This is particularly relevant for organisations developing water management programmes and prevention campaigns to prevent new mosquito-borne disease outbreaks in the context of Mozambique.

The abundance of stored water in and around the urban home constitutes a habitat for mosquitoes (Bayona Valderrama et al., 2020). Studies have warned that climate change, in conjunction with urbanisation, could drive the proliferation of Aedes mosquitoes, with major consequences for public health and disease control strategies. Information campaigns on vector control, promoting daily water management practices such as covering containers/tanks where water is stored and periodically washing and brushing storage containers/tanks, must include the different members of the communities who are in charge of water storage activities.

Notes

1 The research project received formal, prospective, approval from the ethics committee at Universidad Eduardo Mondlane. All the participants provided appropriate informed consent through Informed Consent Forms, which they signed and approved.

2 According to the last two household surveys (conducted in 2008 and 2009), 39% of the households in Capo Delgado were poor (INE, 2015). This percentage increased to 44.8% in 2014 (Maquenzi & Feijó, 2019). The Mozambican Institute for Statistics defines consumption poverty, according to a household's capacity to access a set of necessities (targeted at a food basket of 2,150 calories per person per day). All households deprived of such necessities are considered poor.

3 The lieu of men in matrilineal societies differs from that of men in patrilineal ones. The south of Mozambique is a patrilineal society where the man is supposed to pay *lobola* to the bride's family, conferring him rights over his children and wife (Arnfred, 2001, p. 27).

4 FIPAG: *Fundo de Investimento e Património de Abastecimento de Água*. In English: Water Supply Investment and Property Fund.

5 The minimum wage in Mozambique ranges between 74.87 to 251 dollars a month. However, the size of Mozambique's informal economy is estimated to be 34.3% (INE, 2015). Responding to the challenges faced in the absence of formal employment (von Schnitzler, 2017), many engage in informal jobs, that is, is receiving daily payments (without formal contracts).

References

Abílio, A., Abudasse, G., Kampango, A., Candrinho, B., Sitoi, S., Luciano, J., Tembisse, D., Sibindy, S., Gouveia de Almeida, A., Garcia, G., Rocha David, D., Maciel-de-Freitas, R., Gudo, E. (2018). Distribution and breeding sites of Aedes aegypti and Aedes albopictus in 32 urban/peri-urban districts of Mozambique: Implication for assessing the risk of arbovirus outbreaks. *PLOS Neglected Tropical Diseases*, *12*(9), e0006692.

Acevedo-Guerrero, T. (2022). Water with larvae: Hydrological fertility, inequality, and mosquito urbanism. *Environment and Planning E: Nature and Space*, 25148486221099801.

Alda-Vidal, C., Rusca, M., Zwarteveen, M., Schwartz, K., & Pouw, N. (2017). Occupational genders and gendered occupations: The case of water provisioning in Maputo, Mozambique. *Gender, Place e Culture*, 1–17.

Amadiume, I. (1987). *Male daughters, female husbands: Gender and sex in an African society*. Zed Books.

Arnfred, S. (2001). Family forms and gender policy in revolutionary mozambique (1975–1985). In *Travaux et Documents/Centre d'étude d'Afrique Noire. Pessac France Paris?* Centre d'étude d'Afrique noire, Institut d'études politiques de Bordeaux Centre national de la recherche scientifique.

Bayona-Valderrama, A., Acevedo-Guerrero, T., & Artur, C. (2020). Cities with mosquitoes: A political ecology of Aedes aegypti's habitats. *Water Alternatives*, *14*(1), 186–203.

Beaty, B. J., Black, W. C. IV, Eisen, L., Flores, A., García-Rejón, J., Loroño-Pino, M., & Saavedra-Rodriguez, K. (2016). *The intensifying storm: Domestication of Aedes aegypti, urbanization of arboviruses, and emerging insecticide resistance*. In Forum on Microbial Threats; Board on Global Health; Health and Medicine Division; National Academies of Sciences, Engineering, and Medicine. Global Health Impacts of Vector-Borne Diseases: Workshop Summary. https://www.ncbi.nlm.nih.gov/books/NBK390432/

Bonate, L. (2006). Matriliny, Islam and gender in Northern Mozambique. *Journal of Religion in Africa*, *36*(2), 139–166.

Carney, J. (1993). Converting the wetlands, engendering the environment: The intersection of gender with agrarian change in The Gambia. *Economic Geography*, *69*(4), 329–348.

Chacha, B. K. (2004). *Traversing gender and colonial madness: Same-sex relationships, customary law and change in Tanzania, 1890–1990(3)*. CODESRIA Gender Series: 129–151.

Cock, J. (1980). *Maids & Madams: A study in the politics of exploitation*. Ravan Press.

García-Betancourt, T., Higuera-Mendieta, D. R., González-Uribe, C., Cortés, S., & Quintero, J. (2015). Understanding water storage practices of urban residents of an endemic Dengue area in Colombia: Perceptions, rationale and socio-demographic characteristics. *PLoS ONE*, *10*(6), 1–19.

Geffray, C. (2000). *Nem Pai Nem Mãe: Crítica Do Parentesco: O Caso Macua*. Caminho.
Gorman-Murray, A. (2013). Urban homebodies: Embodiment, masculinity, and domesticity in inner Sydney. *Geographical Research*, *51*(2), 137–144.
Houweling, E. V. (2016). "A good wife brings her husband bath water": Gender roles and water practices in Nampula, Mozambique. *Society & Natural Resources*, *29*(9), 1065–1078.
Humbane, J. (2018). *Empregados do quintal (male domestic workers) in Nampula city: Domestic work, masculinities and matrilineality* [Master Thesis, Department of Anthropology and Sociology, Arts Faculty, University of Western Cape, Cape Town].
Instituto Nacional de Estatísticas (INE). (2015). *Relatório Final do Inquérito ao Orçamento Familiar – IOF 2014/15*. Maputo.
Kjellén, M., & McGranahan, G. (2006). Informal water vendors and the urban poor. *Human Settlements Discussion Paper Series*, 1–26.
Le Roux, T. (1999). "Home Home where the children are": A qualitative study of migratory domestic workers in Mmotla Village, South Africa. In J. H. Momsen (Ed.), *Gender, migration, and domestic service* (pp. 179–190). Routledge.
Macuácua, N. (2013). *Estatísticas do distrito cidade de Pemba*. Instituto Nacional de Estatística.
Maquenzi, J., & Feijó, J. (2019). *Pobreza, Desigualdades e Conflitos No Norte de Cabo Delgado. 76*. Observatório do Meio Rural.
Medeiros, E. (2007). *Os Senhores da Floresta: Ritos de iniciação dos rapazes Macuas e Lómuès*. Universidade de Coimbra.
Meneses, M. P. (2010). O "Indígena" Africano e o Colono "Europeu": A construção da diferença por processos legais. *E-cadernos Indígenas, Nativos, Nações: Identidades, Classificações e Estado*, 7, 68–93.
Miles, M. (1999). Working in the city: The case of migrant women in Swaziland's domestic service sector. In J. H. Momsen (Ed.), *Gender, migration, and domestic service*, (pp. 191–206). Routledge.
Nading, A. (2014). *Mosquito trails: Ecology, health, and the politics of entanglement*. University of California Press.
Nanda, S. (1990). *Neither man nor woman: The Hijras of India*. Wadsworth.
Oludele, J., Lesko, B., Mahumane Gundane, I., de Bruycker-Nogueira, F., Muianga, A., Ali, S., Mula, F., Chelene, I., Falk, K. I., Barreto Dos Santos, F., & Gudo, E. S. (2017). Dengue virus serotype 2 established in northern Mozambique (2015–2016). *American Journal* of *Tropical Medicine* and *Hygiene*, *97*(5), 1418–1422. doi:10.4269/ajtmh.17-0317
O'Reilly, K. (2016). From toilet insecurity to toilet security: Creating safe sanitation for women and girls. *Wiley Interdisciplinary Reviews: Water*, *3*(1).
Osório, C. (2006). Sociedade matrilinear em Nampula: estamos a falar do passado? *Outras Vozes*, *16*.
Overgaard, H. J., Olano, V. A., Jaramillo, J. F., et al. (2017). A cross-sectional survey of aedes aegypti immature abundance in urban and rural household containers in central Colombia. *Parasites & Vectors*, *10*(1), 56.
Oyewùmí, O. (1997). *The invention of women: Making an African sense of western gender discourses*. University of Minnesota Press.
Oyewùmí, O. (1998). De-confounding gender: Feminist theorizing and western culture, a comment on hawkesworth's "confounding gender". *Signs*, *23*(4), 1049–1062.
Pape, J. (1993). Still serving the tea: Domestic workers in Zimbabwe 1980–90. *Journal of Southern African Studies*, *19*(3), 387–404.
Penvenne, J. M. (1995). *African workers and colonial racism: Mozambican strategies and struggles in LourenÁo Marques, 1877–1962*. James Currey.
Quartararo, C., & Falcinelli, D. (2013). Not only men but also migrants in non-traditional occupations. *International Review of Sociology*, *23*(2), 363–378.
Raimundo, I. (2008). The interaction of gender and migration household relations in Rural and Urban Mozambique. In E. Uchendu (Ed.), *Masculinities in contemporary Africa* (Vol. 7). African Books Collective.
Rezeanu, C.-I. (2015). The relationship between domestic space and gender identity: Some signs of emergence of alternative domestic femininity and masculinity. *Journal of Comparative Research in Anthropology and Sociology*, *6*(2), 21.

Sultana, F. (2011). Suffering for water, suffering from water: Emotional geographies of resource access, control and conflict. *Geoforum*, *42*, 163–172.

Truelove, Y. (2011). (Re-)Conceptualizing water inequality in Delhi, India through a feminist political ecology framework. *Geoforum*, *42*(2), 143–152.

Von Schnitzler, A. (2017). *Democracy's infrastructure: Techno-politics and Protest After Apartheid*. Princeton University Press.

Watson-Franke, M.-B. (1992). Masculinity and the "matrilineal puzzle". *Anthopos*, *87*(4/6), 475–488.

Zamparoni, V. D. (1998). *Entre Narros & Mulungos: Colonialismo e paisagem social em Lourenç o Marques c. 1890–c.1940*. Universidade de São Paulo.

Zamparoni, V. D. (1999). Gênero e Trabalho Doméstico Numa Sociedade Colonial: Lourenço Marques, c. 1900–1940. *Afro-Ásia*, *23*, 145–172.

Zuin, V., Ortolano, L., Alvarinho, M., Russel, K., Thebo, A., Muximpua, O., & Davis, J. (2011). Water supply services for Africa's urban poor: The role of resale. *Journal of Water and Health*, *9*(4), 773–784.

Zwarteveen, M., Ahmed, S., & Gautam, R. (2012). *Diverting the flow: Gender equity and water in South Asia*. Zuuban.

PART 5

New feminist futures

25

HOW WATER CHANGES (EVERY)THINGS

A feminist study of how 'water worlds' shape processes of rural agrarian transformations in Maharashtra, India

Arianna Tozzi and Irene Leonardelli

Introduction

In the semi-arid regions of Maharashtra, India, water is central to processes of agrarian transformation and in articulating gender relations (Krishna and Kulkarni, 2019). Driven by discourses of naturalized water scarcity (Mehta, 2007), infrastructure projects in the Indian countryside have historically delineated a pathway of agrarian transformation as the transition towards an irrigated model of commercial agriculture (Bharucha, 2019; Shah et al., 2021). This narrow focus on supply augmentation is evidenced by the turn towards wastewater reuse schemes as the silver bullet to both stabilize agricultural production under a changing climate and increase farmers income (Jamwal et al., 2014; Zhang & Shen, 2019), with little attention paid to the state and quality of return flows. Yet, in a monsoon-dependent landscape like Maharashtra, where water comes first and foremost in the form of precipitation, the persistent "hydrocracy" (Joy & Janakarajan, 2018, p. 3) of the Indian water sector has neglected the richness of agriculture knowledges and livelihood practices tied to the dynamics of monsoonal rainfall (Tozzi et al., 2022).

Reducing water to a measurable substance to be harnessed and put to productive uses, this 'irrigation at all costs' mentality is problematic as much as misguided, as it erases multiple forms of gendered labor taking place across the productive and reproductive divide (Harris, 2009). More crucially, it ignores how farming practices and livelihood dynamics are articulated through and shaped by the different forms water takes as it moves across a landscape (Sultana, 2011). Whether abundant or limited, precipitating or flowing, visible or invisible, pure or contaminated, water in its multiple material forms brings together different actors (human and otherwise) (Bakker, 2012; Barnes, 2013) whose configuration plays a role when it comes to articulating gender-water relations in a place.

In this chapter we think with the relational materiality of water – as co-emergent between water's biophysical characteristics and sociocultural situatedness (Krzywoszynska & Marchesi, 2020) – to reveal how this relationality matters to the unfolding of gender dynamics in specific contexts (Figure 25.1). Bringing together feminist-informed agrarian studies (Harris, 2009;

DOI: 10.4324/9781003100379-31

Figure 25.1 Women farmers in rural Maharashtra.
Source: Illustration by Sara Filippi Plotegher.

Sultana, 2011) and post-humanist approaches (Haraway, 2003; Puig de la Bellacasa, 2017; Tsing, 2015) we examine processes of agrarian transformation through the lens of a transition between different forms of waters moving across the landscape and probe the shift in gender labor relations this transition entails. In doing so, we use the lens of 'water worlds' which, following Barnes and Alatout (2012), we describe as assemblages of practices, knowledges, technologies, histories and ideas, brought together around specific manifestations of water. Taking relationality seriously therefore reflects our commitment to go beyond seeing water as a predetermined substance or a substrate upon which social relations leave their marks (see Linton, 2010). Instead, we conceptualize water as a 'socio-bio-physical' element (Krzywoszynska & Marchesi, 2020, p. 194), an agential substance whose material affordances play a role in shaping the elements, knowledges and practices different waters entangle.

Reflecting on the transformation of the Indian agrarian landscape from a predominantly rainfed to an irrigated 'water world', we use rainwater and wastewater as our two watery manifestations and follow a series of more-than-human elements these worlds bring together as our analytical entry point. Focusing on soils, animals, tools, wells, seeds and moisture as they are variously enrolled through this transition, we interrogate how gendered knowledges, farming practices and labor relations are reassembled as a result. Our objective is to reflect on the concrete political possibilities behind what Tsing (2013) calls 'critical descriptions' – narrations of socialites that are more than just human – to thicken understandings of the gender and intersectional dynamics that imbue processes of rural agrarian transformation in the Indian countryside and beyond.

The remaining part of the chapter is structured as follows; first we introduce the empirical cases the chapter draws upon and our methodology. After describing the literature that informs our thinking, we focus on three more-than-human elements that become significant in articulating gender relations in our two water worlds. These are: (1) water solutes and sediments (2) soil organisms and (3) goats. In the concluding section we reflect on how a more-than-human attentiveness enriches feminist analyses of rural agrarian transformation. We suggest an avenue for feminist post-humanist scholarship to move beyond metaphorical sensitivity to our watery interconnections (Neimanis, 2017) and engage politically with the obligations of our more-than-human relatedness.

Presenting our water worlds and methodology

Inspired by feminist methodologies that let small stories speak to larger issues and concerns (Gibson-Graham, 2014), we draw on ethnographic research from two rural areas of Maharashtra (Figure 25.2). The first is the Arvi sub-district in the rainfed region of Vidarbha.

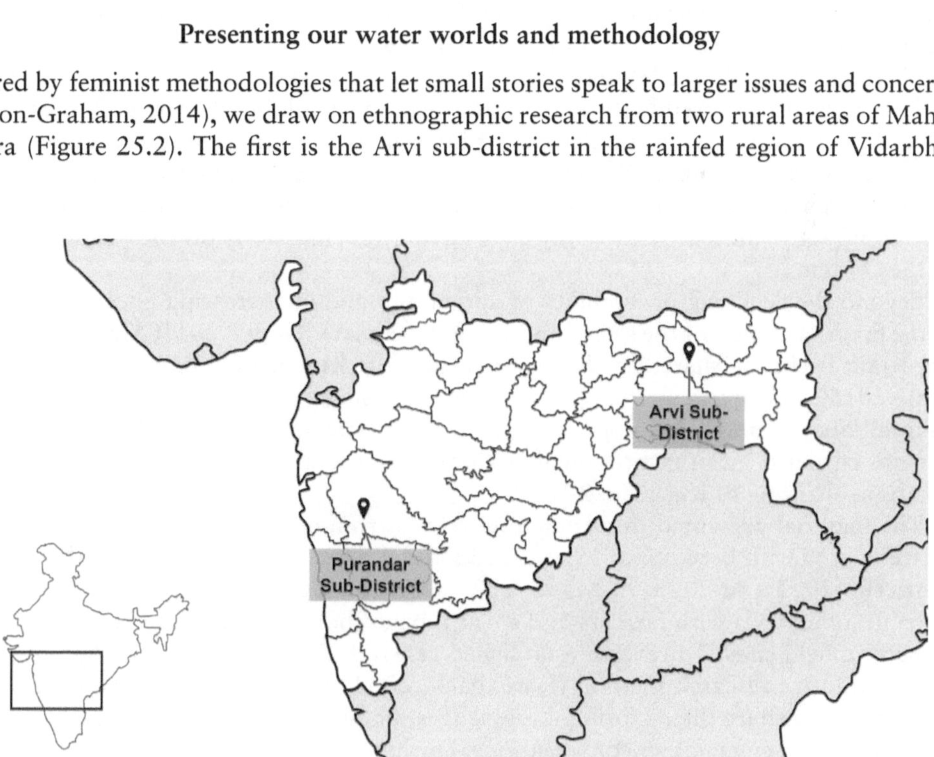

Figure 25.2 Map of Maharashtra indicating the two sub-districts where this research was conducted.
Source: The Authors.

This area lacks surface irrigation, and while few farmers owning land closer to the valley have access to groundwater, the majority of smallholders are completely dependent on the monsoonal rainfall which they use to cultivate cotton and soybean (as cash crops) together with pulses. Outside of the monsoon season unirrigated farmers (both men and women) work as laborers in irrigated fields or migrate to cities for employment. Over the years, policies pushing for the extension of Green Revolution technologies, like high yielding seeds for selected crops dependent on the use of fertilizers and irrigation have driven a situation of agrarian crisis (Reddy & Mishra, 2010). Rampant levels of rural debt incurred to purchase costly farming inputs (Sethi, 2021) and the damaged ecologies caused by overuse of chemicals needed to sustain production in increasingly depleted soils (Patel, 2013) bear witness to this state of distress.

The second case is Pravah,[1] a village in the Purandar sub-district which, from the early 2000s, has been served by the Purandar Lift Irrigation Scheme. Built by the Government of Maharashtra, this scheme consists of an infrastructure that transports mostly untreated water – or 'wastewater', as the government defines it (Government of Maharashtra [GoM], 2018) – from the city of Pune to 60 drought-prone villages in the Purandar sub-district (Chiwhane et al., 2015; Jagtap & Manivanan, 2019). The 'Purandar water' – as farmers call the wastewater delivered by the infrastructure – is purchased by farmers and stored in open ponds, where it percolates into the aquifer, recharging (as well as contaminating) wells before being used for irrigation. Before the implementation of this scheme, the agricultural season was limited to the monsoon months, after which farmers used to move to work in nearby villages and towns. Women would be mostly employed as agricultural laborers, while men often worked in local industries. Beyond making agriculture possible throughout the year, the availability of irrigation has shifted the cropping pattern from food crops for household consumption to commercial crops, mostly flowers and onions. While men continue working in industries in nearby towns, it is women who increasingly perform all farming tasks (including irrigating crops), suggesting how the infrastructure project is having repercussions on the gender dynamics of the community.

Beyond the declining profitability of agriculture and its increasing uncertainty because of the financialization of the sector and changing climate (Gupta, 2017; Matthan, 2023) an important backdrop shared by these two cases is the feminization of agriculture (Pattnaik et al., 2018). This phenomenon describes the increasingly central role women play as agricultural laborers and cultivators, while their identity as farmers and access to land rights remains curtailed by an institutionally patriarchal system with strong class and caste connotations (Krishna & Kulkarni, 2019).

The material presented in this chapter comes from ethnographic research conducted by Irene in Pravah between 2019 and 2022 and Arianna in two villages in the Arvi sub-district in 2022 and 2023. In March 2022 we went together to Pravah and conducted in-depth interviews with farmers and spent time exchanging ideas and reflections on our respective field sites. To reverse established research dynamics, we used 'guided walks' as a qualitative research method (Kusenbach, 2003) asking farmers to show us around their land and share their farming knowledges, or simply participated in people's everyday lives whilst engaging in conversations. During our time in the field, we took detailed field notes and pictures to document farming practices and day-to-day dealings with different waters.

Gender and water in agrarian contexts: exploring invisibilized labors

Feminist scholars working in agrarian contexts have explored how gender and other intersecting axes of social differences are (per)formed in relation to historically and spatially situated environments (Harris, 2006; Nightingale, 2006; Sultana, 2009). Hierarchies of power, they argue, shape how people relate to and are knowledgeable of their environments, which in turn affects dynamics of access, control and use of ecological resources. Feminist water scholars in particular have emphasized how gender relations and water practices are interwoven (Bossenbroek & Zwarteveen, 2018; Harris, 2006; Krishna & Kulkarni, 2019; O'Reilly, 2006; Sultana, 2009) proposing the notion of 'waterscape' (Baviskar, 2007) to reflect on the co-constitution of water and social relations of production and reproduction in simultaneously material and discursive ways.

In the context of water infrastructure projects, Carney (1993) for example describes how an irrigation scheme in a wetland region in The Gambia went hand in hand with changes in gender roles. She shows how, on the one hand, the project increased the work burden for women and reaffirmed a patriarchal system of male-headed landholding, while on the other, it led to women renegotiating access to resources, as they refused to work on irrigated fields without proper remuneration. Harris (2006) adds further texture to our understanding of gender as a relation that plays out in historically and geographically situated ways. Examining the transition to irrigated agriculture in southeastern Turkey, she suggests that rapid socioenvironmental changes like those triggered by the advent of irrigation rearticulate embodied experiences of social identities across gender, class and ethnicity lines. State planners and gender analysts, Harris (2006) argues, should take more seriously how the enactment of intersectional identities are impacted by changes in water availability and quality introduced by infrastructure projects.

This is particularly true in rural Maharashtra where, despite women performing much of the daily work in the farm, the idea of a 'farmer' reproduces the imaginary of a higher caste male landowner (Agarwal, 2003). Women, particularly from scheduled castes and tribes,[2] with limited financial resources, who are single or widowed, remain excluded or silenced, both institutionally and socially (Bhat, 2016; Krishna & Kulkarni, 2019) due to patriarchal norms that shape the gendered division of labor within and outside the home. In fact, while most men oversee market activities and manage farming incomes, women perform most of the domestic/reproductive tasks – or unpaid care labor. Further, even when playing a role in cultivating commercial crops, women's work still gains little public recognition (Ahmed & Zwarteveen, 2012).

Thus, while neoliberally oriented land and water reforms may have marginally improved women's participation in male-dominated spaces (Ahlers & Zwarteveen, 2009), for Krishna and Kulkarni (2019) there is still an urgent need to recognize and value their labors. For this reason, while we speak alongside a long feminist tradition that showed how identities are not fixed but 'in the making' (Sundberg, 2004), we mobilize the lens of 'women farmers' as an empirical entry point and a pragmatic guide for our political goal as feminist scholars aiming to zoom into what is at the margins. As an analytical prism, 'women farmers' therefore helps us unfold the invisibilized dimension of gendered labor relations, while theoretically adhering to an understanding of identities as a fluid 'performative accomplishment' (Butler, 1990, p. 162) that intersects with multiple axes of difference.

When it comes to water itself, a critical review of feminist scholarship in agrarian contexts also reveals some limitations. In fact, while enriching understandings of the operation of gender as a relational accomplishment that is shaped by (and in turn shapes) different waterscapes, an engagement with water as a substance that makes a difference has remained marginal (for exceptions see Barnes, 2013; Savelli et al., 2022). In this regard, feminist posthumanist scholars (Ballestero, 2019; Cortesi, 2021) have begun interrogating how water is not passive and inert but rather vibrant and agential (Bennett, 2010; Strang, 2014) – an unruly substance whose material affordances matter in weaving specific water worlds (Barnes & Alatout, 2012). Troubling a chemical ontology of water as H_2O molecules (Cortesi, 2021), ethnographically informed studies have shown how water is an inherently relational element whose biophysical characteristics and socio-historical significance *together* play a role in shaping human-water relations in specific places. Examining flood-control infrastructures in Bihar, India, Cortesi (2021) for example questions prevailing understandings of rivers as 'a matter of only water' (p. 870). By showing the ontological inseparability between water and the sediments rivers contain, Cortesi's analysis explains the failures of embankments designed to seal off the land from its periodic inundations, as farmers have historically breached these barriers to let nutrient rich waters enter their fields.

When it comes to gender relations and livelihood dynamics, engaging with the relational materiality of water and other ecological elements also broadens questions around invisibilized labors, as it redefines labor as a co-constituted practice of 'working with' the more-than-human world (Battistoni, 2017; Besky & Blanchette, 2019; Krzywoszynska, 2020). Rather than being imposed from above, processes of agrarian transformations therefore entail changes that enroll more-than-human elements as active participants in transforming agrarian environments. Barnes (2013) for example shows how by reducing water to crop intake alone, the idea of 'virtual water flows' erases the labors of both people who bring water to the field, and that of water itself, as it alters the ecological fabric of the landscape it flows through. Through the case of rice cultivation in Egypt, Barnes (2013) highlights the work water does as it washes salts from the soil, thereby maintaining the fertility of the land. Building from this literature, we argue that inquiries into transforming waterscapes should listen more carefully to these more-than-human labors as co-constituted through assemblages of people, waters, soils, animals, whose shape determines how gender and intersectional dynamics unfold in the rural world.

In this spirit, we now turn to three elements, asking how attending to the worlds gathered around water's different manifestations – as rainfall or wastewater – allows us to better understand the gendered labor relations enrolled into processes of agrarian transformation and the more-than-human entanglements that support them.

Solutes, sediments and other things waters are made of

Thinking with Camargo and Cortesi's invitation to see water not as 'a chemical compound, but an ecosystem, a habitat, a mean for the connections and transportation of other elements and organisms' (2019, p. 3), we begin by looking at the material life of water itself. Particularly, we explore how farmer's knowledge and everyday practices are influenced by their interactions with waters' lively ecology and reflect on the gendered connotations of these interactions.

Starting from rainwater, throughout history, farmer's knowledge of the monsoon has problematized understandings of rainfall as an uncontaminated, colorless and odorless substance. Colonial accounts of rainfed areas of Maharashtra are punctuated with descriptions of how farmers have been valuing rainwater through its consistency (light or heavy), temperature (hot or cold) and color to infer its beneficial or harmful effect on crops (Voelcker,

1893). These knowledges have traveled to the present as unirrigated farmers in Vidarbha gave us rich descriptions of how solutes contained in rainwater influenced their farming practices. Hiralal for instance would welcome rainwater flowing into his field from the forest uphill building small channels to direct the rains into his land. "I do it because the rainwater is rich with decomposed leaves that are beneficial to my soil", he explained when we asked the reason for the structures built around his land. On the other hand, Vinita, who had been practicing chemical-free farming for many years, lamented her continued efforts to divert the '*bad rainwater*' entering her field from adjacent farms using chemical inputs. "That water is full of chemicals; it damages the softness of the soil I worked so hard to build", she told us. During the monsoon season she would therefore be busy constructing structures around her land to keep polluted water at bay while trying to harness as much rainwater coming from the sky by building obstacles to arrest its flow and let it percolate into the ground.

Similarly, in Pravah, farmers are responsive to the contaminants the Purandar water is made of, which they know is having a harmful impact on the landscape's ecology. As women increasingly take charge of irrigating the fields, they are developing rich experiential knowledges of the effects pollutants and other hazardous materials have on the quality of the soils and the differences of irrigating crops with different kinds of waters (Leonardelli et al., 2023). Busy with weeding work, Shandidi told us:

> We know that rainwater is better than the Purandar water because of the weeds. Before, there was only one type of grass but nowadays, because of the contaminants of the Purandar water we see at least three different types of grass. This means that we have much more weeding work to do.

A task predominantly (if not solely) performed by women, weeding is carried out manually with a sickle, carefully moving up and down the land crouched to the ground to uproot grasses that may harm the crops and take up nutrients and moisture from the soil. The invisible pollutants brought by the Purandar water interact with soil's ecology, triggering the growth of weeds as well as the proliferation of insects and bacteria – ultimately making this gendered work more challenging.

Other women farmers lamented how sediments carried by the Purandar water ended up clogging the drip irrigation systems they installed to increase the precision of water application to their crops. Padma, for instance, told us how she needs to closely follow the entire irrigation process, walking along the drip lines to make sure the outlets are working properly, promptly unclogging them with a safety pin so that her flowers do not get spoiled. Further adding to her labor, at the start of every cropping cycle the drip lines need to be washed with an acid lotion to remove any sediments that may have remained from the previous season. She explained this process: "There are 'so many things' in the Purandar water that block the drip lines! So every year before planting flowers we [her and her husband] need to use this acid to wash the pipes, otherwise they won't work". When, intrigued by her answer, we asked what the 'things' blocking the drips were, she explained that they were algae and other sediments suspended in the Purandar water. While men are in charge of installing the drip lines, it is women whose labor of maintenance of the technology has been made more time consuming by these suspended sediments the Purandar water is made of.

Engaging with the ways both rainwater and the Purandar water are never *only* water, these examples show how gender relations and labor dynamics are contingent upon the ecological fabric of the water itself. Zooming into the 'nonhuman universes' (Camargo & Cortesi, 2019,

p. 4) waters are made of has therefore revealed the seeming paradox whereby, in a drought-prone unirrigated region, rainwater is kept from entering one's land because of the contaminants it brings. Further, exploring water's lively ecology has also suggested that we reconsider the value of metrics like efficiency and productivity used to evaluate the performances of irrigation infrastructures (Boelens & Vos, 2012; Wanvoeke et al., 2015). As the drip irrigation case indicates, the efficiency of these technological 'solutions' is all but neutral as it legitimizes some actions – in this case the precise delivery of water to the crops – at the expense of others – the labor of women having to manually clean the outlet (see also Tozzi, 2021). Yet, rather than abandoning these metrics because they are necessarily at fault, together with other feminist scholars (Henry, 2018) we call for alternative modes of accounting that broaden which works get counted and valued in specific contexts. Crucially, as the next example will show, these unaccounted labors are also never just the prerogative of humans.

The organisms of soils

Understanding water's ecology as made of combinations of elements brings a parallel shift in the ways we engage with soils as living substances – 'dynamic ecologies' (Krzywoszynska & Marchesi, 2020, p. 194) continuously (re)made through multispecies interactions, including with human who depend on it. Moving away from seeing soils as containers for plant's nutrients (Marchesi, 2020), this relational understanding brings into focus what Puig de la Bellacasa (2014, p. 2) describes the 'working quality' of soils as 'bio-infrastructure' underneath our feet. By calling attention to all those invisibilized more-than-human labors that compose soil's ecosystems (Krzywoszynska, 2020), this definition highlights the importance of attending to this lively subterranean infrastructure and its role in sustaining livable relationalities above the ground.

In the context of rainfed farming, we found that soil's bio-infrastructural character was manifested through its hydrological function, a coordinated effort between human and soil organisms to deliver and maintain sufficient moisture to plants roots (Tozzi, 2024). When we asked Priya, a farmer in Vidarbha, how she managed to raise her cotton crop only with the monsoon rains she replied:

> It all depends on "her" she said referring to the soil. See, black soil retains much more water than red soil and the red soil is better than the 'binghi' soil, which has a lot of stones. Our land has a mix of red and binghi soils so we need to prepare it carefully. Before the start of the monsoon, I clean the surface from stones deposited during the summer, apply cow dung and turn it a little so that nutrients reach deeper into the ground. Then I build obstacles in the field so that when the rain comes it goes deeper in the soil. The trick is to make sure that the soil maintains a good amount of moisture close to the roots of the plants, so I try to help her [the soil] with that.

In an unirrigated land – according to Priya – it is the inseparability of water from soil (that is moisture), combined with the care she mobilizes to make water, soil and roots hold together as one that guides her everyday farming practices.

The hydrological labor of soils is therefore inextricable from farmer's embodied knowledges and practices of devising appropriate cropping patterns, re-ploughing residues to recycle nutrients, reducing tillage and increasing land cover to avoid evaporation and protect microbial life. Veena explained:

> In the black soil, I alternate one year of cotton and one year of soybean and in between I plant rows of tur [green gram], oilseed, or lentils. This is because I balance between what is taken by the main crop and what is provided back by other crops. In the red soil instead, I plant jowar [sorghum] together with mung and arrange them into alternate rows. The jowar is good for red soils because even when there are long dry spells it has good chances to survive since its roots reach different depths.

When asked why she coupled jowar with mung, Veena replied, "Jowar grows first and being taller helps the mung grow on it. Then I cut the jowar, leaving space for the mung to develop on its own". This configuration also creates competition for moisture closer to the surface, stimulating jowar's roots to reach deeper into the ground, thereby strengthening the plant's ability to withstand increasingly frequent dry spells. These processes of adaptation that develop through close and repeated interactions between human and soil organisms are crucial to sustain rainfed agrarian environments in increasingly uncertain climates.

At the same time, the case of Pravah reveals how the hydrological work that takes place within soils is not always desirable: its permeability allows elements that are harmful to the landscape's ecology and people's wellbeing to pass through. From the various ponds where the Purandar water is delivered the solution of water-and-contaminants percolates through the ground, reaching the aquifer, where it travels to public and private wells across the landscape. Soils not only transport contaminated water but also filter it of pollutants, delineating a complex mesh of contamination across sources. Aware of this purifying work carried out by soil organisms, a common practice adopted by farmers is that of letting the Purandar water percolate from the ponds into the soil until it recharges private wells and only then use it to irrigate their crops. "The longer Purandar water percolates through soil, the more it gets purified", Aparna told us while irrigating her onions. "This is why we avoid using Purandar water straight from the pond". Yet, this filtering effort is not seamless, as manifested in a white layer ('*shaar*' in Marathi) that forms over the land and which women recognize as a sign of contamination and a reminder of the slow but steady loss of fertility of their land.

The aliveness of soils therefore suggests that the (waste)waterscape of Pravah is not delineated by (nor does it end at) pipe's outlets dotted around the village. Rather it emerges through the ways in which the contaminated water from the city interacts with the hydrological labors of soil bio-infrastructure, alongside the labor of humans as they filter and contain contaminants, producing a gradation of polluted sources women farmers must learn to navigate (Leonardelli et al., 2023). Where does the infrastructure end and the irrigated land begin? What counts as infrastructural work and what as farming practices? These are questions with which irrigation engineers and planners must engage.

As the water moving across the landscape changes, so does the labor performed by soil organisms, and also the knowledges that farmers (especially women) develop in their everyday dealings within the land. From a coordinated set of practices to maintain sufficient moisture for the plants to grow, to its ambivalent role transporting but also filtering contaminated water, engaging with soils as a bio-infrastructure reveals the relationality of laboring practices as distributed undertakings. Better characterizing the co-constituted character of these imperceptible works may therefore allow us to understand who and what is involved in processes of rural agrarian transformation, what role they play in this transition, and how they may be better accounted for. As the next section will reveal, beyond humans and soils, animals get involved in these transformations too.

Goats

This final section speaks to scholarship on interspecies relations in the agrarian world (Galvin, 2018) to tell the story of how the relationship between the Dhangar pastoralist community of Pravah and their goats was reconfigured by the Purandar water. While not directly addressing gender relations, through this narration we hope to shine a light on the intersectional impact of water infrastructure projects beyond the farmers who are their primary targets, discussing the effect of the Purandar Lift Irrigation Scheme and its polluted water on a pastoralist community and their companion species (Haraway, 2003).

While the social and economic status of the Dhangar community varies across India,[3] they can be described as a herding caste living a traditional nomadic lifestyle. Their livelihood practices integrate pastoralism with agriculture in a combination that is closely tied to the rhythms and dynamics of the monsoon. Having acquired small patches of land one kilometer away from Pravah before the arrival of the Purandar water, the Dhangars used to spend the monsoon season cultivating traditional crops such as bajra (millet), jowar (sorghum) and spices predominantly used for household consumption. As soon as the monsoon harvest ended and the land was cleared, they would set off with their goats and return to Pravah only the following rainy season. While roaming around, they would set up camps close to communities with whom they developed relations over the years, often through sales or exchange of animals and other products. Yet, since the Purandar water arrived, a few Dhangar households decided to abandon their nomadic lifestyle and settle in Pravah. Aided by the Purandar water, those who settled are now diversifying their livelihoods, combining commercial agriculture with animal husbandry.

Chatting with Kastur – a Dhangar man – we started understanding how things have changed for people from his community. "I have just made an agreement with the farmer who owns this land" he told us while his goats grazed on a recently harvested onion plot:

> My goats can graze here for a few days eating the residues from the harvest and in return they fertilize the soil with their excrement. It is good for the next cropping season. We [the Dhangar people] are doing this with many other farmers in Pravah:

Yet, goat's fertilizing efforts have not always been welcome. Shristi, an elderly Dhangar woman, remembers how farmers used to be hostile towards their goats, scaring them away as they would graze on their cultivated plots and spoil the crops. Now as the size of the cultivated farmland has increased and farmers rotate cultivated with fallow lands, pasture and fodder are available throughout the year, resulting in new relationships of exchange between Pravah's farmers, Dhangar people and their animals.

Highlighting the work that goats do as they manure the land, the case of the Dhangar community further highlights the impossibility of defining labor as a prerogative of any single life form – human or otherwise (Battistoni, 2017; Besky & Blanchette, 2019). In fact, as the soil of Pravah loses fertility and hardens because of continued use of contaminated water and chemical inputs, it is the labor of goats alongside that of pastoralists that becomes central to sustaining Pravah's troubled ecology. Yet goats and other animals are also paying a price for this contamination, as they drink the Purandar water directly from the ponds. Kastur shared his worries with us: "Since our goats started drinking the Purandar water, they have started falling sick more and more often". The situation is different from the time when Purandar water was not around: "Tanks would get filled by the rainwater and goats would drink that. Now all the water is contaminated, and our animals are suffering", he continued. In this regard, anthropological studies of multispecies relations in the agrarian world have questioned domestication narratives that suggest that animals are brought into the social world for purely utilitarian motivations. By contrast, these relations are multifaceted and better described as processes of co-becoming, through which humans and animals build relations of 'significant otherness' with one another (Haraway, 2003).

Further, while the relation between the Dhangar pastoralists and farmers from other castes in Pravah has improved, there are implications for the increased labor of women who are now required to both tend to the goats (though mostly alongside men) and perform most of the farming work. This is particularly true for older women, who are most likely to remain in the village with the grandchildren and are finding it difficult to grapple with the knowledge required for practicing sedentary, commercial agriculture.

Looking at goats and the Dhanghar pastoralists has therefore delineated a broader picture of the impact of the Purandar Lift Irrigation Scheme: it was because of the Purandar water that the Dhangars decided to settle in Pravah, diversifying their livelihoods by combining market-oriented agriculture with animal husbandry. As they now remain in the village, their work and that of their goats has become indispensable to sustaining Pravah's agricultural landscape amid increasingly contaminated soil. As processes of agrarian transformation reach beyond agriculture lands, approaching infrastructural planning with a more-than-human sensitivity enriches current analysis, as it directs attention towards the larger intersectional impact these projects have on communities that are not considered the direct 'beneficiaries' in the mind of planners, demanding that we account for them and their companion species in the future.

Conclusions

Examining processes of rural agrarian transformation through the lens of a transition between two forms of water, this paper focused on three more-than-human elements as analytical entry points to interrogate how gendered practices and labor relations are reconfigured through this transition. Taking seriously the relational materiality of water as a socio-bio-physical element whose character emerges at the intersection between water's lively ecology and socio-historical significance (Krzywoszynska & Marchesi, 2020) we brought feminist agrarian scholarship in conversation with post-human literature to explore the gendered connotation of water's relationality. In doing so, we contribute to feminist studies on water in two important ways.

First, by reconceptualizing labor as a more-than-human practice (Besky & Blanchette, 2019), a post-human lens has sharpened our analysis of the ways agrarian environments are reshaped by infrastructure projects aiming to 'develop' the rural countryside. The successes of government-sponsored plans like the Purandar Lift Irrigation Scheme in fact rest on the continued erasures of gendered labors and of the more-than-human world these projects enroll in. In official documents, the Government of Maharashtra advertises the Purandar Lift Irrigation Scheme as an 'efficient plan', providing farmers in drought-prone areas with the opportunity 'to change their cropping patterns to suit market demand' (Government of Maharashtra (GoM), 2018). Describing an idealized win-win situation, these narratives focus exclusively on the productive aspects of the scheme, without addressing how the Purandar water materializes differently for farmers, pastoralists and other members of the community, as it rearticulates their relations with the monsoon, soils and animals in ways that alter their day-to-day practices and livelihood dynamics.

Inspired by a mode of analysis elsewhere referred to as 'obliqueness', we were motivated by the recognition that 'what and who to include or exclude from narratives is never innocent, but the product of distinct political choices that are part of specific enactments of the world' (Leonardelli et al., 2022, p. 4). Echoing Battistoni (2017), employing an expanded understanding of labor as a collective undertaking is therefore a deliberate choice to narrate against the grain of neoliberal fairy tales, and reflect on how barely visible more-than-human labors may be 'reciprocated, compensated or sustained in recognition of their contributions our shared worlds' (p. 23). In a monsoon-dependent landscape like Maharashtra, this requires moving away from processes of agrarian transformations as a pathway towards an irrigated model of agriculture, engaging with the rainfall as the form of water around which relations of production and reproduction are articulated (Tozzi, 2024). Similarly, as wastewater transfer schemes are increasingly adopted as 'solutions' to address a seemingly naturalized condition of scarcity, rethinking metrics of efficiency and productivity to account for the work people and their ecologies perform together is paramount.

Finally, grounding our analysis of water's relational materiality within the complexities of specific waterscapes was an attempt to move feminist post-human literature beyond a metaphorical sensitivity of our watery interrelations (Neimanis, 2017). In this respect, we reflected on how thick descriptions of water worlds and the more-than-human assemblages they bring together can inform political analyses with concrete repercussions for water governance and infrastructural planning. As the Purandar Lift Irrigation Scheme suggests, the contaminated water the infrastructure carries go beyond canals and pipelines, extending into soils, aquifers, animals and gendered bodies all of which are reshaped through its flow. Going forward, it is by attending to the specificities of these configurations, analyzing what (and to whom) opportunities and harms are brought by transforming agrarian landscapes that we may find a place to bridge feminist struggles with environmental concerns.

Acknowledgements

This chapter would not have been possible without the farmers who shared their stories, knowledge and time with us. Many thanks also to the staff of the Society for Promoting Participative Ecosystem Management (SOPPECOM), and of the Revitalising Rainfed Agriculture Network, particularly to Seema Kulkarni, Sneha Bhat and Tarak Kate for their support during our fieldwork and the many conversations which shaped our thinking and writing. We would also like to acknowledge Isha Thorat, Janhvi Nimbarte and Yugandhara Khode, who worked with us as English-Marathi translators.

Notes

1 All names of people and dwelling places are pseudonymized, names of rivers and infrastructure are not changed.

2 Scheduled Castes and Scheduled Tribes are designated groups of people recognized by the Indian Constitutions and which comprise of the most disadvantaged socio-economic groups in India (Ministry of Social Justice and Empowerment, 2022b, 2022a).

3 The story we narrate here pertains only to those whom we met in the area of Pravah, who appeared to be relatively less well-off than members of other castes in the area (e.g., the Maratha caste).

References

Agarwal, B. (2003). Gender and land rights revisited: Exploring new prospects via the state, family and market. *Journal of Agrarian Change*, *3*(1–2), 184–224. https://doi.org/10.1111/1471-0366.00054

Ahlers, R., & Zwarteveen, M. (2009). The water question in feminism: Water control and gender inequities in a neo-liberal era. *Gender, Place & Culture*, *16*(4), 409–426. https://doi.org/10.1080/09663690903003926

Ahmed, S., & Zwarteveen, M. Z. (2012). Gender and water in South Asia: Revisiting perspectives, policies and practice. In M. Zwarteveen, S. Ahmed & S. R. Gautam (Eds.), *Diverting the flow: Gender equity and water in South Asia* (pp. 3–30). Zubaan.

Bakker, K. (2012). Water: Political, biopolitical, material. *Social Studies of Science*, *42*(4), 616–623. https://doi.org/10.1177/0306312712441396

Ballestero, A. (2019). The anthropology of water. *Annual Review of Anthropology*, *48*(1), 405–421. https://doi.org/10.1146/annurev-anthro-102218-011428

Barnes, J. (2013). Water, water everywhere but not a drop to drink: The false promise of virtual water. *Critique of Anthropology*, *33*(4), 371–389. https://doi.org/10.1177/0308275X13499382

Barnes, J., & Alatout, S. (2012). Water worlds: Introduction to the special issue of Social Studies of Science. *Social Studies of Science*, *42*(4), 483–488. https://doi.org/10.1177/0306312712448524

Battistoni, A. (2017). Bringing in the work of Nature: From natural capital to hybrid labor. *Political Theory*, *45*(1), 5–31. https://doi.org/10.1177/0090591716638389

Baviskar, A. (2007). *Waterscapes: The cultural politics of a natural resource*. Permanent Black.

Bennett, J. (2010). *Vibrant matter: A political ecology of things*. Duke University Press.

Besky, S., & Blanchette, A. (2019). *How nature works: Rethinking labor on a troubled planet*. University of New Mexico Press.

Bharucha, Z. P. (2019, May). This is what Nature has become: Tracing climate and water narratives in India's rainfed drylands. *Geoforum*, *101*, 285–293. https://doi.org/10.1016/j.geoforum.2018.09.032

Bhat, S. (2016). Deserted and widowed women's struggles for land and livelihood: A case from Maharashtra. In B. Fernandez, M. Gopal & O. Ruthven (Eds.), *Land, labour and livelihoods* (pp. 73–88). Palgrave Macmillan.

Boelens, R., & Vos, J. (2012). The danger of naturalizing water policy concepts: Water productivity and efficiency discourses from field irrigation to virtual water trade. *Agricultural Water Management*, *108*, 16–26. https://doi.org/10.1016/j.agwat.2011.06.013

Bossenbroek, L., & Zwarteveen, M. (2018). New spaces for water justice? Groundwater extraction and changing gendered subjectivities in Morocco's Saïss region. In R. Boelens, T. Perrault & J. Vos (Eds.), *Water justice* (pp. 330–345). Cambridge University Press.

Butler, J. (1990). *Gender trouble: Feminism and the subversion of identity*. Routledge.

Camargo, A., & Cortesi, L. (2019). Flooding water and society. *Wiley Interdisciplinary Reviews: Water*, *6*(5), 1–9. https://doi.org/10.1002/wat2.1374

Carney, J. (1993). Converting the wetlands, engendering the environment: The intersection of gender with agrarian change in the Gambia. *Economic Geography*, *69*(4), 329–348. https://doi.org/10.2307/143593

Chiwhane, P. A., Shastri, S. S., & Parbat, D. K. (2015). Use of Polluted river water after the treatment for irrigation purpose case study: Purandar lift irrigation scheme. *International Journal of Modern Trends in Engineering and Research*, *2*(2), 80–87.

Cortesi, L. (2021). An ontology of water and land in North Bihar, India. *Journal of the Royal Anthropological Institute*, *27*(4), 870–889. https://doi.org/10.1111/1467-9655.13611

Galvin, S. (2018). Interspecies relations and agrarian worlds. *Annual Review of Anthropology*, *47*, 233–249. https://doi.org/10.1146/annurev-anthro-102317-050232

Gibson-Graham, J. K. (2014). Rethinking the economy with thick description and weak theory. *Current Anthropology*, *55*(Suppl. 9). https://doi.org/10.1086/676646

Government of Maharasta (GoM). (2018). *Purandar lift irrigation scheme Dist. Pune*. https://kypipe.com/new_stuff/Purandhar%20Report%20(T).pdf

Gupta, A. (2017). Farming as speculative activity: The ecological basis of farmers' suicides in India. In U. Heise, J. Christensen & M. Nieman (Eds.), *The Routledge companion to the environmental humanities* (pp. 201–209). Routledge.

Haraway, D. (2003). *The companion species manifesto: Dogs, people, and significant otherness*. Prickly Paradigm Press.

Harris, L. M. (2006). Irrigation, gender, and social geographies of the changing waterscapes of southeastern Anatolia. *Environment and Planning D: Society and Space*, *24*(2), 187–213. https://doi.org/10.1068/d03k

Harris, L. M. (2009). Gender and emergent water governance: Comparative overview of neoliberalized natures and gender dimensions of privatization, devolution and marketization. *Gender, Place and Culture*, *16*(4), 387–408. https://doi.org/10.1080/09663690903003918

Henry, C. (2018). The abstraction of care: What work counts? *Antipode*, *50*(2), 340–358. https://doi.org/10.1111/anti.12354

Jagtap, S. S., & Manivanan, R. (2019). Water pollution status of Mula-Mutha rivers in Pune city: Review. *International Journal of Trend in Scientific Research and Development (IJTSRD)*, *4*(1).

Jamwal, P., Thomas, B. K., Lele, S., & Srinivasan, A. (2014). *Addressing water stress through wastewater reuse: Complexities and challenges in Bangalore*. India. 5th Global Forum on Urban Resilience & Adaptation, Bonn, Germany, 29–31 May 2014. http://hdl.handle.net/10625/53874

Joy, K., & Janakarajan, S. (2018). *India's water futures: Emergent ideas and pathways*. Sage Publications Inc.

Krishna, S., & Kulkarni, S. (2019). Gender and water: Why we need alternatives to alternative discourses. In K. J. Joy & S. Janakarajan (Eds.), *India's water futures: Emergent ideas and pathways* (pp. 235–252). Routledge India.

Krzywoszynska, A. (2020). Nonhuman labor and the making of resources making soils a resource through microbial labor. *Environmental Humanities*, *12*(1), 227–249. https://doi.org/10.1215/22011919-8142319

Krzywoszynska, A., & Marchesi, G. (2020). Toward a relational materiality of soils introduction. *Environmental Humanities*, *12*(1), 190–204. https://doi.org/10.1215/22011919-8142297

Kusenbach, M. (2003). Street phenomenology the go-along as ethnographic research tool. *Ethnography*, *4*(3), 455–485. https://doi.org/10.1177/146613810343007

Leonardelli, I., Kemerink-Seyoum, J., Kulkarni, S., Bhat, S., & Zwarteveen, M. (2023). A feminist analysis of women farmers navigating groundwater qualities in Maharashtra, India. *Water Alternatives*, *16*(1), 134–152.

Leonardelli, I., Zwarteveen, M., & Kemerink-Seyoum, J. (2022). Obliqueness as a feminist mode of analysing waterscapes: Learning to think with overflows. *Environment and Planning E Nature and Space*, *6*(2), 863–879.

Linton, J. (2010). *What is water?: The history of a modern abstraction*. University of Chicago Press.

Marchesi, G. (2020). Justus von Liebig makes the world: Soil properties and social change in the nineteenth century. *Environmental Humanities*, *12*(1), 205–226. https://doi.org/10.1215/22011919-8142308

Matthan, T. (2023). Beyond bad weather: Climates of uncertainty in rural India. *The Journal of Peasant Studies*, *50*(1), 114–135. https://doi.org/10.1080/03066150.2022.2116316

Mehta, L. (2007). Whose scarcity? Whose property? The case of water in western India. *Land Use Policy*, *24*(4), 654–663. https://doi.org/10.1016/j.landusepol.2006.05.009

Ministry of Social Justice and Empowerment. (2022a). *List of scheduled tribes*. Government of India. https://web.archive.org/web/20131107225208/http://censusindia.gov.in/Tables_Published/SCST/ST Lists.pdf

Ministry of Social Justice and Empowerment. (2022b). *Scheduled caste welfare – List of scheduled castes*. Government of India. https://web.archive.org/web/20120913050030/http://socialjustice.nic.in/sclist.php

Neimanis, A. (2017). *Bodies of water: Posthuman feminist phenomenology*. Bloomsbury Academic.

Nightingale, A. (2006). The nature of gender: Work, gender, and environment. *Environment and Planning D: Society and Space*, *24*(2), 165–185. https://doi.org/10.1068/D01K

O'Reilly, K. (2006). "Traditional" women, "modern" water: Linking gender and commodification in Rajasthan, India. *Geoforum*, *37*(6), 958–972. https://doi.org/10.1016/j.geoforum.2006.05.008

Patel, R. (2013). The long green revolution. *Journal of Peasant Studies*, *40*(1), 1–63. https://doi.org/10.1080/03066150.2012.719224
Pattnaik, I., Lahiri-Dutt, K., Lockie, S., & Pritchard, B. (2018). The feminization of agriculture or the feminization of agrarian distress? Tracking the trajectory of women in agriculture in India. *Journal of the Asia Pacific Economy*, *23*(1), 138–155. https://doi.org/10.1080/13547860.2017.1394569
Puig de la Bellacasa, M. (2014). Encountering bioinfrastructure: Ecological struggles and the sciences of soil. *Social Epistemology*, *28*(1), 26–40. https://doi.org/10.1080/02691728.2013.862879
Puig de la Bellacasa, M. (2017). *Matters of care: Speculative ethics in more than human worlds*. University of Minnesota Press.
Reddy, D. N., & Mishra, S. (2010). *Agrarian crisis in India*. Oxford University Press.
Savelli, E., Rusca, M., Cloke, H., Flügel, T. J., Karriem, A., & Di Baldassarre, G. (2022). All dried up: The materiality of drought in Ladismith, South Africa. *Environment and Planning E: Nature and Space*. https://doi.org/10.1177/25148486221126617
Sethi, A. (2021). Terms of trade and the cost of cotton: The paradox of commercial agriculture in India. *The Journal of Peasant Studies*, *48*(7), 1397–1408. https://doi.org/10.1080/03066150.2021.1996356
Shah, S. H., Harris, L. M., Johnson, M. S., & Wittman, H. (2021). A "drought-free" Maharashtra? Politicising water conservation for rain-dependent agriculture. *Water Alternatives*, *14*(2), 573–596.
Strang, V. (2014). Fluid consistencies. Material relationality in human engagements with water. *Archaeological Dialogues*, *21*(2), 133–150.
Sultana, F. (2009). Fluid lives: Subjectivities, gender and water in rural Bangladesh. *Gender, Place and Culture*, *16*(4), 427–444. https://doi.org/10.1080/09663690903003942
Sultana, F. (2011). Suffering for water, suffering from water: Emotional geographies of resource access, control and conflict. *Geoforum*, *42*(2), 163–172. https://doi.org/10.1016/j.geoforum.2010.12.002
Sundberg, J. (2004). Identities in the making: Conservation, gender and race in the Maya Biosphere Reserve, Guatemala. *Gender, Place & Culture*, *11*(1), 43–66. https://doi.org/10.1080/0966369042000188549
Tozzi, A. (2021). Reimagining climate-informed development: From "matters of fact" to "matters of care". *The Geographical Journal*, *187*(2), 155–166. https://doi.org/10.1111/geoj.12390
Tozzi, A. (2024). Reimagining the governance of water from the ground up: On the 'worlding-practices' of grassroots movements building alternative 'water worlds'. *Environment and Planning E: Nature and Space*.
Tozzi, A., Bouzarovski, S., & Henry, C. (2022). Colonizing the rains: Disentangling more-than-human technopolitics of drought protection in the archive. *Geoforum*, *135*, 12–24. https://doi.org/10.1016/j.geoforum.2022.07.011
Tsing, A. (2013). More-than-human sociality: A call for critical description. In K. Hastrup (Ed.), *Anthropology and nature* (pp. 27–42). Routledge.
Tsing, A. (2015). *The mushroom at the end of the world*. Princeton University Press.
Voelcker, J. A. (1893). *Report on the improvement of Indian agriculture*. Eyre and Spottiswoode.
Wanvoeke, J., Venot, J.-P., Zwarteveen, M., & de Fraiture, C. (2015). Performing the success of an innovation: The case of smallholder drip irrigation in Burkina Faso. *Water International*, *40*(3), 432–445. https://doi.org/10.1080/02508060.2015.1010364
Zhang, Y., & Shen, Y. (2019). Wastewater irrigation: Past, present, and future. *Wiley Interdisciplinary Reviews: Water*, *6*(3), e1234. https://doi.org/10.1002/wat2.1234

26

BEYOND *WATER JUSTICE* AND *WATER SECURITY*

Debates on water, women, and climate change in Latin America

Catalina Quiroga and Anyi Castelblanco

Introduction

Climate change has increased water scarcity and deepened inequalities in water access throughout the Latin American region. According to the latest IPCC report, in all scenarios for the Latin American and Caribbean regions, floods will increase due to the degradation of critical ecosystems such as glaciers or wetlands, key ecosystems for water provision. This would significantly affect the daily life of agricultural and fisher communities (IPCC, 2022). Furthermore, according to estimates by the Food and Agriculture Organization (FAO), 32% of the world's renewable water resources can be found in Latin America, and climate change jeopardizes its availability, quality, and quality (FAO & UNEP, 2020). Consequently, the relationship between water and climate change becomes a critical dimension for analysis.

Moving beyond data about risk, water insecurity, loss, and damage, Latin American social movements, in joint work with politically committed academics, have proposed that it is necessary to examine the connection among waters, women's bodies, and territories[1] to understand the impact climate change will have on water.

Scholarship has described how climate change transforms water availability, quality, and quantity and impacts women's daily life. Women, who are generally responsible for house water provision, are differentially impacted by climate change (Agarwal, 1992; Harris, 2009; Sultana, 2018; Ulloa, 2015). Thus, women, especially the most impoverished, are more vulnerable to water scarcity and, intersecting with other inequalities such as social class or race, have fewer options to participate in the transformations and policies to guarantee a different relationship with water for the production and reproduction of life (Ahlers & Zwarteveen, 2009; Caretta & Rothrock, 2021; Coles, 2009; Harris, 2009; Sultana, 2018).

It is in this context of worsening water scarcity and gender inequality that we embark on an exploration of the knowledge produced by social movements and politically committed academia. This chapter focuses on the encounters and disputes between three concepts: *water justice, water-body-territory*, and *water security*.

We argue, following the *water-body-territory* concept, that thinking of water as an integral part of the relationships between territories and the bodies of women is essential to analyzing causes, consequences, and forms of adaptation and mitigation to climate

DOI: 10.4324/9781003100379-32

change. To contextualize this approach, it is necessary to recognize other concepts that have emerged in other arenas and are related to the debate, specifically the ideas around *water security* and *water justice*. We review the ways in which knowledge is being produced at different scales on the relationships between water, women, and climate change, in various and sometimes contradictory ways.

Our interest in this topic stems from our own personal and professional experience. Anyi Castelblanco has been part of the social mobilization in defense of water and territories in Colombia for more than ten years. From her work as an activist researcher, Anyi has worked with different experiences in Colombia in the defense of water against mining, researched on water management, and the impacts of bottled water. Catalina Quiroga has conducted activist research in the Colombian Caribbean, where various forms of extractivism produce inequalities that affect the daily lives of women, children, Indigenous, Afro-descendant, and peasant populations.

Building on feminist principles, we both believe that research should be a collaborative process that allows for dialogues between different actors. This means that in previous experiences, we have included the principles of scholar-activism (Fals Borda, 1999; Borras, 2016; Fals Borda, 2015; Leyva & Speed, 2008) and research by demand (Segato, 2015), which aim to think about local research agendas and contribute to the explanation and analysis of the socio-environmental inequalities in collaborative ways. We believe that the concepts that arise from social mobilization and the situated reflection of women in the Latin American region are key to understanding the relationships between water, gender, and climate change in a complex way. We want to highlight the voice of women and social movements because they have generated initiatives to manage and live with water, initiatives we believe can create new ways of living in the future.

This analysis is done in response to Sultana's assertion that: "feminist scholars can add much to the ongoing debates in the climate change and adaptation literature, explicating the textured ways that space, place, identities, and lived experiences are intersected by a range of processes and social relation" (Sultana, 2018, p. 4). We propose that analyzing the relationships between gender and water that emerge from social and community leadership in Latin America in contexts of climate crisis can provide key inputs to think about the structural causes of climate change and possible ways to counteract it. To do this, we analyze the concepts that operate in policy recommendation reports produced by key multilateral organizations, as well as those that are embraced by (and emerge from) the work of women-led social movements and activist researchers in Latin America.

This chapter has five sections. The first sets out some reference points on the relationship between water and gender in the climate crisis context. The second presents the main postulates of the concept of *water security*. The third explores the discussions associated with work on *water justice*. The fourth section highlights the contributions made by territorial and communitarian feminism, especially the concept of *water-body-territory*. Finally, we close with conclusions.

Water and gender in times of climate crisis: the relation between water, gender, and climate change

Water has been understood from different cultures and paradigms, which define relationships, interactions, management, and actions at different levels. These paradigms can be summarized in four ways: those that see water as a commodity and "see" it as a resource to tame and manage (Ulloa et al., 2020; Linton, 2010; Villa Fontecha, 2012); those that see water as a right and focus on guaranteeing access for humanity (Amigos de la Tierra América Latina y el Caribe (ATALC), 2018; Cantor & Emel, 2018); those that understand water as a common good and as part of life for both humans and non-humans (Tacha, 2022; de Llobatera, 2004; Woster, 2008), and finally, those who understand water as a non-human entity with its own rights (Ulloa et al., 2020; Ulloa, 2020).

The debate around *water security* is usually related to the first paradigm and is associated with the Integrated Water Resource Management (IWRM) model, which multilateral agencies have generally championed. This model seeks to conserve water for economic purposes and tends to think of water as isolated from the ecosystems where it emerges from and flows. Similarly, the debate around the human right to water also seeks to guarantee its access with quantifiable minimums (Valdés de Hoyos & Uribe Arzate, 2016). In this sense, it focuses on ensuring availability from a quantifiable approach, highlighting the relationship between beneficiaries and providers.

In contrast, the visions of water as commons or as a non-human entity with agency understand the dynamics and connections between water and life and focus on fostering its integrity throughout its cycle, thinking about all the beings that depend on it, not only humans (Tacha, 2022; Ulloa, 2020). Debates around *water justice* and *water-body-territory* draw on these paradigms and are embraced by social movements and women in the region.

Considering these water-related paradigms, it is relevant to review feminist scholarship. Most studies suggest that human relationships with water are gendered and affect women unevenly, in different contexts. Researchers worldwide have analyzed how women are excluded from decision-making around the construction of irrigation systems (Caretta, 2015b, 2015a; Caretta & Rothrock, 2021). Other studies have focused on the relationships between water and women as they intend to gain access to drinking water and sanitation systems (Delbene-Lezama, 2019; Salinas Mulder, 2017). Finally, other studies have considered the relationship between water and women from a perspective that integrates health and disease management, for instance, in the pandemic context (Fernández, 2021; Guilcamaigua Pastuña, 2021)

Notably, in Latin America, the debate around the relationship between water and women focuses on the analysis of the implementation of extractive projects that include mining (Colectivo CASA, 2015, 2018; Gaitán Ortiz, 2020), hydrocarbons (Roa Avendaño et al., 2017), dams (Duarte-Abadía et al., 2015; Lobos Castro, 2021), agribusiness (Bolados García et al., 2018; Bravo & Fragkou, 2019), and land grabbing processes (Pena, 2022). These studies are concerned with how women, given their role in reproductive labor, are affected by the lack of water access for the production and reproduction of life. These studies call out the need to attend to the experiences of women, as centering their experience opens space to recognize more nuances in the impacts of climate change on the daily and political life of the region.

Delving into the relationship between water, women, and climate change, Sultana (2018) argues that, because climate change affects the availability of and access to water, any discussion about this process must consider how the relationships between water and gender develop and transform more broadly. In attempting to overcome the supremacy of the paradigm that understands water as a commodity, it is relevant to study how climate change transforms water availability and how this affects women's access and relationship to water. In Latin America, women relate with water in productive, reproductive, and political ways. They are not only in charge of the work inside the home, but they are also part of the "productive" sector, and, in addition, are often involved in communitarian and political spheres

(Salinas & Becker, 2022). For this reason, we are interested in analyzing the proposals born from territorial and community feminisms as an expression of the political action taken by women in their daily lives.

This chapter diverges from the approaches that, amidst climate change and water problems, place women as victims or caregivers. Following Arora-Jonsson, we argue that a causal reading between climate change and vulnerability diverts attention from power relations and inequalities produced by institutional policies and in discourses on climate change, women, and water (2011). We show how reports produced by multilateral agencies have frequently documented the specific impacts of climate change on water access and availability for women without delving into other related aspects such as inequality in the production of knowledge or local proposals to overcome water injustices (Ahlers & Zwarteveen, 2009).

Thus, by exploring the relationships between water, women, and climate change, we aim to understand how productive, reproductive, and political roles are produced, and how in the interaction of these dimensions, local knowledge emerges. We argue that this local knowledge allows us to recognize in a complex way how water and gender inequality is produced in contexts of climate change (Carrillo Rodríguez, 2020; Guilcamaigua Pastuña, 2021).

Water security: water as commodity and multilateral agencies

The concept of *water security* was proposed at the Second World Water Forum in The Hague in 2000. Since then, different policy recommendations have adopted the idea to characterize what should be included in definitions of water access. In Latin America, this concept is tied to development agendas, and, more recently, associated with solutions to the effects of climate change (Peña, 2016; Urquiza & Billi, 2020). Multilateral development agencies build on water security on their public policy implementation guides, which in turn are used by governments at the local level (see Table 26.1). National governments in Latin America must use these guides to present projects in the search of development funding.

Table 26.1 Operationalization of Water Security by Multilateral Agencies

Institution	*Water Security Concept*
UNICEF (2021) and UNWater (2013)	The ability of a population to safeguard sustainable access to adequate quantities of water of acceptable quality to support livelihoods, human well-being, and socio-economic development, to ensure protection against waterborne pollution and related disasters, and to preserve ecosystems in a climate of peace and political stability. Water insecurity occurs when some or all of these needs cannot be met *(own translation)* (UNICEF, 2021; UNWater, 2013).
Paris (2020)	(i) a characteristic of the population, which leads the global community to local action, (ii) refers not only to water emergencies due to extreme events but also includes the condition of health and includes ecosystem services in the hydrological cycle, (iii) guides the definition of public policies on the water issue and defines criteria to establish goals and provides indicators *(own translation)* (Paris, 2020).

Source: Author's own elaboration.

The *water security* concept derived from debates around the IWRM framework. As previously stated, it understands water as a quantifiable resource which can be governed through clear distribution rules (Lautze & Manthrithilake, 2012; UNEP, 2017). In practice, IWRM proposes an approach to the management and handling process in which different actors are involved to define water management's limits, forms, and rules (Miralles-Wilhelm et al., 2022).

In its strategic document to guarantee *water security* in Latin America, the Inter-American Development Bank (IDB) states that:

> The main challenges that Water Insecurity strategies must face in the region go through the complex interaction between all the sectors that use the resource, the high vulnerability of the region to natural disasters of hydric origin, poor access to water and sanitation services, institutional weakness, and poor water infrastructure *(own translation)*.
>
> *(Bretas et al., 2020)*

The same document, which is also the guide for the preparation of projects for loans and technical support, mentions as possible solutions to *water insecurity*: (i) planning of water infrastructure, (ii) flow regulation, (iii) management, preservation, and economic valuation of ecosystem services, and (iv) management of floods and natural disaster risks (Bretas et al., 2020).

These recommendations present a growing concern about water availability, especially the occurrence of "natural disasters" in the context of climate change, without addressing its causes. Faced with worsening problems related to water access in the context of climate change, international banks propose solutions associated with water management for the future, emphasizing the construction of infrastructure. They do not propose solutions to structural processes associated, for example, with extractivism in the region. Instead, the idea of *water security* is focused on guaranteeing measurable minimums so that society can sustain productive activities. The concept of *water security* also focuses on the need to preserve key ecosystems for the maintenance of natural water cycles.

It is important to mention that these concepts are changing, including new elements, and adapting to new water management models. For example, the Valuing Water Initiative, championed by the United Nations and other national governments worldwide, recognizes the relationships between water, sustainable human development, ecosystem health, and economic prosperity (The Valuing Water Initiative, 2020). Although this initiative highlights gender in its approach as well as the relationship between water management and peacebuilding, it does not depart from quantifying water as a manageable resource. It proposes a measurement model based on numerical variables and specific rules that obscure other relationships with the water, such as Indigenous communities' visions of the interconnection between water and territories. In relation to this initiative, the idea of "values" can be used at the local level to demand greater participation. Still, new questions emerge, such as which values will be incorporated or listened to more than others?

In the 1977 United Nations Water Conference held in Mar Del Plata, Argentina, multilateral agencies mentioned that women have a leading role in the collection, management, and water use of households, and are most affected by transformations in water supply. Thirty-five years later, in 2012, the United Nations Framework Convention on Climate Change (UNFCCC) included the premise that climate change can worsen the conditions of women's water access.

Three main arguments or images are developed in relation to women in the reviewed policy documents. In the first argument, women are presented as subjects that need special attention because they are the poorest of the poor. Second, women are the protagonists in the presentation of risk-associated mortality data; that is, women are the most affected during natural disasters caused by climate change. These two arguments are mainly developed by preparing background information and presenting the situation associated with water insecurity. The third argument concerns the idea that women are more aware of the environment. Therefore, they are called to take responsibility for implementing daily changes (Arora-Jonsson, 2011).

These inclusions of gender in water policy recommendations cannot be seen outside of the political work carried out by women and social movements worldwide (Marcha Mundial de las Mujeres, 2020; Sultana, 2018). Nonetheless, even when the feminist movement worldwide has made it possible to include these debates in public policy, this process must be read critically. Some authors argue that the production of neoliberal policies instrumentalizes gender discourse and the role of women as political actors. The process of instrumentalization uses key concepts for mobilization in policy writing, but the meaning of these concepts is transformed and often reduced to sporadic mentions within public policies (Montealegre, 2020; Fosado, 2019). Fosado (2019) argues that this makes social processes invisible, erases the historical trajectories of constructing concepts, and could deepen existing inequalities.

The instrumentalization of gender has led to portrayals of women as victims or caregivers within water governance and environmental policies related to climate change (Arora-Jonsson, 2011, 2014). Within multilateral agency documents, expressions such as "most affected" or "worsening of conditions" reoccur, deepening the dynamic of victimization. However, open dialogues and understanding of structural causes of gender inequality are strikingly absent.

Even though policy documents drawing on the concept of water security do mention that the problem of water scarcity is a consequence of climate change, they do not delve into how climate change is produced as a global phenomenon, its causes, and the relationships that led us to the crisis. Regarding the relationship between water, women, and climate change, the concept of *water security* used in recommendations for Latin America understands women solely as individuals deprived of water access (Arora-Jonsson, 2014). In this sense, women tend to be seen as victims who have suffered the effects of transformations in water availability.

Water justice, activist research, and social movements in Latin America

Latin American social movements, in connection with activist research, have been articulating struggles and local knowledges in the region in order to understand the territorial and water management dynamics behind the policies production (Gudynas, 2020). Latin American activist researchers have supported these struggles by making visible the concepts that rise from social movements and by systematizing the communities processes (Fernández, 2021; Hoetmer, 2021; Isch, 2021). On occasions, academic sectors have engaged in Participatory Action Research, seeking to transform local realities (Fals Borda, 1999, 2015). Moreover, actors from social movements have been actively involved in academic research, just as several academics have decided to be part of the social movements.

Table 26.2 General Characteristics of the Water Justice Concept

Water Justice's Characteristics	*How the Discussion Between Social Movements and Academia has Explained the Characteristic*
WJ recognizes water as an actor that relates to others within the ecosystem, including social relationships.	Water is both an end to struggles, as well as a method and an actor. The defense of ecosystems has also generated ecosystems of movement, social relations, meanings, and horizons of a different future *(Own translation)* (Hoetmer, 2021, p. 164)
WJ recognizes the various ways inequalities and injustices are produced, including in the debate the need to think about the production of knowledge and local forms of knowledge production.	Understanding how water injustices are embedded and localized and looking for possible ways to remedy them, [. . .] This implies a recognition of diversity and plurality – in points of view, knowledge, rights systems, ideas and norms on equity, etc. – without adopting a position of cultural relativism or denying the general similarities across specific cases of injustice *(own translation)* (Boelens, 2021, p. 67)
WJ recognizes the material, cultural, political, and ecological relationships in the context of exclusion in water access.	The interactive social and academic effort to critically explore the production, allocation and management of water knowledge and combine struggles against water-related forms of material dispossession, cultural discrimination, political exclusion, and ecological destruction rooted contexts *(own translation)* (Boelens, 2021, p. 74)

Source: Author's own elaboration.

The concept of *water justice*, which emerges from political ecology,[2] entails analyzing water accumulation processes in terms of class, gender, and ethnicity (Fernández, 2021; Isch, 2021; Saldi, 2021). This concept incorporates a historical and structural reading where axes of power such as capitalism, colonialism and patriarchy produce forms of unequal water access (see Table 26.2) (Cruz Hernández, 2020; Díaz Lozano et al., 2021; Marcha Mundial de las Mujeres, 2020).

Following these ideas, Isch (2021) has identified four angles of *water justice* analysis: environmental, economic-distributive, social power construction, and social movement. Although it has been easier to define *water injustice* within the framework of the dynamics of territorial appropriation and the commodification of life, political ecology has played a fundamental role in understanding the dynamics related to water in the region and in the articulation of academic work with the production of diverse knowledge. In addition, academics and social movements working in Latin America have defined the category of hydrosocial territory[3] to understand power relations around waters. This category recognizes the political character of the territories by studying water use from the praxis of daily life (Ávila-García, 2016; Boelens et al., 2017; Swyngedouw & Zuñiga, 2018)

Scholarship drawing on the *water justice* concept proposes that climate change is felt mainly through water (Doornbos, 2011). For this reason, issues related to food sovereignty, the infrastructure to adapt to climate change, and access to drinking water and water quality are central issues for water justice defenders (*justicieros y justicieras hídricas*).[4]

> Water struggles are also struggles to exercise the right to autonomously define water problems and decide on the directions that solutions will take. They are associated with cultural meanings and (ethnic) identity and with ideals and ideas about what humanity and development mean *(own translation).*
>
> *(Boelens et al., 2011, p. 20)*

Along these same lines, debates around *water justice* reflect critically on the concept of development, arguing that the imposition of a development model in Latin America corresponds to a colonial one (Fernández, 2021; Moreano et al., 2022; Ulloa, 2016). Therefore, local communities generate ways of life far from the imposed development model and propose other relationships with nature and water. Still, their proposals are often made invisible, denied, and silenced – even more so when they are enunciated from the voice of women.

> We are at a fundamental historical moment because we, Indigenous peoples, are proposing alternatives to the global civilizational crisis, seeking forms of resistance against the effects of capitalism in its neoliberal form: against colonialism, patriarchy, and globalization. Now other people are looking at our proposals as a different way of living that gives rise to taking up the original paradigm of *Buen Vivir* (Good Living). I think they are profound processes of full consciousness, of remembering how people at some points have had the opportunity to harmonize for life *(own translation).*
>
> *(Cabnal, 2013b, p. web)*

Building on the concept of *water justice*, social movements argue that the problem of water scarcity is not only an effect of climate change but stems from a crisis of civilization resulting from the dominant development model.

> Water scarcity has been socially constructed in the heat of a *Maldesarrollo* (bad development) model. Extractivism has meant an excessive demand for water for productive uses to the detriment of water access for ecosystems and human beings. Communities must compete for access to water with other productive activities, such as mining, agro-export, real estate or hydroelectricity generation *(own translation).*
>
> *(Salinas & Becker, 2022, p. 8)*

Water scarcity is borne of a predatory relationship with the environment, which is in turn the result of extractive processes and the exploitation of territories and lives, especially of those beings who have been victims of unequal, capitalist, patriarchal, and colonial relationships. Following the postulates of *water justice*, it is essential to transform our relationships with water.

Finally, *water justice* makes different water management alternatives visible, often based on communal, local, feminist, and biocentric ideals. Although social movements have led processes of resistance and caring for territory, it is mainly women who have assumed the role of water justice defenders, "*justicieras hídricas*", in the territories, carrying the work of collaborative care and ensuring access to safe drinking water. In the following section, we document the contributions of feminism to the *water justice* debate.

Water-body-territory: a proposal to think about the future of Latin America

The reflections and actions of feminism in Latin America have played a critical role in the struggles for the defense of territories and water. From different parts of the region, Indigenous, peasant, and Afro-descendant women propose various ways of thinking about the relationship between water, women, and climate change. This last section presents the contributions of these feminist initiatives to debates between *water security* and *water justice*.

We argue in this section that the contributions from feminism allow the integration of new elements to the idea of *water justice* while at the same time problematizing how the *water security* concept is understood. Finally, this section is a commitment to provide new elements to the debate that have resulted from community action in the Latin American region.

Feminist proposals are based on the idea that climate change results from predatory ways humans relate to nature, especially to water (Cabnal, 2010, 2013a; Márquez, 2018). The deepening of the extractivist model and patriarchal relations within the territories has generated:

> Death regimes [that] gain space and commodify common goods (land, water, seed) as if they could be measured in monetary value. This leads us to the planet's collapse due to a patriarchal logic that looks at nature as if it were at the service of people and justifies policies of dispossession *(own translation)*.
>
> *(Marcha Mundial de las Mujeres,*[5] *2020)*

Additionally, in its Latin American chapter, the Marcha Mundial de Mujeres integrates a critique of approaches that see water as a commodity. This critique notes that the deepening of the capitalist, colonialist, and patriarchal model has led to women being seen as in the service of men. For this reason, their struggles for the defense of nature often go unnoticed or are conceived as claims conditioned to the role of care. Faced with this problem, the women of the Marcha Mundial de Mujeres propose that it is necessary to overcome visions that pigeonhole women in jobs related to caring as if this were a natural part of the division of labor. On the contrary, the collective proposes to make visible the different spaces where women contribute to the creation of futures, for example, spaces for political discussion or educational and management spaces (Marcha Mundial de las Mujeres, 2020).

Tied to this discussion, women's organizations throughout Latin America have requested to be taken into account in spaces for participation and decision-making, to guarantee the preservation of life in a highly violent contexts where women defenders of territories and water have been victims of persecution by armed actors (Red Latinoamericana de Mujeres Defensoras de Derechos Sociales y Ambientales, 2018). Political participation by some of these groups goes beyond assistance in decision-making spaces. Women's groups claim that to integrate debates associated with justice into water governance, it is necessary to understand that we must transform how we relate to nature in general (Colectivo CASA, 2015, 2018; Salinas & Becker, 2022).

The Urgent Action Fund documents the local strategies of women who demand transformation in our relationship to nature and highlights the role of *the body* in the debate (Fondo de Acción Urgente, 2020). The contributions of academics and activists concerning the concept of *body-territory* allow us to advance in understanding the body, the territories, and their waters as part of an inseparable ontological relationship (Colectivo Miradas Críticas del Territorio desde el Feminismo, 2017; Zaragocin & Caretta, 2021).

Feminist proposals, in addition to providing a different understanding that integrates bodies into the framework of the defense of territories and water, advance a situated analysis of the struggles against the extractive model and its relationship with the development model.

> From the women's movement, we recover, defend and care for water, seeds, land, harmonious relationships, affection, care, and ancestral practices for life. The defense of the sustainability of life is based on another way of thinking and doing economy: one in which all knowledge and types of work are valued, the contribution of women is recognized, and the cycles and times of nature are respected *(own translation)*.
>
> *(Marcha Mundial de las Mujeres, 2020)*

The *water-body-territory* concept has been developed and embraced by social movements and activist researchers. This concept aims to understand the relationship between these elements as a hybrid that interconnects historical, cultural, social, and economic dimensions in women's daily practices (Lobos Castro, 2021). This concept was first developed from the experience of women against mining and in favor of water justice in Chile. Including the body in the discussion of water and territories allows a comprehensive reading of the problems in water access that range from the scale of the body and everyday context to how definitions are taken at other scales (i.e., national, regional, or global scale).

These concepts, *body-territory* and *water-body-territory* are based on the local reality of many women affected by the extractive model. This, however, is the reality of much of the world, and feminist social movements state that extractivism is directly related to the climate crisis. Lobos Castro (2021) continues:

> This *water-body-territory* triad emerges as a tangle of threads of struggle entangled and intertwined by rural women in an attempt to resist embodied dispossession, and in turn, opens subjective dimensions of re-existence (. . .) An inseparable triad that inspires threads of re-existence, that is, daily emanations of the struggle for the common rural women, bodies that intentionally thread territories from the felt and signified corporeality with the flows of suffering or liberated Andean waters *(own translation)*.
>
> *(Lobos Castro, 2021, p. 114)*

Reflections around the *water-body-territory* concept contribute to the debates on the relationship between water, women, and climate change relationships in three ways. First, integrating the body into the discussion questions the idea of finding a solution to the climate crisis through a management model. Second, given that these concepts were born from the mobilization and research of women activists and academics in Latin America, it indicates that women are an important part of the search for solutions to the environmental crisis. Third, methodologically, these concepts propose solutions based on the contextualized dialogue that recognizes daily life as an essential part of the search for answers to the crisis.

Conclusions

This chapter focused on how social mobilization and activist research conceptualize the relations between water, women, and climate change in conversation with hegemonic concepts such as *water security*. Working with the concepts of *water justice* and *water-body-territory*, social movements and research activists have integrated new variables and forms of production of knowledge to understand water injustice in the context of climate change.

We have shown how debates around *water security* do not delve into the production of scarcity, thus erasing responsibilities in how inequalities come to exist. For their part, debates around *water justice*, which have engaged researchers and social movements in Latin America, understand water as part of the territories and the guarantees of a dignified life. Within this framework, the causes of the environmental crisis, especially for the region, are directly linked to extractive projects. Faced with the inclusion of gender issues in the two debates, we find that the concept of *water security* tends to reproduce stereotypes associated with the division of labor and risks reproducing images of women as victims of climate change. Three main arguments are developed within the policies: first, that women need special attention because they are the poorest of the poor; second, because they have a higher mortality rate during natural calamities caused by climate change, and third because women are more aware of the environment (Arora-Jonsson, 2011).

Although the concept of *water justice* incorporates gender and understands it in the context of inequalities and power relations, many times gender is related to individual efforts. In consequence, the gender focus is not central to the debate. In any case, it is important to mention that activists and social movements who propose the concept of water justice have proposed different reflections that allow us to understand the relationship between water, gender, and climate change in a complex way. Still, they have advanced in producing other explanations involving other actors in the origin of the environmental crisis.

Finally, it is important to highlight that the Latin American feminist debate evidences the double or triple task of women in caring for the territory and social life. Women bear the consequences of water insecurity in their daily lives, which makes it difficult for them to participate in decision-making spaces on water management. In this sense, although women are highly vulnerable, it is no less accurate that their experiences in administering, rationing, and distributing water have made them experts in different adaptation strategies in contexts of water scarcity. Therefore, paying attention to their experiences and knowledge can create more effective and sustainable public policies (Salinas & Becker, 2022)

Finally, we emphasize that the water-body-territory concept expands Latin American contributions to understanding the relationships between water, women, and climate change. It is important to highlight that both the postulates of *water justice* and the contributions of feminism are interesting alternatives which move beyond the idea of water resources management. We insist that it is necessary to integrate the voices of women in the debate about water in climate change policies and recommendations because they allow us to transcend the logic of water insecurity, raising questions about the production of scarcity and proposing ways of caring for water within the defense of bodies and territories.

Notes

1 According to the Brazilian geographer Fernandes, "Territory is the space appropriated by a specific social relationship that produces and maintains it based on a form of power" (Fernandes, 2008). We understand *territorio* as a social construction produced from the conflictive relations that develop over it; the territory is the materialization of a social group's intention and, therefore, is loaded with identities and power relations. It is characterized by fluidity, where borders are diluted in relation to other territorial constructions.

2 Political ecology is an (in)discipline that is born from the relationships between academia and communities (Solíz et al., 2019) and analyses the relationship of the human being with nature, questions the reality and identity of the region and dialogues with Marxist heritage to understand the complex configuration of society-nature relationships (Alimonda, 2006). In this field, power relations play a fundamental role since political ecology demonstrate how environmental management mirrors the ways in which nature is interpreted and understood. Power not only refers to the political government but also to the relationships that are part of the social fabrication, appropriation, and control of nature by different socio-political agents (Ávila-García, 2016; Fernández, 2021).
3 Hydrosocial territories are conceived as politically, socially, and environmentally constructed and produced spaces, created and recreated by the interaction between water flows, human practices, hydraulic technologies, biophysical elements, ecological relationships, basin management, social structures, institutions, and legal-administrative systems (Ávila-García, 2016; Boelens et al., 2017; Swyngedouw and Zuñiga, 2018).
4 It is a concept in Spanish that integrates all the people who fight to achieve water justice in Latin America.
5 The World March of Women (WMM) is an international movement of feminist actions, founded in 2000. This space is made up of grassroots organizations that carry out actions to eliminate the causes that originate poverty and violence against women. The process has proposed principles and values and produces collaborative documents between several organizations.

References

Agarwal, B. (1992). The gender and environment debate: Lessons from India. *Feminist Studies*, *18*(1), 119–158. https://doi.org/10.2307/3178217

Ahlers, R., & Zwarteveen, M. (2009). The water question in feminism: Water control and gender inequities in a neo-liberal era. *Gender, Place and Culture*, *16*(4), 409–426. https://doi.org/10.1080/09663690903003926

Alimonda, H. (2006). *Los tormentos de la materia. Aportes para una ecología política latinoamericana*. CLACSO. http://biblioteca.clacso.edu.ar/ar/libros/grupos/hali/hali.html

Amigos de la Tierra América Latina y el Caribe (ATALC). (2018). *Política pública y derecho fundamental al agua en América Latina y el Caribe – Amigos de la Tierra de América Latina y el Caribe*. https://atalc.org/2018/12/10/politica-publica-y-derecho-fundamental-al-agua-en-america-latina-y-el-caribe/

Arora-Jonsson, S. (2011). Virtue and vulnerability: Discourses on women, gender and climate change. *Global Environmental Change*, *21*(2), 744–751. https://doi.org/10.1016/j.gloenvcha.2011.01.005

Arora-Jonsson, S. (2014). Forty years of gender research and environmental policy: Where do we stand? *Women's Studies International Forum*, *47*(PB), 295–308. https://doi.org/10.1016/j.wsif.2014.02.009

Ávila-García, P. (2016). Hacia una ecología política del agua en Latinoamérica. *Revista de Estudios Sociales*, *55*, Article 55.

Boelens, R. (2021). Una introducción a la Justicia hídrica. In A. Guzmán León (Ed.), *Justicia Hídrica: Una mirada desde América Latina* (pp. 63–81). Centro Bartolomé de las Casas.

Boelens, R., Cremers, L., & Zwarteveen, M. (Eds.). (2011). *Justicia Hídrica. Acumulación de Agua, Conflictos y Acción Social*. Instituto de Estudios Peruanos.

Boelens, R., Hoogesteger, J., Swyngedouw, E., Vos, J., & Wester, P. (2017). Territorios hidrosociales: Una perspectiva desde la ecología política. In C. S. Villamizar & F. A. Pizarro (Eds.), *Recursos, vínculos y territorios. Inflexiones transversales en torno al agua*. Editorial de la Universidad Nacionalde Rosario.

Bolados García, P., Henríquez Olguín, F., Ceruti Mahn, C., Cuevas, A. S., Bolados García, P., Henríquez Olguín, F., Ceruti Mahn, C., & Cuevas, A. S. (2018). La eco-geo-política del agua: Una propuesta desde los territorios en las luchas por la recuperación del agua en la provincia de Petorca (Zona central de Chile). *Revista Rupturas*, *8*(1), 159–191. https://doi.org/10.22458/rr.v8i1.1977

Borras, S. M. (2016). *Land politics, agrarian movements and scholar-activism*. Transnational Institute. www.tni.org/en/publication/land-politics-agrarian-movements-and-scholar-activism

Bravo, L., & Fragkou, M. C. (2019). Escasez hídrica, género, y cultura mapuche. Un análisis desde la ecología política feminista. *Polis (Santiago)*, *18*(54), 64–88. https://doi.org/10.32735/s0718-6568/2019-n54-1401

Bretas, F., Casanova, G., Crisman, T., Embid, A., Martin, L., Miralles, F., & Muñoz, R. (2020). Agua para el futuro: Estrategia de seguridad hídrica para América Latina y el Caribe. In Banco Interameticano del Desarrollo (Ed.), *Agua para el futuro: Estrategia de seguridad hídrica para América Latina y el Caribe*. https://doi.org/10.18235/0002816

Cabnal, L. (2010). Acercamiento a la construcción de la propuesta de pensamiento epistémico de las mujeres indígenas feministas comunitarias de Abya Yala. In Minervas Colectivo de Mujeres (Eds.), *Momento de Paro Tiempo de Rebelión* (pp. 116–143). Minervas Ediciones, una editorial propia.

Cabnal, L. (2013a). Defender un territorio de la minería sin defender a las mujeres de la violencia sexual es una incoherencia. Entrevista realizada por Tamara de Gracia y Diego Jiménez. *Diagonal* (pp. 2–7). https://www.diagonalperiodico.net/global/defender-territorio-la-mineria-sin-defender-cuerpos-mujeres-la-violencia-sexual-es

Cabnal, L. (2013b, May 23). *También hay que luchar contra el machismo indígena*. El Ciudadano. www.elciudadano.com/entrevistas/tambien-hay-que-luchar-contra-el-machismo-indigena/05/23/

Cantor, A., & Emel, J. (2018). New water regimes: An editorial. *Resources*, *7*(2), Article 2. https://doi.org/10.3390/resources7020025

Caretta, M. A. (2015a). *East African Hydropatriarchies*. Holmberg.

Caretta, M. A. (2015b). Hydropatriarchies and landesque capital: A local gender contract analysis of two smallholder irrigation systems in East Africa. *Geographical Journal*, *181*(4), 388–400. https://doi.org/10.1111/geoj.12102

Caretta, M. A., & Rothrock, A. (2021). Water and gender. *Oxford Research Encyclopedias, Education, August*, 23–26. https://doi.org/10.4324/9781849776301-6

Carrillo Rodríguez, E. C. (2020). *Cuerpos-Agua: Defensa y cuidado del territorio a través de la experiencia de las mujeres de la Escuela Campesina de Chapacual, Nariño*. Pontificia Universidad Javeriana. http://repository.javeriana.edu.co/handle/10554/49785

Colectivo CASA. (2015). *La tierra es nuestra madre, el agua es nuestra leche*. https://colectivocasa.org.bo/wp-content/uploads/2020/09/renamat-cartilla-el-agua-es-nuestra-leche.pdf

Colectivo CASA. (2018). *Despojo de agua por mineria y el impacto en las mujeres. Diagnostico del despojo de fuentes de agua por mineria estatal y transnacional*. www.redlatinoamericanademujeres.org/wp-content/uploads/2021/05/cartilla3.pdf

Colectivo Miradas Críticas del Territorio desde el Feminismo. (2017). Mapeando el cuerpo-territorio: Guía metodológica para mujeres que defienden sus territorios. *Geopauta*, *4*(4).

Coles, A. (2009). Domestic water provision and gender roles in drylands. *HAGAR Studies in Culture, Polity and Identities*, *9*(1), 175–191.

Cruz Hernández, D. T. (2020). En un rincón de la frontera se teje insurgencia. Territorios encarnados ante la (re)patriarcalización. *Ecología Política. Cuadernos de Debate Internacional*, *60*, 16–23. https://doi.org/10.53368/ep60macep01

Delbene-Lezama, L. (2019). *Una mirada ecofeminista a la gestión del agua en Uruguay. Las bases materiales que sostienen la vida: Perspectivas ecofeministas*. COTIDIANO MUJER.

de Llobatera, N. V. (2004). La problemática del agua. *Ecología política*, *27*, 101–120.

Díaz Lozano, J., Cruz Hernández, D. T., Magalhaes, L., y Pasero, V. (Coords.). (2021). *Fronteras y cuerpos contra el Capital: Insurgencias feministas y populares en Abya Yala*. Ciudad Autónoma de Buenos Aires: El Colectivo; México: Bajo Tierra. (Chico Mendes). En Memoria Académica. www.memoria.fahce.unlp.edu.ar/libros/pm.5132/pm.5132.pdf

Doornbos, B. Z. (2011). Justicia Hídrica: Por qué y cómo considerar el cambio climático en el análisis de la distribución del agua? In R. A. Boelens, L. Cremers & M. Z. Zwarteveen (Eds.), *Justicia Hídrica: Acumulación, conflicto y acción social* (pp. 135–154). Justicia Hídrica, IEP Instituto de Estudios Peruanos, Fondo Editorial PUCP.

Duarte-Abadía, B., Boelens, R., & Roa-Avendaño, T. (2015). Hydropower, encroachment and the re-patterning of hydrosocial territory: The case of hidrosogamoso in Colombia. *Human Organization*, *74*(3), 243–254. https://doi.org/10.17730/0018-7259-74.3.243

Fals Borda, O. (1999). Orígenes universales y retos actuales de la IAP. *Análisis Político*, *38*, 73–90.

Fals Borda, O. (2015). *Una sociología sentipensante para América Latina*. CLACSO.

FAO, & UNEP. (2020). The state of the world's forests 2020. In *The state of the world's forests 2020* (p. 214). FAO and UNEP. https://doi.org/10.4060/CA8642EN

Fernandes, M. B. (2008). Sobre la tipología de los territorios. *Revista digital para estudiantes de geografía y ciencias sociales*. https://web.ua.es/es/giecryal/documentos/documentos839/docs/bernardo-tipologia-de-territorios-espanol.pdf

Fernández, S. (2021). Los ecofeminismos territoriales frente a las injusticias hídricas: Hacia un horizonte de imaginaciones socio-ecológicas en América Latina (Abya Yala). In R. A. Boelens, L. Cremers & M. Z. Zwarteveen (Eds.), *Justicia Hídrica. Una mirada desde América Latina* (pp. 187–205). Centro Bartolomé de las Casas. www.conicet.gov.ar/new_scp/detalle.php?keywords=&id=32829&capitulos=yes&detalles=yes&capit_id=10227390

Fondo de Acción Urgente. (2020). *Extractivismos, pandemia y otros mundos posibles: Recuperación económica y alternativas ante los extractivismos y pandemia desde las defensoras del territorio en América Latina y el Caribe.* Extractivismos, pandemia y otros mundos posibles. Retrieved March 22, 2023, from https://territorios.fondoaccionurgente.org.co/

Fosado, E. (2019). La lógica de género en la configuración del discurso climático: Un análisis del campo institucional (1994–2015). *Géneros*, *26*(26). https://ru.crim.unam.mx/handle/123456789/411

Gaitán Ortiz, L. M. (2020). El agua, un anhelo permanente: La minería y sus efectos territoriales sobre el agua en la comunidad afrodescendiente de Patilla, La Guajira, Colombia. *Repositorio institucional – GRADE.* https://repositorio.grade.org.pe/handle/20.500.12820/583

Gudynas, E. (2020). *Extractivisms: Politics, economy and ecology.* Fernwood Publishing.

Guilcamaigua Pastuña, G. (2021). *"El agua es nuestro cuerpo": Una mirada desde el feminismo decolonial a la forma como viven las mujeres los problemas del territorio – cuerpo – agua.* Flacso Ecuador. http://repositorio.flacsoandes.edu.ec/handle/10469/17495

Harris, L. M. (2009). Gender and emergent water governance: Comparative overview of neoliberalized natures and gender dimensions of privatization, devolution and marketization. *Gender, Place and Culture*, *16*(4), 387–408. https://doi.org/10.1080/09663690903003918

Hoetmer, R. (2021). La Justicia Hídrica desde los movimientos sociales: Mirada sobre el territorio, conflicto y alternativas. In R. A. Boelens, L. Cremers & M. Z. Zwarteveen (Eds.), *Justicia Hídrica: Una mirada desde América Latina* (pp. 163–181). Centro de Estudios Regionales Andinos Bartolomé de Las Casas.

IPCC. (2022). Summary for policymakers. In H.-O. Pörtner, D. C. Roberts, E. S. Poloczanska, K. Mintenbeck, M. Tignor, A. Alegría, M. Craig, S. Langsdorf, S. Löschke, V. Möller & A. Okem (Eds.), *Climate change 2022: Impacts, adaptation, and vulnerability.* Contribution of Working Group II to the Sixth Assessment Report of the Intergovernmental Panel on Climate Change (pp. 3–33). Cambridge University Press. https://doi.org/10.1017/9781009325844.001.

Isch, E. (2021). Justicia hídrica y ecología política: Una relación académica y militante. In R. A. Boelens, L. Cremers & M. Z. Zwarteveen (Eds.), *Justicia Hídrica: Una mirada desde América Latina, Justicia Hídrica* (pp. 137–157). Centro de Estudios Regionales Andinos Bartolomé de Las Casas.

Lautze, J., & Manthrithilake, H. (2012). Water security: Old concepts, new package, what value?: Water security: Old concepts, new package, what value? *Natural Resources Forum*, *36*(2), 76–87. https://doi.org/10.1111/j.1477-8947.2012.01448.x

Leyva, X., & Speed, S. (2008). Hacia la investigación descolonizada: Nuestra experiencia de co-labor. In X. Leyva, A. Burguete & S. Speed (Eds.), *Gobernar (en) la diversidad: Experiencias indígenas desde América Latina. Hacia la investigación de colabor* (pp. 34–59). CIESAS, FLACSO Ecuador y FLACSO Guatemala.

Linton, J. (2010). *What is water?: The history of a modern abstraction.* University of British Columbia Press.

Lobos Castro, F. (2021). Agua-cuerpo-territorio. Las cicatrices y reexistencias de las mujeres rurales en el Maule Sur precordillerano de Chile. *Ecología Política*, *61*, 112–116.

Marcha Mundial de las Mujeres. (2020). *Documento regional Américas: Quinta acción internacional de la marcha mundial de las muejres* (p. 34). https://marchemondiale.org/index.php/action-areas/el-bien-comun-y-los-servicios-publicos/?lang=es

Márquez, F. (Director). (2018, May 2). *Francia Márquez acceptance speech, 2018 Goldman environmental prize.* www.youtube.com/watch?v=DF8NTVV-thc

Miralles-Wilhelm, F., Sanchez-Maldonado, R., & Muñoz-Castillo, R. (2022). Assessing water security through a set of consistent metrics and application to water funds in Latin America. *Structural Engineer*, *9*(1).

Montealegre, D. (2020). Aportes y críticas feministas sobre la incorporación de las mujeres y el enfoque de género en el desarrollo. *Trabajo Social*, *22*(1). https://doi.org/10.15446/ts.v22n1.80484

Moreano Venegas, M. E., Lang, M., & Ruales Jurado, G. (2022). *Perspectivas de justicia climática desde los feminismos latinoamericanos y otros sures* [Working Paper]. Universidad Andina Simón Bolívar, Sede Ecuador. http://repositorio.uasb.edu.ec/handle/10644/8683

Paris, M. (2020). *La seguridad hídrica y los Objetivos de Desarrollo Sostenible Manual de capacitación para tomadores de decisión.* UNESCO Office Montevideo and Regional Bureau for Science in Latin America and the Caribbean. https://unesdoc.unesco.org/ark:/48223/pf0000374917.locale=es

Peña, H. (2016). *Desafíos de la seguridad hídrica en América Latina y el Caribe.* CEPAL. https://repositorio.cepal.org/handle/11362/40074

Pena, M. (2022). Conflicto hídrico y defensa territorial: Mujeres en el Movimiento Campesino de Santiago del Estero, Argentina. *Íconos – Revista de Ciencias Sociales*, *73*, Article 73. https://doi.org/10.17141/iconos.73.2022.5236

Red Latinoamericana de Mujeres Defensoras de Derechos Sociales y Ambientales. (2018). *Informe sobre la situación de riesgo y criminalización de las defensoras del medioambiente en América Latina.* Red Latinoamericana de Mujeres Defensoras de Derechos Sociales y Ambientales. www.redlatinoamericanademujeres.org/wp-content/uploads/2021/05/cartilla6.pdf

Roa Avendaño, T., Roa, M. C., Toloza, J., & Navas, M. (2017). *Como el agua y el aceite. Conflictos socioambientales por la extracción petrolera.* Centro Nacional Salud, Ambiente y Trabajo, CENSAT Agua Viva. www.business-humanrights.org/es/%C3%BAltimas-noticias/como-el-agua-y-el-aceite-conflictos-socioambientales-por-la-extracci%C3%B3n-petrolera/

Saldi, L. (2021). Reconociendo paisajes hídricos en entornos significativos. Preguntas, conceptos y abordajes metodológicos. In R. A. Boelens, L. Cremers & M. Z. Zwarteveen (Eds.), *Justicia Hídrica: Una Mirada desde América Latina* (pp. 89–105). Centro de Estudios Regionales Andinos Bartolomé de Las Casas.

Salinas, M., & Becker, I. (2022). *Guardianas del agua: (In)seguridad hídrica en la vida cotidiana de las mujeres.* Fundación Heinrich Böll.

Salinas Mulder, S. (2017). Mujeres, género y Gestión del Agua en los Tiempos del Vivir Bien. In *Elemento de vida, el agua en el desarrollo, la cultura y la sociedad.* Revista cooperación suiza en Bolivia 2016–2017. www.academia.edu/35463131/ART%C3%8DCULO_MUJERES_G%C3%89NERO_Y_GESTI%C3%93N_DEL_AGUA_EN_LOS_TIEMPOS_DEL_VIVIR_BIEN_

Segato, R. L. (2015). Género y colonialidad: Del patriarcado comunitario de baja intensidad al patriarcado colonial moderno de alta intensidad. In *La Crítica de la Colonialidad en ocho Ensayos y una Antropología por Demanda* (pp. 69–99). Prometeo Libros.

Solíz, M. F., Yepez, M., Valencia Velasco, M., & Solíz Carrión, R. F. (2019). *Reciclar sin recicladoras es basura: El retorno de las brujas.* GAIA Global Alliance for Incinerator Alternatives.

Sultana, F. (2018). Gender and water in a changing climate: Challenges and opportunities. In C. Fröhlich, G. Gioli, R. Cremades & H. Myrttinen (Eds.), *Water security across the gender divide* (pp. 17–33). Springer.

Swyngedouw, E., & Zuñiga Copano, M. (2018). Politizando las ecologías políticas urbanas. *Investigaciones Geográficas*, *56*, 153–167.

Tacha, V. (2022). *Conflictos por el agua causados por la industria de bebidas y productos comestibles ultraprocesados. Estudio de caso de la empresa Postobón.* Corporación Colectivo de Abogados José Alvear Restrepo.

Ulloa, A. (2015). Políticas globales- nacionales-locales de cambio climático y sus incidencias en desigualdades de género y etnicidad. In IDEAs, CODESRIA, & CLACSO (Eds.), *Colección Documentos del Sur. Documentos de Trabajo no. 12.* http://biblioteca.clacso.edu.ar/clacso/sur-sur/20150424030524/OPsursur-Ulloa.pdf

Ulloa, A. (2016). Feminismos territoriales en América Latina: Defensas de la vida frente a los extractivismos. *Nómadas*, *45*, 123–139. https://doi.org/10.30578/NOMADAS.N45A8

Ulloa, A. (2020). The rights of the Wayúu people and water in the context of mining in La Guajira, Colombia: Demands of relational water justice. *Human Geography*, *13*(1), 6–15. https://doi.org/10.1177/1942778620910894

Ulloa, A., Damonte, G., Quiroga, C., & Navarro, D. (2020). *Gobernanzas plurales del agua: Formas diversas de concepción, relación, accesos, manejos y derechos del agua en contextos de gran minería en Colombia y el Perú.* GRADE. www.grade.org.pe/publicaciones/gobernanzas-plurales-del-agua-formas-diversas-de-concepcion-relacion-accesos-manejos-y-derechos-del-agua-en-contextos-de-gran-mineria-en-colombia-y-el-peru/

UNEP. (2017, September 22). *What is integrated water resources management?* UNEP – UN Environment Programme. www.unep.org/explore-topics/disasters-conflicts/where-we-work/sudan/what-integrated-water-resources-management

UNICEF. (2021). *Water security for all: The world is in a water crisis, and children's lives and futures are at risk*. United Nations Children's Fund. www.unicef.org/media/95241/file/water-security-for-all.pdf

UNWater. (2013). *What is water security? Infographic*. UN-Water. www.unwater.org/publications/what-water-security-infographic

Urquiza, A., & Billi, M. (2020). *Seguridad hídrica y energética en América Latina y el Caribe: Definición y aproximación territorial para el análisis de brechas y riesgos de la población*. CEPAL. https://repositorio.cepal.org/handle/11362/46408

Valdés de Hoyos, E. I. P., & Uribe Arzate, E. (2016). El derecho humano al agua. Una cuestión de interpretación o de reconocimiento. *Cuestiones constitucionales*, *34*, 3–25.

The Valuing Water Initiative. (2020). *Better decisions impacting water*. Valuing Water Initiative. https://valuingwaterinitiative.org/

Villa Fontecha, G. H. (2012). A Propósito de la Gestión del Agua en el Mundo Contemporaneo. Un Enfoque Biopolítico. *Análisis Político*, *25*(74), 109–133.

Woster, D. (2008). *Transformaciones de la Tierra*. Coscoroba Ed.

Zaragocin, S., & Caretta, M. A. (2021). Cuerpo-territorio: A decolonial feminist geographical method for the study of embodiment. *Annals of the American Association of Geographers*, *111*(5), 1503–1518. https://doi.org/10.1080/24694452.2020.1812370

27

BEYOND MATERIAL DIMENSIONS OF WATER INSECURITY

Gendered subjectivities, senses of community, and renewed political possibilities

Evelyn Arriagada, Leila M. Harris and Dacotah-Victoria Splichalova

Introduction: gender and water insecurities, the literature to date and moving forward

There is a vast and growing literature that connects water insecurity and governance to gender dynamics. For instance, women are more likely to experience water-related stress in contexts of insecurity (Wutich & Ragsdale, 2008) and are likely to be more affected by poor water quality or contamination (e.g., see discussion of the Flint water crisis, Radonic & Jacob, 2021). Work has also highlighted ways that shifting notions of masculinity and femininity can be bound up with water access and infrastructures (Harris, 2006). Beyond binary gender conceptions, intersectional categories are also meaningful – whether Indigeneity, impoverishment, caste, or class (see, for instance, Latchmore et al., 2018 and O'Gorman, 2021 on dynamics related to Indigenous water insecurity in Canada). Recent contributions also consider trans, non-binary, or non-normative gender relations and experiences in relation to water insecurities (Brewis et al., 2024; Mukherjee et al., 2021).

Here, our aim is to think expansively about both "water" and "gender", highlighting key insights that invite retheorization, from discussions of ontologies of water and kinship to post-structuralist, queer, trans, nonbinary, and intersectional theories and approaches. Taken together, we ask: what are key gender and water connections and insights that are revealed through broadened engagement with these categories? This exploration highlights issues beyond the material, offering key starting points to 'think otherwise' about gender-water dynamics and intersections.

Pushing the limits: relational waters, non-materiality, and reconceptualizing gender

Moving towards a relational understanding of water

A growing literature engages with ontology, epistemology, and axiology, asking questions such as: what is water? How do we come to know and understand water? And how do we

DOI: 10.4324/9781003100379-33

value and relate to water? Among them, Linton (2010) exposes the place-based, techno-scientific, and disciplinary biases behind assumed 'universal' notions of water. He challenges modern and Western abstractions of water that position it as distinct from humans, and as something to be controlled or managed (e.g., through damming of rivers, or treatment, see also Yates et al., 2017; Boelens et al., 2022). Other relational conceptions exist that understand water as embedded in territory, local ecologies, and political, social, and cultural dynamics (see concepts such as "waterscape" or the "hydrosocial cycle", Linton & Budds, 2014, "hydrosocial territories" or "riverhood", Boelens et al., 2016, 2022).

While such approaches stress human-water linkages, it is clear that Indigenous ontologies and epistemologies have *long* emphasized notions of interconnection and mutuality – situating water and aquatic species as kin, or centering on notions of respect, reciprocity, or love (Wilson & Inkster, 2018; Daigle, 2018; van Horn et al., 2022; McGregor, 2015). Attending to multiple ontologies of water, we can situate water not as an inert thing, nor as a chemical compound, but as a set of relationships, a living being, or with a spirit imbued with lifeforce (Ibid, see also Craft & King, 2021; Yates et al., 2017). These interventions allow us to understand that there are multiple and vastly different "waterworlds". This extends beyond varied cultural understandings of water to the more fundamental notion of multiple ontologies of water – water *is fundamentally different*. In turn, axiologies – or how we engage with or value water – might vary or shift considerably, moving from control or management toward care, awe, or sense of responsibility (Ibid.). In parallel, feminist and humanities scholars also emphasize the need to rethink our connections to water in ways that foster strengthened senses of connection, mutuality, and care.

Together, our expansive approach to water ties together multiple strands and traditions that invite us to theorize and extend our approaches – highlighting the idea that water is part of, and connected to, myriad relationships (see also Jepson et al., 2017; McGregor, 2012; Krause & Strang, 2016). Attending to the non-material dimensions of water through these lenses helps to foreground and enliven the complex emotions, senses of community, belonging, wonder, senses of injustice, fairness, or values linked with waterways and their changing conditions (see also Boelens et al., 2022; Norgaard, 2017).

Beyond normative gender: intersectional, queer, and trans perspectives

While the section above highlights several pivots that move us towards more relational understandings of water-society, here we provide a brief overview of interventions that necessitate a fundamental rethinking of gender. We then move from these unsettled conceptual foundations to several starting points that help to chart gender-water linkages *otherwise*. While not all elements of this exploration are new (as noted, senses of mutuality with local ecologies as well as more fluid or nonbinary notions of gender have long been in evidence, including for many Indigenous communities), we nonetheless consider these foundations to offer revisions and reconceptualization that are both necessary and timely.

Feminist theory has largely moved beyond binary, universal, and (hetero) normative conceptions of gender. In the 1980s and 1990s, Black feminists expressed caution with gender approaches that erase the complexities of race, class, location, or other key considerations (Crenshaw, 1991; Collins, 1998). Intersectionality offers the insight that not all women and men experience phenomena similarly, rather, specific marginalities emerge at the intersection of race, class, caste, location, or other relevant operations of difference (see Truelove, 2019; Mollett & Faria, 2013; Shah et al., 2023 for recent study of intersectional dynamics vis-a-vis water insecurity). Linked with studies of water and waterscapes, work has highlighted that despite flux and variability, the gender binary is at times made to appear as natural and stable in relation to complex labors and waterscape practices (Harris, 2006), or to suggest that caste, class, and other dynamics intertwine to condition complex relationships to water uses or infrastructures (O'Leary, 2019).

Queer and trans theorists have similarly highlighted the gaps, erasures, or indeed violence, that occurs when there is an insistence on male and female (or masculinity and femininity) as the most salient categories, further marginalizing trans or non-normative bodies and experiences. Indeed, assumptions about cisgendered and heteronormative experiences are rife in many mainstream explorations of gender and water. While the need to think about trans and queer bodies and experiences has been signaled for some time (see Hawkins et al., 2011), only very recently have we seen work that explicitly considers trans, non-binary, and gender-diverse peoples and experiences (Brewis et al., 2024; Mukherjee et al., 2021; Hazard, 2022).

Theoretically, interventions from poststructuralist and queer theory question the notion that gender categories are fixed or given, instead seeking to understand them as sedimented in relation to ongoing and specific discourses and practices (Butler, 1990, 1993). Also pushing at key concepts from a queer and trans perspective, Neimanis has usefully extended notions of corporality, including the notion of "gestationality" (Chandler & Neimanis, 2013; Neimanis, 2013, 2017), to de-center the relationship between water and life-giving associated with the female womb. Conceived as "a facilitative mode of being, but one that is *not necessarily* tied to the female human" (Neimanis, 2017, pp. 68–69, emphasis added), water is gestational in the sense of giving (us) life "in both the most banal and sacred of the terms (Neimanis, 2013, p. 30)". Neimanis emphasizes: "we are all watery, then we all harbor the potential of watery gestationality within our corporeal selves (Neimanis, 2017, p. 69)". Hazard's (2022) recent intervention, *Underflows*, also engages queer and trans concepts, methods, and bodies, to refigure how we research and engage with rivers and water bodies. To do so invites new possibilities for what is livable, both for trans and queer lives, and for beavers, salmon, and riverbodies – attending to "hidden flows and their movements, excesses, and relations" (2022, p. 9), in ways that challenge what is invisibilized through colonialism, capitalism, and mainstream river management.

Such interventions invite a radical rethinking of these very categories and their associated assumptions. What emerges when we enable possibilities for thinking otherwise, beyond common framings and understandings? In so doing, what new types of insights, politics, and hydro-social futures might emerge?

Towards reimaging and broadening gender-water connections

As we have suggested, there are several contributions which clearly make progress in ways that are consistent with the goals of this piece. We review such key insights, then consider several terrains of analysis that might emerge from continuing to push these analytics.

Regarding intersectionality, building on contributions by theorists such as Mollett and Faria (2013), it is clear that, at times, privileging gender as a key category of difference sidesteps and minimizes other key axes of difference and inequality, including those related to caste, class, race, or rurality. Despite clear progress, there is an ongoing need for work on water insecurity and governance to adopt more intersectional perspectives (Shah et al., 2023; Nightingale, 2011; Harris, 2008; Sultana, 2020). O'Leary (2019) highlights the ways that water use and access in relatively impoverished settlements of Delhi provides pathways that both enable and restrict class mobilities among low income and low caste individuals. Similarly, Mawani (2022) highlights the ways that religion structures urban planning and ongoing struggles of Muslim communities to access water in Ahmedabad, India, while Ranganathan (2022) explores intersections of caste, class, religion, and gender to analyze property, infrastructure, and labor regimes in urban Bangalore. The imperative is not to set gender aside, but to engage gender in relation to other key inequities, and to analyze differentiated conditions that structure power and marginality as central to water-society dynamics.

Many scholars also critically interrogate the male/female gender binary. As Vinyeta et al. (2015) argue, notions of gender diversity and their associated roles have long varied among Indigenous communities in North America (see also Simpson, 2017). They argue that the imposition of the gender binary through Western perspectives and colonial practices had detrimental consequences for communities, waterways, and socio-ecological resilience. Explaining this, Vinyeta and collaborators document the unique roles of many women, non-binary, and two-spirit individuals with respect to water protection in Indigenous contexts. The colonial imposition of binary gender categories led to a disconnect with associated practices, including those that historically protected waterways. Barker (2019) also describes the roles and responsibilities around water in Indigenous communities not only for cis women but also for "women *and women-identified individuals*" (p. 4, emphasis added). The Michi Saagiig Nishnaabeg scholar Leanne Betasamosake Simpson (2017) goes further, asserting that land-based Indigenous philosophies are often more open to different possibilities of being in the world. Simpson asserts: "Our thought systems within grounded normativity are fluid, dynamic, and responsive. . . . They also come from the land – the land that provides endless examples of queerness and diverse sexualities and genders" (Simpson, 2017, p. 122). To reinforce this, Simpson quotes Indigenous scholar Alex Wilson (Opaskwayak Cree Nation) to emphasize that in Cree language the relationship with sacred entities – including water – are genderless and focused on kinship: "In Cree, land (*aski*) is not gendered. . . . Same for water. It's not gender but it has a spirit of life and it's fluid" (Wilson, 2015. Quoted in Simpson, 2017, p. 121). Simpson uses the example of a community-based tool produced by the Women's Earth Alliance and the Native Youth Sexual Health Network (2016)[1] to argue that:

> We all have a relationship to creation and that these relationships are not tied to certain body parts. It centers on the idea that creating life comes in many forms, not just from the womb, and it creates a space where all genders can have valuable, ethical, consensual, meaningful, and reciprocal relationships with all aspects of creation.
>
> *(Simpson, 2017, p. 121)*

Many theorists point out that concepts and approaches that cite and sediment the gender binary exclude trans, non-binary, or non-normative and gender-queer bodies and practices. Seymour (2017, pp. 258–259), examines transphobic and cisgender assumptions behind environmental research and interrogates narratives that associate poisoning the environment/water and consequences for gender ambiguity as the main focus, instead of cancer and other health issues. Seymour asks: "To what extent have transphobia, cisnormativity, and cissexism informed our past and present understandings of the nonhuman natural world"?

As an alternative, the author proposes a "trans ecologies framework" as a nimbler and fluid way of conceiving these relationships (Seymour, 2017). Hazard (2022) provides a compelling illustration of collaborative research that seeks to fundamentally refashion Indigenous, queer, and trans relationships with rivers, opening up fluid possibilities for ways of living, being, and thriving in the world (recall also trans gestationality from Neimanis, previously). Adding to these contributions, several recent works offer empirical insights regarding the situation of water insecurity for trans, queer, and non-gender conforming communities (Mukherjee et al., 2021; Boyce et al., 2018). Even with such examples, there remains an overriding tendency in the gender and water literature to reinforce the gender binary, or to focus exclusively on women (Brewis et al., 2024; Davies et al., 2023).

These contributions radically rethink water-gender dynamics, moving beyond binarized notions and challenging ways of thinking and being imposed and reinforced through settler colonialism and Western science. One possible alternative is to stress *deep connections* between humans and (sacred) waters. These realities serve to re-define and expand the ways in which we conceive ourselves, our bodies, and our relationships and engagements with multiple waters. As we explore further in the next section, some of these ideas intersect with recent developments in post-structural and posthumanist feminist theories, emphasizing the fluidity and (intra/inter) connection among diverse bodies and the ethical commitments derived from them (Alaimo, 2010; Braidotti, 2022; Neimanis, 2013, 2017). This meshes with our attention to the broad non-material dimensions of water insecurity, including values, spiritual relations, and broad social-political dynamics. This orientation is necessary to fully realize the insights offered by a broadened gender + perspective.

The sections that follow invite critical rethinking to problematize and expand gendered notions of the self; deepen, reveal, and expand human connections to each other, places, and diverse ecologies and enable revised and reinvigorated political engagements and possibilities.

Problematizing and expanding gendered notions of self

Feminist and critical scholars (particularly ecofeminist, de/post-colonial, Indigenous, and posthumanist authors) have pointed out that anthropocentrism, patriarchy and colonialism characterize many dominant traditions in Western thought. In such traditions the "ideal self" is often implicitly (or explicitly) defined as a male, white, and rational against which others (humans and non-human beings) are defined by contraposition or negation (Butler, 1990, 1993; de Beauvoir, 1949; Plumwood, 1993; Simpson, 2017). Within these dominant traditions, personal identities and subjectivities correspond with self-contained individuals, whose bodies are materially separated from other entities and from social-natural environments. As critical feminist and posthumanist scholars have argued, those understandings offer very limited possibilities of being in the world, often constraining women's, as well as trans and queer people's, identities, aspirations, and relationships with other human and non-human beings, including broader ecologies or waterways (Hazard, 2022). All such debates beg the question: *is it possible that broader understandings of gender dynamics, and of waters, can help us to problematize and expand notions of the self?*

Several scholars working in Feminist Political Ecology, critical geography, and environmental justice traditions have stressed that sexed, gendered, and racialized subjectivities are relationally constructed through everyday connections, practices, and struggles over natures, including water or polluted landscapes and urban spaces (e.g., notions of racialized space, Pulido et al., 1996). Feminist political ecologists such as Leila Harris (2006, 2015) and Farhana Sultana (2009) draw upon the work of Judith Butler (1990, 1993) to explain how certain notions of gendered subjectivities and associated intersectional identities are often reinforced or challenged by changes in water quality, location, access, and control. For Harris (2006), shifting irrigation practices or new cropping patterns, constitute shifting senses of the self and of difference/otherness (with importance for everyday lives and livelihoods). Specifically, water and changing agro-ecologies are part and parcel of what cites and reproduces the "factness" of gender, sex, and ethnic difference (see also discussion of "embodied subjectivities" as engaged by Sultana, 2009). Other contemporary works in FPE consider attachments to certain water bodies as expanded notions of the self. According to Yaka (2019, p. 363), our material bodies but also "[o]ur self-knowledge, consciousness and subjectivity emerge out of perceptual engagement with our environment in many sensory and affective ways, through which the human body extends into the world."

Insight from Indigenous elders and scholars, as well as feminist and queer interventions, have also been critical in expanding dominant understandings of the human self in relation to land, waters, and non-human beings. Leanne Betasamosake Simpson (2017) argues that acknowledging "queer Indigeneity" offers a form of "radical resurgence" that allows Indigenous peoples to reconnect to land-based and sacred relationships to "build strong societies of individuals who are functioning as their best selves" (Simpson, 2017, p. 122). She reinforces: "Queer Indigeneity . . . is about a web of supportive, reciprocal, generative relationships that we often do not have names for in English and that exist outside of the hierarchy and the imagination of heteropatriarchy" (Simpson, 2017, p. 134). These ideas are bolstered by insights from other Indigenous thinkers that have emphasized kin relations with humans and non-human others (Todd, 2017; van Horn et al., 2021).

Feminist humanist philosopher Astrida Neimanis (2013) suggests we reframe our understandings of our bodies, and ourselves, as part of a reformulated "watery subjectivity", a feminist and embodied politics of location that takes seriously the situatedness of our bodies, and the mutuality of water worlds with our lives and well-being. Neimanis asks what it would mean to fundamentally understand "that we are all bodies of water", given that we are all primarily made up of water, our bodies are porous, and water passes through us and back to the environment. Watery subjectivities thus critically reframe the ways that we are also fundamentally connected to each other, and to the waterways and the diverse species of plants and animals that depend on them. Moving beyond individuated subjectivity, and the boundary processes that affirm self-contained bodies, they offer that "attention to water's material capacities informs a new way of thinking about subjectivity in collective rather than individualistic terms . . . [in ways that are] also concerned for the well-being of water under neo-liberal capitalist regimes" (Neimanis, 2013, p. 34). This recognition, Neimanis argues, can trigger commitments to take care of planetary waters, and as such can help to foster and sediment more socially just and ecologically responsible relations, including those aligned with a feminist ethic and practice of care and connection (see also De la Cadena, 2015 for similar focus on the ways that subjectivities and waters are co-constituted).

Engaging enriched sense of place and belonging connecting humans and more-than-human others

Refashioned understandings of water and gender provide a much stronger basis from which to refuse other common binaries, including human-non-human, or nature-culture. Here we offer several other extensions from feminist, posthumanist, poststructuralist, and

Indigenous thought that highlight senses of belonging and connection to place, landscape, and multi-species kin.

As noted above, Neimanis offers an emphasis on ways that water subjectivity invites a politics of location that connects bodies to broader communities and ecologies, reconstituting subjectivity and ethical-political commitments. Yaka (2019) highlights that knowledge and subjectivity are always emergent in relation to sense and affect with the surrounding environment, offering another framework to consider the sensory experiences and fundamental inseparability between humans, place, territory, and more-than-human others. Explaining the political commitments of Eastern Black Sea women against small scale hydroelectric plants (HPPs), Yaka explains that these relational subjectivities emerge through sensorial experiences with water, wherein, "[s]ubjectivity . . . is formed through a relational ontology of sentience and intelligence, within a tactile universe" (2017, p. 883). One need not go far to find a similar emphasis on connections to salmon, beaver, or waterways held by many Indigenous groups (van Horn, 2021). For these communities, lifeways, knowledges, and well-being have always been understood as inseparable from the more-than-human world.

Other key offerings have stressed how water-related connections with more than human beings are emergent through corporeal, sensorial, and emotional experiences. For instance, Martina Caretta, Sofia Zaragocin, and others (Caretta et al., 2020) examine how female water activists in Ecuador and United States deploy the notion "*cuerpo-territorio*" to understand the "inseparable ontological relationships between body and territory: what is experienced by the body is simultaneously experienced by territory" (Zaragocin & Caretta, 2020, p. 2). Returning to Neimanis, understanding ourselves as "bodies of water" not only shifts senses of the self, but as noted in the previous examples, fosters senses of mutuality, belonging, and connection with specific ecologies, including those trees, animals, and other humans who similarly depend on the same waterways. As stated by Neimanis (2013), new possibilities might emerge that connect humans and waterscapes, with possibilities for novel and reinvigorated subjectivities responsibly attuned to other watery bodies, within global flows of power.

Some authors also examine how water justice struggles produce community bonds and new ways to signify and practice water justice. Clark (2019) examines how water justice activists in Johannesburg (South Africa), Michigan (United States) and Dublin (Ireland) enact and expand definitions of the human right to water as "a process of commoning". Women activists especially highlight that the right to water goes beyond individual needs, conceiving of it as relational and communal: enhancing community bonds and triggering empowerment of dispossessed groups while encouraging politicization. In Michigan, female activists also recall a deep connection with the Detroit River as a place to go "for spiritual healing" (Clark, 2019, p. 87; see also Botero-Mesa and Roca-Servat, 2019 for similar processes in an urban neighborhood in Medellín, Colombia).

Davis's work on the Magdalena River echoes the sense of a river as an extraordinary source of knowledge. He demonstrates how the river/water shapes conceptions of belonging, identity, and community through place-making that brings to life varied river-dwelling societies over space and time (Davis, 2020). This notion of place attachment is accompanied by ambivalent and complicated feelings of alienation and disconnection, as slavery and drug trafficking also exist along the river. Still, it is clear that the currents and flows of the Magdalena inextricably shape individuals' and communities' identities, attachments, and detachments.

Together, we can trace emergent, and exciting, refashioned notions of the self, bodies, individuals, and collectives. As suggested, shifting ethics, political commitments and activism all coincide with these shifts. Our last section explores these shifting ethics, political commitments, and activism more fully.

Political engagements and possibilities

Water access, quality, and infrastructure are deeply connected with issues such as everyday encounters between citizens and states, senses of inclusion and exclusion from political "imagined" communities, expanded notions of water justice (including issues of social and cultural recognition and participation in decision-making processes), as well as alternative forms of water governance, political autonomy, and self-determination. Gendered categories and relations remain central to these multi-scalar dynamics of power and situated intersectional categories of identity and difference.

A vast literature documents myriad examples of Indigenous, peasant, racialized, and urban women mobilizing to protect forests or water bodies, from massive demonstrations, such as the paradigmatic Cochabamba war in Bolivia (Bustamante et al., 2005; Laurie, 2011), to more subtle forms of "everyday" and "slow" resistance (Hayman et al., 2015; Jenkins, 2017, see also Norman, 2018). Scholars highlight linkages between these politics and material needs, complex connections between women+ and water, as well as perceived senses of intersectional injustices. These politics might emerge from material considerations such as divisions of labor in changing environments, as well as from bodily, sensorial, and emotional experiences rooted in deep relationalities with water and nature – as highlighted previously. The praxis from Grandmother Josephine Mandamin and other Anishinaabe women through Mother Earth Water Walks (MEWW) (McGregor, 2015) or the Women's Council, which enhanced the Nibi Declaration of Treaty #3 (Craft and King, 2021), are remarkable examples of enacting ethical and political commitments rooted in deep responsibilities to waterways. The experiences documented by Yaka (2017) in Turkey, and Caretta et al. (2020) in Ecuador and the United States, again show how expanded notions of the self which encompass non-human water bodies often push diverse women to mobilize and resist extractivist activities and ongoing assaults on the health of waterways (see also Buechler & Hanson, 2015; Johnson et al., 2020; Bustamante et al., 2005; Drew, 2014; Hayman et al., 2015; Jenkins, 2017, Boelens et al., 2022).

Such interventions not only cast a light on why women or Indigenous communities might often be at the forefront of resistance movements but also invite critical attention to issues of politics and social dynamics consistent with our interest in broadened conceptualization of gender, water, and non-material dimensions of water insecurity. Other relevant works have discussed "hydraulic or hydro-citizenship" (Anand, 2011) or ways that lack of access to safe and affordable water not only threatens health and survival, but offers challenge to formal notions of citizenship, political belonging, or state legitimacy (Harris, 2008). From this literature, we learn that for disenfranchised communities, water insecurity is connected to broader experiences of injustice, including denial of individual and community rights, or reinforcing a sense of exclusion to participate in broader political communities (Lemanski, 2019). In the case of Flint, Michigan, community members linked their sense of injustice to lead contamination of water and the denial of their knowledge of associated health challenges to the broader politics of injustices experienced by African American communities in the United States (see Pauli, 2019 on carceral injustice and violence against those communities as highlighted by Black Lives Matter and similar movements).

Nascent literature also documents how, in the context of water insecurity, gendered forms of citizenship, and water justice emerge while communities and individuals contest what is considered formal or institutional participation, rights, and inclusion. For instance, Arriagada et al. (2022) document how female leaders of impoverished rural communities in Chile moved from experiencing water and environmental "suffering" (cf. Auyero & Swistun, 2009; Sultana, 2011) to enacting diverse forms of "lived environmental citizenship". Other researchers in the Global South show how women engage in water-related organizations and build diverse roles as community leaders and (informal) authorities based on their lived experiences and knowledges of securing, distributing, and taking care of waters (Botero-Mesa & Roca-Servat, 2019; Bustamante et al., 2005; Ennis-McMillan, 2005; Clark, 2019; Zaragocin & Caretta, 2020). Such examples provide empirical evidence to inform the broadened philosophical and conceptual debates concerning gender-water subjectivities, including senses of connection to place, territory, and ecology, and linked political implications.

Conclusions and future directions

All told, these conceptual, theoretical, methodological, and empirical currents move us far beyond simplistic notions of gender-water connections, including those that have been easily cast aside as overly facile or as reinforcing marginalities in problematic and exclusionary ways. Instead, we offer numerous examples of concepts and frameworks that provide the inspiration to rethink these gender-water dynamics, especially in ways that enable inclusive ethical-political accountability, responsibility, Indigenous resurgence, and care of diverse water bodies/bodies of water. Considered together, ideas and interventions such as "gestationality" (Neimanis, 2013, 2017), "confluence" (Barker, 2019), "transcorporeality" (Alaimo, 2010), "trans environments" (Seymour, 2017), and kinship (van Horn et al., 2021), can serve as powerful inspiration to re-imagine broader *feminist intersectional and posthumanist* solidarities, alliances and coalitions based on affinities rather than those based on singular or fixed identities (Rocheleau, 1995; Sundberg, 2017).

As such, we must simultaneously be attentive to uneven gendered/intersectional experiences of water insecurities/injustices, to the meaning of the water crisis for human and non-human beings, and to those social and environmental discourses and practices that reinforce the exclusion of non-conforming bodies and communities. We also understand that the effort to refine and expand our notion of what is meant by "gender" and "water" offers a number of key conceptual and political resources. Our chapter aims to trace and highlight these insights and pathways to enable nuanced attunement and possibilities for thinking *with* waterbodies. Doing so allows us to reflect on how to live with, and foster, the health of waterways and of each other – enabling diverse ways of being in the world, while nurturing and caring for these continuities. These are crucial starting points for reinvigorated gender theory and also for work on water insecurity and governance. Maintaining focus on the ontological, epistemological, and axiological concerns regarding water (what water is, and how we come to know and value water) and critically extending our understandings of and approaches to gender together provide a key nexus to foster new relationships, emotional-affective relations, and possibilities in the face of ever-changing and increasingly precarious socio-hydrologies.

Note

1 Native Youth Sexual Health Network and Women's Earth Alliance (2016). Violence on the Land, Violence on Our Bodies: Building an Indigenous Response to Environmental Violence, 5–6. Available here: http://landbodydefense.org/uploads/files/VLVBReportToolkit2016.pdf?

References

Alaimo, S. (2010). *Bodily natures: Science, environment and the material self*. Indiana University Press.

Anand, N. (2011). Pressure: The politechnics of water supply in Mumbai. *Cultural Anthropology*, *26*(4), 542–563. https://doi.org/10.1111/j.1548-1360.2011.01111.x

Arriagada, E., Garcés, A., Maillet, A., Viveros, K., & Zambra, A. (2022). Lived environmental citizenship through intersectional lenses: The experience of female community leaders in rural Chile. *Journal of Rural Studies*, *94*, 353–365. https://doi.org/10.1016/j.jrurstud.2022.07.007

Auyero, J., & Swistun, D. (2009). *Flammable: Environmental suffering in an Argentine shantytown*. Oxford University Press.

Barker, J. (2019). Confluence: Water as an analytic of Indigenous Feminisms. *American Indian Culture and Research Journal*, *43*(3), 1–40. https://doi.org/10.17953/aicrj.43.3.barker

Boelens, R., Escobar, A., Bakker, K., Hommes, L., Swyngedouw, E., Hogenboom, B., Huijbens, E. H., Jackson, S., Vos, J., Harris, L. M., Joy, K. J., de Castro, F., Duarte-Abadía, B., Tubino de Souza, D., Lotz-Sisitka, H., Hernández-Mora, N., Martínez-Alier, J., Roca-Servat, D., Perreault, T., . . . Wantzen, K. M. (2022). Riverhood: Political ecologies of socionature commoning and translocal struggles for water justice. *The Journal of Peasant Studies*. https://doi.org/10.1080/03066150.2022.2120810

Boelens, R., Hoogesteger, J., Swyngedouw, E., Vos, J., & Wester, P. (2016). Hydrosocial territories: A political ecology perspective. *Water International*, *41*(1), 1–14. https://doi.org/10.1080/02508060.2016.1134898

Botero-Mesa, M., & Roca-Servat, D. (2019). Water rights and everyday ch'ixi practices in the Barrio El Faro in Medellín, Colombia. *Water*, *11*(10), 2062. https://doi.org/10.3390/w11102062

Boyce, P., Brown, S., Cavill, S., Chaukekar, S., Chisenga, B., Dash, M., Dasgupta, R., De La Brosse, N., Dhall, P., & Fisher, J. (2018). Transgender inclusive sanitation-insights from South Asia. *Waterlines*, *37*(2), 102–117. https://doi.org/10.3362/1756-3488.18-00004

Braidotti, R. (2022). *Posthumanist feminism*. Cambridge University Press.

Brewis, A., DuBois, L. Z., Wutich, A., Adams, E. A., Dickin, S., Elliott, S. J., Empinotti, V. L., Harris, L. M., Ilboudo Nébié, E., & Korzenevica, M. (2024). Gender identities, water insecurity, and risk: Re-theorizing the connections for a gender-inclusive toolkit for water insecurity research. *WIREs Water*, *11*(2), e1685. https://doi.org/10.1002/wat2.1685

Buechler, S., & Hanson, A.-M. (Eds.). (2015). *A political ecology of women, water, and global environmental change*. Routledge.

Bustamante, R., Peredo, E., & Udaeta, M. E. (2005). Women in the "water war" in the Cochabamba Valleys. In V. Bennett, S. Dávila-Poblete & N. Rico (Eds.), *Opposing currents: The politics of water and gender in Latin America* (pp. 72–89). University of Pittsburgh Press.

Butler, J. (1990). *Gender Trouble feminism and the subversion of identity*. Routledge.

Butler, J. (1993). *Bodies that matter: On the discursive limits of 'Sex'*. Routledge.

Caretta, M. A., Zaragocin, S., Turley, B., & Orellana, K. T. (2020). Women's organizing against extractivism: Toward a decolonial multi-sited analysis. *Human Geography*, *13*(1), 49–59. https://doi.org/10.1177/1942778620910898

Chandler, M., & Neimanis, A. (2013). Water and gestationality: What flows beneath ethics. In C. Chen, L. MacLeod & A. Neimanis (Eds.), *Thinking with water* (pp. 61–83). McGill-Queen's University Press.

Clark, C. (2019). Water justice struggles as a process of commoning. *Community Development Journal*, *54*(1), 80–99. https://doi.org/10.1093/cdj/bsy052

Collins, P. H. (1998). Learning from the outsider within: The sociological significance of Black feminist thought. *Social Problems*, *33*(6), S14–S32. https://doi.org/10.2307/800672

Craft, A., & King, L. (2021). Building the treaty #3 Nibi Declaration using an Anishinaabe methodology of ceremony, language and engagement. *Water*, *13*(4), 532. https://doi.org/10.3390/w13040532

Crenshaw, K. (1991). Mapping the margins: Intersectionality, identity politics, and violence against women of color. *Stanford Law Review*, *43*(6), 1241–1299. https://doi.org/10.2307/1229039

Daigle, M. (2018). Embodying kinship responsibilities in & through nipi (water). In J. Ellis (Ed.), *Water rites: Reimagining water in the west* (pp. 1–11). University of Calgary Press.

Davies, G., Arriagada Oyarzún, E., & Harris, L. (2023, forthcoming). Reframing gendered intersections in water and development. In S. Hellberg, A. Swain, J. Ojendal & F. Soderbaum (Eds.), *Routledge handbook on water and development*. Routledge.

Davis, W. (2020). *Magdalena: River of dreams*. Knopf Canada.

de Beauvoir, S. (1949). *The second sex*. Penguin Books.

De la Cadena, M. (2015). *Earth beings: Ecologies of practice across Andean worlds*. Duke University Press.

Drew, G. (2014). Mountain women, dams, and the gendered dimensions of environmental protest in the Garhwal Himalaya. *Mountain Research and Development*, *34*(3), 235–242. www.jstor.org/stable/mounresedeve.34.3.235

Ennis-McMillan, M. C. (2005). Women, equity and household water management in the Valley of Mexico. In V. Bennett, S. Dávila-Poblete & N. Rico (Eds.), *Opposing currents: The politics of water and gender in Latin America* (pp. 137–153). University of Pittsburgh Press.

Harris, L. M. (2006). Irrigation, gender, and social geographies of the changing waterscapes of Southeastern Anatolia. *Environment and Planning D: Society and Space*, *24*(2), 187–213. https://doi.org/10.1068/d03k

Harris, L. M. (2008). Water rich, resource poor: Intersections of gender, poverty, and vulnerability in newly irrigated areas of Southeastern Turkey. *World Development*, *36*(12), 2643–2662. https://doi.org/10.1016/j.worlddev.2008.03.004

Harris, L. M. (2015). Hegemonic waters and rethinking natures otherwise. In W. Harcourt & I. L. Nelson (Eds.), *Practicing feminist political ecologies: Moving beyond the green economy* (pp. 157–181). Zed Books.

Hawkins, R., Ojeda, D., Asher, K., Baptiste, B., Harris, L., Mollet, S., Nightingale, A., Rocheleau, D., Seagar, J., & Sultana, F. (2011). Gender and environment: Critical tradition and new challenges. *Environment and Planning D: Society and Space*, *29*(2), 237–253. https://doi.org/10.1068/d16810

Hayman, E., Wedge, M., Gooshú, A., James, C., & Tláa, G. (2015). Storytelling water north of the future Héen Kas'él'ti Xoo 1 (among the ragged lakes): Collaborative water research with Carcross/Tagish First Nation, Yukon Territory, Canada. In S. Buechler & A. Hanson (Eds.), *A political ecology of women, water and global environmental change* (pp. 185–205). Routledge.

Hazard, C. W. (2022). *Underflows: Queer and trans ecologies of river justice*. University of Washington Press.

Jenkins, K. (2017). Women anti-mining activists' narratives of everyday resistance in the Andes: Staying put and carrying on in Peru and Ecuador. *Gender, Place & Culture*, *24*(10), 1441–1459. https://doi.org/10.1080/0966369X.2017.1387102

Jepson, W., Budds, J., Eichelberger, L., Harris, L., Norman, E., O'Reilly, K., Pearson, A., Shah, S., Shinn, J., Staddon, C., Stoler, J., Wutich, A., & Young, S. (2017). Advancing human capabilities for water security: A relational approach. *Water Security*, *1*, 46–52. https://doi.org/10.1016/j.wasec.2017.07.001

Johnson, A., Zalik, A., Mollet, S., Sultana, F., Havice, E., Osborne, T., Valdivia, G., Lu, F., & Billo, E. (2020). Extraction, entanglements, and (im)materialities: Reflections on the methods and methodologies of natural resource industries fieldwork. *Environment and Planning E*, *4*(2), 383–428. https://doi.org/10.1177/2514848620907470

Krause, F., & Strang, V. (2016). Thinking relationships through water. *Society & Natural Resources*, *29*, 633–638. https://doi.org/10.1080/08941920.2016.1151714

Latchmore, T., Shuster-Wallace, C. J., Longboat, D. R., Dickson-Anderson, S. E., & Majury, A. (2018). Critical elements for local Indigenous water security in Canada: A narrative review. *Journal of Water and Health*, *16*(6), 893–903. https://doi.org/10.2166/wh.2018.107

Laurie, N. (2011). Gender water networks: Femininity and masculinity in water politics in Bolivia. *International Journal of Urban and Regional Research*, *35*(1), 172–188. https://doi.org/10.1111/j.1468-2427.2010.00962.x

Lemanski, C. (2019). *Citizenship and infrastructure: Practices and identities of citizenship and the state*. Routledge.
Linton, J. (2010). *What is water? The history of a modern abstraction*. UBC Press.
Linton, J., & Budds, J. (2014). The hydrosocial cycle: Defining and mobilizing a relational-dialectical approach to water. *Geoforum*, *57*, 170–180. https://doi.org/10.1016/j.geoforum.2013.10.008
Mawani, V. (2022). "Making do": Religious segregation and everyday water struggles. *Environment and Planning E: Nature and Space*. https://doi.org/10.1177/25148486221086481
McGregor, D. (2012). Traditional knowledge: Considerations for protecting water in Ontario. *International Indigenous Policy Journal*, *3*(3), 11. https://doi.org/10.18584/iipj.2012.3.3.11
McGregor, D. (2015). Indigenous women, water justice and Zaagidowin (Love). *Canadian Woman Studies*, *30*(2/3), 71–78.
Mollett, S., & Faria, C. (2013). Messing with gender in feminist political ecology. *Geoforum*, *45*, 116–125. https://doi.org/10.1016/j.geoforum.2012.10.009
Mukherjee, S., Sundberg, T., & Schütt, B. (2021). Issues, dimensions and approaches of assessing urban water security in developing and emerging countries: An inclusive perspective. In P. K. Sikdar (Ed.), *Environmental management: Issues and concerns in developing countries* (pp. 151–184). Springer.
Neimanis, A. (2013). Feminist subjectivity, watered. *Feminist Review*, *103*(1), 23–41. www.jstor.org/stable/41819667
Neimanis, A. (2017). *Bodies of water*. Bloomsbury Academic.
Nightingale, A. (2011). Bounding difference: Intersectionality and the material production of gender, caste, class and environment in Nepal. *Geoforum*, *42*(2), 153–162. https://doi.org/10.1016/j.geoforum.2010.03.004
Norgaard, K. M. (2017). Emotional impacts of environmental decline: What can Native cosmologies teach sociology about emotions and environmental justice? *Theory and Society*, *46*(6), 463–495. www.jstor.org/stable/44981877
Norman, E. (2018). Toward a global water ethic: Learning from Indigenous communities. *Ethics and International Affairs*, *32*(2), 237–247. https://doi.org/10.1017/S0892679418000333
O'Gorman, M. (2021). Mental and physical health impacts of water/sanitation infrastructure in First Nations communities in Canada: An analysis of the Regional Health Survey. *World Development*, *145*, 105517. https://doi.org/10.1016/j.worlddev.2021.105517
O'Leary, H. (2019). Conspicuous reserves: Ideologies of water consumption and the performance of class. *Economic Anthropology*, *6*(4), 195–207. https://doi.org/10.1002/sea2.12150
Pauli, B. J. (2019). *Flint fights back: Environmental justice and democracy in the Flint water crisis*. MIT Press.
Plumwood, V. (1993). *Feminism and the mastery of nature*. Routledge.
Pulido, L., Sidawi, S., & Vos, R. O. (1996). An archaeology of environmental racism in Los Angeles. *Urban Geography*, *17*(5), 419–439. https://doi.org/10.2747/0272-3638.17.5.419
Radonic, L., & Jacob, C. E. (2021). Examining the cracks in Universal Water Coverage: Women document the burdens of household water insecurity. *Water Alternatives*, *14*(1), 60–78.
Ranganathan, M. (2022). Towards a political ecology of caste and the city. *Journal of Urban Technology*. https://doi.org/10.1080/10630732.2021.2007203
Rocheleau, D. (1995). Maps, numbers, text, and context: Mixing methods in feminist political ecology. *The Professional Geographer*, *47*(4), 458–466. https://doi.org/10.1111/j.0033-0124.1995.458_h.x
Seymour, N. (2017). Transgender environments. In S. MacGregor (Ed.), *Routledge handbook of gender and environment*. Routledge.
Shah, S., Harris, L. M., Menghwani, V., Stoler, J., Brewis, A., Miller, J. D., Workman, C. L., Adams, E. A., Pearson, A. L., Hagaman, A., Wutich, A., & Young, S. (2023). Variations in household water affordability and water insecurity: An intersectional perspective from 18 low- and middle-income countries *Environment and Planning F*. Online First.
Simpson, L. B. (2017). *As we have always done: Indigenous freedom through radical resistance*. University of Minnesota Press.
Sultana, F. (2009). Fluid lives: Subjectivities, gender and water in rural Bangladesh. *Gender, Place and Culture: A Journal of Feminist Geography*, *16*(4), 427–444. https://doi.org/10.1080/09663690903003942
Sultana, F. (2011). Suffering for water, suffering from water: Emotional geographies of resource access, control and conflict. *Geoforum*, *42*(2), 163–172. https://doi.org/10.1016/j.geoforum.2010.12.002

Sultana, F. (2020). Embodied intersectionalities of urban citizenship. *Annals of the American Association of Geographers, 110*(5), 1407–1424. https://doi.org/10.1080/24694452.2020.1715193

Sundberg, J. (2017). Feminist political ecology. In D. Richardson, N. Castree, M. F. Goodchild, A. Kobayashi, W. Liu & R. A. Marston (Eds.), *International encyclopedia of geography: People, the earth, environment and technology*. Wiley-Blackwell.

Todd, Z. (2017). Fish, Kin and Hope: Tending to water violations in amiskwaciwâskahikan and Treaty Six Territory. *Afterall: A Journal of Art, Context and Enquiry*, *43*, 102–107. https://doi.org/10.1086/692559.

Truelove, Y. (2019). Gray Zones: The everyday practices and governance of water beyond the network. *Annals of the American Association of Geographers*, *106*(6), 1758–1774. https://doi.org/10.1080/24694452.2019.1581598

Van Horn, G., Kimmerer, R. W., & Hausdoerffer, J. (2021). *Kinship: Belonging in a world of relations*. Center for Humans and Nature.

Vinyeta, K., Whyte, K., & Lynn, K. (2015). *Climate change through an intersectional lens: Gendered vulnerability and resilience in indigenous communities in the United States*. [online]. U.S. Department of Agriculture, Forest Service, Pacific Northwest. https://papers.ssrn.com/sol3/papers.cfm?abstract_id=2770089

Wilson, A. (2015). *Violence on the land, violence on our bodies building an indigenous response to environmental violence. A toolkit by Native Youth Sexual Health Network and Women's Earth Alliance*. http://landbodydefense.org/uploads/files/VLVBToolkit_2016.pdf

Wilson, N. J., & Inkster, J. (2018). Respecting water: Indigenous water governance, ontologies, and the politics of kinship on the ground. *Environment and Planning E: Nature and Space*, *1*(4), 516–538. https://doi.org/10.1177/2514848618789378

Wutich, A., & Ragsdale, K. (2008). Water insecurity and emotional distress: Coping with supply, access, and seasonal variability of water in a Bolivian squatter settlement. *Social Science & Medicine*, *67*, 2116–2125. https://doi.org/10.1016/j.socscimed.2008.09.042

Yaka, Ö. (2017). A feminist-phenomenology of women's activism against hydropower plants in Turkey's Eastern Black Sea region. *Gender, Place & Culture*, *24*(6), 869–889. https://doi.org/10.1080/0966369X.2017.1340873

Yaka, Ö. (2019). Rethinking justice: Struggles for environmental commons and the notion of socio-ecological justice. *Antipode*, *51*(1), 353–372. https://doi.org/10.1111/anti.12422

Yates, J. S., Harris, L. M., & Wilson, N. J. (2017). Multiple ontologies of water: Politics, conflict and implications for governance. *Environment and Planning D: Society and Space*, *35*(5), 797–815. https://doi.org/10.1177/0263775817700395

Zaragocin, S., & Caretta, M. A. (2020). Cuerpo-territorio: A decolonial feminist geographical method for the study of embodiment. *Annals of the American Association of Geographers*, *111*(5), 1503–1518. https://doi.org/10.1080/24694452.2020.1812370

28
SEMÁ:TH X_Ó:TSA
Fringe natures as decolonial feminist-queer-trans water imaginaries

Madeline Donald and Astrida Neimanis

Introduction

In November 2021, a series of atmospheric rivers caused devastating flooding in Lower Mainland British Columbia and of the Sumas Valley specifically. Characterized as a major disaster, this 2021 event also brought the return of Semá:th X_ó:tsa (the Sto:lo name for a large body of water in the Fraser Valley) in its lake form. This seasonal iteration of Semá:th X_ó:tsa had been mostly repressed for a century by major colonial drainage engineering that turned this place into farmland, but an extreme form of the lake was reactivated in 2021 by the river in the sky. How might we understand such dramatic climate change events *not* as errant catastrophes, but rather in relation to colonial histories and land/water imaginaries that attempt to fix bodies of water like Semá:th X_ó:tsa into categories they refuse? In this paper, we argue that how we *imagine* bodies of water – what they are and how we think they should act in relation to other bodies of water, including humans – is a key part of water policy and governance. More livable futures in the Sumas Valley thus demand perforating the settler colonial land/water imaginary.

This chapter proceeds first by rehearsing the settler colonial transformation of Semá:th X_ó:tsa from a fecund body of water of great seasonal variation in depth, size and temperature, into an ostensibly controlled agricultural prairie. In noting the colonial environmental imaginaries that buttress these actions, this section provides the necessary context for current or future governance actions but also reminds us that settler colonialism is itself also a mode of water governance, often with undesirable consequences for lands, waters and existing communities. Following this, we outline an anticolonial, feminist-queer-trans theoretical framework that helps us to imagine Semá:th X_ó:tsa otherwise. This begins with a hydrofeminist understanding of bodies of water (Neimanis, 2017) that highlights the porous, shifting and interconnected being of water bodies. While maintaining that all water needs a body, that is, some semi-permeable membrane of containment, we aver that rigid categories of containment can lead to harmful paradigms of water governance with devastating results. Instead of inflexible divides between land and water, the alternative imaginary requires attention to marginality and border crossings. The next section therefore turns to queer and trans ecological theory developed by Cleo Wölfle Hazard, Eva

DOI: 10.4324/9781003100379-34

Hayward, and others, which inspires an imaginary of water bodies that are curious about boundaries, rather than rigidly opposed to them. Attuned to this motile ambivalence, we then return to Semá:th X_ó:tsa and read it through the figuration of "fringe natures" as our paper's key conceptual intervention. To conclude, we suggest that in a time of increasing climate instability, "fringe natures" such as Semá:th X_ó:tsa should be imagined as part of an expanded littoral zone, where forcible maintenance of fixed binary categories can no longer hold. In this expanded littoral zone, we might articulate modes of water governance in which not only Semá:th X_ó:tsa might flourish, but so too might other anticolonial, feminist, and queer-trans modes of being.

Before continuing, though, we stress that we do not seek to "import" theories about gender and sexuality from a gender studies context into a water governance one as "mere" metaphor or conceptual analogy. Following the diverse work of scholars such as Kim TallBear, Michelle Murphy, Catriona Sandilands, and Farhana Sultana, as well as Wölfle Hazard, we argue that gendered embodiment (and how we understand/control/claim it), good, life-sustaining ecological relationships between humans and other ecological bodies, and anticolonial support to Indigenous ways of knowing and LANDBACK movements are all materially and imaginatively connected. Settler violences done unto land are always gendered processes that distort and disappear Indigenous understandings and enactments of gender (e.g., Whetung, 2019) while often reimposing binaristic, "Master Model" (Plumwood, 1993) understandings on both human bodies and places. As Michi Saagiig Nishnaabeg scholar Leanne Betasamoskae Simpson (2017) notes, drawing on the work of Kanien'keha:ka scholar Audra Simpson, the sovereignty of Indigenous women's, queer and trans peoples' bodies under their own various political systems was a major threat to settler states imposition of colonial systems, and these bodies were therefore targeted for death and disappearance. In Simpson's words, "you simply cannot ignore the fact that performance of settler governmentality in Canada and the United States is strategically gendered" (p. 118). To this, we add Wölfle Hazard's (2022) understanding of settler control of waterways "as a central force of heteropatriarchy" (p. 130).

As Wölfle Hazard (2022) further asserts, concepts of trans (gender/species) embodiment bring "attention to how the vicissitudes of settler legacy and present human need shape the water body and transform the water body" (p. 104). This attention, "together with the various species and elementals in the watershed," can help "shape the possible presents and futures of that water body" towards renewed ecological possibility (pp. 104–105). Using feminist-queer-trans frameworks to rethink watershed governance thus serves as a reminder, first, that governance is always gendered even if in non-obvious ways, and second, that what we do to watersheds (e.g., through policy and governance) is informed by our always gendered understandings of embodiment and power. Feminist-queer-trans-theory is thus materially relevant to imaginaries of water governance.[1]

Semá:th X_ó:tsa: from Semá:th to settler colonial water governance

Sumas valley, unceded and occupied Stó:lō territory,[2] lies between two mountains, Sumas to the northwest and Vedder to the southeast, and within the Fraser River watershed.[3] The Semá:th are the Stó:lō group most closely enmeshed with this valley (Bolton & Daly, 2013), a valley whose mountains seem to end abruptly when they meet its floor, as if a large body spent a long time flattening the land in between them.

Today the valley's prairie bottom hosts a meaningful proportion of the commercial agricultural production in the province (British Columbia Ministry of Agriculture, 2017). Eggs are laid in long metal buildings by large numbers of chickens; cows graze on the grasses nourished by the valley's soil; blueberries grow obediently in rows; and potatoes ready themselves for dinner tables province wide. Before the chickens, cows, valley-bottom blueberries, and potatoes, there was a lake. This lake-place Semá:th X_ó:tsa, varied in size throughout the year, and at high water in late spring stretched the expanse of the valley, from the base of one mountain to the other (Reimer, 2018). (We refer to this "lake" as such only tentatively, because as we detail next, this name is directly related to settler colonial imaginaries of water governance). Water would stop in Semá:th X_ó:tsa on its way to the Fraser River, gathering and pooling, welcoming and nourishing a diverse array of fish, birds, insects, soil, humans, and more (Cameron, 1997).

The Sumas valley has nourished humans since long-ago glaciers receded, leaving a valley in their wake (Reimer, 2018). Stó:lō communities fished, hunted, frequented berry patches, and root digging grounds, and procured materials for shelter, tools, adornment, and trade from up and down the valley (Bolton & Daly, 2013). The lake's annual variation in extent, depth, and temperature played a major part in the ebb and flow of Stó:lō livelihood activities and served as a regional gathering place in both summer and winter (Silver, n.d.).

When European settlers moved into the valley – first in small numbers for travel-though, and then in masculine droves to hunt the exuberant communities of waterfowl who would stop at the lake mid-migration to rest and dine – they quickly noticed the rate and extravagance with which the grasses grew in the littoral zone when the freshet receded in midsummer (Cameron, 1997; Reimer, 2018). Cows, a staple in European food and agri-cultures, would eat well on such grasses. In a manner familiar to modes and methods of colonial land relations, this thought was taken to the extreme. If this grass is good, surely more would be better, and the grass appeared when the lake receded (Cameron, 1997).

With aspirations to establish settlements and feed livestock, human kin, and customers elsewhere, European farmers moved into Semá:th territory. These settlers overrode the needs, knowledge, and opinions of the Semá:th in their desire to farm productively and profitably. Laura Cameron (1997) quotes an exchange[4] between Chief Ned, the male chief of the Sumas band, and "the commission"[5]:

Q: (commission): Could there be any land reclaimed here by dyking?

A: (Chief Ned): I could not say. I am against the dyking because that will mean more starvation for us.

Q: Why do you think that you would be starved out if this land were dyked?

A: Because the lake is one of the greatest spawning grounds there is and this dyking would cut it off and that way would cut off our fish supply.

The disregard for the needs and knowledge of the Semá:th people has echoed throughout history. In 2008 James Murton explained that "[w]ith the approval of the landowners and of *nearly everybody else who mattered in the province*, the Land Settlement Board was planning to drain Sumas Lake" (p. 92, emphasis added).

A multi-year campaign to empty the valley bottom of the lake began.

When the settlers had reached an agreement amongst themselves that the lake would be drained, dikes, canals, and pumps were put in place to divert the shallow, widespread lake waters to the Sumas River, which would flow neatly into the Fraser at the north end of Sumas Mountain (Cameron, 1997). In 1924 a lake was manipulated, shifted from the valley bottom towards a riverine flow – a flow along which now runs the trans-Canada highway, flanked by industrial outpost buildings and strip-mall-style shopping opportunities. Behind these, approaching the would-be lake bottom, lie farms.

As currently practiced, agricultural life in Semá:th X_ó:tsa is rife with paradox. Rich soils created by centuries of seasonal sediment deposition resulting from ebbing lake and river waters attracted early-arrival agriculturalists (Reimer, 2018). Now, for over 100 years – a very short story in the long life of Semá:th X_ó:tsa – generations have cultivated livelihoods in this prairie-between-mountains. Despite efforts towards governance and control, the lake maintains its prominence in the valley. Pumps run constantly to move water out of the valley-bottom and into the river, flood risk is high in most parts of the prairie, and anthropocentric infrastructure periodically fails to match the water's determination (University of the Fraser Valley Library, 2022). Or, as explained by Tyler Olsen of the Abbotsford News (2016), "the work of keeping Sumas Prairie dry will never be finished."

Since Semá:th X_ó:tsa was re-placed, many floods have threatened agricultural lifeways in the valley. In addition to increasingly erratic climatological phenomena, clear cut logging and high-intensity forest fires create hillsides off which precipitation can flow ferociously, overwhelming human abilities to maintain the valley's prairie bottom. One such flood occurred in November of 2021 with the arrival of a series of tropical atmospheric rivers that dumped themselves as rain while unseasonably warm temperatures melted early-season snow off recently scorched mountainsides.

Despite paradox and intermittent devastation, the ongoing maintenance of a farmable valley bottom has become an obvious necessity to many, unimaginable otherwise: "It's our entire livelihood. We'll come back stronger" (Sandhu, in McSheffrey, 2021). Catastrophic climate change events may belie this conviction. Can we reimagine Semá:th X_ó:tsa in a way that both looks back at its fluctuating liveliness but also looks forward to climate events that must be lived with?

In this chapter we are thinking with Semá:th X_ó:tsa – a glacier turned lake, turned prairie and, for several weeks in November 2021, back to lake again – to ask what alternative decolonial and feminist-queer-trans conceptual understandings of places/ecologies might better accommodate more livable futures.

Imagining Semá:th X_ó:tsa otherwise

Bodies of water

Human bodies as bodies of water are akin to geophysical bodies of water – rivers, lakes, clouds, aquifers- we are all comprised mostly of water. This is a material fact, and a shared ontology. This watery constitution is what enables such bodies to move, grow, change, and gestate other bodies which, too, are mostly watery. When we start to pay attention to the wateriness of bodies, we begin to deduce their watery "onto-logics" – that is, not only that we *are* watery but also that we share *modes* (or logics) of being and relating as bodies of water (Neimanis, 2017, p. 95). Onto-logics of watery bodies thus pertain to what they are, how they act in the world, and what is needed to allow them to fulfill their ecological responsibilities. Two key watery onto-logics that Neimanis elaborates are inter-permeation-exchange across and between bodies – and differentiation, whereby through these exchanges water is continually becoming different. Alerted to these onto-logics we understand that bodies of water *inherently and ontologically* expand, shrink, shift, and transform. As a human, for example, you may be approximately sixteen times bigger than you were fifty years ago, but you are still the same person; a "lake" on a floodplain, similarly, is still that "lake," even as the precise location of its edge shifts with the seasons and the weather. Geological time may eventually render the category of "lake" nonsensical (as glacial ice sheets form, or continents separate and join), but bodies of water have a metastable flexibility – remaining what they are even as they push and pull at their borders – until they are annihilated entirely (Deleuze & Guattari, 1987). Until then, in order for these transformations to happen, bodies borrow from and bequeath their waters to other bodies, in a wondrous, complex, cascading and nesting series of multibeing hydrological cycles (Neimanis, 2017, pp. 65–67).

Paying attention to these onto-logics means appreciating the subtle ambivalence of watery bodiedness: although continuously changing, a body of water can still hold coherence as a body. Put otherwise, these watery onto-logics reveal bodies of water as porous and permeable, but this does not mean they are all just one big amorphous flow. A recognition that bodies of water shift and morph *as a fundamental aspect of their being* is not a submission to total dissolution. The bodies-of-water concept also recognizes the imperative of what physicist and feminist theorist Karen Barad (2007) has called "agential separability" (p. 140). According to Barad, material bodies appear as distinct only when we "cut" them into separate bodies for pragmatic purposes. Barad calls this "agential cutting" because this separation is an active and changeable process. Sometimes these cuts are made by human language ("this is the lake but this here is the river that flows into it"); sometimes these cuts are made by non-human material processes (a slow, centuries-long shift turns the river into a separate pool); and they are often a matter of scale (a river and a lake can become one body if we understand both as part of a watershed).[6] The key point here is that relata do not pre-exist relations; material bodies come into meaningful existence through the intra-action of materiality and the discourses or processes in which they are made intelligible. Again, while the borders of water within an entire river basin system and those of a single farm's irrigation system are both permeable and contingent, from each perspective different borders would be salient. To make such cuts is to affirm agential separability; this separability is a necessary part of responsive and relational meaning making in the world. Barad further underscores that responsivity to changing environments – circumstantially, cyclically, and over longer durations – can also be a matter of a body's survival. As Barad (2007) notes, "intelligibility and materiality are not fixed aspects of the world but rather intertwined agential performance" (p. 376).

In other words, this agential separability – or discernments that distinguish between bodies of water – are practical and necessary, both for the body itself and for those who are in relation to it. We need to know if a body of water can be traversed with a canoe, for example, or whether it is shallow or dispersed enough to ride a bike across. "It matters which cuts are enacted," Barad (2007) reminds us, because "different cuts enact different materialized becomings" (p. 361). As Neimanis puts it, all bodies need water, but concomitantly, all water needs a body, or some semi-permeable membrane of containment such that it can persist as a being or entity (such as a human body, a lake). All water needs a body *in order to be meaningful in the world*, even if that container (whether discursive or material or likely both) is temporary or contingent. This containment generates difference, and differentiation is what allows the proliferation or multiplication of bodies of water. These onto-logics exist in both complementarity and tension; interpermeation allows bodies to retain a metastability, but this exchange is also what enables bodies to become something new. Living well with other bodies of water requires neither fixity nor total dissolution of borders, but rather curiosity about shifting embodiment in response to time, space, and other bodies.

Attuning to these onto-logics helps us understand what enables bodies to flourish and fulfill the roles they have in relation to other bodies of water. While these onto-logics are articulated by Neimanis within a Western feminist context, they can also be anticolonial in their dismantling of dominating[7] white colonial and heteropatriarchal ontologies of bodies. As we detail below, understanding water's onto-logics can be directly relevant for water governance seeking to counter the effects of colonial imaginaries. Before turning back to Semá:th X_ó:tsa, though, we augment these hydrofeminist onto-logics with explicitly queer and trans ecological insights.

Transing water bodies

Water's hydrofeminist onto-logics suggest that sustaining relationships with other bodies of water requires neither a rigid fixity of boundaries, nor their total refusal. This imaginary can be helpfully supplemented by queer and trans ecological theories that offer more nuanced understandings of boundary transgression, transformation, and a rejection of binaristic thinking and being. Here, trans theory is not an alternative to feminist hydro-logics, but an amplification and specific attunement to some of its tenets. Queer trans feminist STS scholar Cleo Wölfle Hazard (2022) invites us to "consider how trans bodies are like bodies of water." He continues: "Bodies of water are not containable. They overflow their banks in floods, or they seep away and go dry: the boundary is always a matter of contingency" (p. 103). This section brings closer attention to that boundary work, and the movements, places, and affects it elicits.

Acknowledging the great diversity and even disagreement that is internal to trans theory (March, 2021), many of its insights can usefully inform an analysis of environmental questions and water governance specifically. These concepts appear in the emerging subfield of trans ecologies, but their roots reach back across several decades of trans scholarship. In the first place, trans theory draws our attention to bodies and places *of and as* movement. Trans is to "move across." As Eve Hayward (2008) explains, trans is "a prefix weighted with across, beyond, through (into another state or place – elsewhere)" (p. 68). Susan Stryker (2008) similarly points out that "trans" means "the movement across a socially imposed boundary away from an unchosen starting place – rather than any particular destination or mode of transition" (p. 1). In other words, trans theories invite us to perceive movement, or transitioning, as ontologically significant. The relationship of trans movement to borders is neither to categorically affirm or denounce them, but to be curious about them, including how they might be useful, what work they might do, as well as how they can be violently enforced. Trans ecological theories, specifically, remind us that non-human worlds are and have always been home to transitioning – moving, changing, shifting. This work underscores that transitioning, or transing – not simply as a means or a temporary phase on the way to a more important or more stable state, but as a way of being in itself – is not pathological, but entirely common, and "natural" (see also Freyne, 2020; Seymour, 2020). The desire to perceive stable states (land *or* water, for example) may be related to Western ontologies of binary thinking that are linked to (and even anchored in) how these systems of thought understand gendered bodies (Plumwood, 2003). For Wölfle Hazard (2022), "the watershed body" (which comprises the extended and connected embodiment of multiple watery bodies) is always "a kind of body-in-transition" (p. 110).

Trans ecological thinking reminds us that trans places are also contact zones (Bedford, 2020). As Wölfle Hazard's "watershed body" also reveals, transing bodies are always transing in relation to other bodies; from a queer trans ecological standpoint, transition is not a journey to individual fulfillment as much as an incitement to "trans community cocreation." He continues: "Becoming a member of a transfigured watershed body requires that one act in ways that foster good relations" (2022, p. 128). Transing always happens in relation to other bodies, and in ways that reconfigure how those bodies fit together, in place. In these lively contact zones of transing, binaries are blurred and the in-between is enlivened. Sometimes known as ecotones, such spaces are often ecologically and evolutionarily significant (see Krall, 1994). Culturally speaking, these lively zones can also be where sedimented modes of thinking are loosened, and new imaginaries can take hold. Here, trans ecologies emphasize what eco- and hydrofeminist thought has similarly suggested about in-between zones (see also Mortimer-Sandilands, 2004; Neimanis, 2012), but reminds us that these insights are also related to, verified by, and deepened through lived experiences of trans people. Littoral zones, where land and water meet, are one such ecotone, and are particularly salient for water governance.

This liveliness of transitional spaces is also affectively significant. For example, as Taylor Coyne (2022), drawing on research about eels and urban waterways, notes: "The margins are, for queer eels and for queer folk, safe places – places of liberation and pleasure" (n.p.). Of course, this is not universally true; as trans geographer Loren March underscores, unstable borders and encounters that cross them can also be "disorienting" (2021, p. 466) and "potentially deadly." But Coyne's point can nonetheless remind us that transitioning, and transitional spaces, can be life-affirming and joyful. Bedford (2020) reminds that "'contact zones' and 'borderlands' . . . have the potential to be reclaimed as sites of belonging for those cast as 'impure' within the social body" (p. 2) – and the ecological one, we argue. Even when located in urban and suburban zones, such margins are still uncontainable in their evasion of normative strictures (in this case, of sociality and sexuality). These are place-times at or on the fringe. These sentiments are echoed in the marginal queer trans multispecies watershed body that Wölfle Hazard (2022) describes, that brings "river lovers of all genders" (p. 110) together in a "yearning against settler colonial affects" (p. 116). Locating affirmation in fringey transition might shift attention to what is still possible and cocreated in such "yearning against," rather than reinforcing violent and pathological associations with certain bodies or lifeways.[8]

In other words, transing is natural, and can also be joyful and affirmative; ecological transition zones can be similarly abundant and life-affirming, just as anticolonial "marginalized" worlds that counter heteropatriarchal settler colonial imaginaries and support Indigenous life and land relations are full of joy and affirmation. While the dangers and violences of marginality or liminality are real, they are not inherent to these bodies-spaces themselves, but rather linked to how they are imagined and treated from dominating colonial or heteropatriarchal positions. To find affirmation in these fringey margins does not mark a return to an Edenic or pure state. As bodies of water that enact a logic of continual differentiation, transing is border play across bodies (e.g., land and water) but also across times – looking back and forwards at once. As Wölfle Hazard (2022), citing the work of Hayward, asserts, there is a "vertiginous joy at being unable either to turn back or to continue as before; new ways, transways, transselves are required of us here." (p. 105). In Hayward's (2008) words, "to be *trans-* is to be transcending or surpassing particular impositions, whether empirical, rhetorical, or aesthetic" (p. 68, italics in original). Trans ecologies can draw our attention to transitioning environments not as pathological ones in need of correction or stabilization, but as teachers for different ways of living in relation.

Aswim in hydrofeminist onto-logics and queer and trans understandings of the possibilities afforded by bodies in transition, we offer a reading of Semá:th X_ó:tsa following what Cleo Wölfle Hazard (2022) calls a "trans theory of watershed embodiment" (p. 134). We support Wölfle Hazard's claim that such a theory can "sharpen and focus these narratives of unsettling projects" (p. 134) to act in solidarity with Indigenous nations (in this case, the Stol:lo) working towards LANDBACK. We also propose that this can suggest a way to understand what Semá:th X_ó:tsa actually *is* – neither Sumas Lake nor Sumas Prairie – and open to different governance perspectives and approaches better fit for the climate instability we must learn to live with.

Semá:th X_ó:tsa as fringe nature

"Semá:th X_ó:tsa" does not exist in the settler water governance imaginary. While this Sto:lo name already holds within it a necessary contingency of terrestrial-aquatic categories, in settler languages such as English, neither "lake" nor "prairie" will do, loaded as these words are with fixed separability. Indeed, as Reimer (2018) suggests, settler imaginaries of this place were guided by Christian-influenced discourses of nature according to "two fixed elements: water and land" (p. 152). Reimer elaborates what this perspective would have supported: "Water that was silted up with land, or land that had too much water on it, was faulty and had to be fixed. This is what was wrong with the aging, shallow Sumas Lake and its surrounding wetlands" (p. 152). According to such a view, this shapeshifting body could not fulfill its "proper purpose" as "fertile agricultural lands" (p. 152). Reclaiming the land takes on the literal language of redemption in the settler imaginary, emphasized in the 1894 words of Presbyterian minister John Locke: "If this fertile land . . . ever comes to what God or nature intended, there must be a strong and united effort made to reclaim it from its watery grave" (cited in Reimer, 2018, p. 152). Yet, as Reimer bewilderedly inquires: "how can you reclaim something that was never there" (p. 151)?

Evoking Hayward's (2008) reminder that "trans . . . does the now familiar work of suggesting the unclassifiable" (p. 68), we propose that bringing queer and trans ecologies to the question of Semá:th X_ó:tsa might enact a necessary curiosity around these settler classification and containment paradigms that have proved devastating in so many ways. What if, countering the idea that this place "needed to be redeemed" (Reimer, 2018, p. 151), we become curious about such ambivalent or fringey spaces as sites of pleasure and possibility? What if fringey ecotones were instead places of ecological and social significance, as Coyne above suggests, where a watershed body, as Wölfle Hazard muses, could be "most joyous, bountiful and most vibrantly itself" (p. 129)? Guided by these questions, we propose "fringe natures" as an operational concept for telling an alternative story of Semá:th X_ó:tsa. We suggest the image of the fringe can hold marginality, transition, movement, and alternative sociality in a way that might encourage different approaches to governance and policy making.

"Fringe natures" is thus a figuration, following the development of figurations offered by feminist theorists Rosi Braidotti (2011) and Donna Haraway (1997). A figuration is a real, material phenomenon, which holds explanatory and propositional agency for inaugurating different kinds of relationships (Neimanis, 2017, pp. 5–6). In this sense, even as they have conceptual, associative, and suggestive power, figurations are always more-than-metaphorical. In our case, the "fringe" is a real environmental phenomenon that refers to the uppermost section of a littoral zone, that is, where terrestrial and aquatic habitats meet (see Environmental Encyclopedia, 2018). We thus use the term "fringe nature" to refer to the ambiguous and contingent land/water space of Semá:th X_ó:tsa with a degree of ecological soundness. This lake-made prairie-become lake again is an actual place where land and water come together. Rejecting the fixity of "either/or," Semá:th X_ó:tsa transcends land/water, surface water/groundwater, and atmospheric/terrestrial with a full-bodied "yes and," an affirmation and addition that holds multiple and trans-capable possibilities for relating to the flows that facilitate our lives. Contextualized by seasons, inter-species relations, and stories, Semá:th X_ó:tsa and uncountable generations of Stol:lo thriving as participants of this place show that it is possible to live well with such vicissitude, within these literal "fringe natures."

Semá:th X_ó:tsa is a transing place-body, always in movement, consistently becoming anew in a dance of seasons, weather, climate, and community relations. "A river or lake is not contained by its shoreline," nor are shorelines containable in their defining two dimensions, "but seep around underground, hidden or revealed through relations to other entities" (Wölfle-Erskine & Cole, 2015, p. 304). This inherent border curiosity of watery bodies is repressed by the insistence that these bodies must be governed according to settler imaginaries that treat a drained and fallibly engineered floodplain as though this is the only way it could ever be. On increasingly frequent occasions, infrastructural restrictions on waterways cease to perform the function for which they were engineered: dykes and levees seep and break, containment mechanisms fail to contain. Even though "all water needs a body" and these bodies are constantly subject to agential cutting in Barad's terms, in the context of climate change rubbing up against the anthropogenic insistence on the fixity of these separations, the usefulness of this particular agential cut overflows and seeps out, flooding the territory in ways that exacerbate relations with other bodies of water. As humans living as visitors in this unceded land in the context of an increasingly erratic climate, we are – as pronounced by B.C. Public Safety Minister Mike Farnworth in November of 2021 – "in uncharted territory when it comes to these storms." Perhaps it is the ideal of charting that needs to be reconsidered, for to shift between wetter and drier states does not mean it must "pick a side" but rather reveals it as "naturally" aligning with water's onto-logics.

As noted, figurations also draw attention to epistemological positioning, and hold symbolic, imaginative power that connects a real phenomenon to broader social, ecological, and political questions. We thus invoke the concept of fringe natures moreover to point to the ways in which ecological spaces such as littoral zones – where land and water meet, mingle, and transition – are marginalized within apparatuses of knowing and imagining. They exist at the frayed edges of what is securely knowable and categorizable (i.e., land *or* water) and often at the margins of our policy and governance imaginations too. In their characteristic "submerged, floating, or emergent vegetation" (Environmental Encyclopedia), fringe natures inspire us to ask about which voices are not-yet audible.[9] While Semá:th X_ó:tsa comprises fringe natures in the ecological sense, it is also a fringe nature in the submergence of Sto:lo knowledges within water governance decision-making in this place. To say that these knowledges are "fringe," however, is not to relegate them to a minority status as an essential or fixed position. Note that in the definition above, fringe natures are also "emergent." In this, they hold possibility – a possibility whose realization depends on shifting the imaginary of Semá:th X_ó:tsa to respond to *what is*. In other words, marginal or fringe may not index "minority" in the quantitative sense; the marginal refers to what is submerged and emergent in response to the given context.

Conclusion: water governance for an expanded littoral zone

In the context of colonial climate catastrophe, we need to shift what knowledges and figures are relegated to the margins. Returning to Karen Barad's (2007) concept of agential separability, we note that what a body is, and what it becomes as it transitions, depends on how our discourses "cut" it. If the discourse demands land or water as the only "proper" categories, then Semá:th X_ó:tsa is either lake or prairie. We are led to imagine that most everything must be land or water. In this dominating discourse, there is no way to explain a tulip field below two feet of water other than aberrant, catastrophic, and spectacular. As we move more deeply into the wilds of climate change, however, aberration becomes increasingly familiar. Each fire season is worse than most have ever seen, and we continually confront above average temperatures, or rivers, upon rivers, upon rivers falling from the sky. When the aberrant increases in frequency, shifting towards the quotidian, perhaps the fringe becomes the norm; the world of atmospheric rivers and devastating floods are part of an expanding fringe that is not to be romanticized or aestheticized through academic conceptual wordsmithing, but responded to through governance and policy that meet this transition aptly. Living with these exigencies is not just imaginative work (although it is this as well); it can also be enacted through governance and policy changes both small and large.[10]

This is an expanded littoral zone, encompassing far more than the river's edge as a fixed line on a map, and instead extends across entire valleys, and vertically too, up into a riverine sky and down into the hyporheic zone.[11] In this expanded littoral zone, where water and land are always commingling, reconfiguring the watershed body anew, water governance must learn from, with, and within these fringe natures. For humans to live well with Semá:th X_ó:tsa's amphibious ecology requires situated attention to the changing conditions of climate and weather variability. "[F]ocusing on the shifting affordances of wet, dry, and in-between environments," which the Semá:th have extensive in-place experience with, will allow future governance to recognize the "spectrum of possibilities and oppositions that emerge in relation to particular human activities and projects" (Krause, 2017, pp. 405–406). This is the stochastic world we need to learn to live in with compassion, because warm temperatures *will* melt snow that has fallen on scorched hillsides and water *will* insist that it "*belongs to itself*" (Maracle, 2017, p. 37, italics in original), that it governs itself, despite the efforts of settler colonialism to fix it within impermeable containers and either/or categories. Wölfle Hazard claims that "once they are deployed to orderly ends, queer and trans are no longer themselves and they are no longer ours" (p. 105) (that is to say, of and belonging to queer and trans people). But perhaps in a feminist queer transing imaginary, the watershed body is never simply ours, to be controlled. This is the water that water governance must reckon with.

Notes

1 These theoretical frames also insist that human queer and trans bodies directly belong in and make valuable contributions to these ecologically oriented debates. As trans geographer March (2021) confirms, "utilizing queer (usually trans*) subjects as abstracted, deconstructive conceptual tools or as a metaphor for liminality" (p. 461) is problematic, when their lived experiences are not taken into account in the production of this scholarship (see also Seymour, 2020). Cleo Wölfle Hazard (2022, pp. 80–102) also notes that having queer trans scientists on a river restoration project, for example, brings new perspectives and knowledges. Our research for this chapter did not directly involve transgender or queer watershed researchers of Semá:th X_ó:tsa, but it takes place in the context of our research lab, The FEELed Lab (www.thefeeledlab.ca) that advocates for and practices the inclusion of queer and trans people and their lived knowledges in environmental humanities research.

2 "Stó:lō traditional territory extends from Yale to Langley, BC" (Stó:lō Service Agency, n.d.). Halq'eméylem (alternate spellings include Hul'q'umi'num', Halq'eméylem, and hən̓q̓əmin̓əm̓) is the language of this land (Silver et al., 2020; First People's Map of BC, n.d.).

3 These are their present day anglicized colonial names.

4 These interviews have been described by historian Paul Thompson (1988) as "a peculiarly intimidating form of interview, in which the lone informant is confronted by the whole committee" (p. 105, cited in Cameron, 1997, p. 26).

5 Royal Commission on Indian Affairs for the Province of BC 1913–1915.

6 For a rich discussion on scale as a matter of what relationships matter, see Liboiron (2021).

7 We use the gerund "dominating" here as opposed to the more often used adjective "dominant" to highlight the ongoing practice and perpetuation of settler colonial ontologies.

8 This can relate to Eve Tuck's (2009) warning to scholars writing about Indigenous communities to resist telling damage stories and to focus instead on the life-affirming ways of being that persist in those communities, in spite of settler colonialism.

9 See Macarena Gomez "Submerged Perspectives" in *The Extractive Zone*. See Sophie Chao on emergence within ecological spaces.

10 For example, Sarah Whatmore's (2013, p. 43) research on flooding in the United Kingdom reveals how such exigencies can be addressed with small policy changes related to building, as reflected by a research collaborator who notes:

> [I]n the past, people had houses with solid floors and if it flooded you swept it out, but now of course we have got your electrics and carpets, and you put your wooden floors in. Why do you do it when you know you are going to be flooded?
>
> *(MP, 11/09/2007)*

11 The hyporheic zone (from the Greek hypo, under, and rheos, flow (Wölfle Hazard, 2022, p. 8) includes concepts such as undercurrents, subsurface flows, groundwater, and aquifers. See Wölfle Hazard (2022) for elaboration on this dimension.

References

Barad, K. (2007). *Meeting the universe halfway: Quantum physics and the entanglement of matter and meaning*. Duke University Press.

B.C. Public Safety Minister Mike Farnworth provides update on flooding situation. (2021, November 24). Cable Public Affairs Channel. www.youtube.com/watch?v=WP1oznYoXT0

Bedford, A. (2020). Introduction: Transecology – (Re)claiming the natural, belonging, intimacy, and impurity. In D. A. Vakoch (Ed.), *Transecology: Transgender perspectives on environment and nature* (pp. 1–16). Routledge.

Bolton, R. P., & Daly, R. (2013). *Xwelíqwiya: The life of a Stó:lō Matriarch*. Athabasca University Press.

Braidotti, R. (2011). *Nomadic subjects: Embodiment and sexual difference in contemporary feminist theory* (2nd ed.). Columbia University Press.

British Columbia Ministry of Agriculture. (2017). *Agriculture in brief: Fraser valley regional district, 2016*. https://www2.gov.bc.ca/assets/gov/farming-natural-resources-and-industry/agriculture-and-seafood/statistics/census/census-2016/aginbrief_2016_fraser_valley.pdf

Cameron, L. (1997). *Openings: A meditation on history, method, and Sumas Lake*. McGill-Queen's Press-MQUP.

Coyne, T. (2022, June 22). *Queering the Swamp. Network in Canadian History & Environment | Nouvelle initiative Canadienne en histoire de l'environnement website*. https://niche-canada.org/2022/06/27/queering-the-swamp/

Deleuze, G., & Guattari, F. (1987). *A thousand plateaus: Capitalism and schizophrenia*. Bloomsbury Publishing.

Environmental Encyclopedia. (2018). *Littoral Zone*. Encyclopedia.com. www.encyclopedia.com/earth-and-environment/geology-and-oceanography/geology-and-oceanography/littoral-zone#3404800887

First Peoples Map of BC. (n.d.). Retrieved June 13, 2022, from https://maps.fpcc.ca/languages

Freyne, G. G. (2020). Transgender: An expanded view of the ecological self. In *Transecology: Transgender perspectives on environment and nature* (pp. 174–189). Routledge.

Haraway, D. J. (1997). *Modest_Witness@Second_Millennium.FemaleMan©_Meets_OncoMouseTM*. Routledge.

Hayward, E. (2008). More lessons from a starfish: Prefixial flesh and transspeciated selves. *Women's Studies Quarterly*, *36*(3/4), 64–85.

Krall, F. R. (1994). *Ecotone: Wayfaring on the margins*. State University of New York Press.

Krause, F. (2017). Towards an amphibious anthropology of delta life. *Human Ecology*, *45*(3), 403–408. https://doi.org/10.1007/s10745-017-9902-9

Liboiron, M. (2021). *Pollution is colonialism*. Duke University Press.

Maracle, L. (2017). Water. In D. Christian & R. Wong (Eds.), *Downstream: Reimagining water*. Wilfrid Laurier University Press.

March, L. (2021). Queer and trans* geographies of liminality: A literature review. *Progress in Human Geography*, *45*(3), 455–471. https://doi.org/10.1177/0309132520913111

McSheffrey, E. (Director). (2021, November 26). BC floods: The financial and emotional impact on the province's farmers. *Global News*. https://globalnews.ca/news/8406360/farming-impact-bc-flood/

Mortimer-Sandilands, C. (2004). The marginal world. In J. A. Wainwright (Ed.), *Every grain of sand: Canadian perspectives on ecology and environment* (pp. 45–54). Wilfird Laurier University Press.

Murton, J. (2008). Creating order: The liberals, the landowners, and the draining of Sumas Lake, British Columbia. *Environmental History*, *13*(1), 92–125.

Neimanis, A. (2012). Hydrofeminism: Or, on becoming a body of water. In H. Gunkel, C. Nigianni & F. Soderback (Eds.), *Undutiful daughters: Mobilizing future concepts, bodies and subjectivities in feminist thought and practice* (pp. 96–115). Palgrave Macmillan.

Neimanis, A. (2017). *Bodies of water: Posthuman feminist phenomenology*. Bloomsbury Publishing.

Olsen, T. (2016, November 19). Keeping the lake at bay in Abbotsford: How an engineering marvel defeats gravity to keep Sumas Prairie dry. *The Abbotsford News*. www.abbynews.com/news/keeping-the-lake-at-bay-in-abbotsford/

Plumwood, V. (1993). *Feminism and the mastery of nature*. Routledge.

Plumwood, V. (2003). Animals and ecology: Towards a better integration. *Unpublished Article*. https://openresearch-repository.anu.edu.au/handle/1885/41767

Reimer, C. (2018). *Before we lost the lake: A natural and human history of sumas valley*. Catlin Press.

Seymour, N. (2020). "Good animals": The past, present, and futures of trans ecology. In D. A. Vakoch (Ed.), *Transecology: Transgender perspectives on environment and nature* (pp. 190–204). Routledge.

Silver, C. D. (n.d.). *Chief's message*. Retrieved June 13, 2022, from Sumas First Nation website: www.sumasfirstnation.com/

Silver, T. C., Victor, X. C., Foulds, K., & Schneider, L. (2020). *Semá:th Xó:tsa: Sts'ólemeqwelh Sxó:tsa Great-Gramma's Lake*. The reach Gallery Museum.

Simpson, L. B. (2017). *As we have always done: Indigenous freedom through radical resistance*. University of Minnesota Press.

Stó:lō Service Agency. (n.d.). *Welcome*. Stolo Nation. www.stolonation.bc.ca/

Stryker, S. (2008). *Transgender history: The roots of today's revolutions*. Seal Press.

Thompson, P. (1988). *The voice of the past: Oral history*. Oxford University Press.
Tuck, E. (2009). Suspending damage: A letter to communities. *Harvard Educational Review*, *79*(3), 409–428. https://doi.org/10.17763/haer.79.3.n0016675661t3n15
University of the Fraser Valley Library. (2022). *Sumas Prairie: Failed dikes, flooded farmland: Semá:th X_ó:tsa*. https://libguides.ufv.ca/sumas_prairie_flooding/sumas_lake
Whatmore, S. J. (2013). Earthly powers and affective environments: An ontological politics of flood risk. *Theory, Culture & Society*, *30*(7–8), 33–50. https://doi.org/10.1177/0263276413480949
Whetung, M. (2019). (En)gendering shoreline law: Nishnaabeg relational politics along the Trent Severn Waterway. *Global Environmental Politics*, *19*(3), 16–32. https://doi.org/10.1162/glep_a_00513
Wölfle-Erskine, C., & Cole, J. (2015). Transfiguring the Anthropocene: Stochastic reimaginings of human-beaver worlds. *Transgender Studies Quarterly*, *2*(2), 297–316. https://doi.org/10.1215/23289252-2867625
Wölfle Hazard, C. (2022). *Underflows: Queer trans ecologies and river justice*. University of Washington Press.

CONCLUDING REFLECTIONS – FUTURE DIRECTIONS FOR FEMINIST WATER RESEARCH

This collection brings together varied reflections and studies, providing many inspiring examples of hydrofeminism. The book presents feminist water scholarship as a flourishing field of studies where multiple methods, theoretical lenses, and approaches coexist – even if sometimes in tension. We hope that the diversity of analyses and stories will serve as inspiration for new generations of feminist water scholars, activists, and managers, whetting their appetite to further enrich and expand it. In this short concluding chapter, we would like to draw on the book's content to highlight some key issues for future feminist water research, focusing on what we think is important to address when moving forward.

First of all, in a world in which capitalist and neoliberal reforms externalize forms of accountability and control, also in water, it remains important to continue exposing the structural features of inequalities in water access, use, and knowledge. Languages of productivity and efficiency continue to be dominant in the water sector, while prices and profits continue to be the determining criteria for water allocation. The implication is that those with the power to invest will be prioritized when accessing water or making decisions about it. Indeed, with markets and market logics shaping where and to whom water flows, intersecting structural differences – of class, ethnicity caste, gender, religion, ability, and other – that co-determine incomes and wealth will also importantly shape how water, and water-related benefits and risks, are distributed. In addition, and as many have noted, there is a complex causal relation between the depletion and pollution of aquifers, rivers, and lakes on the one hand and processes of social differentiation on the other. The systematic undervaluation of the work needed to look after and care for people and nature is one important thread here. Much of this work of care and repair is done by (some) women, as well as by those – peasant farmers figuring prominently among them – belonging to lower classes and castes. When processes of dispossession disrupt their direct connections to the waters they rely on for their livelihoods, they are likely to become less engaged in the day-to-day efforts needed to protect, maintain, restore, and heal water bodies and water infrastructures. This sets in motion a downward spiral, in which an ever-widening gap between those with responsibilities (and powers) for sustaining and those benefitting from, waters and watery ecosystems lies at the basis of

DOI: 10.4324/9781003100379-35

systematic neglect. This will in turn further endanger the lives and wellbeing of many – humans and more-than-humans.

Exposing and learning to recognize how water-based dispossessions and forms of exploitation structurally produce gendered forms of differentiation and marginalization is needed and important. Yet, feminist activism and scholarship does not stop at this. It is equally important to bring forward and narrate stories of hope, stories that use the present-day existence of (more) sustainable and equitable waterworlds as an inspiration for new ways of thinking and doing water. How to (re-)imagine and (re-)create waters and ways of living with water, and what are the strategies, alliances, collaborations, and forms of organization needed at different levels to realize more feminist water futures? The book provides testimony of the emergence and flourishing of a diverse range of hydrofeminisms, each linked to and grounded in their own practices and concerns. Reading the different chapters of the book serves as a reminder that the political urgency to create connections and alliances between those different feminisms should not result in erasing the differences between them. After all, each waterscape is unique; it consists of its own specific human and more-than-human relations that need to be carefully attended to. Waters and the relations they give rise to will always overflow the categories – urban and rural, irrigation and domestic, surface and ground, clean and polluted – used to connect and compare different place and time-based realities. Waters will also flow in and out of these categories, with what happens to water in one place carrying implications for water quality and quantity in other places, and for other people. Hence, while awareness of connections and connectedness is of particular importance in water, feminist forms of hydrosolidarity need to always remain attentive to specificities.

This also means actively investing in (re-)appreciating the value and importance of other-than-mainstream scientific knowledges, thereby challenging the colonial legacies that continue to shape water policy interventions and water research projects in the Global South. The different chapters of the book suggest different ways of doing this, providing examples of how feminist water research can support marginalized water communities in the Global South as well as in the Global North. How to support cultures of caring for water, and work towards the healing of the well-being of multiple and different water bodies and territories? As it is, colonial legacies and North-South disparities continue to intersect with dominant gendered water professionalisms to co-determine who is considered knowledgeable in terms of water, how much different water experts earn, who is (considered) a water leader, who gets a seat at the table and what topics make it to the water agenda (see also Khandekar et al., 2023). The fact that 'western' science and theories remain dominant in water research, moreover, implies that the experiences that inform this science come to be seen as more relevant and truer than others (see Linton, 2008). Colonial legacies and present-day funding realities also continue to make it much more normal for Northern researchers to carry out research in the Global South (often relying on the underpaid assistance of Southern translators and assistants) than the other way around. Feminist water futures therefore importantly hinge on decolonizing water research and knowledge, something that includes the active facilitation of Global South researchers to study water systems and questions in the Global North. Creating more diversity and inclusion also hinges on transforming institutional cultures, replacing the focus on individual excellence with the pro-active encouragement of collaboration and joint learning. A more caring learning environment for water is needed, one in which there is enough safety to admit to, discuss and correct biases and inequities and where hierarchical divides between research and activism, or between theory and practice, cease to matter. Only in this way can the *feminist intersectional and posthumanist* solidarities, alliances, and coalitions be created and expanded that are needed for more just and equitable water futures.

Although we have tried to include a wide range of feminist water topics, research contexts, and methodologies in the book, the book does not cover everything there is to say about gender/feminism and water. We would, for instance, have liked to include more critical reflections about how in the professional water world, possibilities to thrive and excel are different for differently bodied people. Studies in different parts of the world have noticed that water administrations tend to be characterized by particular hegemonic masculine attitudes and professional cultures that make it difficult for women (as well as for others who do not conform to prevailing ideals of professional masculinity), to climb up in the professional hierarchy (Sehring et al., 2023; Liebrand & Udas, 2017; Rap & Oré, 2017; Zwarteveen, 2008). Similar mechanisms may be at work in water research institutes, where the expertise and authority of (young) female or other-gendered researchers is often not recognized and rewarded in the same way as that of cis-gendered male colleagues. This makes it harder for them to make a career as water researchers. During the "me too movement", some female researchers were brave enough to expose unsafe work environments and dominant masculine cultures (Kemerink-Seyoum, 2017), sharing testimonies that resonate with the experiences of the editors of this book. Yet, the impact of such stories seems to have been temporary. Our conclusion is that it remains important to continue to speak out against hegemonic masculine water cultures, and to be critical and reflexive about the professional water environments we work in.

The book also spends relatively little thought on what it means for gender to have become an almost obligatory theme in water development cooperation, with projects needing to include references to gender in proposals and evaluations. While this has positively increased the legitimacy and space to raise and discuss gender concerns, it also leads to instrumentalization and depoliticization, with gender becoming a mere box to tick for funds to be released and projects to be approved. Often, "women" or "gender" get treated as issues to be integrated in already established water programs or agendas. The effect of this is that the more critical and transformative questions, those that challenge established ways of doing things, remain unaddressed. Different chapters of the book indeed demonstrate how the mention of gender on water policy agendas is no guarantee that feminist change will happen, perhaps on the contrary. We therefore invite feminist scholars to remain vigilant about how gender is mainstreamed in the water sector, in water policies, and agendas. Key questions could be articulated around: What kind of transformative measures are considered to achieve real gender equity? What efforts are made on the ground to effectively address gender differences and inequities in water programs? How to transform existing water policies and agenda's so that they better reflect women's needs and gender concerns (see Zwarteveen, 2006)?

The book as a whole demonstrates not just the diversity and liveliness of feminist water scholarship but also forcefully underscores the continued need for feminist activism in water. In addition, the book's chapters convincingly show the importance of feminist research and action in generating useful entry-points and inspiration for the re-thinking, re-imagining, and re-doing dominant models and modes of water-development and management. By experimenting with and nurturing pluriversal forms of living with water that are rooted in care and conviviality (Singh et al., 2018), feminists indeed have and will continue to have an important role to play in envisioning pathways towards the hydrosocial transformations that are so urgently needed.

References

Kemerink-Seyoum, J. (2017, October 20). #Metoo: A testimony of sexual harassment in the water sector. *FLOWs*. Retrieved March 3, 2024, from https://doi.org/10.58079/orzd

Khandekar, N., Arulingam, I., Joshi, D., Udalagama, U., Chakraborty, S., Ghosh, K., & Pacheco, P. (2023). What will it take to decolonize water science, policy, and practice?: Reflections by young water researchers from the Global South. In *The water dissensus: Water alternatives forum* (15th ed.). https://www.water-alternatives.org/index.php/blog/decolonize

Liebrand, J., & Udas, P. B. (2017). Becoming an engineer or a lady engineer: Exploring professional performance and masculinity in Nepal's department of irrigation. *Engineering Studies*, *9*(2), 120–139. https://doi.org/10.1080/19378629.2017.1345915

Linton, J. (2008). Is the hydrologic cycle sustainable? A historical – geographical critique of a modern concept. *Annals of the Association of American Geographers*, *98*(3), 630–649. https://doi.org/10.1080/00045600802046619

Rap, E., & Oré, M. T. (2017). Engineering masculinities: How higher education genders the water profession in Peru. *Engineering Studies*, *9*(2), 95–119. https://doi.org/10.1080/19378629.2017.1342255

Sehring, J., ter Horst, R., & Zwarteveen, M. (2023). *Gender dynamics in transboundary water governance*. Routledge.

Singh, N., Kulkarni, S., & Pathak Broome, N. (Eds.). (2018). *Ecologies of hope and transformation: Post-development alternatives from India*. Kalpavriksh and SOPPECOM.

Zwarteveen, M. (2006). *Effective gender mainstreaming in water management for sustainable livelihoods: From guidelines to practice* (Both ENDS Working Paper Series, Issue). www.iwmi.cgiar.org/assessment/files_new/research_projects/EffectiveGender_BothEnds.pdf

Zwarteveen, M. (2008). Men, masculinities and water powers in Irrigation. *Water Alternatives*, *1*(1), 111–130.

INDEX

Note: Page numbers in *italics* indicate a figure and and page numbers in **bold** indicate a table on the corresponding page.